POROUS SILICON

From Formation to Application

Optoelectronics, Microelectronics, and Energy Technology Applications, Volume Three

Porous Silicon: From Formation to Application

Porous Silicon: Formation and Properties, Volume One

Porous Silicon: Biomedical and Sensor Applications, Volume Two

Porous Silicon: Optoelectronics, Microelectronics, and Energy Technology Applications, Volume Three

Edited by
Ghenadii Korotcenkov

POROUS SILICON
From Formation to Application

*Optoelectronics, Microelectronics, and
Energy Technology Applications, Volume Three*

CRC Press
Taylor & Francis Group
Boca Raton London New York

CRC Press is an imprint of the
Taylor & Francis Group, an **informa** business

CRC Press
Taylor & Francis Group
6000 Broken Sound Parkway NW, Suite 300
Boca Raton, FL 33487-2742

Printed on acid-free paper
Version Date: 20151104

ISBN: 9781482264586 (Hardback)
ISBN: 9780367575083 (Paperback)

Visit the Taylor & Francis Web site at
http://www.taylorandfrancis.com

and the CRC Press Web site at
http://www.crcpress.com

Contents

Preface...xi
Editor..xiii
Contributors..xv

SECTION I—Optoelectronics and Photoelectronics

Chapter 1 Electroluminescence Devices (LED)..3
Bernard Jacques Gelloz

Chapter 2 Photodetectors Based on Porous Silicon..35
Ghenadii Korotcenkov and Nima Naderi

Chapter 3 PSi-Based Photonic Crystals...51
Gonzalo Recio-Sánchez

Chapter 4 Passive and Active Optical Components for Optoelectronics Based on Porous Silicon..........................77
Raghvendra S. Dubey

SECTION II—Electronics

Chapter 5 Electrical Isolation Applications of PSi in Microelectronics.............................109
Gael Gautier, Jérôme Billoué, and Samuel Menard

Chapter 6 Surface Micromachining (Sacrificial Layer) and Its Applications in Electronic Devices..........................129
Alexey Ivanov and Ulrich Mescheder

Chapter 7 Porous Silicon as Substrate for Epitaxial Films Growth....................................143
Eugene Chubenko, Sergey Redko, Alexey Dolgiy, Hanna Bandarenka, and Vitaly Bondarenko

Chapter 8 Porous Silicon and Cold Cathodes..165
Ghenadii Korotcenkov and Beongki Cho

Chapter 9 Porous Silicon as Host and Template Material for Fabricating Composites and Hybrid Materials..........................183
Eugene Chubenko, Sergey Redko, Alexey Dolgiy, Hanna Bandarenka, Sergey Prischepa, and Vitaly Bondarenko

SECTION III—Energy Technologies

Chapter 10 Porous Si and Si Nanostructures in Photovoltaics..211
Valeriy A. Skryshevsky and Tetyana Nychyporuk

Chapter 11 PSi-Based Betavoltaics 239

Ghenadii Korotcenkov and Vladimir Brinzari

Chapter 12 Porous Silicon in Micro-Fuel Cells 249

Gael Gautier and Ghenadii Korotcenkov

Chapter 13 Hydrogen Generation and Storage in Porous Silicon...273

Valeriy A. Skryshevsky, Vladimir Lysenko, and Sergii Litvinenko

Chapter 14 PSi-Based Microreactors 297

Caitlin Baker and James L. Gole

Chapter 15 Li Batteries with PSi-Based Electrodes 319

Gael Gautier, François Tran-Van, and Thomas Defforge

Chapter 16 PSi-Based Supercapacitors 347

Diana Golodnitsky, Ela Strauss, and Tania Ripenbein

Chapter 17 Porous Silicon as a Material for Thermoelectric Devices 375

Androula G. Nassiopoulou

Chapter 18 Porous Silicon–Based Explosive Devices 387

Monuko du Plessis

Index 403

Contents for Porous Silicon: *Formation and Properties, Volume One*

SECTION I—Introduction

Chapter 1 Porous Silicon Characterization and Application: General View
Ghenadii Korotcenkov

SECTION II—Silicon Porosification

Chapter 2 Fundamentals of Silicon Porosification via Electrochemical Etching
Enrique Quiroga-González and Helmut Föll

Chapter 3 Technology of Si Porous Layer Fabrication Using Anodic Etching: General Scientific and Technical Issues
Ghenadii Korotcenkov and Beongki Cho

Chapter 4 Silicon Porosification: Approaches to PSi Parameters Control
Ghenadii Korotcenkov

SECTION III—Properties and Processing

Chapter 5 Methods of Porous Silicon Parameters Control
Mykola Isaiev, Kateryna Voitenko, Dmitriy Andrusenko, and Roman Burbelo

Chapter 6 Structural and Electrophysical Properties of Porous Silicon
Giampiero Amato

Chapter 7 Luminescent Properties of Porous Silicon
Bernard Jacques Gelloz

Chapter 8 Optical Properties of Porous Silicon
Gilles Lérondel

Chapter 9 Thermal Properties of Porous Silicon
Pascal J. Newby

Chapter 10 Alternative Methods of Silicon Porosification and Properties of These PSi Layers
Ghenadii Korotcenkov and Vladimir Brinzari

Chapter 11 The Mechanism of Metal-Assisted Etching of Silicon
Kurt W. Kolasinski

Chapter 12 Porous Silicon Processing
Ghenadii Korotcenkov and Beongki Cho

Chapter 13 Surface Chemistry of Porous Silicon
Yannick Coffinier and Rabah Boukherroub

Chapter 14 Contacts to Porous Silicon and PSi-Based *p-n* Homo- and Heterojunctions
Jayita Kanungo and Sukumar Basu

Index

Contents for Porous Silicon: *Biomedical and Sensor Applications, Volume Two*

SECTION I—Sensors

Chapter 1   Solid State Gas and Vapor Sensors Based on Porous Silicon
Ghenadii Korotcenkov

Chapter 2   Optimization of PSi-Based Sensors Using IHSAB Principles
James L. Gole and Caitlin Baker

Chapter 3   Porous Silicon-Based Optical Chemical Sensors
Luca De Stefano, Ilaria Rea, Alessandro Caliò, Jane Politi, Monica Terracciano, and Ghenadii Korotcenkov

Chapter 4   PSi-Based Electrochemical Sensors
Yang Zhou, Shaoyuan Li, Xiuhua Chen, and Wenhui Ma

Chapter 5   Porous Silicon-Based Biosensors: Some Features of Design, Functional Activity, and Practical Application
Nicolai F. Starodub

Chapter 6   MEMS-Based Pressure Sensors
Ghenadii Korotcenkov and Beongki Cho

Chapter 7   Micromachined Mechanical Sensors
Ghenadii Korotcenkov and Beongki Cho

Chapter 8   Porous Silicon as a Material for Thermal Insulation in MEMS
Pascal J. Newby

Chapter 9   PSi-Based Microwave Detection
Jonas Gradauskas and Jolanta Stupakova

SECTION II—Auxiliary Devices

Chapter 10   PSi-Based Ultrasound Emitters (Acoustic Emission)
Ghenadii Korotcenkov and Vladimir Brinzari

Chapter 11   Porous Silicon in Micromachining Hotplates Aimed for Sensor Applications
Ghenadii Korotcenkov and Beongki Cho

Chapter 12   PSi-Based Preconcentrators, Filters, and Gas Sources
Ghenadii Korotcenkov, Vladimir Brinzari, and Beongki Cho

Chapter 13 Silicon Nanostructures for Laser Desorption/Ionization Mass Spectrometry
Sergei Alekseev

Chapter 14 PSi Substrates for Raman, Terahertz, and Atomic Absorption Spectroscopy
Ghenadii Korotcenkov and Songhee Han

Chapter 15 PSi-Based Diffusion Membranes
Andras Kovacs and Ulrich Mescheder

Chapter 16 Liquid Microfluidic Devices
Ghenadii Korotcenkov and Beongki Cho

SECTION III—Biomedical Applications

Chapter 17 Biocompatibility and Bioactivity of Porous Silicon
Adel Dalilottojari, Wing Yin Tong, Steven J.P. McInnes, and Nicolas H. Voelcker

Chapter 18 Applications of Porous Silicon Materials in Drug Delivery
Haisheng Peng, Guangtian Wang, Naidan Chang, and Qun Wang

Chapter 19 Tissue Engineering
Pierre-Yves Collart Dutilleul, Csilla Gergely, Frédérique Cunin, and Frédéric Cuisinier

Chapter 20 Use of Porous Silicon for *In Vivo* Imaging Techniques
Igor Komarov and Sergei Alekseev

Index

Preface

In recent decades, porous silicon has been regarded as a means to further increase the functionality of silicon technology. It was found that silicon porosification is a simple and cheap way of nanostructuring and bestowing to silicon properties, which are markedly different from the properties of the bulk material. Because of this, increased interest in porous silicon appeared in various fields, including optoelectronics, microelectronics, photonics, medicine, chemical sensing, and bioengineering. It was established that this nanostructured and biodegradable material has a range of properties, making it ideal for indicated applications. As a result, during the last decade we have been observing an extremely fast evolution of porous Si-based optoelectronics, photonics, microelectronics, sensorics, energy technologies, and biomedical devices and applications. It is predicted that this growth will continue in the near future. However, despite the progress achieved in the field of design of porous silicon–based devices and their applications, it is necessary to note that there is a very limited number of books published in the field of silicon porosification and especially porous silicon (PSi) applications. No doubt such situations should be recognized as unsatisfactory. Thus, it was decided to prepare a set of books devoted to the analysis of the current state of the technology of silicon porosification and the use of these technologies in the development of devices for different applications. While developing the concept of this series, the objective was to collect in one edition information concerning all aspects of the formation and the use of porous silicon. This is of great importance nowadays, due to the speed of technological development and the rate of an appearance of new fields of PSi-based technology applications.

This set, Porous Silicon: From Formation to Application, prepared by an international team of expert contributors, well known in the field of porous silicon study and having high qualifications, represents the most recent progress in the field of porous silicon and gives a fascinating report on the state-of-the-art in silicon porosification and the valuable perspective one can expect in the near future.

The set is divided into three books by their content. Chapters in *Porous Silicon: Formation and Properties, Volume One* focus on the fundamentals and practical aspects of silicon porosification by anodization and the properties of porous silicon, including electrical, luminescence, optical, thermal properties, and contact phenomena. Processing of porous silicon, including drying, storage, oxidation, etching, filling, and functionalizing, are also discussed in this book. Alternative methods of silicon porosification using chemical stain and vapor etching, reactive ion etching, spark processing, and so on are analyzed as well. *Porous Silicon: Biomedical and Sensor Applications, Volume Two* describes applications of porous silicon in bioengineering and various sensors such as gas sensors, biosensors, pressure sensors, optical sensors, microwave detectors, mechanical sensors, etc. The chapters in this book present a comprehensive review of the fabrication, parameters, and applications of these devices. PSi-based auxiliary devices such as hotplates, membranes, matrices for various spectroscopies, and catalysis are discussed as well. Analysis of various biomedical applications of porous silicon including drug delivery, tissue engineering, and *in vivo* imaging can also be found in this book. No doubt, porous silicon is rapidly attracting increasing interest in this field due to its unique properties. For example, the pores of the material and surface chemistry can be manipulated to change the rate of drug release from hours to months.

Finally, *Porous Silicon: Optoelectronics, Microelectronics, and Energy Technology Applications, Volume Three* highlights porous silicon applications in opto- and microelectronics, photonics, and micromachining. Features of fabrication and performances of photonic crystals, fuel cells, elements of integral optoelectronics, solar cells, LED, batteries, cold cathodes, hydrogen generation and storage, PSi-based composites, etc. are analyzed in this volume.

I believe that we have prepared useful books that could be considered a real handbook encyclopedia of porous silicon, where each reader might find the answers to most questions related to the formation, properties, and applications of porous silicon in practically all possible fields. Previously published books do not provide such an opportunity. Recently, several interesting books became available to readers, such as *Porous Silicon in Practice: Preparation, Characterization and Applications* by M.J. Sailor (Wiley-VCH 2011), *Porous Silicon for Biomedical Applications* by Santos H.A. (ed.) (Woodhead Publishing Limited 2014), and *Handbook of Porous Silicon*, by Canham L. (ed.) (Springer 2015). However, in *Porous Silicon in Practice: Preparation, Characterization and Applications*, M.J. Sailor describes mainly features of silicon porosification by electrochemical etching without any analysis of the correlation between parameters of porosification and properties of porous silicon. In the *Handbook of Porous Silicon* most attention was paid to the properties of porous silicon and, as well as in the book of M.J. Sailor, the consideration of PSi applications in devices was brief. At the same time, *Porous Silicon for Biomedical Applications* focuses on only the analysis of biomedical applications of porous silicon. I hope that our books, which cover all of the above-mentioned fields and provide a more detailed analysis of PSi advantages and disadvantages for practically all possible applications, will also be of interest to the reader. Our books contain a great number of various figures and tables with necessary information. These books will be a technical resource and indispensable guide for all those involved in the research, development, and application of porous silicon in various areas of science and technology.

From my point of view, our set will be of interest to scientists and researchers, either working or planning to start activity in the field of materials science focused on multifunctional porous silicon and porous silicon–based semiconductor devices. It also could be useful for those who want to find out more about the unusual properties of porous materials and about possible areas of their application. I am confident that these books will be interesting for practicing engineers or project managers working in industries and national laboratories who intend to design various porous silicon–based devices, but don't know how to do it. They might help select an optimal technology of silicon porosification and device fabrication. With many references to the vast resources of recently published literature on the subject, these books can serve as a significant and insightful source of valuable information and provide scientists and engineers with new insights for better understanding of the process of silicon porosification, for designing new porous silicon–based technology, and for improving performances of various devices fabricated using porous silicon.

I believe that these books can be of interest to university students, post docs, and professors, providing a comprehensive introduction to the field of porous silicon application. The structure of these books may serve as a basis for courses in the field of material science, semiconductor devices, chemical engineering, electronics, bioengineering, and environmental control. Graduate students may also find the books useful in their research and for understanding that porous silicon is a promising multifunctional material.

Finally, I thank all contributing authors who have been involved in the creation of these books. I am also thankful that they agreed to participate in this project and for their efforts in the preparation of these chapters. Without their participation, this project would have not been possible.

I also express my gratitude to Gwangju Institute of Science and Technology, Gwangju, Korea, which invited me and gave me the ability to prepare these books for publication, and especially to Professor Beongki Cho for his fruitful cooperation. Many thanks to the Ministry of Science, ICT, and Future Planning (MSIP) of the Republic of Korea for supporting my research. I am also grateful to my family and my wife, who always support me in all undertakings.

Editor

Ghenadii Korotcenkov earned his PhD in physics and the technology of semiconductor materials and devices from Technical University of Moldova in 1976 and his DrSci degree in the physics of semiconductors and dielectrics from the Academy of Science of Moldova in 1990 (Highest Qualification Committee of the USSR, Moscow). He has more than 40 years of experience as a teacher and scientific researcher. He was a leader of a gas sensor group and manager of various national and international scientific and engineering projects carried out in the Laboratory of Micro- and Optoelectronics, Technical University of Moldova. In particular, during 2000–2007 his scientific team was involved in eight international projects financed by EC (INCO-Copernicus and INTAS Programs), United States (CRDF, CRDF-MRDA Programs), and NATO (LG Program). In 2007–2008, he was an invited scientist at the Korea Institute of Energy Research (Daejeon) in the Brain Pool Program. Since 2008, Dr. Korotcenkov has been a research professor in the Department of Materials Science and Engineering at Gwangju Institute of Science and Technology (GIST) in Korea.

Specialists from the former Soviet Union know Dr. Korotcenkov's research results in the field of study of Schottky barriers, MOS structures, native oxides, and photoreceivers on the base of III-Vs compounds very well. His present scientific interests include material sciences, focusing on metal oxide film deposition and characterization, surface science, porous materials, and gas sensor design. Dr. Korotcenkov is the author or editor of 29 books and special issues, including the 11-volume Chemical Sensors series published by Momentum Press, the 10-volume Chemical Sensors series published by Harbin Institute of Technology Press, China, and the 2-volume *Handbook of Gas Sensor Materials* published by Springer. He has published 17 review papers, 19 book chapters, and more than 200 peer-reviewed articles (h-factor = 33 [Scopus] and h = 38 [Google scholar citation]). A citation average for his papers, included in Scopus, is higher than 25. He is a holder of 18 patents. In most papers, Dr. Korotcenkov is the first author. He has presented more than 200 reports on national and international conferences, and was the co-organizer of several conferences. His research activities were honored by an award of the Supreme Council of Science and Advanced Technology of the Republic of Moldova (2004), a prize of the Presidents of Ukrainian, Belarus and Moldovan Academies of Sciences (2003), a Senior Research Excellence Award of the Technical University of Moldova (2001, 2003, 2005), a fellowship from International Research Exchange Board (1998), and the National Youth Prize of the Republic of Moldova (1980), among others.

Contributors

Caitlin Baker
Georgia Institute of Technology
Atlanta, Georgia

Hanna Bandarenka
Belarussian State University of Informatics
    and Radioelectronics
Minsk, Belarus

Jérôme Billoué
Université François Rabelais de Tours
Tours, France

Vitaly Bondarenko
Belarussian State University of Informatics
    and Radioelectronics
Minsk, Belarus

Vladimir Brinzari
State University of Moldova
Chisinau, Republic of Moldova

Beongki Cho
Gwangju Institute of Science and Technology
Gwangju, Republic of Korea

Eugene Chubenko
Belarussian State University of Informatics
    and Radioelectronics
Minsk, Belarus

Thomas Defforge
Université François Rabelais de Tours
Tours, France

Alexey Dolgiy
Belarussian State University of Informatics
    and Radioelectronics
Minsk, Belarus

Raghvendra S. Dubey
Swarnandhra College of Engineering and Technology
Seetharampuram, India

Monuko du Plessis
University of Pretoria
Pretoria, South Africa

Gael Gautier
Université François Rabelais de Tours
Tours, France

Bernard Jacques Gelloz
Nagoya University
Nagoya, Japan

James L. Gole
Georgia Institute of Technology
Atlanta, Georgia

Diana Golodnitsky
Tel Aviv University
Tel Aviv, Israel

Alexey Ivanov
Furtwangen University
Furtwangen, Germany

Ghenadii Korotcenkov
Gwangju Institute of Science and Technology
Gwangju, Republic of Korea

Sergii Litvinenko
Taras Shevchenko National University of Kyiv
Kyiv, Ukraine

Vladimir Lysenko
Université de Lyon
Villeurbanne, France

Samuel Menard
ST Microelectronics
Tours, France

Ulrich Mescheder
Furtwangen University
Furtwangen, Germany

Nima Naderi
Materials and Energy Research Center (MERC)
Karaj, Iran

Androula G. Nassiopoulou
IMEL/NCSR Demokritos
Institute of Nanoscience and Nanotechnology
Athens, Greece

Tetyana Nychyporuk
Université de Lyon
Villeurbanne, France

Sergey Prischepa
Belarussian State University of Informatics
    and Radioelectronics
Minsk, Belarus

Gonzalo Recio-Sánchez
Universidad Católica de Temuco
Temuco, Chile

Sergey Redko
Belarussian State University of Informatics
    and Radioelectronics
Minsk, Belarus

Tania Ripenbein
Tel Aviv University
Tel Aviv, Israel

Valeriy A. Skryshevsky
Taras Shevchenko National University of Kyiv
Kyiv, Ukraine

Ela Strauss
Israel Ministry of Science, Technology and Space
Jerusalem, Israel

François Tran-Van
Université François Rabelais de Tours
Tours, France

I

Optoelectronics and Photoelectronics

Electroluminescence Devices (LED)

Bernard Jacques Gelloz

CONTENTS

1.1	Introduction	4
1.2	Background	5
1.3	Electrolytic Systems	6
	1.3.1 Anodic Polarization of PSi	6
	1.3.1.1 Anodic Oxidation	6
	1.3.1.2 Other Systems	8
	1.3.2 Cathodic Polarization of PSi with Persulfate Ions	8
1.4	Charge Carrier Transport in Electrolytic Systems and Differences with Solid-State Devices	10
	1.4.1 Charge Carrier Transport in Electrolytic Systems	10
	1.4.2 Charge Carrier Transport in Solid-State Devices	12
1.5	Devices Including As-Formed PSi	13
1.6	Porosified p-n Junctions	13
1.7	Partially Oxidized PSi	15
	1.7.1 Chemical and Thermal Oxidation	16
	1.7.2 Electrochemical Oxidation	16
	1.7.3 Oxidation by High-Pressure Water Vapor Annealing	18
1.8	PSi Impregnated by Another Material	18
	1.8.1 Incorporation of Metals	20
	1.8.2 Incorporation of Polymers	20

	1.8.3	Incorporation of Inert Materials	21
1.9		Influence of Device Top Electrode	21
	1.9.1	Metals and Conductive Oxides as Top Electrodes	22
	1.9.2	Compact PSi between Active PSi and Top Electrode	23
	1.9.3	Other Materials between Active PSi and Top Electrode	24
1.10		EL Stabilization by Capping and Surface Passivation	26
	1.10.1	Capping PSi	26
	1.10.2	PSi Surface Modification	26
1.11		Emission Spectrum, Speed, and Integration	27
	1.11.1	Tuning and Narrowing the EL Spectrum with Microcavities	27
	1.11.2	EL Modulation Speed	28
	1.11.3	Integration of PSi EL	29
1.12		Conclusion	30
		References	31

1.1 INTRODUCTION

Silicon (Si) technology overwhelmingly dominates microelectronics. However, Si is a very poor light emitting material because of its indirect bandgap. Furthermore, its bandgap of ~1.1 eV at room temperature would limit its use to the near infrared range. As a result, other light emitting materials, such as compound semiconductors, or organic molecules have been chosen for optoelectronic applications such as optical communication, display, and lighting.

On-chip Si-based light emitters are very much desired to overcome the so-called electrical interconnect bottleneck in microelectronics. The display community, as well as other arenas involving lighting, would also benefit from a cheap, efficient, stable, and integrated Si-based visible light emitter.

Nanostructured Si, such as porous silicon (PSi), could be a solution. Indeed, carrier localization, relaxation of k-conservation rule, and bandgap enlargement enable efficient and visible light emission from nanocrystalline silicon. Tremendous efforts have been devoted to the development of PSi electroluminescence (EL) for more than two decades, starting from the discovery of PSi luminescence (Canham 1990). However, all of these efforts have failed to deliver practical EL devices. The problems of PSi EL are mainly low charge carrier injection efficiency and poor stability.

The research activities in PSi EL have considerably slowed down in recent years. However, the hope may be revived by recent advances in luminescence efficiency and stability. Indeed, rather high photoluminescence (PL) external quantum efficiency (EQE) has been achieved from PSi: 23% (Gelloz and Koshida 2005; Gelloz et al. 2005a). The stability of both PL and EL (Gelloz et al. 2006) was also dramatically improved.

This chapter first presents the background of PSi EL, including a few definitions, the problems of PSi EL, as well as the state of the art and requirements in terms of efficiency, stability, and so on. Then, the EL of PSi in contact with different types of liquid electrolytes is discussed. These systems were the first where EL of PSi could be observed. Furthermore, they exhibit very attractive EL properties (such as low voltage operation, rather high efficiency, and voltage-tunable emission energy), radically different from those of solid-state PSi EL devices. Next, as a transition toward the presentation of solid-state devices, the charge carrier transport mechanisms in both electrolytic systems and conventional solid-state devices are discussed. Then, many types of noticeable PSi solid-state EL devices are presented, using a classification based on the main device structures and main processes involved. Finally, further techniques proposed to improve the EL characteristics (including tuning the emission spectrum), efficiency, and stability are presented. Progress in integration is also discussed.

1.2 BACKGROUND

EL is the result of radiative recombination of electrons and holes electrically injected into the Si nanocrystals in PSi layers. The EQE of an EL device is defined as the number of photons emitted outside its structure divided by the number of electrons flowing through it. The external power efficiency (EPE) is the optical power of photons emitted outside the device divided by the input electrical power. The EPE is important for applications. It is usually given in percentages, although the industry of lighting and display prefers the lumen/watt.

Various mechanisms can lead to nonradiative recombinations, thus lowering the luminescence efficiency (see Chapter 7 in *Porous Silicon: Formation and Properties*). In particular, trapping by surface states is a significant problem in PSi due to its large surface area. In addition, in an EL device, the balance between electron and hole injection is a very important issue. Besides, luminescent PSi is usually highly resistive due to depletion of free charge carriers because of quantum confinement and inactivation of dopants by surface traps (Lehmann et al. 1995). This makes the charge carrier flow and injection into Si nanocrystals very difficult. This is one reason for the usually low EQE of PSi EL devices.

The characteristics and problems of PSi EL devices may be summarized as follows:

1. Poor efficiency mainly due to leakage (large imbalance of electron and hole injection and carriers flowing through nonluminescent parts of PSi) or poor carrier injection efficiency in the smaller parts (light-emitting parts) of PSi.
2. Poor stability due to surface oxidation or defect generation on storage in air and during device operation.
3. Difficulty obtaining green and blue EL due to interface levels within the bandgap (Wolkin et al. 1999); see Chapter 7 in *Porous Silicon: Formation and Properties*. Thus, PSi EL is currently mostly limited to red–orange emission. In addition, typical emission spectra are quite broad, although some solutions are presented in Section 1.11.1.
4. Usually high operating voltages due to the high series resistance of PSi. This is detrimental to the EPE.
5. Rather long response times (switching times), mostly in the microsecond range, limited by radiative lifetimes and capacitive effects, although particular device structures may shorten the response time (as discussed in Section 1.11.2) possibly down to the nanosecond range using Auger effect (Carreras et al. 2008).

However, recent progress in PL and EL efficiency (Gelloz and Koshida 2000, 2005; Gelloz et al. 2005a) and stability (Gelloz and Koshida 2005; Gelloz et al. 2003b, 2005a, 2006) is very encouraging. Table 1.1 summarizes the requirements and status of PSi devices in terms of EQE, time response, and stability. The minimum requirement for the EPE is ~1% for the least demanding applications such as displays. Optical communications are much more demanding (~10%).

TABLE 1.1 Requirements and Status of PSi EL

Application	EPE (%)	Modulation Speed (MHz)	Peak Emission	Spectral Width	Stability
Displays	>1	>0.001	Red, green, blue	100 meV	10^5 h
Interconnects	>10	>100	–	–	>10^5 h
PSi EL	<0.4	~1	Red	~220 meV	s to weeks

Note: EPE refers to external power efficiency.

TABLE 1.2   EL Characteristics of Porous Silicon Contacted by a Liquid Electrolyte

Polarization Condition	Electrolytic System and EL Properties
Anodic polarization	Electrochemical oxidation in acidic solution.
Holes injected into PSi from Si substrate	Electrons injected from oxidation of silicon (Chazalviel and Ozanam 1992).
	EL triggered by injection of holes into luminescent nanocrystals; blue-shift of EL during oxidation. EL limited in time because Si nanocrystals are irreversibly oxidized (Billat 1996; Bsiesy et al. 1991; Cantin et al. 1996; Halimaoui et al. 1991; Hory et al. 1995).
	Oxidation of species in acidic electrolyte.
	Electrons injected from oxidation of Methylviologen (Kooij et al. 1995) or formic acid (Green et al. 1995).
	EL triggered by injection of holes into luminescent nanocrystals; not stable due to simultaneous oxidation of silicon under anodic polarization.
Cathodic polarization	Reduction of persulfate ions in acidic electrolyte.
Electrons injected into PSi from Si substrate (Gelloz and Bsiesy 1998; Peter and Wielgosz 1996)	Holes injected from reduction of $S_2O_8^{2-}$ (Bressers et al. 1992; Canham et al. 1992; Noguchi et al. 1999; Peter et al. 1996; Romestain et al. 1995).
	EL triggered by injection of electrons into luminescent nanocrystals; EL spectrum is voltage-tunable (Romestain et al. 1995). Reversible; stability affected by ion consumption, slow oxidation of silicon surface and possibly evolution of the location of charge exchanges (Noguchi et al. 1999; Saren et al. 2002). Reaction kinetics affects the EL characteristics (Gelloz and Bsiesy 1998; Gelloz et al. 1996).

1.3 ELECTROLYTIC SYSTEMS

The first demonstration of EL from PSi was performed with PSi impregnated with an acidic aqueous electrolyte (Halimaoui et al. 1991). The electrolyte has to be at least slightly acidic because silicon can be chemically etched by bases. In such EL electrolytic systems, one type of carrier is injected electrically from the silicon substrate and the other one is injected in PSi because of a chemical reaction. Table 1.2 summarizes the characteristics of the different electrolytic systems that have been studied.

1.3.1 ANODIC POLARIZATION OF PSi

Next, we consider the case where holes are injected into PSi from the substrate by using anodic polarization of PSi.

1.3.1.1 ANODIC OXIDATION

EL from PSi was first demonstrated using anodic oxidation of p-type PSi of high porosity (Halimaoui et al. 1991). Anodic oxidation, also called electrochemical oxidation (ECO), may be performed either under potentiostatic (constant applied potential) or galvanostatic (constant applied current) conditions, with PSi in contact with an aqueous conductive electrolyte. In most cases, aqueous HCl or H_2SO_4 were used as electrolyte. During ECO PSi is oxidized following the reaction $Si + 2H_2O + 4h^+ \rightarrow SiO_2 + 4H^+$, where holes ($h^+$) are supplied by the substrate.

The galvanostatic mode is usually preferred because it turned out that an important parameter in the anodic oxidation process is the total charge exchanged Q in the experiment, and Q is simply related to the electrical current I by $Q = I.t$, where t is the experiment time (Billat 1996; Bsiesy et al. 1991; Cantin et al. 1996; Halimaoui et al. 1991; Hory et al. 1995; Ligeon et al. 1993).

Figure 1.1 shows the evolution of the potential and EL intensity as a function of time in the case of lightly doped PSi (Ligeon et al. 1993). At first, the potential increases very slowly as the surface of the PSi skeleton is oxidized. The initial step of Si oxidation is hole injection. These holes flow through the paths of lowest energy in PSi. Thus, they flow through the largest parts of PSi, where no quantum confinement takes place. The oxidation is therefore selective in size. However, as

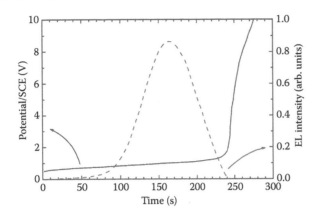

FIGURE 1.1 Potential and electroluminescence intensity as a function of time for a PSi layer of porosity 65%, and thickness 5 µm, during anodic oxidation at 10 mA/cm². The Si substrate was p-type. SCE refers to saturated calomel electrode. (From Ligeon M. et al., *J. Appl. Phys.* 74, 1265–1271, 1993. Copyright 1993: AIP. With permission.)

the experiment proceeds, the formation of oxide at the surface makes charge-exchange between silicon and the electrolyte more difficult. Then, the potential must correspondingly be increased in order to keep the current constant. Hole injection in levels of higher energy in PSi is then achieved. There is a period of time when injection occurs in crystallites small enough to be luminescent. EL can be observed during this period. EL reaches a maximum when carrier injection in confined crystallites is optimal. Afterward, the EL decreases and eventually vanishes, when, in the final stage of ECO, the potential increases very sharply. At this stage, the oxide grown at the PSi/Si substrate interface prevents further injection of holes into PSi and constitutes a high-resistance film between the Si substrate and the electrolyte. No hole injection into PSi then means no EL.

The total charge exchanged at the final stage Q_0 corresponds to the amount of Si oxidized at the surface of PSi (four charges needed to oxidize one Si atom). It is a constant for a given PSi layer. Thus, a higher current density will lead to a shorter total time. In addition, increasing the PSi layer thickness proportionally increases the amount of Si to oxidize and Q_0 is proportional to the thickness.

The EL spectrum blueshifts during ECO due to the fact that hole injection occurs in higher energy levels as the potential increases. Such experiment is highly reproducible and controllable. Bright EL, visible by the naked eye in daylight, at potentials below 2 V can be achieved. However, ECO is nonreversible because PSi is oxidized in the process and severely limited in time. Thus, this type of EL cannot be used in applications.

ECO can also be carried out under potentiostatic mode. In this case, the maximum energy level for hole injection is limited by the applied potential. The current density (and thus the EL intensity) decrease as time elapses due to the growing oxide (Billat et al. 1995).

It is interesting to notice a difference between lightly doped PSi (detailed above) and heavily doped PSi, resulting from a difference in PSi morphology/structure. The latter includes large Si structures inside PSi. The final sharp increase in potential is not seen because these large structures cannot be oxidized enough to break the electrical contact between PSi and the Si substrate. As a result, the oxidation can process further, accompanied by a progressive increase of the potential (Billat 1996; Billat et al. 1995), as shown in Figure 1.2. In this case, the reason for the EL extinction can be explained by the fact that the extent of the oxidation becomes so high that electrical contact between nonconfined PSi and confined crystallites is broken. However, a deterioration of the nanocrystals surface passivation at high potentials (>2 V) has also been suggested (Billat 1996; Billat et al. 1995).

The observation of EL in ECO may seem surprising as only holes are electrically injected into PSi. In fact, the oxidation of one Si atom involves four holes. It turns out that in the oxidation reaction, one electron is injected into the conduction band of PSi. Consequently, EL can be observed, attributed to radiative recombination of holes coming from the Si substrate with

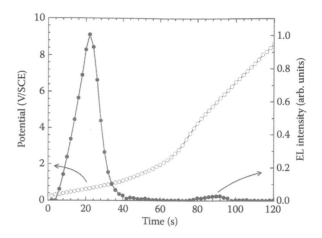

FIGURE 1.2   Potential and EL intensity as a function of time for a p⁺-type porous layer of porosity 80% and thickness 1 μm, during anodic oxidation at 5 mA/cm² in H_2SO_4 (1 M).

electrons resulting from the oxidation of Si atoms at the inner surface of PSi (Chazalviel and Ozanam 1992).

1.3.1.2 OTHER SYSTEMS

Anodic polarization means the substrate supplies holes. For EL, electrons must then be injected via a chemical reaction. ECO is completely nonreversible. However, it is possible to inject electrons in PSi using another specific electrochemical reaction, hopefully without affecting the PSi structure too much. In this case, the electrolyte contains an electro-active molecule chosen for its ability to inject electron at high energy so that this electron will be injected in the conduction band of PSi. Here, an oxidation reaction involving the exchange of two charge carriers with the working electrode (Si) is required. In a first step, a hole from PSi triggers the oxidation reaction, which then proceeds with the injection of an electron in the conduction band of PSi.

Methylviologen (Kooij et al. 1995) and formic acid (Green et al. 1995) oxidation lead to electron injection into PSi, after capture of holes provided by the substrate. EL was indeed observed in both experiments. However, this type of EL is not stable because simultaneous PSi oxidation taking place under anodic polarization could not be avoided.

Since the injection of holes by the substrate unavoidably leads to PSi oxidation, research efforts shifted toward injection of electrons by the substrate (cathodic polarization) and injection of holes by an electro-active species present in the electrolyte. These results are presented in the following section.

1.3.2 CATHODIC POLARIZATION OF PSi WITH PERSULFATE IONS

We now consider the case where electrons are electrically injected into PSi from the substrate by using cathodic polarization of PSi. Species capable of injecting high-energy holes in the valence band of PSi are rare. Only persulfate $\left(S_2O_8^{2-}\right)$ ions are suitable for this purpose, without deteriorating much PSi by secondary chemical reactions. The reduction of the $S_2O_8^{2-}$ ion is a two-step process. In the first step, an electron from silicon is consumed leading to the formation of a highly reactive radical. In the second step, the radical relaxation to stable SO_4^{2-} involves the injection of a hole into the silicon valence band (Memming 1969). The reaction itself does not oxidize PSi.

Since the two types of charge carriers can be injected in Si nanocrystals, EL can be generated in PSi (Bressers et al. 1992; Canham et al. 1992). The PSi electrode is not consumed during the process and the EL is more stable than that resulting from ECO. However, the stability is still affected by persulfate ion consumption and slow chemical oxidation of the PSi surface in the aqueous electrolyte.

Since the silicon substrate must supply electrons, mainly n-type PSi has been investigated with persulfate ions. However, EL could also be observed with heavily doped p-type PSi, which is also conductive under reverse bias (Gelloz et al. 1996), and with lightly doped PSi using back-side (Peter and Wielgosz 1996) or front-side (Gelloz and Bsiesy 1997; Gelloz et al. 1999a) illumination to generate electrons in the substrate.

An attractive feature of this type of EL is the very low voltage required because it is governed by the redox potential of persulfate ions. Thus, EL can be observed at voltages typically in the 1–4 V range. An even more attractive feature is the voltage-induced spectral shift of the EL (Bsiesy et al. 1993, 1994; Bsiesy and Vial 1996; Noguchi et al. 1999; Peter et al. 1996; Romestain et al. 1995; Saren et al. 2002). Indeed, as seen in Figure 1.3, the EL peak position blueshifts as the potential is rendered more cathodic (potential difference between Si and electrolyte increased). As in the case of ECO, injection in higher energy levels lead to emission of photons of higher energies as the voltage is increased. However, looking closer at the phenomenon, the spectral shift does not occur only by activation of higher and higher emission energies, but also by simultaneous quenching of the lowest emission energies by Auger effect (Romestain et al. 1995). Injection of one electron into a luminescent nanocrystal can trigger EL. However, if two or more electrons are injected into a nanocrystal, fast nonradiative Auger recombination dominates, resulting in no EL emission. This mechanism has been confirmed by the energy-selective quenching of PL triggered by selective electron injection (Bsiesy and Vial 1996; Gelloz et al. 2003a).

The dynamics of the EL of PSi with persulfate ions, as well as that of the PL quenching of PSi, has been investigated (Gelloz et al. 1997, 2003a). The response time was shown to be limited by the dynamic of charge carrier injection into PSi nanocrystals. The time constant was proportional to the Helmoltz capacitance at the electrolyte/PSi interface and the resistance of the electrolyte. The Helmoltz capacitance is proportional to the inner surface area of PSi, which can be very large (~300 m^2/cm^3). Thus, the EL response time was very long, typically several tens of milliseconds. The effect of the electrolyte resistivity and PSi thickness confirmed the proposed

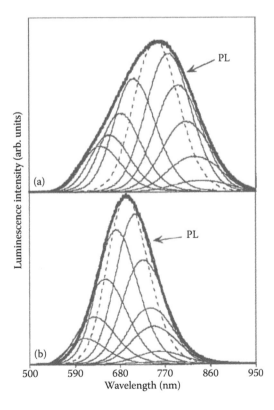

FIGURE 1.3 Comparison of PL and EL spectra for two samples of different porosity. The bold lines refer to the PL spectra. The non-normalized EL spectra are represented using a fine line. For the sake of comparison, the scale of the PL intensity is adjusted to the same peak intensity as that of the EL. (From Bsiesy A. and Vial J. C., *J. Lumin.* 70, 310–319, 1996. Copyright 1996: Elsevier. With permission.)

mechanism. For the PL quenching, an asymmetry in the response time can be obtained in particular conditions where the dynamic of carrier accumulation into PSi and depletion from PSi are different (Gelloz et al. 2003a).

These electrolytic EL systems are not suitable for applications due to the technologic difficulties related to the liquid electrodes and the stability issues.

1.4 CHARGE CARRIER TRANSPORT IN ELECTROLYTIC SYSTEMS AND DIFFERENCES WITH SOLID-STATE DEVICES

Even though the EL of electrolytic systems is rather efficient, achieved at low voltages, and exhibits large voltage-tunable spectral shifts, the EL of solid-state devices is weak in all aspects. Indeed, as we will in subsequent sections, it is usually low-efficiency, unstable, and requires rather high voltages. One very big difference between the two types of systems is the charge carrier transport and conduction in PSi. In this section, the conduction of electrolytic systems is presented first and then that of solid-state devices is briefly discussed.

1.4.1 CHARGE CARRIER TRANSPORT IN ELECTROLYTIC SYSTEMS

In electrolytic systems, conductive liquid impregnates completely the PSi structure and electrically short circuits the highly resistive PSi layer. There are two questions: what are the conditions for carrier accumulation into PSi and what is the transport mechanism in PSi?

The first question is addressed first. Figure 1.4 shows the different types of conductions available, depending on what limits the flow of carriers (Bsiesy et al. 1996; Gelloz et al. 1996). The

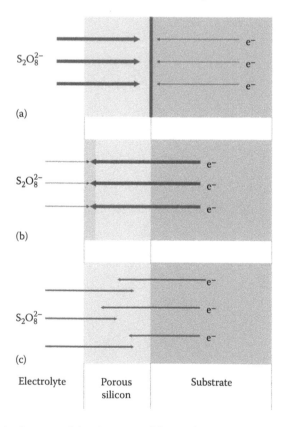

FIGURE 1.4   Schematic diagram of the three possible conduction regimes of PSi/electrolyte system illustrated for the case of persulfate ions. Electrons or holes limited conduction (a), limitation by the mass transport in the electrolyte (b), and by the electrochemical reaction rate (c). (From Bsiesy A. et al., *J. Appl. Phys.* 79, 2513–2516, 1996. Copyright 1996: AIP. With permission.)

current can be limited by either of three mechanisms: (1) the supply of carriers by the substrate (regime 1). The carriers are consumed immediately as they are generated at the Si/PSi interface by ions in large numbers. They do not penetrate into PSi, and consequently no EL can be generated. This is actually what is taking place during the formation of PSi when the electrolyte contains HF. (2) The supply of ions by diffusion in the electrolyte (regime 2). EL is possible only within a thin layer at the top of the PSi layer because all ions inside PSi have been consumed. (3) The kinetic of the electrochemical reaction (regime 3). EL is possible within the entire PSi layer. This is the case for ECO, or when PSi is partially oxidized in order to lower the charge exchange kinetics.

Figure 1.5 shows an experiment where a transition from regime 1 to regime 2 is observed *in situ* (Bsiesy et al. 1996; Gelloz et al. 1996). The P+ substrate can generate electrons at the Si/PSi interface, but at a limited rate. During the first part of the experiment, a plateau of current and no EL were observed. This corresponds to consumption of electrons at the Si/PSi interface. Then the current decreases and EL can be observed. This happens when all the ions at the bottom of the PSi layer have been consumed, and electrons could penetrate into PSi. This stage depends on the amount of ions in solution (i.e., ion concentration), as seen in the figure. After a short transition, the charge exchanges and then proceeds only at the top of PSi.

We now address the question of the carrier transport mechanism in PSi. When charge carriers can penetrate into PSi, their transport was shown to be mostly a diffusion process, supported by the fact that no significant electric field is set across PSi (the electrolyte electrically short-circuits PSi) (Gelloz and Bsiesy 1997, 1998; Gelloz et al. 1999a). This mechanism was also supported by the observation that carriers photo-generated in PSi are not electrically active and by PL quenching being triggered only by carrier accumulation independently of the electric field set across the system (Gelloz et al. 1999a).

Therefore, a key feature responsible for the high efficiency and low operating voltages is the absence of any large voltage drop in the liquid-impregnated porous skeleton. Another important characteristic of electrolytic EL is the fact that both electrons and holes can be efficiently injected into Si nanocrystals in part or the whole thickness of PSi. In the case of cathodic polarization, electrons are accumulated in PSi, and holes are injected from the electrolyte, in a parallel manner because the liquid impregnates completely the porous network. If only persulfate ions react, the balance of electron and hole injection is one to one, which is perfect. In practice, a supporting electrolyte has to be added to the system (e.g., H_2SO_4) for stability and parallel reduction of hydrogen reduces the EL efficiency. The efficiency is also affected by charge exchanges taking place in nonconfined parts of PSi.

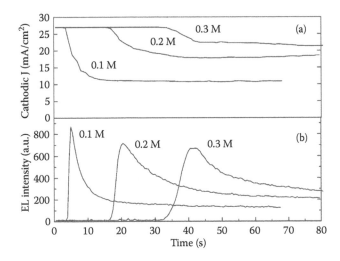

FIGURE 1.5 Time-evolution of the cathodic current density (a) and the EL signal (b) for a 4-μm-thick p+-type PSi layer polarized at –1.6 V/SCE. The different characteristics correspond to different concentrations (in mol/l) of $S_2O_8^{2-}$ in the electrolyte. (From Bsiesy A. et al., *J. Appl. Phys.* 79, 2513–2516, 1996. Copyright 1996: AIP. With permission.)

1.4.2 CHARGE CARRIER TRANSPORT IN SOLID-STATE DEVICES

In solid-state devices, high voltages are necessary in order to get a significant current through PSi because of the very high series resistance of the layers. The resistivity of PSi is usually greater than 10^5 Ω·cm at 300 K. As a result, high electric fields are set across PSi. Locally, the electric field can be high enough to separate the electron-hole pairs, thus seriously limiting the EL efficiency (Oguro et al. 1997). This is a major difference with the electrolytic configuration.

Another great difference is the charge carrier injection mechanism. In dry PSi, the charge carrier supply in Si nanocrystals occurs in series and not in parallel as carriers flow from electrode toward the other. Furthermore, the charge carriers flow preferentially through the lowest resistance paths, which include all the nonluminescent parts of PSi. This effect reduces considerably the efficiency by inducing a large leakage current.

The electron-hole pair generation mechanism in most solid-state PSi-based EL devices is due to impact processes via accelerated electrons through the highly resistive PSi skeleton, as illustrated in Figure 1.6 (Oguro et al. 1997). In contrast, in wet PSi, both electrons and holes are efficiently injected without the need for high electric fields across PSi. This effect is another reason for the typically low efficiency and stability of solid-state PSi EL.

Another difference with electrolytic systems is the quality of the electrode contacting PSi. With the liquid, a very smooth and intimate contact results in an ideal contact interface, whereas in solid-state devices, a thin solid electrode is deposited onto PSi. The contact is mostly physical rather than chemical and is not smooth and continuous due to the roughness of PSi. As a result, the contact itself is a major cause of instability and operation problems in many solid-state devices. This issue is discussed in Section 1.9.

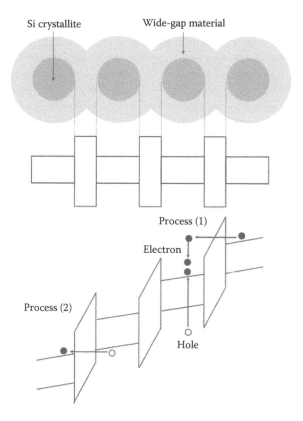

FIGURE 1.6   Simplified schematic band diagram of a quantum wire in PSi under thermal equilibrium, and band structure of the Si nanostructure system under a strong electric field. Two possible high-field effects are illustrated; process 1 (electron-hole-pair generation by energetic hot electrons) and process 2 (tunneling injection of electrons from the valence band). (From Oguro T. et al., *J. Appl. Phys.* 81, 3, 1407–1412, 1997. Copyright 1997: AIP. With permission.)

TABLE 1.3   Some Characteristics of Most Devices Including a Single PSi Layer

Contact	Structure	Posttreatment	EL Threshold V-mA/cm²	Stability	Emission Peak (nm)	Highest Efficiency EQE-EPE (%)	Ref.
Au	p(D)		4–50		680	10^{-3}–	Koshida and Koyama 1992
ITO	p(D)		<2–<10	>5 h	580		Namavar et al. 1992
Au or ITO	n(L)		–1	Minutes	700		Kozlowski et al. 1992
Au	n⁺(L)		–0.1		680	0.05–	Oguro et al. 1997
Au	Oxide-free PSi from p(D)	L+H₂ exposure for 12 hours			430		Mizuno and Koshida 1999

Note: D and L mean that anodization was conducted in the dark and under illumination, respectively. ECO stands for electrochemical oxidation. EQE and EPE are external quantum efficiency and external power efficiency, respectively.

1.5 DEVICES INCLUDING AS-FORMED PSi

Most devices in this section are listed in Table 1.3. These devices are based on Si/PSi/top contact junctions, where the top contact is usually either gold or indium tin oxide (ITO). The best results are usually obtained with ITO because it offers much better transparency than gold and fair conductivity (also see Section 1.9.1).

The first true demonstration of injection mode solid-state EL based on PSi was reported in 1992 (Koshida and Koyama 1992). The efficiency of this type of devices, without any further treatment, is usually low, that is, $<10^{-3}$% (Koshida and Koyama 1992; Kozlowski et al. 1992; Maruska et al. 1992; Namavar et al. 1992), except in one case where 0.05% was claimed (Oguro et al. 1997). Operating voltages for EL visible in daylight are rather high, usually greater than 10 V.

Noticeably, blue EL has been demonstrated (Mizuno and Koshida 1999) using an as-formed PSi layer formed from p-type Si, with a gold top contact. PSi was thinned by using photochemical etching in HF. The efficiency of this device was very low. Blue emission is difficult to get for several reasons: (1) the fabrication of mechanically stable layers with a high density of small nanocrystals is difficult, (2) keeping the nanocrystals oxide-free (to avoid the luminescence red-shift due to introduction of surface states by oxygen) is not an easy task, and (3) charge carrier injection into high energy levels is difficult to obtain. The important result of this report (Mizuno and Koshida 1999) is the demonstration of the possibility of RGB LEDs based on Si.

1.6 POROSIFIED p-n JUNCTIONS

Most devices in this section are listed in Table 1.4. It was hoped that devices including a PSi layer formed from a p-n junction could allow injection of holes and electrons from the two different electrodes. However, it is unclear whether a true p-n junction can exist in PSi. Indeed, free carriers are mostly absent from PSi, even when the bulk Si initial doping level is high, and should be trapped by surface states (Grosman and Ortega 1997). As for p-type Si, it seems that dopant atoms are still present in PSi at concentrations similar to bulk Si, but are passivated (Grosman and Ortega 1997).

The use of p⁺-n junctions has provided rather good results. Boron atoms were incorporated into an n-type substrate either by implantation or by diffusion. After PSi was formed, the top contact was deposited onto the p⁺ side. Devices in which a p-n junction is porosified has shown better efficiencies than devices including one Si type only and without any postanodization treatment. p⁺-n⁻ (Lalic and Linnros 1996a,b; Lang et al. 1997; Linnros and Lalic 1995; Loni et al. 1995; Nishimura et al. 1998; Steiner et al. 1993), n⁺-p⁻ (Peng and Fauchet 1995), n⁺-p⁺ (Chen et al. 1993), and p⁺-n⁺ (Gelloz et al. 1999b) junctions have been studied.

A noticeable device used a porosified n-p⁺ junction (Lalic and Linnros 1996a,b; Linnros and Lalic 1995). The junction was at a depth of 0.25 µm. Anodization was performed until the total PSi thickness was in the range of 20–60 µm. PSi from the n-type part is composed of a nanoporous layer at the top and a macroporous underlying layer, as illustrated in Figure 1.7. EL originated

TABLE 1.4   Some Characteristics of Most Devices Based on Porosified p-n Junction

Contact	Structure	Posttreatment	EL Threshold V-mA/cm²	Stability	Emission Peak (nm)	Highest Efficiency EQE-EPE (%)	Ref.
Au	n⁺p⁺p(D)		−<600	>6 h	640	0.18−	Chen et al. 1993
Au	p⁺n(L)		1.7−0.1	80 h	700	10^{-2}−	Steiner et al. 1993
ITO	p⁺n(L)	1 min L	$2.3-10^{-3}$	Hours	600	0.18−	Loni et al. 1995
Au	p⁺n(L)			Seconds	630	0.16−0.016	Linnros and Lalic 1995
Au	p and p⁺n(L)	2 min exposure in 10% HNO₃	5−		650	0.01−	Peng and Fauchet 1995
Au	p⁺n(L)		3−1	Seconds	670−780	0.2−	Lalic and Linnros 1996a,b
Au	p⁺n(L)		$>10-10^{-3}$		690	0.8−0.07	Nishimura et al. 1998
ITO	p⁺n⁺(L)	ECO	$5-1.5 \times 10^{-4}$	Hours	650	1.1−0.08	Gelloz et al. 1999b

Note: D and L mean that anodization has been conducted in the dark and under illumination, respectively. ECO stands for electrochemical oxidation. EQE and EPE are external quantum efficiency and external power efficiency, respectively.

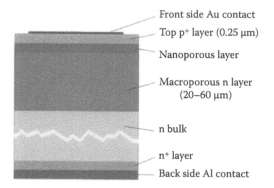

Front side Au contact
Top p⁺ layer (0.25 µm)
Nanoporous layer
Macroporous n layer (20−60 µm)
n bulk
n⁺ layer
Back side Al contact

FIGURE 1.7 Schematic presentation of a PSi diode structure based on a porosified p⁺n junction. (From Lalic N. and Linnros J., *J. Appl. Phys.* 80, 10, 5971–5977, 1996. Copyright 1996: AIP. With permission.)

from the n-type part. The device showed high series resistance (due to high PSi thickness), inducing high operating voltages, and perhaps shunting the p-n behavior. It was pulse-operated. The best device showed an EQE of about 0.2%.

In another device (Nishimura et al. 1998) using a porosified p-n junction, an EQE as high as 0.8% was reported under pulsed operation, but for high voltages and very low EL intensities.

With a thin p⁺-n porosified junction (Loni et al. 1995; Simons et al. 1997), calculations supported the fact that a p-n junction was existing in the porosified layer in the device, even though the boron-implanted atoms were not annealed for electrical activation. The EL was believed to take place in the p⁺-n junction. Due to the thin PSi layer used (400 nm), their fresh device was CW-operated at rather low voltages, below 6 V. The best EQE was 0.18% (Canham et al. 1996; Cox et al. 1999; Loni et al. 1995; Simons et al. 1997) at 4–6 V and an EQE of about 0.1% was reproducibly achieved. The typical output of the device is shown in Figure 1.8. The EQE was highly dependent on the boron dose. Best performance was found for a dose of about 10^{16} cm⁻². The temperature was found very important for good reproducibility.

Another interesting device has been fabricated by anodizing an n⁺-p-⁺p substrate (Chen et al. 1993). PSi was formed from the n⁺ side and extended up to the p region. The top contact was Au. EL could be seen within a forward voltage of 5–10 V and a current density of about 600 mA/cm². Rectifying properties were very good. The light originated from a PSi region a few micrometers below the top contact-PSi interface. The authors then concluded that the EL had to be the result of electron-hole recombination at the porous n⁺-p⁺ junction.

Finally, a record EQE of 1.1% (with an EPE of 0.08%) was reported using a porosified p⁺-n⁺ junction (Gelloz et al. 1999b). The PSi layer was about 20 µm. This device also suffered from high

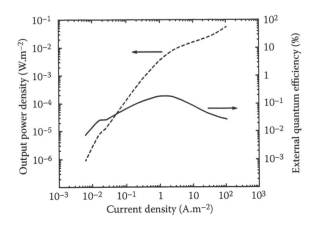

FIGURE 1.8   Efficiency and output of a device based on porosified thin p⁺n junction. (From Canham L. T. et al., *Appl. Surf. Sci.* 102, 436–441, 1996. Copyright 1996: Elsevier. With permission.)

voltage operation. The high EQE was attributed mainly to the postanodization ECO of PSi (see Section 1.7.2).

1.7 PARTIALLY OXIDIZED PSi

Most devices in this section are listed in Table 1.5. Also, see oxidized devices in Table 1.4. Partially oxidizing PSi is useful for three reasons: (1) it reduces the sizes of the Si network, inducing an enhancement of the PL efficiency (via enhanced charge carrier localization or creation of new luminescent nanocrystals), (2) it usually leads to a reduction of the leakage current because low-resistance nonluminescent regions are reduced, and (3) it generally enhances the PSi stability by a surface passivation effect.

PSi oxidation has been conducted under various conditions: chemically (Kozlowski et al. 1995), thermally (Fauchet et al. 1997; Tsybeskov et al. 1995, 1996), and electrochemically (ECO) (Gelloz and Koshida 1999, 2000; Gelloz et al. 1998, 1999b; Pavesi et al. 1999).

TABLE 1.5   Some Characteristics of Most Devices Based on Partially Oxidized PSi

Contact	Structure	Posttreatment	EL Threshold V-mA/cm²	Stability	Emission Peak (nm)	Highest Efficiency EQE-EPE (%)	Ref.
Au	n(L)	H₂O₂ oxidation			650–750		Kozlowski et al. 1995
Au	n(UV)	H₂O₂ oxidation		>7 h	460–550		
Al-poly Si	p⁺p(D)	Anneal in N₂ or in 10% O₂ in N₂	1.5–2	1 month	620–770	0.1–	Fauchet et al. 1997; Tsybeskov et al. 1995, 1996
ITO	n⁺(L)	ECO	$3.5–4 \times 10^{-4}$		640	0.51–0.05	Gelloz et al. 1999b
ITO	n⁺(L)	ECO	$3–10^{-4}$	Hours	640	0.21–0.02	Gelloz et al. 1998
ITO	n⁺(D)n⁺(L)	ECO	$2–1.8 \times 10^{-3}$	Hours	680	0.5–0.2	Gelloz and Koshida 1999
Al/n⁺	p(L)	ECO		>1 week			Pavesi et al. 1999
ITO	n⁺(D)n⁺(L)	ECO	$2.2–7 \times 10^{-4}$	Days, EQE is stable	680	1.07–0.37	Gelloz and Koshida 2000
ITO	n⁺p	HWA	$2–10^{-4}$	Stable	700	10^{-3}–	Gelloz and Koshida 2006
ITO	n⁺	ECO+HWA	$2–10^{-4}$	Stable	820	–	Gelloz et al. 2006

Note: D and L mean that anodization has been conducted in the dark and under illumination, respectively. ECO stands for electrochemical oxidation. EQE and EPE mean external quantum efficiency and external power efficiency, respectively.

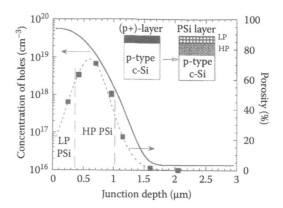

FIGURE 1.9 Initial hole concentration and resulting porosity profile of part of an LED. The inset shows how the p+/p c-Si substrate structure is transformed by anodization into a low porosity (LP) top layer and a high porosity (HP) active layer on the p c-Si substrate. (From Fauchet P. M. et al., *Thin Solid films* 297, 254–260, 1997. Copyright 1997: Elsevier. With permission.)

1.7.1 CHEMICAL AND THERMAL OXIDATION

The most noticeable devices are now briefly discussed. With n-type PSi, oxidation in H_2O_2/water/ ethanol mixture (Kozlowski et al. 1995) stabilized the EL at about 50% of the initial value for several hours. EQE was found to increase during operation.

One group has oxidized PSi using thermal oxidation in the range 800–900°C, either in N_2 (nominally) or in N_2 with 10% O_2 (Fauchet et al. 1997; Tsybeskov et al. 1995, 1996). The device consisted of a porosified p⁺-p⁻-type junction, with the characteristics shown in Figure 1.9. The top contact was a 300-nm thick n⁺ type poly-Si deposited onto the p⁺ side. The resulting device had several advantages. First, the stability was rather good, with weeks of stable EL (Tsybeskov et al. 1995). Second, the EQE (0.1%) was relatively high (Fauchet et al. 1997; Tsybeskov et al. 1995, 1996). Third, the EL could be modulated by a square wave current pulse with frequencies greater than 1 MHz (Tsybeskov et al. 1996). Finally, a bipolar device fully compatible with conventional Si microelectronic processing was demonstrated (Hirschman et al. 1996; Tsybeskov et al. 1996). The high stability was due to the replacement of the fragile hydrogen covering the initial PSi surface by a thin Si oxide layer. The increase in response speed may be related to a different recombination mechanism involving the oxide rather than the interior of the Si nanocrystals.

Blue and green EL have been reported from partially oxidized PSi (Mimura et al. 1996). Blue EL was obtained from PSiC layers and green EL from PSi formed under UV illumination. In this latter case, the PL and EL spectra were different, and thought to have different origins. The efficiency was very low (<10^{-5}%).

1.7.2 ELECTROCHEMICAL OXIDATION

Postanodization of partial ECO of PSi has been studied in a view to increase the efficiency and stability (Gelloz et al. 1998). PSi includes nonluminescent weakly confined Si regions through which a large leakage current flows. This is one of the reasons for the low EQE of PSi-based EL. The objective of ECO is to selectively decrease the size of the nonconfined Si skeleton without affecting much the confined Si nanocrystals, in order to reduce the leakage current without damaging the luminescence. In addition, ECO performed in appropriate conditions optimizes charge carrier injection into luminescent crystallites.

When oxidation is performed up to the maximum of EL during ECO, the nonemissive coarser regions of PSi have been significantly oxidized and therefore their sizes have been much reduced, whereas low dimensional luminescent nanocrystals have only been slightly oxidized and have been well preserved. As a result, ECO decreases the leakage current by several orders of magnitude. Thermal or chemical oxidation could not lead to such an enhancement because it occurs almost uniformly on the whole internal surface of PSi.

ECO also enhances the luminescence homogeneity. If PSi is not uniform, ECO acts as a kind of healing treatment as far as it tends to homogenize the size distribution along the conductive paths. The EL during ECO is also a unique probe of PSi homogeneity.

The strong effect of ECO on the EL efficiency was first demonstrated in 1998 (Gelloz et al. 1998), with an obtained EQE of 0.21% (EPE of 0.02%) with a single PSi layer made from n$^+$ (100) Si. EQE of 0.51% and 1.1% (EPE of 0.08%) were then obtained with n$^+$ (111) and p$^+$n$^+$ (111) (p$^+$ on the top side) PSi layers, respectively (Gelloz et al. 1999b). However, the operating voltages of these devices were still high, exceeding 10 V. An optimized device, which remains to date one of the best PSi EL devices, operating below 5 V, with an EQE of 1.07% and an EPE of 0.37% was later demonstrated (Gelloz and Koshida 2000). The modulation speed was about 33 kHz. Strictly controlled conditions were very important for good reproducibility. In particular, the temperature was maintained at 0°C during PSi formation of the active layer. An even lower voltage was achieved with a device also including a PSi layer oxidized using ECO (Gelloz and Koshida 2004).

Figure 1.10 shows the EL intensity and the current density as a function of voltage, for two devices (Gelloz and Koshida 2000). They have been prepared in the same conditions, except that one has been subjected to ECO and the other one was not oxidized. The EL of the oxidized device could be seen with the unaided eye in room lighting at operating voltages below 5 V. Since ECO induced a decrease of the current density of about 3 orders of magnitude and a tenfold enhancement of EL intensity, the EQE of the device was increased by more than four orders of magnitude. The dramatic enhancement of EQE of PSi EL due to ECO has been confirmed by another group (Pavesi et al. 1999).

ECO also improves the EL stability (Gelloz and Koshida 2000; Gelloz et al. 1998, 1999b). The optimized device discussed previously (Gelloz and Koshida 2000) showed no loss of efficiency during operation, and after one-month storage in air. However, the PSi layer still was slowly oxidized during operation and storage in air. The enhanced stability provided by ECO was confirmed by another group (Pavesi et al. 1999). No degradation of EL intensity during several days could be seen. In this case, the conditions used for the ECO have probably led to the growth of an oxide layer, which has replaced the initial hydrogen passivation.

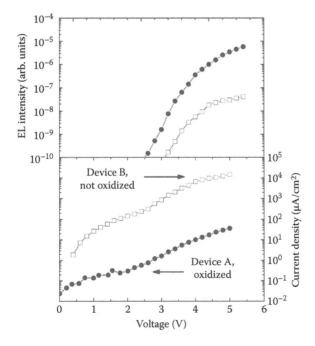

FIGURE 1.10 EL intensity and current density as a function of applied voltage for an anodically oxidized (device A) and a not-oxidized device (device B). (From Gelloz B. and Koshida N., *J. Appl. Phys.* 88(7), 4319–4324, 2000. Copyright 2000: AIP. With permission.)

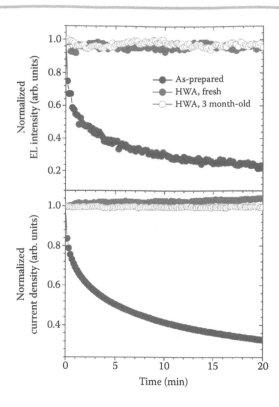

FIGURE 1.11 Normalized EL intensity and normalized current density versus continuous operation time two devices. The two devices were identical except one was treated by HWA. The as-prepared device and HWA-treated device were driven at reverse bias voltages of 10 V and 15 V, respectively. (From Gelloz B. et al., *Appl. Phys. Lett.* 89, 191103, 2006. Copyright 2006: AIP. With permission.)

1.7.3 OXIDATION BY HIGH-PRESSURE WATER VAPOR ANNEALING

A treatment of PSi (Gelloz and Koshida 2005; Gelloz et al. 2005a) based on high-pressure water vapor annealing (HWA) has been proposed to enhance the efficiency and the stability of PSi PL (see Chapter 7 in *Porous Silicon: Formation and Properties*). A record high PL EQE of 23% has been reported, with an outstanding stability, using a pressure of 2.6 MPa at 260°C. The extremely high efficiency is a result of high exciton localization in Si nanocrystals and much reduced non-radiative defect density at the Si/SiO$_2$ interface, the stress in the oxide being very much relaxed compared to that of conventional partially oxidized PSi.

Application of this technique in EL (Gelloz and Koshida 2006b; Gelloz et al. 2006) led to very stable EL emission (Gelloz et al. 2006). The EL was absolutely stable on operation (Figure 1.11) and also on storage in air for several months (Figure 1.12). It is the most stable PSi EL device reported to date. HWA tends to increase the diode conductivity, probably owing to the marked reduction of trapping centers in PSi. The diode was operated below 10 V. More work is needed to improve its efficiency.

1.8 PSi IMPREGNATED BY ANOTHER MATERIAL

Most devices in this section are listed in Table 1.6. In order to emulate the efficient electrolytic EL, impregnation of PSi by a conductive material, either a metal or a polymer, has been considered.

It is worth noticing that good impregnation of a material much more conductive than PSi could short-circuit the PSi layer, resulting in low carrier injection into Si nanocrystals. Such a situation would be similar to regime 1 described in Section 1.4.1 for liquid contact (Bsiesy et al. 1996; Gelloz et al. 1996), in the sense that the contact would prevent carrier injection from the Si substrate into PSi. Solid-state emulation of regime 2 or, even better, regime 3 of the liquid contacted systems would be a promising approach for enhancing the efficiency.

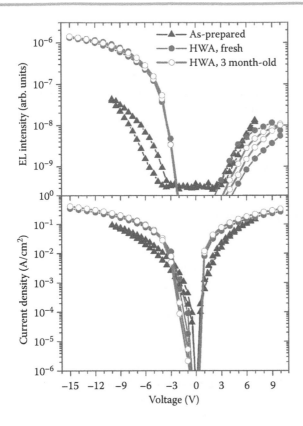

FIGURE 1.12   Current-voltage curves of the two devices shown in Figure 1.11, and the corresponding EL intensities. The as-prepared and HWA-treated devices were operated just after fabrication (fresh). Measurement of the HWA-treated device was conducted again after 3-months storage in air. (From Gelloz B. et al., *Appl. Phys. Lett.* 89, 191103, 2006. Copyright 2006: AIP. With permission.)

TABLE 1.6   Some Characteristics of Most Devices Based on PSi Impregnated by Another Material

Contact	Structure	Posttreatment	EL Threshold V-mA/cm²	Stability	Emission Peak (nm)	Highest Efficiency EQE-EPE (%)	Ref.
Au	p(D)	Polypyrrole electro-deposition	2–<0.01		590		Koshida et al. 1993, 1994
Au	n⁺(D)	PANI chemical deposition	3–400		800		Bsiesy et al. 1995
Au	n(L)	Polyaniline chemical deposition	–500		790		Halliday et al. 1996
Au	n(L:UV)	In electroplating	–0.1	Hours	455–480	0.01–	Lang et al. 1997; Steiner et al. 1994, 1996
Au	n(L:UV)	Ga electroplating		Hours	520		Steiner et al. 1996
Au	n(L:UV)	Sn electroplating		Hours	550	0.0005–	Lang et al. 1997; Steiner et al. 1996
Au	n(L:UV)	Sb electroplating		Hours	700–750	0.0001–	
Au	n(L:UV)	Al electroplating	–0.1	Hours	480	0.005–	Lang et al. 1997

Note: D and L mean that anodization has been conducted in the dark and under illumination, respectively. EQE and EPE mean external quantum efficiency and external power efficiency, respectively.

1.8.1 INCORPORATION OF METALS

The incorporation of In (Lang et al. 1997; Steiner et al. 1994, 1996), Al (Lang et al. 1997), Sn (Lang et al. 1997; Steiner et al. 1996), and Sb (Lang et al. 1997; Steiner et al. 1996) into PSi pores has been investigated by electrochemical techniques. The best device was obtained with In, with an EQE of 0.01% in ac conditions (Lang et al. 1997). In plating increased the EQE by a factor of 150, but PSi was also partially oxidized by the deposition process. The mechanism by which the efficiency was increased by In electroplating was not fully understood as the oxidation was probably also contributing to the enhancement. This device emitted blue (480 nm) light. This EL may as well be oxide-related rather than the result of exciton recombination in Si nanocrystals.

1.8.2 INCORPORATION OF POLYMERS

As for polymers, polypyrrole (Koshida et al. 1993, 1994) and polyaniline (Bsiesy et al. 1995; Halliday et al. 1996) have been used. Efficiency is usually enhanced by the polymer impregnation, but the attempt to emulate completely the liquid contact could not be achieved in any case, the efficiency remaining low.

One group (Koshida et al. 1993, 1994) studied a device in which polypyrrole had been electrochemically deposited into p-type PSi pores and a gold contact deposited onto PSi. The current-voltage characteristics and the voltage and current dependence of EL were significantly improved in comparison with a control device. Figure 1.13 shows that the EL intensity as a function of injection power tends to saturate for the gold electrode whereas it stays fairly linear for the polymer electrode because of the difference in potential drop across the devices. The EPE was improved by a factor of 3 by polymer incorporation.

Another group (Halliday et al. 1996) has chemically deposited polyaniline (PANI) into and onto PSi made from n-type Si. The polymer acted as a hole injector. EL was obtained at 0.5 A/cm². Another group (Bsiesy et al. 1995) deposited PANI into n⁺ PSi by chemical oxidation of aniline by persulfate ions. Optimal results were obtained with deposition of two layers of PANI. More layers of PANI reduced the EL intensity. The test device without polymer showed a voltage threshold of EL of about 9 V. When two PANI layers were deposited in PSi, the voltage threshold of EL became 3 V and EL intensity increases with a much higher slope than when no polymer was present in PSi, as shown in Figure 1.14. However, when more PANI layers were incorporated,

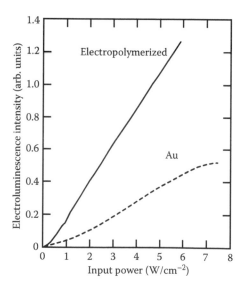

FIGURE 1.13   EL intensity of PSi diodes with a polymerized contact (solid curve) and a thin Au contact (dashed curve) as a function of injected electrical power. (From Koshida N. et al., *Appl. Phys. Lett.* 63, 2655–2657, 1993. Copyright 1993: AIP. With permission.)

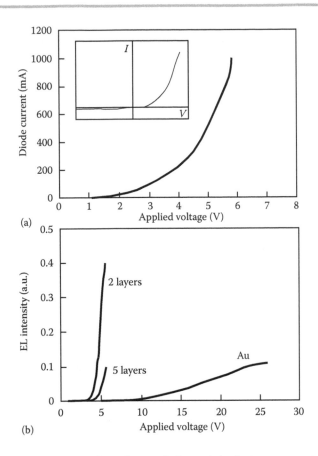

(a)

(b)

FIGURE 1.14   (a) *I-V* characteristic of the forwardly biased diode. Inset: complete *I-V* characteristic; (b) EL-V characteristics (i) of PSi diodes contacted with 2 and 5 PAN1 layers and (ii) of a PSi diode contacted by an evaporated semitransparent thin Au film. (From Bsiesy A. et al., *Thin Solid Films* 255, 43–48, 1995. Copyright 1995: Elsevier. With permission.)

the performance of the diode dropped, even though it was still better than without polymer. The EL intensity was found six times lower than that from the liquid junction cell. The current-voltage characteristics and the voltage and current dependence of EL were significantly improved in comparison with a control sample including a gold top contact. In another device by another group (Li et al. 1994) the EL intensity was increased compared to a control device.

1.8.3 INCORPORATION OF INERT MATERIALS

Filling of the pores with materials that are inert from the electrical and optical point of view has been proposed to enhance the mechanical stability of the PSi skeleton (Fauchet et al. 1997). Various polymers have been tried, such as PVC, polyamide, PMMA, polystyrene, and polypropylene. The hardness of the resulting nanocomposite increased with the density of the base polymer. Maximum increase in hardness achieved was 50%.

1.9 INFLUENCE OF DEVICE TOP ELECTRODE

Several devices in this paragraph are listed in Table 1.7. Oxidized devices can be found in Table 1.5. Several top contact configurations have been investigated. An ideal contact should be transparent to visible light in order to guarantee maximum light extraction, be mechanically strong, and resistant to stress induced by heat.

TABLE 1.7   Some Characteristics of Several Devices in Which a Particular Top Contact Has Been Used

Contact	Structure	Posttreatment	EL Threshold V-mA/cm²	Stability	Emission Peak (nm)	Highest Efficiency EQE-EPE (%)	Ref.
ITO/n-type SiC	p(D)	KOH dip					Futagi et al. 1992
Al/Al₂O₃	n or n⁺ poly		5–400	>1 month		–0.01	Lazarouk et al. 1996a,b
Au	n poly Si (L)				860–930	0.04–	Kozlowski et al. 1996
Au	p⁺n(L)			Minutes	700	0.02–	Simons et al. 1997
ITO	p⁺n(L)			Hours	700	0.2–	
Au-Al-pin (a-Si:H)	np(D)				3 peaks: 455, 590, 670	0.13–	Chen et al. 1997a
Au-Al-npin (a-Si:H)	p(D)		6–				Chen et al. 1997b
Au-Al-npnin (a-Si:H)	p(D)		3.6–				
ITO/a-C	n⁺(D)n⁺(L)	ECO	0.5–	Minutes	Voltage-tunable	–0.35	Gelloz and Koshida 2004

Note: D and L mean that anodization has been conducted in the dark and under illumination, respectively. ECO stands for electrochemical oxidation. EQE and EPE mean external quantum efficiency and external power efficiency, respectively.

1.9.1 METALS AND CONDUCTIVE OXIDES AS TOP ELECTRODES

Most devices include a thin semitransparent gold or ITO top contact. The two types of top electrodes have been compared (Simons et al. 1997). Devices with a semitransparent Au contact were much less stable in air than those with an ITO contact because of the gold layer being more permeable to air than the ITO layer. The devices contacted by gold were oxidized and degraded much faster than those contacted by ITO were. The device with ITO exhibited an efficiency ten times higher than with gold. These results can be explained partly because gold electrodes are typically 10–20 nm thick (to be semitransparent), whereas ITO can be more than 100 nm thick and therefore mechanically much stronger and impermeable to air.

Some devices include Al top contacts. In one case (Lazarouk et al. 1996a,b), aluminum contact was deposited onto PSi. Then, parts of the Al layer were electrochemically oxidized into Al₂O₃ to create transparent windows through which EL was observed. The device structure is shown in Figure 1.15. The stability was reported to be more than a month. However, the efficiency was limited by the fact that a significant part of the light could not be extracted. Moreover, the color emitted by the device was white, leading the conclusion that the EL probably originated from some oxide-related centers rather than from the nanocrystals cores.

EL colors from red to blue have been obtained by variation of the contact metal (Kozlowski et al. 1996; Lang et al. 1997; Steiner et al. 1994, 1996). In/Au, Al/Au, Ga/Au, Sn/Au, and Sb/Au contacts lead to EL emission at 455, 455, 520, 555, and 700 nm, respectively, as shown in Figure 1.16 (Lang et al. 1997). The metals were also impregnated into PSi. However, the oxide in the PSi layers is believed to be responsible for the short wavelength EL emission.

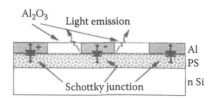

FIGURE 1.15   Scheme of a device presenting an Al/Al₂O₃ top layer. (From Lazarouk S. et al., *Appl. Phys. Lett.* 68, 1646–1648, 1996. Copyright 1996: AIP. With permission.)

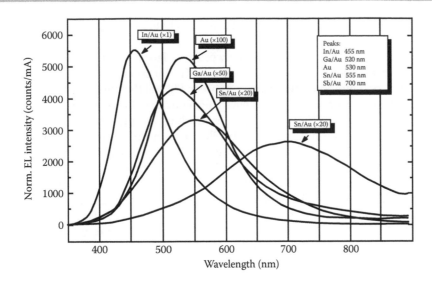

FIGURE 1.16 Spectra taken from various metal compositions. The intensity of the EL is normalized by the current, which flows through the device, so that some information on the efficiency is provided. (From Steiner P. et al., *Thin Solid Films* 276, 159–163, 1996. Copyright 1996: Elsevier. With permission.)

1.9.2 COMPACT PSi BETWEEN ACTIVE PSi AND TOP ELECTRODE

If the top contact is directly deposited onto the rough and uneven surface of a high-porosity PSi layer, conduction peculiarities may arise and reproducibility can be bad (Gelloz and Koshida 1999). To solve these problems, some authors (Fauchet et al. 1997; Gelloz and Koshida 1999, 2000; Gelloz et al. 1999b; Tsybeskov et al. 1995, 1996) have included a superficial compact porous layer between the optically active porous layer and the top contact. This superficial layer provides a better electrical contact and greater mechanical stability to the device.

In one case (Gelloz and Koshida 1999, 2000), already discussed in Section 1.7.2, the superficial layer consisted of a PSi layer which surface was still exhibiting the mirror property of the silicon substrate and which was more compact than the underlying active porous layer. The device structure is shown in Figure 1.17. In some other cases, a p⁺ PSi superficial layer (Fauchet et al. 1997; Gelloz et al. 1999b; Tsybeskov et al. 1995, 1996) was used, as shown earlier in Figure 1.9. PSi formed from p⁺ Si showed much better mechanical stability than PSi made from other Si substrates. Therefore, such a layer should be a good choice as a buffer layer between the top contact and the active PSi layer.

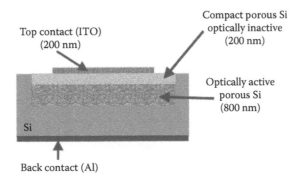

FIGURE 1.17 Cross-sectional view of a device which structure is ITO/superficial compact PSi/optically active PSi. Not to scale. Substrate is n⁺ type Si. This device can be operated both under forward and reverse bias. Such a device has been shown to exhibit an external quantum efficiency of more than 1% and a record of 0.37% of external power efficiency below 6 V.

1.9.3 OTHER MATERIALS BETWEEN ACTIVE PSi AND TOP ELECTRODE

A 300-nm thick n⁺ poly-Si layer has been used between the Al top contact and PSi (Fauchet et al. 1997). This device has already been discussed in Section 1.7.1. The incorporation of the poly-Si layer reduces the surface states concentration at the interface between PSi and the top contact (Fauchet et al. 1997).

Structures that are more sophisticated have been considered in a view to enhance the carrier injection into PSi. One group has studied devices in which a p-i-n (Chen et al. 1997a), an n-p-i-n (Chen et al. 1997c), an n-i-p-n (Chen et al. 1997b), or an n-i-n-p-n (Chen et al. 1997b) an-Si:H multilayer structure has been deposited onto PSi. The n-i-n-p-n device had a lower threshold voltage for EL detection (3.6 V) than the n-i-p-n LED (6 V) (Chen et al. 1997b), which in turn was better than PSi LED (>10 V). The a-Si:H multilayer structure was believed to enhance the carrier injection into PSi. Red-orange EL could be seen by the naked eye in the dark (Chen et al. 1997b). Brightness of about 30–50 cd/m² at 600 mA/cm² and an EL efficiency of 0.13% have been reported (Chen et al. 1997a). The p-i-n PSi LED (Chen et al. 1997a) was said to be voltage-tunable between 30 and 90 V, as a result of three peaks present in the EL spectrum.

A device with microcrystalline SiC deposited onto PSi was also reported (Futagi et al. 1992). The diode structure was p-type Si/PSi/n-type SiC/ITO. The EL efficiency was very low. The EL was observed by the naked eye in the current range from 200 mA to 619 mA (contact area equals 1 mm²), at a voltage above 20 V.

One group (Gelloz and Koshida 2004) studied the effect of a few nanometer thick amorphous carbon layers deposited (by sputtering) onto PSi before the deposition of the ITO top contact in order to enhance the stability and reproducibility of the operation of an optimized device operating a low voltage. Visible EL in room lighting was obtained at an operating voltage as low as 3 V from n⁺-Si/ECO-treated thin PSi (600 nm thick)/a-C/ITO junctions. Figure 1.18 shows the *I-V*

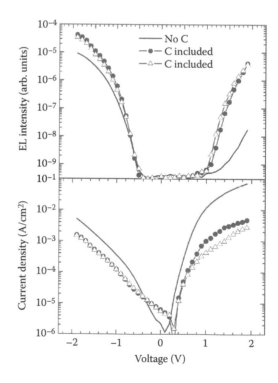

FIGURE 1.18   EL intensity and current density versus voltage for three devices. One does not have any carbon film whereas the two others do. The data shown represent a voltage scan from 2 V to –2 V. In this case, the voltage scan is not started at 0 V. The reason why the current density is not 0 at zero voltage could be explained by some charging effect. The two devices with a carbon film were prepared in the same conditions, showing the very good reproducibility achieved. (From Gelloz B. and Koshida N., *Jpn. J. Appl. Phys.* 43(4B), 1981–1985, 2004. Copyright 2004: The Japan Society of Applied Physics. With permission.)

and EL-V characteristics. A brightness of 3 Cd/m² was achieved at 3 V. The EL voltage threshold was below 1 V and about −0.5 V under forward and reverse operation, respectively. The carbon film enhanced the stability and the EL efficiency. In addition, the reproducibility from device to device was very much improved by the carbon film, as seen in Figure 1.18. The enhancement in stability was attributed to the capping of PSi by the carbon film and the high chemical stability of carbon and Si-C bonds, which should prevent PSi oxidation. The carbon film acts as an efficient mechanical and electrical buffer layer between PSi and ITO, resulting in enhanced mechanical, electrical, and chemical stability of the top contact. Figure 1.19 shows that the EL peak wavelength was continuously voltage-tunable between 700 nm to 630 nm for a voltage ranging from 2 to 5 V, a property that is not directly related to the carbon film. The EL peak voltage was linear in both forward and reverse operation conditions as shown in Figure 1.20. This behavior is unique in PSi-based EL. It originates from the field-induced EL generation mechanism combined with high efficiency and size distribution of Si nanocrystals (Gelloz and Koshida 2004).

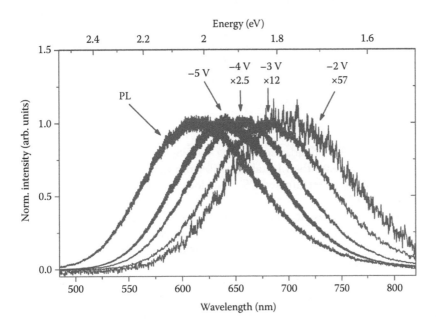

FIGURE 1.19 Normalized PL spectrum and EL spectra at different reverse voltages for an optimized device with an a-C buffer layer between ITO and PSi. (From Gelloz B. and Koshida N., *Jpn. J. Appl. Phys.* 43(4B), 1981–1985, 2004. Copyright 2004: The Japan Society of Applied Physics. With permission.)

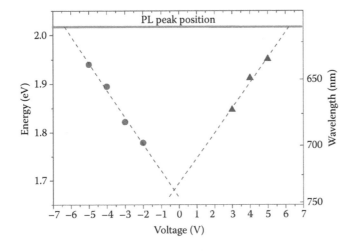

FIGURE 1.20 EL peak positions at different voltages for a device with an a-C buffer layer between ITO and PSi, corresponding to Figure 1.19.

1.10 EL STABILIZATION BY CAPPING AND SURFACE PASSIVATION

1.10.1 CAPPING PSi

Layer capping and encapsulation techniques have been proposed in order to prevent the contamination of PSi by all sorts of molecules present in ambient atmosphere. Especially, PSi should be protected from water to avoid the ineluctable oxidation that occurs when it is left in air.

The top contact itself could act as a partial capping layer. For example, it was shown (Simons et al. 1997) that semitransparent gold layers are much more permeable to air than thicker ITO layers. More complete capping of PSi has been achieved by Lazarouk et al. (1996b) using an aluminum layer which parts were electrochemically oxidized into Al_2O_3 to create transparent windows (Figure 1.15). The EL was stable for more than a month.

The capping of ECO-treated PSi by electron cyclotron resonance sputtered SiO_2 films was studied (Koshida et al. 2001). The device used was that developed in the same group (Gelloz and Koshida 2000). The stabilization of the EQE was becoming better when the capping layer was made thicker. Two deposition modes were studied: the oxide and metal modes. The metal mode led to much better stabilization than the oxide mode because it exhibits lower permeability to water molecules. The microscopic defects in SiO_2 films deposited by the metal mode possibly acted as antidiffusion trapping sites for water molecules.

1.10.2 PSi SURFACE MODIFICATION

The replacement of the Si-H bonds terminating the PSi surface by more stable bonds has been proposed in order to increase the EL stability. EL from deuterium-terminated PSi exhibits better stability than hydrogen-terminated PSi, but does not solve the problem for the long run (Matsumoto et al. 1997).

A simple and rather successful approach to stabilize the PSi PL has been the replacement of most Si-H bonds at the surface of PSi by stable Si-C bonds using hydrosylilation reactions either catalyzed by Lewis acids or thermally activated (Buriak 2002). The latter process has been successfully implemented for the enhancement of the EL stability (Gelloz et al. 2003b, 2005b). The surface modification was performed by thermal reaction of the PSi surface with 1-decene, ethyl-undecylenate, n-caprinaldehyde, and undecylenic acid at about 100°C.

Figure 1.21 shows the EL intensity and the current density as a function of time for a not modified reference device and devices including PSi layers modified using different molecules,

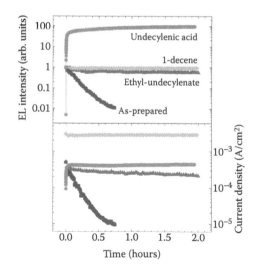

FIGURE 1.21 Current density and normalized EL intensity as a function of time for a reference device and devices treated with various organic molecules. Applied voltages were 20 V, 15 V, 10 V, and 40 V for as-prepared, ethyl-undecylenate, 1-decene, and undecylenic acid, respectively. (From Gelloz B. et al., *Appl. Phys. Lett.* 83, 2342–2344, 2003. Copyright 2003. AIP. With permission.)

under constant voltage. Modified devices exhibit very good stability whereas the reference device degrades rather quickly. The efficiency was somewhat lowered by the modification using 1-decene, ethyl-undecylenate, and n-caprinaldehyde, but was well preserved when using undecylenic acid (Gelloz et al. 2005b). The enhanced stability was attributed to a better stability of the Si-C or Si-O bonds and by the fact that the long organic groups represented a hydrophobic physical barrier that prevented water molecules and other contaminants from accessing the PSi surface, thus preventing PSi oxidation.

As already discussed in Section 1.7.3, and more thoroughly in Chapter 7 in *Porous Silicon: Formation and Properties*, HWA is a more recent technique that provides very stable PL (Gelloz and Koshida 2005; Gelloz et al. 2005a). It also gives very stable EL, as shown earlier in Figure 1.11 (Gelloz and Koshida 2006b; Gelloz et al. 2006). It is a very promising low temperature process. Furthermore, a combination of HWA and hydrosylilation reactions has been investigated (Gelloz and Koshida 2006a). The results depend on whether HWA is performed before or after hydrosylilation. Performing both treatments together leads to better surface passivation and higher PL intensity.

HWA has been used to get efficient blue PL (Gelloz et al. 2009). Recently, blue phosphorescence has been reported (Gelloz and Koshida 2009) (see Chapter 7 in *Porous Silicon: Formation and Properties*).

1.11 EMISSION SPECTRUM, SPEED, AND INTEGRATION

1.11.1 TUNING AND NARROWING THE EL SPECTRUM WITH MICROCAVITIES

Making a microcavity by inserting a luminescent PSi layer between two reflective media offers three main advantages. First, a significant PL line narrowing is achievable in this way (Pavesi et al. 1996). Second, high luminescence directionality can be achieved (Chan and Fauchet 1999; Pavesi et al. 1996). Finally, although not experimentally demonstrated yet, a reduction of the EL response time is potentially achievable (Pavesi et al. 1996) because the PL lifetime may be reduced in this way (Purcell effect).

Practically, the tuning of the EL emission has been achieved by placing the luminescent PSi layer between two multilayer Bragg reflectors, as illustrated in Figure 1.22, thus realizing a

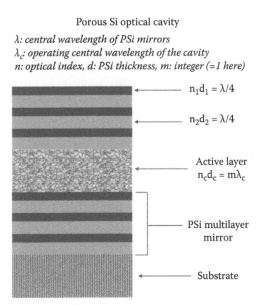

FIGURE 1.22 Tuning of EL emission spectrum is possible by placing the luminescent PSi layer in between two Bragg reflectors, thus obtaining a Fabry–Pérot resonator. A high optical line narrowing is achievable. In this example, the mirrors consist of two Bragg reflectors made of three periods of PSi layers of alternating refractive index. Subscripts 1 and 2 refer to the low porosity PSi layer (high refractive index), and the high porosity PSi layer (low refractive index), respectively.

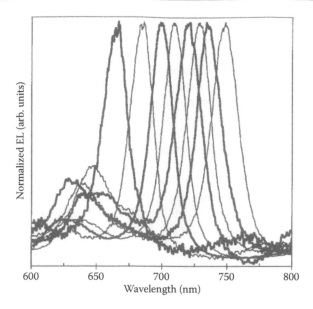

FIGURE 1.23   EL from oxidized PSi microcavity resonators with varying active layer porosity. The devices were reverse-biased at about 100 V. (From Chan S. and Facuchet P. M., *Appl. Phys. Lett.* 75, 274–276, 1999. Copyright 1999: American Institute of Physics. With permission.)

Fabry–Pérot resonator. The Bragg reflectors consisted of PSi layers of alternating refractive index (Pavesi et al. 1996). In another case, a silver top contact was used as both top reflector and top electrode, and a reduction of the EL FWHM by a factor of 3 was achieved (Araki et al. 1996a). The FWHM was about 100 meV for the emission energy of 1.8 eV. The same group has demonstrated the possible tuning of the PL emission from 1.5 eV to 2.2 eV (Araki et al. 1996b). Narrow spectra (10–40 meV in FWHM) are possible by using this approach, compared to the wide typical FWHM of PSi PL ($\approx$0.25 eV).

Another group (Chan and Fauchet 1999) demonstrated narrow and tunable EL, depending on the anodization parameters, using an active layer sandwiched between two Bragg reflectors. The substrate was p$^+$ Si. With 6 periods per mirror, the devices were typically operated at a reverse bias as high as 100 V due to the thick total PSi layer involved. The FWHMs were about 50 meV. The EL could be tuned from 1.65 eV to 1.85 eV by changing the porosity of the active layer from 76% to 94%, as shown in Figure 1.23. In addition, a high directionality of the EL emission was observed in these devices, the emission being concentrated within a 30° cone around the main axis. Since PSi made from p$^+$ Si is usually not highly luminescent, these devices were not efficient. The problem is that good quality microcavities are difficult to get with low doped PSi and n-type PSi.

1.11.2 EL MODULATION SPEED

The decay and rise times of all PSi devices are below the millisecond, which is enough for display applications. However, the modulation speed of all devices was always below the GHz, making application in optical interconnects very challenging. It is usually influenced by the carrier mobility in PSi, the radiative recombination processes, and charge trapping. The PL decay following pulsed excitation is usually in the microsecond range (Bisi et al. 2000; Canham et al. 1996; Cullis et al. 1997; Smith and Collins 1992).

The EL modulation speed was typically found to be of the order of about tens of microseconds (Cox et al. 1999; Gelloz and Koshida 2000; Kozlowski et al. 1996; Lalic and Linnros 1996a,b; Linnros and Lalic 1995; Peng and Fauchet 1995). However, one group has reported an efficient (EQE of 0.1%) device based on partially oxidized PSi that can be modulated at a frequency greater than 1 MHz (Tsybeskov et al. 1996). The high response speed may be due to a recombination mechanism that does not involve the interior of the Si nanocrystals. Modulation frequency of 200 MHz has been reported (Balucani et al. 1998) for a device showing lower efficiency and which

EL probably does not originate from recombination of excitons in Si nanocrystals. The device speed was limited by the junction capacitance.

A reduction of the EL response time is potentially possible by using PSi in a microcavity (Pavesi et al. 1996) (see Section 1.11.1). Another promising route for faster devices is the use of blue luminescence because the PL lifetime decreases down to the nanosecond regime for blue emission (Mizuno and Koshida 1999). Making use of fast Auger recombination (Carreras et al. 2008) or plasmonic systems may be promising too.

1.11.3 INTEGRATION OF PSi EL

Besides EL devices, useful optical devices such as optical waveguides (Takahashi and Koshida 1999), optical cavities (Berger et al. 1997; Frohnhoff and Berger 1994), and nonvolatile memories (Ueno and Koshida 1998, 1999) based on PSi have been fabricated on silicon substrates by simple processing. Optical nonlinearity in PSi has been demonstrated in a Fabry–Pérot resonator, leading to the availability of PSi for optical switches and optical logic gates (Takahashi et al. 2000). Most of these phenomena have only been observed separately, on different substrates. However, progress has been made in the integration of some of these functions, especially the EL.

Figure 1.24 shows an integrated bipolar device fully compatible with conventional Si microelectronic processing (Hirschman et al. 1996). The EL device was based on thermally oxidized PSi. The driving transistor, connected in the common-emitter configuration, could modulate the light emission by amplifying a small base input signal and controlling the current flow through the EL device. The device could be turned on and off by applying a small current pulse to the base of the bipolar transistor. Arrays of such integrated structures have also been fabricated.

EL from polycrystalline Si and its possible integration in large-area applications have also been demonstrated (Koshida et al. 1998). Porous polycrystalline Si diodes were shown to operate with efficiencies comparable to that of conventional crystalline PSi devices. The EL mechanism is believed to be the same in both cases. It has also been confirmed that porous polycrystalline EL devices can be driven by a poly-Si-based switching TFT (Koshida et al. 1998).

The compatibility of PSi with silicon-on-insulator technology has been demonstrated (El-Bahar and Nemirovsky 2000). The formation of PSi using silicon-on-insulator substrates has been done using an alternating current electrochemical process. The characteristics of the resulting PSi layers were similar to that of conventional PSi.

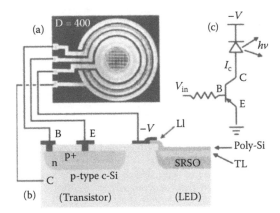

FIGURE 1.24   Micrograph of an integrated LED/bipolar-transistor structure (a) along with the cross-section (b) and equivalent circuit (c). The bipolar transistor is identified by the concentric emitter (E), base (B), and collector (C) terminals. The cross-section is taken through the center of the structure and can be mirrored on the right edge because of the symmetric design. LI and TL refer to "local interconnect" and "transition layer," respectively. SRSO refers to "silicon-rich silicon oxide," and is in fact oxidized PSi. (From Hirschman K. D. et al., *Nature* 384(6607), 338–341, 1996. Copyright 1996: NPG. With permission.)

Furthermore, a PSi EL device was fabricated using a process compatible with an industrial bipolar plus complementary MOS plus diffusion MOS technology. The device was based on a p/n$^+$ junction (Barillaro et al. 2001).

1.12 CONCLUSION

The visible EL from PSi has been extensively studied, in a view to achieving sufficient efficiency, brightness, and stability for applications. However, this goal has not been achieved yet and research efforts have significantly dropped in recent years.

Figure 1.25 shows the evolution of the efficiency of PSi devices since 1990. The 1% mark of EQE was reached (Gelloz and Koshida 2000) in 2000, but no new records have been reported since then. The EPE has reached about 0.4% (Gelloz and Koshida 2000).

The speed of PSi-based LEDs is enough for display purposes but is far too low for interconnects. Thus, the most likely application of PSi-based LED would be in display areas, lighting, and other devices where speed is not an issue.

The major breakthroughs in the field have been the use of microcavities to narrow and tune the emission spectrum, ECO to enhance the efficiency (Gelloz and Koshida 2000; Gelloz et al. 1998), chemical modification of the surface with organic molecules (Gelloz et al. 2003b, 2005b), and HWA (Gelloz and Koshida 2005, 2006b; Gelloz et al. 2005a, 2006) to enhance the stability. HWA also leads to efficient red and blue PL (Gelloz et al. 2009) and blue phosphorescence (Gelloz and Koshida 2009) and may be promising for the realization of multicolor stable and efficient PSi EL devices.

The PSi structure is extremely sensible on many parameters, rendering the reproducibility of the EL characteristics sometimes quite difficult. For instance, a rather small variation in the doping level of the substrate could lead to very different EL results even though all the other fabrication conditions would be the same. Illumination (power, directionality, spectrum of emission) is another very important and sensible fabrication condition that is often used. Temperature should also be very well controlled.

The rather good results obtained recently in EL stability (Gelloz et al. 2006) and PL efficiency (Gelloz et al. 2005a) may raise again the hope for practical PSi EL.

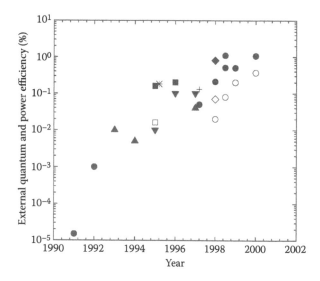

FIGURE 1.25 Selected values of external quantum and power efficiencies. The same symbol shape refers to the same research group. Solid and hollow symbols refer to quantum and power efficiencies, respectively. ●○ (Gelloz and Koshida 1999, 2000; Gelloz et al. 1998, 1999; Koshida and Koyama 1992; Oguro et al. 1997), ▼ (Kozlowski et al. 1992, 1996; Lang et al. 1997; Steiner et al. 1993, 1994), * (Loni et al. 1995), ■□ (Lalic and Linnros 1996a,b; Linnros and Lalic 1995), ◆◇ (Nishimura et al. 1998), ▲ (Fauchet et al. 1997; Peng and Fauchet 1995; Tsybeskov et al. 1996), + (Chen et al. 1997a). Most devices appearing in this figure are also listed in the tables of this chapter.

REFERENCES

Araki, M., Koyama, H., and Koshida, N. (1996a). Controlled electroluminescence spectra of porous silicon diodes with a vertical optical cavity. *Appl. Phys. Lett.* **69**(20), 2956–2958.

Araki, M., Koyama, H., and Koshida, N. (1996b). Precisely tuned emission from porous silicon vertical optical cavity in the visible region. *J. Appl. Phys.* **80**, 4841–4844.

Balucani, M., La Monica, S., and Ferrari, A. (1998). 200 mhz optical signal modulation from a porous silicon light emitting device. *Appl. Phys. Lett.* **72**(6), 639–640.

Barillaro, G., Diligenti, A., Pieri, F., Fuso, F., and Allegrini, M. (2001). Integrated porous-silicon light-emitting diodes: A fabrication process using graded doping profiles. *Appl. Phys. Lett.* **78**(26), 4154–4156.

Berger, M.G., ArensFischer, R., Thonissen, M. et al. (1997). Dielectric filters made of ps: Advanced performance by oxidation and new layer structures. *Thin Solid Films* **297**(1–2), 237–240.

Billat, S. (1996). Electroluminescence of heavily doped p-type porous silicon under electrochemical oxidation in galvanostatic regime. *J. Electrochem. Soc.* **143**(3), 1055–1061.

Billat, S., Gaspard, F., Herino, R. et al. (1995). Electroluminescence of heavily-doped p-type porous silicon under electrochemical oxidation in the potentiostatic regime. *Thin Solid Films* **263**(2), 238–242.

Bisi, O., Ossicini, S., and Pavesi, L. (2000). Porous silicon: A quantum sponge structure for silicon based optoelectronics. *Surf. Sci. Rep.* **38**(1–3), 5–126.

Bressers, P.M.M.C., Knapen, J.W.J., Meulenkamp, E.A., and Kelly, J.J. (1992). Visible-light emission from a porous silicon solution diode. *Appl. Phys. Lett.* **61**(1), 108–110.

Bsiesy, A. and Vial, J.C. (1996). Voltage-tunable photo- and electroluminescence of porous silicon. *J. Lumin.* **70**, 310–319.

Bsiesy, A., Vial, J.C., Gaspard, F. et al. (1991). Photoluminescence of high porosity and of electrochemically oxidized porous silicon layers. *Surf. Sci.* **254**(1–3), 195–200.

Bsiesy, A., Muller, F., Ligeon, M. et al. (1993). Voltage-controlled spectral shift of porous silicon electroluminescence. *Phys. Rev. Lett.* **71**(4), 637–640.

Bsiesy, A., Muller, F., Ligeon, M. et al. (1994). Relation between porous silicon photoluminescence and its voltage-tunable electroluminescence. *Appl. Phys. Lett.* **65**(26), 3371–3373.

Bsiesy, A., Nicolau, Y.F., Ermolieff, A., Muller, F., and Gaspard, F. (1995). Electroluminescence from n(+)-type porous silicon contacted with layer-by-layer deposited polyaniline. *Thin Solid Films* **255**(1–2), 43–48.

Bsiesy, A., Gelloz, B., Gaspard, F., and Muller, F. (1996). Origin of the charge carriers accumulation depletion in porous silicon contacted by a liquid phase. *J. Appl. Phys.* **79**(5), 2513–2516.

Buriak, J.M. (2002). Organometallic chemistry on silicon and germanium surfaces. *Chem. Rev.* **102**(5), 1271–1308.

Canham, L.T. (1990). Silicon quantum wire array fabrication by electrochemical and chemical dissolution of wafers. *Appl. Phys. Lett.* **57**(10), 1046–1048.

Canham, L.T., Leong, W.Y., Beale, M.I.J., Cox, T.I., and Taylor, L. (1992). Efficient visible electroluminescence from highly porous silicon under cathodic bias. *Appl. Phys. Lett.* **61**(21), 2563–2565.

Canham, L.T., Cox, T.I., Loni, A., and Simons, A.J. (1996). Progress towards silicon optoelectronics using porous silicon technology. *Appl. Surf. Sci.* **102**, 436–441.

Cantin, J.L., Schoisswohl, M., Grosman, A. et al. (1996). Anodic oxidation of p- and p(+)-type porous silicon: Surface structural transformations and oxide formation. *Thin Solid Films* **276**(1–2), 76–79.

Carreras, J., Arbiol, J., Garrido, B., Bonafos, C., and Montserrat, J. (2008). Direct modulation of electroluminescence from silicon nanocrystals beyond radiative recombination rates. *Appl. Phys. Lett.* **92**(9).

Chan, S. and Fauchet, P.M. (1999). Tunable, narrow, and directional luminescence from porous silicon light emitting devices. *Appl. Phys. Lett.* **75**(2), 274–276.

Chazalviel, J.N. and Ozanam, F. (1992). Mechanism of electron injection during the anodic oxidation of silicon. *Mater. Res. Soc. Symp. Proc.* **283**, 359–364.

Chen, Z.L., Bosman, G., and Ochoa, R. (1993). Visible-light emission from heavily doped porous silicon homojunction pn diodes. *Appl. Phys. Lett.* **62**(7), 708–710.

Chen, Y.A., Chen, B.F., Tsay, W.C. et al. (1997a). Porous silicon light-emitting diode with tunable color. *Solid-State Electron.* **41**(5), 757–759.

Chen, Y.A., Liang, N.Y., Laih, L.H., Tsay, W.C., Chang, M.N., and Hong, J.W. (1997b). Improvement of electroluminescence characteristics of porous silicon led by using amorphous silicon layers. *Electron. Lett.* **33**(17), 1489–1490.

Chen, Y.A., Liang, N.Y., Laih, L.H., Tsay, W.C., Chang, M.N., and Hong, J.W. (1997c). Improvement of current injection of porous silicon. *Jpn. J. Appl. Phys., Part 1* **36**(3B), 1574–1577.

Cox, T.I., Simons, A.J., Loni, A. et al. (1999). Modulation speed of an efficient porous silicon light emitting device. *J. Appl. Phys.* **86**(5), 2764–2773.

Cullis, A.G., Canham, L.T., and Calcott, P.D.J. (1997). The structural and luminescence properties of porous silicon. *J. Appl. Phys.* **82**(3), 909–965.

El-Bahar, A. and Nemirovsky, Y. (2000). A technique to form a porous silicon layer with no backside contact by alternating current electrochemical process. *Appl. Phys. Lett.* **77**(2), 208–210.

Fauchet, P.M., Tsybeskov, L., Duttagupta, S.P., and Hirschman, K.D. (1997). Stable photoluminescence and electroluminescence from porous silicon. *Thin Solid Films* **297**(1–2), 254–260.

Frohnhoff, S. and Berger, M.G. (1994). Porous silicon superlattices. *Adv. Mater.* **6**(12), 963–965.

Futagi, T., Matsumoto, T., Katsuno, M., Ohta, Y., Mimura, H., and Kitamura, K. (1992). Visible electroluminescence from p-type crystalline silicon porous silicon n-type microcrystalline silicon carbon pn junction diodes. *Jpn. J. Appl. Phys., Part 2* **31**(5B), L616–L618.

Gelloz, B. and Bsiesy, A. (1997). Investigation of the carrier transport process in wet porous silicon by photocurrent measurements: Significance for the luminescence properties. *Electrochem. Soc. Proc.* **97**(7), 92–103.

Gelloz, B. and Bsiesy, A. (1998). Carrier transport mechanisms in porous silicon in contact with a liquid phase: A diffusion process. *Appl. Surf. Sci.* **135**(1–4), 15–22.

Gelloz, B. and Koshida, N. (1999). Enhancing efficiency and stability of porous silicon electroluminescence using electrochemical techniques. *Electrochem. Soc. Proc.* **99**(22), 27–34.

Gelloz, B. and Koshida, N. (2000). Electroluminescence with high and stable quantum efficiency and low threshold voltage from anodically oxidized thin porous silicon diode. *J. Appl. Phys.* **88**(7), 4319–4324.

Gelloz, B. and Koshida, N. (2004). High performance electroluminescence from nanocrystalline silicon with carbon buffer. *Jpn. J. Appl. Phys., Part 1* **43**(4B), 1981–1985.

Gelloz, B. and Koshida, N. (2005). Mechanism of a remarkable enhancement in the light emission from nanocrystalline porous silicon annealed in high-pressure water vapor. *J. Appl. Phys.* **98**(1), 123509.

Gelloz, B. and Koshida, N. (2006a). Highly efficient and stable photoluminescence of nanocrystalline porous silicon by combination of chemical modification and oxidation under high pressure. *Jpn. J. Appl. Phys., Part 1*, in press.

Gelloz, B. and Koshida, N. (2006b). Highly enhanced efficiency and stability of photo- and electro-luminescence of nano-crystalline porous silicon by high-pressure water vapor annealing. *Jpn. J. Appl. Phys., Part 1* **45**(4B), 3462–3465.

Gelloz, B. and Koshida, N. (2009). Long-lived blue phosphorescence of oxidized and annealed nanocrystalline silicon. *Appl. Phys. Lett.* **94**, 201903.

Gelloz, B., Bsiesy, A., Gaspard, F., and Muller, F. (1996). Conduction in porous silicon contacted by a liquid phase. *Thin Solid Films* **276**(1–2), 175–178.

Gelloz, B., Bsiesy, A., Gaspard, F. et al. (1997). Charge carrier dynamics in wet porous silicon. *Electrochem. Soc. Proc.* **97**(7), 422–437.

Gelloz, B., Nakagawa, T., and Koshida, N. (1998). Enhancement of the quantum efficiency and stability of electroluminescence from porous silicon by anodic passivation. *Appl. Phys. Lett.* **73**(14), 2021–2023.

Gelloz, B., Bsiesy, A., and Herino, R. (1999a). Light-induced porous silicon photoluminescence quenching. *J. Lumin.* **82**(3), 205–211.

Gelloz, B., Nakagawa, T., and Koshida, N. (1999b). Enhancing the external quantum efficiency of porous silicon leds beyond 1% by a post-anodization electrochemical oxidation. *Mater. Res. Soc. Symp. Proc.* **536**, 15–20.

Gelloz, B., Bsiesy, A., and Herino, R. (2003a). Electrically induced luminescence quenching in p(+)-type and anodically oxidized n-type wet porous silicon. *J. Appl. Phys.* **94**(4), 2381–2389.

Gelloz, B., Sano, H., Boukherroub, R., Wayner, D.D.M., Lockwood, D.J., and Koshida, N. (2003b). Stabilization of porous silicon electroluminescence by surface passivation with controlled covalent bonds. *Appl. Phys. Lett.* **83**(12), 2342–2344.

Gelloz, B., Kojima, A., and Koshida, N. (2005a). Highly efficient and stable luminescence of nanocrystalline porous silicon treated by high-pressure water vapor annealing. *Appl. Phys. Lett.* **87**(3), 031107.

Gelloz, B., Sano, H., Boukhrerroub, R., Wayner, D.D.M., Lockwood, D.J., and Koshida, N. (2005b). Stable electroluminescence from passivated nano-crystalline porous silicon using undecylenic acid. *Phys. Status Solidi C* **2**(9), 3273–3277.

Gelloz, B., Shibata, T., and Koshida, N. (2006). Stable electroluminescence of nanocrystalline silicon device activated by high pressure water vapor annealing. *Appl. Phys. Lett.* **89**, 191103.

Gelloz, B., Mentek, R., and Koshida, N. (2009). Specific blue light emission from nanocrystalline porous Si treated by high-pressure water vapor annealing. *Jpn. J. Appl. Phys., Part 1* **48**(4), 04C119.

Green, W.H., Lee, E.J., Lauerhaas, J.M., Bitner, T.W., and Sailor, M.J. (1995). Electrochemiluminescence from porous silicon in formic-acid liquid-junction cells. *Appl. Phys. Lett.* **67**(10), 1468–1470.

Grosman, A. and Ortega, C. (1997). Properties of porous silicon. In: Canham L.T. (Ed.) *Properties of Porous Silicon*, Emis datareviews series, vol. 18. INSPEC, The Institution of Electrical Engineers, London, pp. 328–335.

Halimaoui, A., Oules, C., Bomchil, G. et al. (1991). Electroluminescence in the visible range during anodic-oxidation of porous silicon films. *Appl. Phys. Lett.* **59**(3), 304–306.

Halliday, D.P., Holland, E.R., Eggleston, J.M., Adams, P.N., Cox, S.E., and Monkman, A.P. (1996). Electroluminescence from porous silicon using a conducting polyaniline contact. *Thin Solid Films* **276**(1–2), 299–302.

Hirschman, K.D., Tsybeskov, L., Duttagupta, S.P., and Fauchet, P.M. (1996). Silicon-based visible light-emitting devices integrated into microelectronic circuits. *Nature* **384**(6607), 338–341.

Hory, M.A., Herino, R., Ligeon, M. et al. (1995). Fourier-transform ir monitoring of porous silicon passivation during posttreatments such as anodic-oxidation and contact with organic-solvents. *Thin Solid Films* **255**(1–2), 200–203.

Kooij, E.S., Despo, R.W., and Kelly, J.J. (1995). Electroluminescence from porous silicon due to electron injection from solution. *Appl. Phys. Lett.* **66**(19), 2552–2554.

Koshida, N. and Koyama, H. (1992). Visible electroluminescence from porous silicon. *Appl. Phys. Lett.* **60**(3), 347–349.

Koshida, N., Koyama, H., Yamamoto, Y., and Collins, G.J. (1993). Visible electroluminescence from porous silicon diodes with an electropolymerized contact. *Appl. Phys. Lett.* **63**(19), 2655–2657.

Koshida, N., Mizuno, H., Koyama, H., and Collins, G.J. (1994). Visible electroluminescence from porous silicon diodes with immersed conducting polymer contacts. *Jpn. J. Appl. Phys. Part 2* **34**, 92–94.

Koshida, N., Takizawa, E., Mizuno, H., Arai, S., Koyama, H., and Sameshima, T. (1998). *MRS Proc.* **486**, 151.

Koshida, N., Kadokura, J., Takahashi, M., and Imai, K. (2001). Stabilization of porous silicon electroluminescence by surface capping with silicon dioxide films. *MRS Proc.* **638**, F18.13.11.

Kozlowski, F., Sauter, M., Steiner, P., Richter, A., Sandmaier, H., and Lang, W. (1992). Electroluminescent performance of porous silicon. *Thin Solid Films* **222**(1–2), 196–199.

Kozlowski, F., Wagenseil, W., Steiner, P., and Lang, W. (1995). *MRS Proc.* **358**, 677.

Kozlowski, F., Sailer, C., Steiner, P., Knoll, B., and Lang, W. (1996). Time-resolved electroluminescence of porous silicon. *Thin Solid Films* **276**(1–2), 164–167.

Lalic, N. and Linnros, J. (1996a). A porous silicon light-emitting diode with a high quantum efficiency during pulsed operation. *Thin Solid Films* **276**(1–2), 155–158.

Lalic, N. and Linnros, J. (1996b). Characterization of a porous silicon diode with efficient and tunable electroluminescence. *J. Appl. Phys.* **80**(10), 5971–5977.

Lang, W., Kozlowski, F., Steiner, P. et al. (1997). Technology and rbs analysis of porous silicon light-emitting diodes. *Thin Solid Films* **297**(1–2), 268–271.

Lazarouk, S., Jaguiro, P., Katsouba, S. et al. (1996a). Visible light from aluminum-porous silicon schottky junctions. *Thin Solid Films* **276**(1–2), 168–170.

Lazarouk, S., Jaguiro, P., Katsouba, S. et al. (1996b). Stable electroluminescence from reverse biased n-type porous silicon-aluminum schottky junction device. *Appl. Phys. Lett.* **68**(15), 2108–2110.

Lehmann, V., Hofmann, F., Moller, F., and Gruning, U. (1995). Resistivity of porous silicon—A surface effect. *Thin Solid Films* **255**(1–2), 20–22.

Li, K.H., Diaz, D.C., He, Y.S., Campbell, J.C., and Tsai, C.C. (1994). Electroluminescence from porous silicon with conducting polymer film contacts. *Appl. Phys. Lett.* **64**(18), 2394–2396.

Ligeon, M., Muller, F., Herino, R. et al. (1993). Analysis of the electroluminescence observed during the anodic-oxidation of porous layers formed on lightly p-doped silicon. *J. Appl. Phys.* **74**(2), 1265–1271.

Linnros, J. and Lalic, N. (1995). High quantum efficiency for a porous silicon light-emitting diode under pulsed operation. *Appl. Phys. Lett.* **66**(22), 3048–3050.

Loni, A., Simons, A.J., Cox, T.I., Calcott, P.D.J., and Canham, L.T. (1995). Electroluminescent porous silicon device with an external quantum efficiency greater-than 0.1-percent under cw operation. *Electron. Lett.* **31**(15), 1288–1289.

Maruska, H.P., Namavar, F., and Kalkhoran, N.M. (1992). Current injection mechanism for porous-silicon transparent surface light-emitting-diodes. *Appl. Phys. Lett.* **61**(11), 1338–1340.

Matsumoto, T., Masumoto, Y., Nakagawa, T., Hashimoto, M., Ueno, K., and Koshida, N. (1997). Electroluminescence from deuterium terminated porous silicon. *Jpn. J. Appl. Phys., Part 2* **36**(8B), L1089–L1091.

Memming, R. (1969). Mechanism of the electrochemical reduction of persulfates and hydrogen peroxide. *J. Electrochem. Soc.* **116**(6), 785–790.

Mimura, H., Matsumoto, T., and Kanemitsu, Y. (1996). Green and blue light emitting devices using Si-based porous materials. *J. Non-Cryst. Solids* **200**, 961–964.

Mizuno, H. and Koshida, N. (1999). Enhancement in efficiency and stability of oxide-free blue emission from porous silicon by surface passivation. *MRS Proc.* **536**, 179–184.

Namavar, F., Maruska, H.P., and Kalkhoran, N.M. (1992). Visible electroluminescence from porous silicon np heterojunction diodes. *Appl. Phys. Lett.* **60**(20), 2514–2516.

Nishimura, K., Nagao, Y., and Ikeda, N. (1998). High external quantum efficiency of electroluminescence from photoanodized porous silicon. *Jpn. J. Appl. Phys., Part 2* **37**(3B), L303–L305.

Noguchi, H., Kondo, T., Murakoshi, K., and Uosaki, K. (1999). Visible electroluminescence from n-type porous silicon/electrolyte solution interfaces: Time-dependent electroluminescence spectra. *J. Electrochem. Soc.* **146**(11), 4166–4171.

Oguro, T., Koyama, H., Ozaki, T., and Koshida, N. (1997). Mechanism of the visible electroluminescence from metal porous silicon n-Si devices. *J. Appl. Phys.* **81**(3), 1407–1412.

Pavesi, L., Guardini, R., and Mazzoleni, C. (1996). Porous silicon resonant cavity light emitting diodes. *Solid State Commun.* **97**(12), 1051–1053.

Pavesi, L., Chierchia, R., Bellutti, P. et al. (1999). Light emitting porous silicon diode based on a silicon/porous silicon heterojunction. *J. Appl. Phys.* **86**(11), 6474–6482.

Peng, C. and Fauchet, P.M. (1995). The frequency-response of porous silicon electroluminescent devices. *Appl. Phys. Lett.* **67**(17), 2515–2517.

Peter, L.M. and Wielgosz, R.I. (1996). Light-induced electroluminescence of porous silicon layers on p-Si in persulfate solution. *Appl. Phys. Lett.* **69**(6), 806–808.

Peter, L.M., Riley, D.J., Wielgosz, R.I. et al. (1996). Mechanisms of luminescence tuning and quenching in porous silicon. *Thin Solid Films* **276**(1–2), 123–129.

Romestain, R., Vial, J.C., Mihalcescu, I., and Bsiesy, A. (1995). Saturation and voltage quenching of the porous silicon luminescence and importance of the auger effect. *Phys. Status Solidi B* **190**(1), 77–84.

Saren, A.A., Kuznetsov, S.N., Pikulev, V.B., Gardin, Y.E., and Gurtov, V.A. (2002). Electroluminescence from porous silicon in the cathodic reduction of persulfate ions: Degree of reversibility of the tuning effect. *Semiconductors* **36**(10), 1184–1187.

Simons, A.J., Cox, T.I., Loni, A., Canham, L.T., and Blacker, R. (1997). Investigation of the mechanisms controlling the stability of a porous silicon electroluminescent device. *Thin Solid Films* **297**(1–2), 281–284.

Smith, R.L. and Collins, S.D. (1992). Porous silicon formation mechanisms. *J. Appl. Phys.* **71**(8), R1–R22.

Steiner, P., Kozlowski, F., and Lang, W. (1993). Light-emitting porous silicon diode with an increased electroluminescence quantum efficiency. *Appl. Phys. Lett.* **62**(21), 2700–2702.

Steiner, P., Kozlowski, F., Wielunski, M., and Lang, W. (1994). Enhanced blue-light emission from an indium-treated porous silicon device. *Jpn. J. Appl. Phys., Part 1* **33**(11), 6075–6077.

Steiner, P., Wiedenhofer, A., Kozlowski, F., and Lang, W. (1996). Influence of different metallic contacts on porous silicon electroluminescence. *Thin Solid Films* **276**(1–2), 159–163.

Takahashi, M. and Koshida, N. (1999). Fabrication and characteristics of three-dimensionally buried porous silicon optical waveguides. *J. Appl. Phys.* **86**(9), 5274–5278.

Takahashi, M., Toriumi, Y., Matsumoto, T., Masumoto, Y., and Koshida, N. (2000). Significant photoinduced refractive index change observed in porous silicon fabry-perot resonators. *Appl. Phys. Lett.* **76**(15), 1990–1992.

Tsybeskov, L., Duttagupta, S.P., and Fauchet, P.M. (1995). Photoluminescence and electroluminescence in partially oxidized porous silicon. *Solid State Commun.* **95**(7), 429–433.

Tsybeskov, L., Duttagupta, S.P., Hirschman, K.D., and Fauchet, P.M. (1996). Stable and efficient electroluminescence from a porous silicon-based bipolar device. *Appl. Phys. Lett.* **68**(15), 2058–2060.

Ueno, K. and Koshida, N. (1998). Negative-resistance effects in light-emitting porous silicon diodes. *Jpn. J. Appl. Phys., Part 1* **37**(3B), 1096–1099.

Ueno, K. and Koshida, N. (1999). Light-emissive nonvolatile memory effects in porous silicon diodes. *Appl. Phys. Lett.* **74**(1), 93–95.

Wolkin, M.V., Jorne, J., Fauchet, P.M., Allan, G., and Delerue, C. (1999). Electronic states and luminescence in porous silicon quantum dots: The role of oxygen. *Phys. Rev. Lett.* **82**(1), 197–200.

Photodetectors Based on Porous Silicon

2

Ghenadii Korotcenkov and Nima Naderi

CONTENTS

2.1   Overview and Background of PSi-Based Photodetectors   36
2.2   Schottky Contact, p-n Junction, and MSM PSi-Based Photodetectors   37
      2.2.1   PSi-Based Photodetectors and Their Disadvantages   38
      2.2.2   Stabilization of PSi-Based Photodetectors   40
2.3   Photodetectors Designed on the Base of PSi-Based Composites   42
2.4   Silicon Photodetectors with PSi-Based Antireflecting Covering and Filters   43
2.5   Hybrid Heterojunction Photodetectors   45
2.6   Summary   47
Acknowledgment   47
References   47

2.1 OVERVIEW AND BACKGROUND OF PSi-BASED PHOTODETECTORS

Photodetectors are fundamentally semiconductor devices that convert optical energy (light) into electrical energy, which is mostly manifested as photocurrent. High-sensitivity and high-speed photodetectors have been widely studied over the past years because of their application in optical communication networks.

The performance of the semiconductor-based photodetectors depends on the distribution and flux of the incident light as well as on electronic parameters of the substrate material, such as doping levels and band structures (Al-Hardan et al. 2011). Based on the application, the performance of photodetectors refers to sensitivity, wavelength selectivity, response, and recovery times (Yang et al. 2008; Casalino et al. 2010).

The responsivity of the detector is defined as the ratio of the electrical output to the optical input for a given wavelength and can be expressed as

$$(2.1) \qquad R = \frac{\text{Photocurrent}(A)}{\text{Power}(W)} = \frac{I_{ph}(A)}{P_{inc}(W)}$$

where I_{ph} is the photocurrent and P_{inc} is the incident optical power on the device. Responsivity of photodetectors indicates the current produced by a certain optical power. Reasonable responsivities are necessary for an acceptable signal-to-noise ratio and for easing the design and realization of the amplifier circuitry that follows after the photodetector. Responsivity is strictly linked to a device's quantum efficiency, a property describing how many carriers per photon are being collected. The quantum efficiency, η ($0 \leq \eta \leq 1$), of a photodetector is defined as the number of carriers collected to produce the photocurrent (I_{ph}) generated per number of incident photons. This parameter is indicated by the following equation:

$$(2.2) \qquad \eta = \frac{I_{ph}}{q} \times \frac{h\nu}{P_{inc}}$$

where q is the quantum of electric charge (charge of one electron) and υ is the frequency of incident light. Thus, the relation between responsivity and quantum efficiency is

$$(2.3) \qquad R = \frac{\eta q}{h\upsilon} = \frac{\eta \lambda(\mu m)}{1.24}$$

For example, in the telecommunications field, a responsivity ≥ 0.1 A/W (Schaub et al. 2001), corresponding to the external quantum efficiencies η of 15%, 10%, and 8% at λ = 850, 1300, and 1550 nm, respectively, is required.

Dark current is also an important parameter for efficient operation because the shot noise, associated with the fluctuations in a measured signal due to the random arrival time of the particles, carrying energy, is generated by a leakage current. The current gain, which is the ratio of photocurrent to dark current (I_{ph}/I_d), can be used to measure photodetection ability. In a typical photodetector, dark currents less than 1 μA are required.

A further requirement of photodetectors is low-voltage operation. It would be desirable to realize devices operating at the same power supply as the CMOS circuitry, that is, bias voltage <5 V and as low as 1 V for advanced CMOS generation. Finally, the shrinking of the photodetector dimensions would allow the integration of photonic components with integrated electronic circuits, enabling interconnection bandwidths that are not limited by the RC time constant or the reliability constraints of the metal lines. It is useful to recall briefly the main performance requirements of integrated photodetectors for near-infrared (NIR) optical applications. Thus, high speed, high responsivity, low dark current, low bias voltage, and small dimensions are appealing properties for a photodetector, especially designed for optoelectronic applications (Casalino et al. 2010).

At present, there are different kinds of devices based on porous silicon that can operate as photodetectors. They are the following: Schottky diodes (Figure 2.1a), *p-n* junctions,

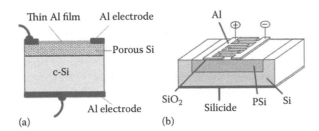

FIGURE 2.1   (a) Cross-sectional view of Schootky barrier type Al/PSi/Si/Al photodetector. (b) Schematic structure of MSM type PSi-based photodiode. (From Rossi A.M. and Bohn H.G., *Phys. Stat. Sol. (a)* 202, (8), 1644, 2005. Copyright 2005: John Wiley & Sons, Inc. With permission.)

heterostructures, and metal–semiconductor–metal (MSM) (Figure 2.1b) structures (Su et al. 2002; Chen et al. 2006; Liu et al. 2006). One should note that Schottky barrier, *p-n* junction, and MSM-based photodetectors were ones of the first optoelectronic devices fabricated on the base of porous silicon (PSi). The MSM photodetectors have been popular in the field of optical communications in the past few years due to their several advantages. These photodetectors consist of two interdigitated metal fingers on a semiconductor that forms two back-to-back connected Schottky contacts (Zebentout et al. 2011). Photons are detected by collecting electric signals produced by photo-excited electrons and holes in the semiconductor, which drift under the electrical field applied between the fingers (Yu and Wie 1993). The metal structure is composed of two contact pads and fingers, which form the active area of the device, as shown in Figure 2.1b. When the active area of the device is under illumination, carriers in the semiconductor absorption layer are generated (electron–hole pairs generation) by incident photons having energy higher than the bandgap energy of the semiconductor. The carriers are transported to the metal contact pads, and a current is detected in the external circuit under the application of an external bias voltage. One of the most outstanding advantageous characteristics is its high response speed, which is a function of the geometry of the structure. It was found that the spectral response of the PSi-based MSM photodetectors was similar to that of an Si PIN photodiode (Yu and Wie 1993).

Analysis of the results of experimental and theoretical studies, conducted in *Porous Silicon: Formation and Properties*, has shown that PSi has the unique ability to change its properties when changing the porosity of the material. These properties have been used in the development of different sensors (*Porous Silicon: Biomedical and Sensor Applications*) and fabrication of various devices for micro and optoelectronics (*Porous Silicon: Optoelectronics, Microelectronics, and Energy Technology Applications*). Numerous attempts to use PSi in the development of photodetectors were also undertaken (Yu et al. 1993; Balagurov et al. 1997; Belyakov et al. 1997; Bisi et al. 2000; Min et al. 2001; Martinez-Duart and Martin-Palma 2002; Liu et al. 2006; Pérez 2007; Tu et al. 2010; Chou et al. 2012). While analyzing these studies, we can distinguish four areas of potential use of PSi in the development of photodetectors. They are as follows:

1. Schottky contact, *p-n* junction, and MSM PSi-based photodetectors
2. Photodetectors designed on the base of PSi-based composites
3. Si photodetectors with PSi-based unreflecting covering and filters
4. Hybrid and heterostructure PSi-based photodetectors

Let us consider these approaches in detail. It should be noted that the electrical properties of Schottky contacts, *p-n* junctions, and heterostructures based on PSi were considered earlier in Chapter 14 in *Porous Silicon: Formation and Properties*.

2.2 SCHOTTKY CONTACT, p-n JUNCTION, AND MSM PSi-BASED PHOTODETECTORS

As it is known, PSi is a material with a variable bandgap. In particular, it was established that for PSi, the bandgap strongly increased compared with the non-porous Si (read Chapters 7 and 8 in *Porous Silicon: Formation and Properties*). According to the quantum confinement theory, this effect takes place due to the decrease in the crystallite size (Diesinger et al. 2000). The ability to tune the optical absorption/emission properties of PSi nanostructures by varying their structure sizes is essential in the bandgap engineering of PSi-based photodetectors. In particular, it is assumed that this property allows creating PSi-based photodetectors with controllable spectral characteristics. For example, the bandgap of PSi may be adjusted for optimum sun light absorption.

2.2.1 PSi-BASED PHOTODETECTORS AND THEIR DISADVANTAGES

The experiment has shown that the use of PSi with different porosity really allows controlling the range of spectral sensitivity (see Figures 2.2a,b and 2.3a).

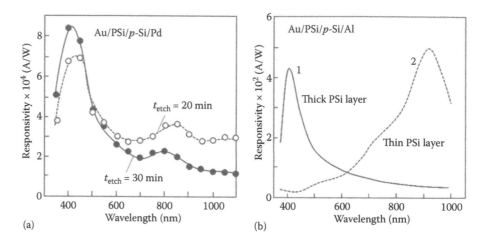

FIGURE 2.2 (a) Spectral sensitivity of Au/PSi/*p*-Si/Pd structures for different formation times of PSi layers. (From Hadjersi T. and Gabouze N., *Opt. Mater.* 30, 865, 2008. Copyright 2007: Elsevier. With permission.) (b) Photosensitivity spectra of Al/PSi/p-Si/Al structures with thick (1) and thin (2) PSi layers. (Data extracted from Svechnikov S.V. et al., *Semicond. Phys. Quantum Electron. Optoelectron.* 1(1), 13, 1998.)

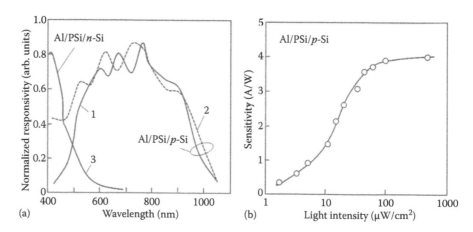

FIGURE 2.3 (a) Normalized photosensitivity spectra measured at 3 V reverse bias for (1) non-oxidized, (3) oxidized Al/PSi/*p*-Si structures with 1.5-μm thick PSi layer, (2) non-oxidized Al/PSi/*n*-Si structure with 1-μm thick PSi layer. Spectra were measured under illumination of uncovered PSi surface (1,3) or semi-transparent Al contact (2); (b) dependence of the photosensitivity on the light intensity measured at 3 V reverse bias, 0.85 μm wavelength for oxidized Al/PSi/*p*-Si structure with 1.3 μm thick PSi layer. (From Balagurov L.A. et al., *Solid-State Electron.* 47, 65, 2003. Copyright 2003: Elsevier. With permission.)

PSi-based photodiodes with close to unity quantum efficiency (Zheng et al. 1992; Tsai et al. 1993; Balagurov et al. 2001a) and modulated barrier photodiodes with more than unity quantum efficiency (Balagurov et al. 1997) were also reported. Moreover, as it was established in Tsai et al. (1993), Lee et al. (1998) and Balagurov et al. (2003), some PSi-based photoresistors had photosensitivity up to 6 A/W (see Figure 2.3b). In addition, it was found that the most interesting structures for photodetector applications were PSi-based photodiode structures fabricated on p-type substrates because they had a high photosensitivity in the near ultraviolet range (Balagurov et al. 2001a). At the same time, experiment has shown that all methods of silicon porosification, including electrochemical anodization, chemical etching, and laser-assisted etching, can be used for PSi-based photodetectors fabrication.

However, the same experiment has shown that these PSi-based photodetectors had high noise (Balagurov et al. 2003; Rossi and Bohn 2005) and low threshold sensitivity due to high back dark current (above 1 $\mu A/cm^2$ at 1–10 V reverse bias), and relatively low sensitivity in the blue spectral range (Krüger et al. 1996). Optimization of the parameters of silicon porosification and device fabrication allows reducing the value of reverse currents and improves operating parameters of photodetectors (Yu and Wie 1993; Ait-Hamouda et al. 2003; Hadjersi and Gabouze 2008; Abd Rahim et al. 2011; Naderi and Hashim 2012a,b). Some results of such optimization are shown in Figure 2.4. It is seen that the anodization parameters are important for the photodetector application, and there are optimal values of current density, as well as anodization time at a given current density, which are required for

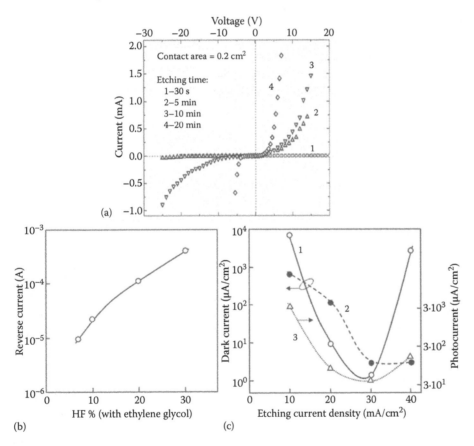

FIGURE 2.4   (a) Current–voltage curves measured from Au/PSi/*p*-Si/Pd structures for different formation times of PSi layers. (From Hadjersi T. and Gabouze N., *Opt. Mater.* 30, 865, 2008. Copyright 2007: Elsevier. With permission.) (b) Variations of reverse current, at constant bias voltage (60 V), with HF concentration during electrochemical etching. (Data extracted from Ait-Hamouda K. et al., *Solar Energy Mater. Solar Cells* 76, 535, 2003.) (c) Etching current density influence on (1, 2) dark current of MSM PSi structures under reverse bias 5 V (1: as formed, 2: after rapid-thermal-oxidized [RTO] [750°C, 50 sec], and rapid-thermal-annealed [RTA] [750°C for 15 sec] processes), and (3) photocurrent as a function of the preparation current density at specific illumination power 200 mW/cm². (Data extracted from Alwan A.M. and Jabbar A.A., *Modern Appl. Sci.* 5(1), 106, 2011.)

achievement of maximal efficiency of the photodetectors. However, it is necessary to recognize that due to the high concentration of structural defects, high concentration of the surface state, and high density of recombination centers, which are formed during silicon porosification, this problem cannot be resolved completely. Therefore, PSi-based photodetectors in comparison with silicon photodetectors tend to have worse parameters. As it is known, bulk photodetectors are perhaps the oldest and the best understood silicon optoelectronic devices. Commercial products operate at wavelengths below 1100 nm, where band-to-band absorption occurs. One of the most important advantages of silicon is that due to the quality of its crystalline material and its excellent passivation properties, very low dark-current photodetectors (PD) can be obtained (Zimmermann 2004).

Temporal instability is another shortcoming of PSi-based photodetectors, especially when they need to be used in harsh environments (Ghosh et al. 2009). It was established that PSi surface was very sensitive to different gases present in ambient air, or to any atomic species being in contact with it. This behavior, being very useful for certain applications such as chemical, gas, and biosensors (read Chapters 1–5 presented in *Porous Silicon: Biomedical and Sensor Applications*), will cause instability in the performance of PSi-based photodetectors. Thus, the degradation of the optical properties of PSi surface after prolonged light irradiation (Mahmoudi et al. 2007a,b) and after long periods of exposure to the atmosphere (Tsybeskov et al. 1995; Huang 1996; Shi et al. 2000) is a barrier for fabricating stable photodetectors on the PSi substrates (Ait-Hamouda et al. 2003; Tuura et al. 2008). Experiment has shown that the instability of physical properties of PSi occurs because of the existence of metastable bonds between silicon and hydrogen atoms (Si–H) at the PSi surface, formed during the preparation process (Huy et al. 2003). Several studies have shown that the silicon hydride species alter very easily in different ambient. This means that the existence of these chemical bonds on the PSi surface can result in deterioration of the physical properties of the PSi layer and instable behavior of the PSi-based photodetectors (Mahmoudi et al. 2007a,b). Therefore, different techniques based on replacing hydrogen terminations with other more stable species were proposed as a solution to this problem (Salonen et al. 2000; Nakamura et al. 2010).

2.2.2 STABILIZATION OF PSi-BASED PHOTODETECTORS

At present, there are several approaches for stabilization of the PSi-based photodetector parameters. However, as a rule, thermal carbonization (Naderi and Hashim 2012c,d) and oxidation (Balagurov et al. 2003; Rossi and Bohn 2005; Lin et al. 2013) are being used for this purpose. Thermal carbonization (TC) technique is based on the replacement of the existing Si–H terminations with the Si–C species. The unique characteristics of the Si–C species, such as chemical inertness and thermal stability (Rittenhouse et al. 2003; Keffous et al. 2007; Wang and Li 2010) are the main motivations for PSi stabilization using the TC method. In this technique, a formation of an ultrathin stabilizing layer on PSi is possible through exposure to acetylene (C_2H_2) gas at high temperatures (Torres-Costa et al. 2008). Here, desorption of the C_2H_2, C_2H, and C_2 species occurs at a temperature of 750°C. At temperatures above 700°C, carbon atoms penetrate into the Si lattice, thereby forming a thin stabilizing layer (Torres-Costa et al. 2008).

The photoelectrical and electrical properties of a typical PSi photodetector before and after thermal carbonization were reported by Naderi and Hashim (2012c,d). The I–V characteristics of fabricated MSM photodetectors based on the PSi and TC-PSi samples are shown in Figure 2.5. The dark current (I_d) is plotted as a baseline. For the metallization of the PSi substrates to fabricate MSM photodetectors, two interdigitated Schottky contacts (electrode) of Ni with four fingers for each electrode (Baharin and Hashim 2007) were deposited onto both porous substrates by a metal mask. For studying the stability of photocurrent (I_{ph}), samples were kept under illumination of low power green laser (532 nm, 5 mW) for 120 min and the I–V measurements were conducted every 30 min.

Experiment has shown that without applying light, the electrical conductivity of TC-PSi was lower than that of PSi, indicating that the carbonized sample exhibited a more resistive nature. The pioneer photocurrent, which was measured immediately after applying light, is shown by a solid line in both diagrams. On exposure to photon, the PSi sample showed a tremendous response to produce a sufficient number of free carriers for enhanced current conduction. It seems that the PSi sample is more sensitive to the initial photons received from the laser radiation compared to the carbonized one. However, the photocurrent of PSi decreased due to the long-term impact of

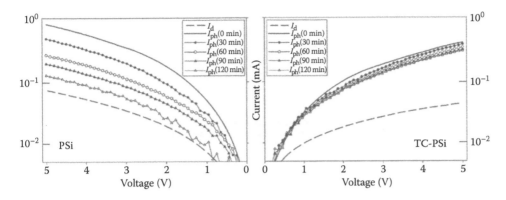

FIGURE 2.5   The *I–V* characteristics of MSM photodetectors based on PSi and TC-PSi under continuous exposure of low power laser radiation at regular intervals of 30 min.

radiation, but this rate of reduction was much lower for carbonized sample even after prolonged exposure to a laser beam (120 min). It indicated the stability of electrical properties of the TC-PSi sample while the initially high current gain of PSi photodetector was diminished exponentially by continuous laser exposure. In other words, the Si–C structure can really act as a protective layer for the silicon walls in the PSi structure. These results are in accordance with results of the photoluminescence (PL) study. For example, analysis of the PL spectra of the PSi samples (Figure 2.6) has shown that a gradual decay (~35%) was observed in the PL spectrum of the PSi sample under a continuous laser radiation, which resulted from the existence of initial metastable bonds on the surface of the bare PSi. This means that these structures are affected by low-power radiation. At the same time, a reduction in the intensity of the PL peak for the TC-PSi sample is considerably smaller. Observed decrease in the PL intensity after carbonization can be explained by a slight decrease in the specific surface area after carbonization due to the attachment of the carbon atoms.

Subsequent studies of the TC-PSi samples have shown that the PL quenching was arrested and reduced to a small amount even after prolonged exposure to laser illumination. This behavior is a good indicator of the formation of a practically stable PSi surface. Thus, TC-PSi layers are acceptable in the manufacturing of PSi-based photodetectors.

The higher surface resistivity of TC-PSi compared to PSi can be explained by a two-dimensional model of PSi that was proposed by Stievenard and Deresmes (1995). The carbonization can increase the concentration of an interface state, which should be accompanied by the effects such as an increase in the concentration of electrons trapped on the surface and widening of the space charge region. From this model, after carbonization, the central channel for the electron transport in the PSi walls becomes narrow and the dark current decreases compared to a PSi-based photodetector.

Numerous experimental studies have shown thermal oxidation of PSi also can be used for stabilization parameters of PSi-based photodiodes (Balagurov et al. 2001a,b, 2003; Rossi and Bohn

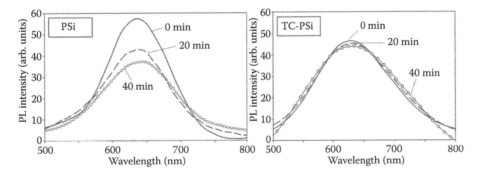

FIGURE 2.6   PL spectra of PSi prior carbonization (PSi) and after that (TC-PSi), under continuous exposure of low power laser radiation for 0, 20, and 40 min.

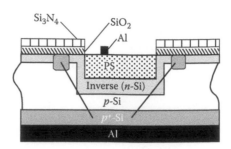

FIGURE 2.7   Schematic view of metal/PSi/p-Si structure with p^+ "stop" ring. (From Balagurov L.A. et al., *Solid-State Electron.* 47, 65, 2003. Copyright 2003: Elsevier. With permission.)

2005). However, it was established that thermal oxidation leads to the formation of a low resistive inversion (n-type) layer in the p-Si substrate adjacent to the PSi/c-Si heterojunction, which drastically increases the active device area, capacitance, noise current, and response time (Balagurov et al. 2001a,b). For resolving this problem, Balagurov et al. (2003) have proposed fabrication technology, which allowed suppressing the reverse current by orders of magnitude. The technology includes formation of the PSi layer in dielectric windows that limits the device area. "Stop" rings (p^+-Si rings) were formed at the edge of the device area (see Figure 2.7). The role of the "stop" ring is to prevent the depletion region from developing too much along the surface. In other words, "stop" rings decrease the spreading of current along the inverse layer formed in the c-Si substrate at its interface with the SiO$_2$ layer. Due to a strong decrease in the back current, low noise metal/PSi/c-Si photodiodes and phototransistor-like structures were fabricated. Balagurov et al. (2003) stated that response time and noise characteristics of these devices were close to those of high-speed c-Si detectors.

However, Rossi and Bohn (2005) believed that the approaches like guard rings did not solve the problem of the high back current in the MSM structures. According to Rossi and Bohn (2005), the use of low-doped substrates is more efficient. In particular, they proposed using (100) oriented B-doped Si with a resistivity of 7–30 Ω·cm as starting material. Backside metallization was achieved by deposition of a 50-nm thick Ti layer and subsequent annealing at 850°C to form conductive Ti-silicide. This silicide is stable at even higher temperatures and can support the subsequent annealing at 850°C in contrast to the Al back contact. The PSi layer was formed by anodic etching in a 20 HF:210 H$_2$O:190 C$_2$H$_5$OH solution at a constant current density of 20 mA/cm^2 for 15 min in the dark. The thickness of the porous layer was about 6 μm. The porous layer was then rapid-thermal-oxidized (850°C, 90 s) to form a 5-nm thick SiO$_2$-layer on top and subsequently rapid-thermal-annealed at 850°C for 15 s in a N$_2$-atmosphere. A 100-nm thick Al interdigital finger structure was then deposited on the top by thermal evaporation to form electrical contacts. A sketch of the resulting layer system is shown in Figure 2.1b. The testing of these devices has shown that a photodetector based on PSi had dark current around 1 μA in reverse bias, and the responsivity was about 2.5 A/W at 400 nm and increased to 5.5 A/W at 800 nm. According to Rossi and Bohn (2005), the avalanche effect is a possible explanation of the gain in designed PSi photodetector.

2.3 PHOTODETECTORS DESIGNED ON THE BASE OF PSi-BASED COMPOSITES

Typically, composite-based photodetectors are fabricated using a pore-filling technique by materials that have a high photosensitivity or specific properties. For example, Abd Rahim et al. (2010) used Ge for this purpose, while Chou et al. (2012) used the filling of the PSi pores by CdSe/CdS/ZnS quantum dots (QDs). Colloidal QDs are semiconductor nanocrystals with tunable optical property depending on their sizes and shapes that can be controlled by the fabrication process. Outstanding optoelectronic and photonic properties, such as electroluminescence, photoluminescence, photovoltaics, absorption, and narrow emission linewidth, make QDs a good candidate for photodetector application (Nayfeh et al. 2004; McDonald et al. 2005; Konstantatos et al. 2006; Shieh et al. 2009; Tu et al. 2010; Tang and Sargent 2011).

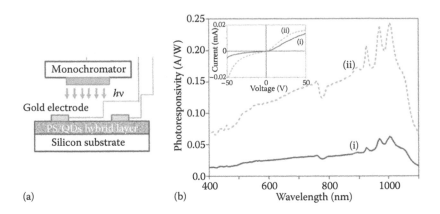

FIGURE 2.8   (a) Schematic setup of spectral photoresponsivity measurement; (b) comparison of spectral photoresponsivity and I-V curve (inset) from undoped (curve i) and QDs doped (curve ii) PSi-MSM photodetector. The anodization time was 25 min. I-V curve was measured in a dark environment. The increase of dark current (inset) indicates that the doped QDs enhance the charge transfer between PSi and electrode. (From Chou C.-M. et al., *Nanoscale Res. Lett.* 7, 291, 2012. Published by Springer as open access.)

Principles of the pore-filling technique were described previously in Chapter 12 in *Porous Silicon: Formation and Properties* and Chapter 9 in this book. As it is shown in these chapters, different methods can be used for pore filling. In particular, Abd Rahim et al. (2010) applied a conventional technique of thermal evaporation, where PSi acted as a patterned substrate. The process was completed by the Ni deposition using thermal evaporation followed by a metal annealing at 400°C for 10 min. In the case of the CdSe/CdS/ZnS QDs, for pore filling, the method of the PSi dipping in the solution of CdSe/CdS/ZnS coreshell colloidal QDs dissolved in toluene following toluene evaporating at room temperature was used (Chou et al. 2012).

The testing of fabricated photodetectors has shown that the Si/PSi-Ge/Ni MSM photodetectors had lower dark currents compared to a control device of PSi. In addition, the device showed enhanced current gain compared to a conventional PSi device, which could be associated with the presence of the Ge nanostructures in the PSi (Abd Rahim et al. 2010). Chou et al. (2012) reported that a PSi metal–semiconductor metal (PSi-MSM) photodetector embedded with colloidal QDs inside the pore layer also demonstrated an improvement of spectral photoresponsivity. The detection efficiency of the QDs/PSi hybrid-MSM photodetector was enhanced five times larger than that of the undoped PSi-MSM photodetector (see Figure 2.8). It was also shown that the photoresponsivity of the QD/PSi hybrid-MSM photodetector depended on the number of layer coatings of QDs and the pore sizes of PSi. It was assumed that the bandgap alignment between PSi (approximately 1.77 eV) and QDs (approximately 1.91 eV) facilitated the photo-induced electron transfer from QDs to PSi, whereby enhancing the photoresponsivity took place.

2.4 SILICON PHOTODETECTORS WITH PSi-BASED ANTIREFLECTING COVERING AND FILTERS

It is known that the optical reflectance from a single-crystalline silicon wafer is ~38% in the visible wavelength range, which is a disadvantage for Si-based photodetectors because in this wavelength interval the solar radiation intensity is maximal. For this reason, antireflective coating layers are mostly coated onto the external surface of silicon-based photodetectors (Balagurov et al. 2003). Experiments and simulations have shown that the PSi layer formed on the surface of silicon could play the same role (Martinez-Duart and Martin-Palma 2002; Salman et al. 2011) (see Table 2.1). It was also established that the refractive index of PSi could vary over a wide range in order to fabricate interference filters (read Chapter 8, *Porous Silicon: Formation and Properties*; Chapter 3, *Porous Silicon: Biomedical and Sensor Applications*; and Chapter 4, this book). Moreover, as it was shown in Chapter 4 in *Porous Silicon: Formation and Properties*, this

TABLE 2.1   **Average Reflectance Values in the Visible Wavelength Range and Beyond (300–900 nm) for Different Silicon and PSi Structures**

Material	Average Reflectance (%)
Monocrystalline silicon	34.5
Polycrystalline silicon	26.8
Electrochemically formed PSi (type A)	9.9
Electrochemically formed PSi (type B)	5.3
Chemically formed PSi	5.0

Source: Martinez-Duart J.M. and Martin-Palma R.J., *Phys. Stat. Sol. (b)* 232 (1), 81, 2002. Copyright 2002: John Wiley & Sons, Inc. With permission.

technique is very cheap because optical systems can be fabricated by anodical etching of pure silicon wafers, without any expensive deposition process. Formation of the PSi interference filters, that is, a controlled stack of layers with different refractive indexes, can be achieved by periodically changing the anodization current density. In addition, a process of fabrication is fast and different kinds of optical components can be fabricated on the same chip very easily by selective area anodization (read Chapter 4, this book). Thus, based on the dimensions of the pores, the structure of PSi can modulate the optical absorption and reflectance properties, and therefore a PSi layer can be used for either antireflective coating or improving optical confinement in silicon photodetectors. Other important advantages of using PSi in optical detection devices are that PSi-based interference filters can be developed to match the desired optical properties, which allow avoiding the use of extra antireflection coatings. In addition, due to use of PSi-based optical filters and variation in the bandgap of PSi, selective and color sensitive photodetectors can be fabricated (Krüger et al. 1996, 1997; Torres-Costa et al. 2007).

Krüger et al. (1996, 1997) have shown that fabrication of the silicon photodetectors with PSi-based filters required standard photolithography processing steps. A cross-section of the Si photodiode with the PSi multilayer stack in the upper, p^+-type part of the p-n-junction fabricated by Krüger et al. (1996, 1997) is shown in Figure 2.9a. This structure implies that during the anodic etch process, the p-n-junction is reverse biased. As the anodization current densities are very

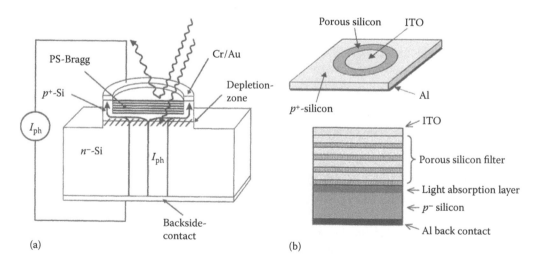

(a)   (b)

FIGURE 2.9   (a) Device geometry of the photodiode with integrated PSi multilayer stack. Devices had backside contact (Cr/AuSb/Au), which was necessary for applying the anodization current for PSi formation, and ohmic contacts (Cr/Au) formed by evaporation on the front side. (From Krüger et al., *Jpn. J. Appl. Phys.* 36, L24, 1997. Copyright 1997: Japan Society of Applied Physics. With permission.) (b) Top and cross-sectional schematization of the PSi-based photodetecting elements designed by Torres-Costa et al. (2007). (From Torres-Costa V. et al., *Mater. Sci. Eng. C* 27, 954, 2007. Copyright 2007: Elsevier. With permission.)

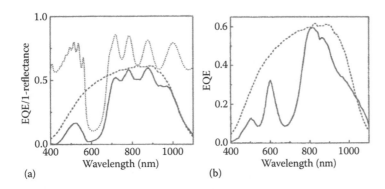

FIGURE 2.10 (a) Spectral characteristics of external quantum efficiency (EQE) of a photodiode with an integrated (HL)^{10} Bragg-reflector (-), compared to a reference diode (--) without PSi, but fabricated on the same wafer, and 1-reflectance (•••) of the Bragg-reflector. The sensitivity of the photodiode with filter is strongly reduced in the spectral range from 550 to 700 nm, and as an undesired effect, oscillations occur between 700 and l000 nm. Both structures correspond very well to the transmission of the Bragg-reflector, which is roughly estimated by 1-reflectivity neglecting the absorption of the PSi. (b) External quantum efficiency (EQE) of a photodiode with a (HL)^4(LU)^4 FP-filter (-) compared to the EQE of a reference diode (--). The filter wavelength is 600 nm. (From Krüger M. et al., *Jpn. J. Appl. Phys.* 36, L24, 1997. Copyright 1997: Japan Society of Applied Physics. With permission.)

high (it was applied 20 mA/cm² for the formation of the H-layer and 400 mA/cm² for the L-layer), the samples had to be illuminated during the etch process. This resulted in degraded filter characteristics compared to filters fabricated on pure substrates without illumination; and this fact requires further investigation. The same approach to design of color-sensitive photodetectors was used by Torres-Costa et al. (2007). Figure 2.9b shows top and cross-sectional schematic views of these devices.

Krüger et al. (1996, 1997) have also shown that the I-V characteristics of the diodes with an integrated PSi filter are nearly the same as the ones of reference diodes fabricated in the same way but without a PSi layer on the top of the sample. At that, the spectral response of the photodiodes can be strongly modified using reflectance filters (Figure 2.10a) or transmission filters (Figure 2.10b). Results presented in these figures testify that after optimization of filter fabrication technology photodetectors, which are sensitive just to a small spectral range, can be designed.

Regarding the PSi used as an antireflection coating, this application of PSi is discussed in detail in Chapters 4 and 10 of this book.

It should be noted that the presence of PSi-based coating on the surface of a p-n junction might also promote the growth of the photodetector sensitive in the ultraviolet region. This approach to design of the UV photodetectors has been realized by Min et al. (2001). To increase detection efficiency in the UV range, the peak spectral response wavelength of the *p-n* junction diode was matched with the peak wavelength of photoluminescence (PL) emitted from a PSi layer. As it is known, PSi has strong enough PL in the spectral range of the Si photosensitivity (read Chapter 7 in *Porous Silicon: Formation and Properties*). In other words, the PSi layer converts UV to visible light, and then the p-n junction beneath the PSi layer detects a light emission from the PSi. In photodetectors fabricated by Min et al. (2001), the PSi layer was formed using chemical (stain) etching. A description of this method can be found in Chapter 10 in *Porous Silicon: Formation and Properties*. Testing of these devices has shown that the PSi-based detectors designed by Min et al. (2001) were much more sensitive to UV light than to red light, while reference diodes without PSi layers showed no sensitivity to UV light. The measured photocurrent increased rapidly with UV power. The differential sensitivity was calculated as 2.91 mA/mW.

2.5 HYBRID HETEROJUNCTION PHOTODETECTORS

For realization of photodiodes integrated in photonics circuits operating at wavelengths beyond 1.1 μm and in the UV region, silicon was not considered the right material (Casalino et al. 2010).

In spectral range >1.1 μm, Si is transparent, while in the UV range a strong absorption in the surface region takes place. Therefore, much research for finding materials and approaches for integrating into silicon electronics of photodetectors with the required spectral sensitivity is being conducted. Studies have shown that the use of PSi makes it possible to solve this problem. It has been found that the lattice parameter of PSi depends on the porosity, and thus through the porosity varying it is possible to select conditions allowing the silicon substrate with PSi buffer layer to grow epitaxial and polycrystalline films of other materials with the required optical properties. The PSi layer in this case plays the role of an elastic matrix with nano-size voids, which compensate for the elastic stresses of the polymorphic layer. The features of this process are discussed in Chapter 7 in this book.

By now, many materials have been already tested as materials for the development of efficient chip-scale hybrid photodetectors integrated on silicon substrates. ZnO (Rajabi et al. 2012; Shabannia et al. 2013; Wu et al. 2013; Keramatnejad et al. 2014; Wu 2014) and SiC (Naderi and Hashim 2013) were used for the development of the UV photodetectors, while InN (Amirhoseiny et al. 2013) and PbTe layers (Belyakov et al. 1997) were used in the development of photodetectors for the IR spectral range. An example of such application is shown in Figure 2.11.

Testing of these devices has shown that the photodetectors have many suitable parameters. As for lead-chalcogenide films deposited on PSi substrates, studies have shown that these layers can serve as a basis for the formation of a wide class of optoelectronic devices operating in the infrared in conjunction with silicon-based readout circuits. Despite a great mismatch in the lattice constants and the temperature expansion coefficients between silicon and lead telluride, the photodiode parameters were similar to those of photodiodes in the orienting substrates (Belyakov et al. 1997).

It should be noted that attempts to develop photodetectors based on nanoscaled 1D and 2D materials on PSi substrate are also present. There are reports on the development of graphene/PSi (Kim et al. 2014) and CNTs/PSi (Suhail 2013) based photodetectors. For example, the detector's structure shown in Figure 2.11 permitted the PSi layer to trap the incident optical radiation and to reduce the reflection coefficient fluctuations of the front face to a value of about 7%. Optoelectronic properties of CNTs make them very interesting components for infrared sensors; CNTs exhibit wide absorbance in the infrared range. At the same time, graphene is an interesting material for fabrication of the carrier collector. Applications of graphene in flexible and transparent electrodes are potentially available because graphene has outstanding physicochemical properties such as transparency over 97%, strong mechanical strength/flexibility, high electrical transportation, and excellent thermal conductivity (Kim et al. 2014). According to Kim et al. (2014), the characteristics of graphene/PSi photodiodes (see Figure 2.12) were governed by typical Schottky diode-like transport of charge carriers at the graphene/PSi junctions, based on bias-dependent variations of the band profiles. The PSi PDs, especially at $P = 60\%$, showed extremely high speed of photoresponse, as short as ~3 μs in t_{decay}, compared to the as-Si PDs. Optimized response speed was ~10 times faster compared to graphene/single-crystalline Si PDs.

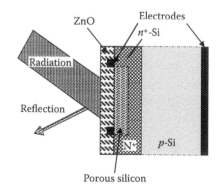

FIGURE 2.11   The structure of ZnO/PSi/c-Si photodiode. (From Ben Achour Z. et al., *Nucl. Instrum. Meth. Phys. Res. A* 579, 1117, 2007. Copyright 2007: Elsevier. With permission.)

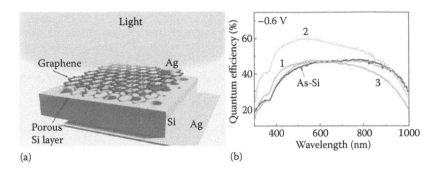

FIGURE 2.12   (a) Schematic diagram of graphene/PSi/*n*-Si photodetector with silver top and bottom electrodes under light illumination; (b) spectral dependence on quantum efficiencies of graphene/PSi (1–3) and graphene/Si-based photodiodes. The samples had the following porosities: (1) *P* = 46.5%, (2) *P* = 60.3%, and (3) *P* = 79.3%. (From Kim J. et al., *ACS Appl. Mater. Interfaces* 6, 20880, 2014. Copyright 2014: American Chemical Society. With permission.)

2.6 SUMMARY

Consideration of devices, conducted in this chapter, shows that PSi is a multifunctional material and, in principle, PSi has great potential for use in photodetectors with the purpose to improve their performance. The PSi layers can be used either for antireflective coating or for improving optical confinement in the silicon photodetectors. Other important advantages of PSi in the optical detection devices are that the optical properties of the PSi surface can be tuned by changing the morphology of pores and the crystallite size (quantum confinement). However, it should be recognized that despite this, PSi-based devices are not present on the photodetector market. Most developments are finished on the stage of prototypes that only demonstrate the capabilities of PSi. Unfortunately, mass production requires high uniformity and reproducibility of the parameters that cannot be realized while using the technology of silicon porosification. There is also the difficulty of this technology incorporation into the traditional process of semiconductor devices manufacturing. Temporal instability of the PSi parameters also hinders its implementation. The optical properties of PSi alter as a function of time of exposure to illumination. The degradation of the PSi properties after prolonged light irradiation and after long periods of exposure to the atmosphere is a strong barrier for fabricating stable photodetectors based on the PSi substrates. By now, a variety of methods aimed for stabilization of the parameters of PSi is being developed. The results showed that the intrinsic instabilities of these materials could be reduced by simple chemical modifications. However, often the stabilization is accompanied by a deterioration of responsivity of the PSi-based photodetectors. All of this suggests that the realization of the full potential of PSi still requires further research and development of new technologies.

ACKNOWLEDGMENT

G.K. is grateful to the Ministry of Science, ICT and Future Planning (MSIP) of the Republic of Korea for supporting his research.

REFERENCES

Abd Rahim, A.F., Hashim, M.R., and Ali, N.K. (2010). Study of Ge embedded inside porous silicon for potential MSM photodetector. *Microelectron. Intern.* 27(3), 154–158.

Abd Rahim, A.F., Hashim, M.R., and Ali, N.K. (2011). High sensitivity of palladium on porous silicon MSM photodetector. *Physica B* 406, 1034–1037.

Ait-Hamouda, K., Gabouze, N., Hadjersi, T. et al. (2003). Influence of solution resistivity and postanodizing treatments of PS films on the electrical and optical properties of metal/PS/Si photodiodes. *Solar Energy Mater. Solar Cells* 76, 535–543.

Al-Hardan, N.H., Abdullah, M.J., Ahmad, H., Aziz, A.A., and Low, L.Y. (2011). Investigation on UV photo-detector behavior of RF-sputtered ZnO by impedance spectroscopy. *Solid-State Electron.* 55(1), 59–63.

Alwan, A.M. and Jabbar, A.A. (2011). Design and fabrication of nanostructures silicon photodiode. *Modern Appl. Sci.* 5(1), 106–112.

Amirhoseiny, M., Hassan, Z., and Ng, S.S. (2013). Fabrication of InN based photodetector using porous silicon buffer layer. *Surf. Eng.* 29(10), 772–777.

Baharin, A. and Hashim, M.R. (2007). Study of electrical characteristics of Ge islands MSM photodetector structure grown on Si substrate using conventional methods. *Semicond. Sci. Technol.* 22(8), 905–910.

Balagurov, L.A., Yarkin, D.G., Petrova, E.A., Orlov, A.F., and Andrushin, S.Ya. (1997). Highly sensitive porous silicon based photodiode structures. *J. Appl. Phys.* 82, 4647–4650.

Balagurov, L.A., Bayliss, S.C., Andrushin, S.Ya. et al. (2001a). Metal/PS/c-Si photodetectors based on unoxidized and oxidized porous silicon. *Solid-State Electron.* 45, 1607–1611.

Balagurov, L.A., Bayliss, S.C., Kasatochkin, V.S., Petrova, E.A., Unal, B., and Yarkin, D.G. (2001b). Transport of carriers in metal/porous silicon/c-Si device structures based on oxidized porous silicon. *J. Appl. Phys.* 90, 4543–4548.

Balagurov, L.A., Bayliss, S.C., Yarkin, D.G. et al. (2003). Low noise photosensitive device structures based on porous silicon. *Solid-State Electron.* 47, 65–69.

Belyakov, L.V., Zakharova, I.B., Zubkova, T.I., Musikhin, S.F., and Rykov, S.A. (1997). Study of PbTe photodiodes on a buffer sublayer of porous silicon. *Semiconductors* 31(1), 76–77.

Ben Achour, Z., Touayar, O., Akkari, E., Bastie, J., Bessais, B., and Ben Brahim, J. (2007). Study and realization of a transfer detector based on porous silicon for radiometric measurements. *Nucl. Instrum. Meth. Phys. Res. A* 579, 1117–1121.

Bisi, O., Ossicini, S., and Pavesi, L. (2000). Porous silicon: A quantum sponge structure for silicon based optoelectronics. *Surf. Sci. Rep.* 38(1–3), 1–126.

Casalino, M., Coppola, G., Iodice, M., Rendina, I., and Sirleto, L. (2010). Near-infrared sub-bandgap all-silicon photodetectors: State of the art and perspectives. *Sensors* 10, 10571–10600.

Chen, X., Yang, W., and Wu, Z. (2006). Visible blind p–i–n ultraviolet photodetector fabricated on 4H-SiC. *Microelectron. Eng.* 83(1), 104–106.

Chou, C.-M., Cho, H.-T., Hsiao, V.K.S., Yong, K.-T., and Law, W.-C. (2012). Quantum dot-doped porous silicon metal–semiconductor metal photodetector. *Nanoscale Res. Lett.* 7, 291.

Diesinger, H., Bsiesy, A., Hérino, R., and Gelloz, B. (2000). Effect of the quantum confinement on the optical absorption of porous silicon, investigated by a new in-situ method. *Mater. Sci. Eng. B* 69–70, 167–170.

Ghosh, R.N., Loloee, R., Isaacs-Smith, T., and Williams, J.R. (2009). High frequency inversion capacitance measurements for 6H-SiC n-MOS capacitors from 450 to 600°C. *Mater. Sci. Forum* 600–603, 739–742.

Hadjersi, T. and Gabouze, N. (2008). Photodetectors based on porous silicon produced by Ag-assisted electroless etching. *Opt. Mater.* 30, 865–869.

Huang, Y.M. (1996). Photoluminescence of copper-doped porous silicon. *Appl. Phys. Lett.* 69(19), 2855–2857.

Huy, B., Binh, P.H., Diep, B.Q., and Luong, P.V. (2003). Effect of ageing on the luminescence intensity and lifetime of porous silicon: Roles of recombination centers. *Physica E* 17, 134–136.

Keffous, A., Bourenane, K., Kechouane, M. et al. (2007). Effect of anodization time on photoluminescence of porous thin SiC layer grown onto silicon. *J. Lumin.* 126(2), 561–565.

Keramatnejad, K., Khorramshahi, F., Khatami, S., and Asl-Soleimani, E. (2014). Optimizing UV detection properties of n-ZnO nanowire/p-Si heterojunction photodetectors by using a porous substrate. *Opt. Quant. Electron.* DOI 10.1007/s11082-014-0032-y

Kim, J., Joo, S.S., Lee, K.W. et al. (2014). Near-ultraviolet-sensitive graphene/porous silicon photodetectors. *ACS Appl. Mater. Interfaces* 6, 20880–20886.

Konstantatos, G., Howard, I., Fischer, A. et al. (2006). Ultrasensitive solution-cast quantum dot photodetectors. *Nature* 442, 180–183.

Krüger, M., Berger, M.G., Marso, M. et al. (1996). Integration of porous silicon interference filters in Si-photodiodes. In: *Proceedings of the 26th European Conference on Solid State Device Research, ESSDERC '96*, Sept. 9–11, Bologna, Italy, pp. 891–894.

Krüger, M., Berger, M.G., Marso, M. et al. (1997). Color-sensitive Si-photodiode using porous silicon interference filters. *Jpn. J. Appl. Phys.* 36, L24–L26.

Lee, M.K., Tseng, Y.C., and Chu, C.H. (1998). A high gain porous silicon metal–semiconductor–metal photodetector through rapid thermal oxidation and rapid thermal annealing. *Appl. Phys. A* 67, 541–543.

Lin, M.-L., Lin, Y.-C., Wu, K.-H., and Huang, C.-P. (2013). Preparation of oxidized nano-porous-silicon thin films for ultra-violet optical-sensing applications. *Thin Solid Films* 529, 275–277.

Liu, X.F., Sun, G.S., Li, J.M. et al. (2006). Visible blind p+–π–n−–n+ ultraviolet photodetectors based on 4H–SiC homoepilayers. *Microelectron. J.* 37(11), 1396–1398.

Mahmoudi, B., Gabouze, N., Guerbous, L., Haddadi, M., Cheraga, H., and Beldjilali, K. (2007a). Photoluminescence response of gas sensor based on CHx/porous silicon-Effect of annealing treatment. *Mater. Sci. Eng. B* 138, 293–297.

Mahmoudi, B., Gabouze, N., Haddadi, M. et al. (2007b). The effect of annealing on the sensing properties of porous silicon gas sensor: Use of screen-printed contacts. *Sens. Actuators B* 123, 680–684.

Martinez-Duart, J.M. and Martin-Palma, R.J. (2002). Photodetectors and solar cells based on porous silicon. *Phys. Stat. Sol. (b)* 232(1), 81–88.

McDonald, S.A., Konstantatos, G., Zhang, S. et al. (2005). Solution-processed PbS quantum dot infrared photodetectors and photovoltaics. *Nat. Mater.* 4, 138–142.

Min, N.-K., Kang, C.-G., Jin, J.-H., Ko, J.-Y., and Kim, S.-K. (2001). Porous silicon-based UV detector. *J. Korean Phys. Soc.* 39, S63–S66.

Naderi, N. and Hashim, M.R. (2012a). A combination of electroless and electrochemical etching methods for enhancing the uniformity of porous silicon substrate for light detection application. *Appl. Surf. Sci.* 258, 6436–6440.

Naderi, N. and Hashim, M.R. (2012b). Effect of surface morphology on electrical properties of electrochemically-etched porous silicon photodetectors. *Int. J. Electrochem. Sci.* 7, 11512–11518.

Naderi, N. and Hashim, M.R. (2012c). Fabrication of silicon carbide thin film as a stabilizing layer for improving the stability of porous silicon photodiodes. *Mater. Sci. Forum* 717–720, 1283–1286.

Naderi, N. and Hashim, M.R. (2012d). Stabilization of photoluminescence properties of silicon nanocrystallites by thermal carbonization of porous silicon. In: *Proceedings of Int. Conference on Enabling Science and Nanotechnology, ESciNano,* 5–7 Jan., Johor Bahru, Malaysia, pp. 1–2.

Naderi, N. and Hashim, M.R. (2013). Porous-shaped silicon carbide ultraviolet photodetectors on porous silicon substrates. *J. Alloys Compounds* 552, 356–362.

Nakamura, T., Ogawa, T., Hosoya, N., and Adachi, S. (2010). Effects of thermal oxidation on the photoluminescence properties of porous silicon. *J. Lumin.* 130(4), 682–687.

Nayfeh, O.M., Rao, S., Smith, A., Therrien, J., and Nayfeh, M.H. (2004). Thin film silicon nanoparticle UV photodetector. *IEEE Photo. Tech. Lett.* 16, 1927–1929.

Pérez, E.X. (2007). Design, Fabrication and characterization of porous silicon multilayer optical devices. PhD thesis, Universitat Rovira i Virgili.

Rajabi, M., Dariani, R.S., and Iraji, Zad, A. (2012). UV photodetection of laterally connected ZnO rods grown on porous silicon substrate. *Sens. Actuators A* 180, 11–14.

Rittenhouse, T.L., Bohn, P.W., and Adesida, I. (2003). Structural and spectroscopic characterization of porous silicon carbide formed by Pt-assisted electroless chemical etching. *Solid State Commun.* 126(5), 245–250.

Rossi, A.M. and Bohn, H.G. (2005). Photodetectors from porous silicon. *Phys. Stat. Sol. (a)* 202(8), 1644–1647.

Salman, K.A., Omar, K., and Hassan, Z. (2011). The effect of etching time of porous silicon on solar cell performance. *Superlatt. Microstructur.* 50(6), 647–658.

Salonen, J., Lehto, V.P., Björkqvist, M., Laine, E., and Niinistö, L. (2000). Studies of thermally-carbonized porous silicon surfaces. *Phys. Stat. Sol. (a)* 182(1), 123–126.

Schaub, J.D., Li, R., Csutak, S.M., and Campbell, J.C. (2001). High-speed monolithic silicon photoreceivers on high resistivity and SOI substrates. *J. Lightw. Tech.* 19, 272–278.

Shabannia, R., Abu Hassan, H., Mahmodi, H., Naderi, N., and Abd, H.R. (2013). ZnO nanorod ultraviolet photodetector on porous silicon substrate. *Semicond. Sci. Technol.* 28, 115007.

Shi, J.X., Zhang, X.X., Gong, M.L., Zhou, J.Y., Cheah, K.W., and Wong, W.K. (2000). Photoluminescence of erbium, zinc, and copper doped porous silicon and a phenomenological model for the metal electrodeposition. *Phys. Stat. Sol. (a)* 182(1), 353–357.

Shieh, J.-M., Yu, W.-C., Huang, J.Y. et al. (2009). Near-infrared silicon quantum dots metal-oxide semiconductor field-effect transistor photodetector. *Appl. Phys. Lett.* 94, 241108.

Stievenard, D. and Deresmes, D. (1995). Are electrical properties of an aluminum-porous silicon junction governed by dangling bonds? *Appl. Phys. Lett.* 67(11), 1570–1572.

Su, Y.-K., Chiou, Y.-Z., Chang, C.-S., Chang, S.-J., Lin, Y.-C., and Chen, J.F. (2002). 4H-SiC metal–semiconductor–metal ultraviolet photodetectors with Ni/ITO electrodes. *Solid-State Electron.* 46(12), 2237–2240.

Suhail, A.M. (2013). Carbon nanotubes—Porous silicon high sensitivity infrared detector. *IJSR – Int. J. Sci. Res.* 2(1), 209–210.

Svechnikov, S.V., Kaganovich, E.B., and Manoilov, E.G. (1998). Photosensitive porous silicon based structures. *Semicond. Phys. Quantum Electron. Optoelectron.* 1(1), 13–17 (Ukraine).

Tang, J. and Sargent, E.H. (2011). Infrared colloidal quantum dots for photovoltaics: Fundamentals and recent progress. *Adv. Mater.* 23, 12–29.

Torres-Costa, V., Martín-Palma, R.J., and Martínez-Duart, J.M. (2007). All-silicon color-sensitive photodetectors in the visible. *Mater. Sci. Eng. C* 27, 954–956.

Torres-Costa, V., Martín-Palma, R.J., Martínez-Duart, J.M., Salonen, J., and Lehto, V.P. (2008). Effective passivation of porous silicon optical devices by thermal carbonization. *J. Appl. Phys.* 103(8), 083124.

Tsai, C., Li, K.H., Campbell, J.C., and Tasch, A.L. (1993). Photodetectors fabricated from rapid thermal oxidized porous silicon. *Appl. Phys. Lett.* 62, 2818–2820.

Tsybeskov, L., Duttagupta, S.P., and M. Fauchet, P.M. (1995). Photoluminescence and electroluminescence in partially oxidized porous silicon. *Solid State Commun.* 95(7), 429–433.

Tu, C.-C., Tang, L., Huang, J., Voutsas, A., and Lin, L.Y. (2010). Solution-processed photodetectors from colloidal silicon nano/micro particle composite. *Opt. Express* 18, 21622–21627.

Tuura, J., Bjorkqvist, M., Salonen, J., and Lehto, V. (2008). Electrically isolated thermally carbonized porous silicon layer for humidity sensing purposes. *Sens. Actuators B* 131, 627–632.

Wang, H.Y. and Li, X.J. (2010). Capacitive humidity-sensitivity of carbonized silicon nanoporous pillar array. *Mater. Lett.* 64(11), 1268–1270.

Wu, K.-H. (2014). Oxidized nano-porous-silicon buffer layers for suppressing the visible photoresponsivity of ZnO ultraviolet photodetectors on Si substrates. *J. Nanomater.* 2014, 756527.

Wu, K.-H., Tang, C.-C. and Lin, S.-C. (2013). Development of ultra-violet sensing devices with Zinc-Oxide thin-films on oxidized nano-porous-silicon substrates. In: *Proceedings of IEEE 8th Conference on Nanotechnology Materials and Devices (NMDC)*, Oct. 6–9, Tainan, pp. 105–107.

Yang, W., Zhang, F., Liu, Z., and Wu, Z. (2008). Effects of annealing on the performance of 4H-SiC metal–semiconductor–metal ultraviolet photodetectors. *Mater. Sci. Semicond. Process.* 11(2), 59–62.

Yu, L.Z. and Wie, C.R. (1993). Study of MSM photodetector fabricated on porous silicon. *Sen. Actuators A* 39, 253–251.

Zebentout, A.D., Bensaad, Z., Zegaoui, M., Aissat, A., and Decoster, D. (2011). Effect of dimensional parameters on the current of MSM photodetector. *Microelectron. J.* 42(8), 1006–1009.

Zheng, J.P., Jiao, K.L., Shen, W.P., Anderson, W.A., and Kwok, H.S. (1992). Highly sensitive photodetector using porous silicon. *Appl. Phys. Lett.* 61, 459–461.

Zimmermann, H. (2004). *Silicon Photonics, Topics in Applied Physics*. Springer-Verlag, New York.

PSi-Based Photonic Crystals

Gonzalo Recio-Sánchez

3

CONTENTS

3.1 Introduction 52

3.2 Working Principles 53

3.3 1D Photonic Crystals 54

    3.3.1 Microcavities 55

    3.3.2 Applications 57

3.4 2D Photonic Crystals 60

    3.4.1 Bulk 2D Photonic Crystals 60

    3.4.2 Defects in 2D Photonic Crystals 63

    3.4.3 Finite 2D Photonic Crystals (Slabs) 65

3.5 3D Photonic Crystals 68

3.6 Applications of 2D and 3D Photonic Crystals 71

3.7 Outlook 71

References 72

3.1 INTRODUCTION

For the last decades, microelectronic technology has dominated several fields such as telecommunications and information processing, playing an important role in almost every aspect of daily life. However, the need for more efficient, higher speed, and lower sized integrated electronic circuits has demanded novel devices that can suppose an alternative to classic electronic computation. For this reason, increased interest is being devoted to other technologies as photonic. In this technology, electrons are substituted by photons to transmit information, allowing the development of faster and more efficient devices.

In this context, the optical analogues to electronic semiconductors, the so-called photonic crystals, have been the subject of intense research efforts for the last few decades (Joannopoulos et al. 1997; Yablonovitch 2001). The idea of photonic crystals was introduced by two independent groups almost three decades ago. Both, Yablonovitch (1987) and John (1987), proposed to inhibit light propagation inside a material by the design of its dielectric constant. In this way, a photonic crystal can be defined as a structure whose dielectric constant periodically varies in one or more spatial directions. That allows classifying the photonic crystals in one, two, and three dimensions (1D, 2D, and 3D) depending on the spatial directions in which the dielectric constant varies periodically. Figure 3.1 shows three different kinds of photonic crystals (1D, 2D, and 3D from left to right) where the dielectric constant varies periodically in one, two, or three spatial directions. The periodical distribution of its dielectric constant can prohibit light propagation of a certain range of electromagnetic wave frequencies inside the crystal.

Since the microelectronic technology has been dominated by silicon, probably becoming the best-known material to humans, it is expected that silicon be the base of this new photonic technology. Furthermore, silicon has interesting optical properties including low absorption in the infrared range and high refractive index, near to 3.5 for this range. Unfortunately, silicon is an indirect gap semiconductor and hence its radiative transitions are low efficiency (around 10^{-6}). This quandary can be solved by using other kinds of direct gap semiconductors such as Ge, GaAs, or InP. However, these materials are expensive and, most importantly, do not allow their direct integration in silicon technology. In this respect, it is desirable to develop a silicon-based or a silicon compatible material with the appropriate photonic properties, which could be easily integrated into the current standard CMOS technology.

Within this context, porous silicon (PSi) is an appropriate material for the development of optical devices such as photonic crystals for different reasons. First, it is based on silicon, and very compatible with current CMOS technology (Hirschman et al. 1996). Another interesting property is its well-known efficient photoluminescence (Canham 1990) and electroluminescence (Koshida and Koyama 1992) at room temperature, which can be tuned by choosing the appropriate condition during the fabrication process. In the same way, the dielectric constant of PSi layers can be controlled from silicon value to almost air by simply changing its porosity (Theib 1996, 1997). Given that their pores and silicon structures can be lower than 20 nm, in the visible and infrared ranges whose wavelengths are greater than these structures, PSi behaves as an effective dielectric medium. Additionally, the possibility to turn PSi into a biocompatible material makes it a very adequate material to generate novel uses for photonic crystals including biomedical applications (Cunin et al. 2002; Meade et al. 2004).

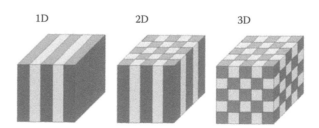

FIGURE 3.1   Simple examples of 1D, 2D, and 3D photonic crystals. Light and dark areas represent materials of different dielectric constant. (Idea from Joannopoulos J.D. et al., *Photonic Crystals— Molding the Flow of Light*. Princeton University Press, Princeton, NJ, 2008.)

In the present chapter, we will focus on advances in PSi-based photonic crystals. In Section 3.2, we will briefly review the theoretical working principles of photonic crystals. Then, we will discuss the main fabrication process of 1D, 2D, and 3D PSi-based photonic crystals, and some of their principal applications.

3.2 WORKING PRINCIPLES

In general, the dispersion relation of light in a dielectric medium relates the frequency and propagation vector of light through its dielectric constant. In a photonic crystal, the periodical distribution of its dielectric constant inhibits the propagation of a certain range of electromagnetic wave frequencies inside the structure. Lightwaves experience periodic perturbations when they propagate through these structures, analogous to electrons in a solid crystal. This analogy suggests that the dispersion of electromagnetic waves in a photonic crystal can be described in terms of photonic band structure (PBS). The frequency ranges over which electromagnetic waves propagation is forbidden are usually called photonic bandgaps. Thus, photonic crystals can control the propagation of light allowing the fabrication of optical devices similar to transistors for the light propagation.

Despite the analogies between electronic waves in semiconductors and electromagnetic waves in photonic crystals, there are significant differences between both. For example, electrons are described by a scalar wavefield whereas the nature of an electromagnetic field is vectorial. Furthermore, electromagnetism inside dielectric materials does not have a fundamental scale, neither in the spatial coordinate nor in the dielectric constant. This makes photonic crystals scalable. Similarly, a same periodic structure will have similar optical properties, independent of the frequency ranges in which it works. For that, a periodic structure in the range of millimeters will work in the microwaves frequency range. If the periodicity is of micrometers, it will work in the infrared frequency range, and so on. Consequently, system unities will be state in function of the lattice parameter (a), which defines the periodicity of the dielectric constant.

The theoretical calculation of light propagation through a photonic crystal has been extensively studied (Sakoda 2005; Busch et al. 2006; Joannopoulos et al. 2008). In general, from Maxwell's equations, which lead the macroscopic electromagnetism, in the absence of external current and sources, the problem is reduced to solve the wave equation subject to the transversality requirement:

(3.1)
$$\nabla \times \left(\frac{1}{\varepsilon(r)} \nabla \times H(r) \right) = \left(\frac{\omega}{c} \right)^2 H(r)$$

where $\varepsilon(r)$ is the macroscopic dielectric function, $H(r)$ is the spatial part of the magnetic field of the photon, ω is its frequency, and c is the speed of light. $H(r)$ and ω are determined by the strength and symmetry properties of $\varepsilon(r)$. If $\varepsilon(r)$ is perfectly periodic, $\varepsilon(r) = \varepsilon(r + a)$, a being the lattice parameter, the solutions are a discrete sequence of frequencies $\omega_n(k)$ characterized by a wavevector k. Each sequence forms a continuous dispersion relation, the so-called "photonic-band" over the Brillouin zone as a function of the wave vector. The collection of all these solutions is termed photonic band structure and provides a complete representation of all possible electromagnetic states in the system.

In order to calculate the photonic band structure of a complex system, novel mathematical and computational developments are needed. Over the last years, different approaches have been proposed for the calculation of photonic band structure. The most commonly used method is the plane wave (Ho et al. 1990) in the frequency domain. Briefly, this method is based on expanding electromagnetism fields as definite-frequency states in some truncation of complex basis and solving the resulting linear eigenproblem. The method has been used with many variations, depending on the choices of basis and eigensolver algorithms (see Soüzer and Haus 1992; Sailor et al. 1998; Dobson 1999; Johnson and Joannopoulos 2001). Other common techniques are the so-called finite-difference time domain (FDTD) algorithms and "transfers matrix" methods. The first method involves the direct simulation of Maxwell's equations over time in a discrete grid

(Chan et al. 1995; Sakoda and Shiroma 1997; Ward and Pendry 2000). The second computes the transfer matrix relating field amplitudes at one end of a unit cell with those at the other at a fixed frequency, applying different methods such as finite-difference, analytical, and so on (Pendry and Mackinnon 1992; Elson and Tran 1996; Chongjun et al. 1997). Both methods allow transmission calculations through finite slabs of photonic materials, and can handle ordered photonic crystals as well as disordered structures, but they require considerable computational resources.

3.3 1D PHOTONIC CRYSTALS

The periodical distribution of dielectric constants in one-dimensional structures such as Bragg reflector and interference filters can constitute 1D photonic crystals. However, they are not referred to as crystal since this name is usually reserved for 2D and 3D structures. As the refractive index of the PSi layer can be changed from a silicon value until almost air depending on the porosity, 1D structures based on PSi are multilayer stacks that alternate PSi layers of different porosities. Normally, these structures interchange PSi layers of high porosity (low refractive index) and low porosity (high refractive index). For this task, different strategies can be used. These structures can be obtained by changing the doping level of crystalline silicon substrate and applying a constant density current in the electrochemical etching (Berger et al. 1994, 1995). However, this method requires the use of complex technological techniques. A simple method results by periodically changing the etch parameters such as the applied current density or light power during the electrochemical etching between two constant values (Loni et al. 1996; Pavesi and Mulloni 1999). This method allows a mayor control in the reproducibility.

Figure 3.2a shows a multilayer stack that interchanges PSi layers of 1 μm of thickness of high and low porosity. A multilayer stack was fabricated by changing the applied current density between 10 mA/cm^2 for 80 s and 100 mA/cm^2 for 15 s during the electrochemical etching (p-type silicon wafer [<100>; ρ = 0.01–0.05 Ω·cm; HF:ethanol 1:2]). Figure 3.2b shows the experimental reflectance spectrum. A high reflectance region can be seen, the so-called stop-band, centered around 1500 cm^{-1}.

This stop-band can be adjusted by changing the thickness of the layers, the number of layers, and their porosity (Mazzoleni and Pavesi 1995; Theib 1997). In this sense, an appropriate design of the structure can give the possibility of tuning the reflectance peak. Figure 3.3 shows simulated reflectance spectra of multilayer stacks which interchange two PSi layers of 500 nm of thickness of high porosity (1.4 of refractive index) and low porosity (2.2 of refractive index) as a function of the number of periods. It can be noticed that as the number of periods increases, the stop-band becomes sharper and larger. In addition, by increasing the number of periods, the reflectance of the stop-bands increases. For that, the number of layers can be optimized taking into account the

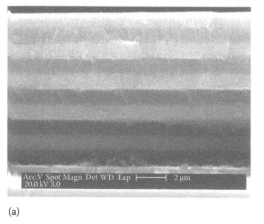

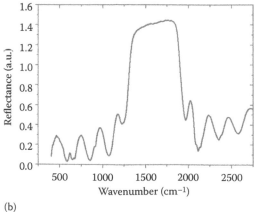

(a)   (b)

FIGURE 3.2   (a) Scanning electron microscope (SEM) image of the cross-section of a PSi multilayer, which alternates PSi layer of 1 μm thickness of low and high porosity. (b) Experimental reflectance spectrum of the PSi multilayer.

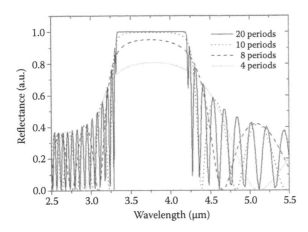

FIGURE 3.3 Simulated reflectance spectra of PSi-based multilayer stack formed by a different number of periods. The refractive index of each layer was set to 1.4 and 2.2, which correspond to typical PSi high and low porosity, respectively.

increase in reflectance for a large number of repetitions and the effects of depth in homogeneity. In addition, the side bands decrease with an increase in the number of periods. Wider stop-bands can be obtained by using random dielectric Bragg reflectors (Pavesi 1997).

Additionally, it can be demonstrated that the half width $\Delta\lambda$ of a stop-band centered at wavelength λ_0 depends on the refractive index of both layers, which form the multilayer stacks (Macleod 1969) as follows:

$$(3.2) \qquad \Delta\lambda = \frac{2\lambda_0}{\pi}\arcsin\left(\frac{n_H - n_L}{n_H + n_L}\right)$$

where n_H and n_L are the refractive index of the low and high porosity layer, respectively. Consequently, the width of the stop-band can be determined by changing the refractive index difference between the high porosity and low porosity layers. A higher difference results in a larger width of the stop-band, as can be observed in Figure 3.4.

Figure 3.5 shows experimental reflectance spectra of different Bragg reflectors in the visible range based on PSi multilayer structures fabricated at different conditions (Theib 1997). The reflectance peak is very tunable over the visible range. The best result was reached for a stop-band centered at a longer wavelength due to an increased absorbance of PSi layers at a shorter wavelength.

Another kind of 1D photonic structure is called rugate filters (Bovard 1993). Rugate filters are structures whose refractive index varies smoothly and periodically in depth. These structures also show stop-band in analogy with Bragg reflectors. However, they have the beneficial effect that there are fewer sidelobes at both sides of the stop-band. In addition, the sidelobes from rugate filters can be suppressed with an anodization function where the refractive index contrast is modulated with a smooth function. The easiest way to obtain a rugate filter based on PSi is applying a sinusoidal current density during the electrochemical etching (Berger et al. 1997; Lorenzo et al. 2005; Salem et al. 2007).

3.3.1 MICROCAVITIES

A planar optical microcavity can be obtained by fabricating a central PSi layer (also named spacer) surrounded by two interference filters (Pavesi 1997). The central PSi layer may have a different porosity than the surrounding layers used for the interference filters (i.e., different dielectric constant) or different thickness. This fact introduces defects in the periodic structure giving the possibility of creating modes inside the bandgap. Special interest has been shown for resonant microcavities in which a quarter wave film is introduced inside two Bragg reflectors resulting in

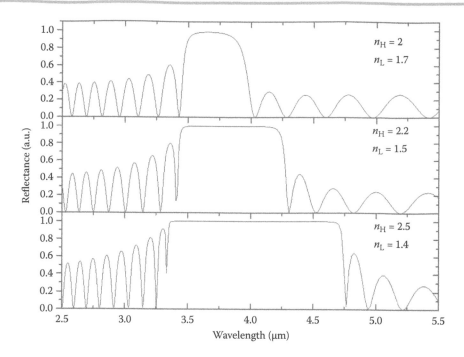

FIGURE 3.4 Simulated reflectance spectra of PSi-based multilayer stack formed by a different mismatch between the high and low porosity layers. The refractive index of each stack is shown in the inset where n_H determines the refractive index of the low porosity layer and n_L the refractive index of the high porosity layer.

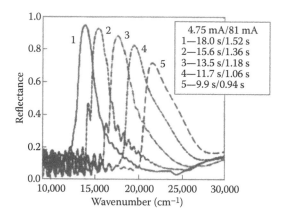

FIGURE 3.5 Experimental reflectance spectra of the PSi Bragg reflector in the visible range fabricated at different conditions. (From Theib W., *Surf. Sci. Rep.*, 29(3–4), 91, 1997. Copyright 1997: Elsevier With permission.)

a defect inside the bandgap at wavelength of λ_0. A squematic representation of this structure is shown in Figure 3.6a where A and B represent a PSi layer of low and high porosity, respectively. In this structure, a transmission range is opened at wavelength λ_0 as can be noticed in the computed reflectance spectrum of Figure 3.6b.

Figure 3.7a shows the cross-section of a PSi microcavity formed by a spacer of 180 nm thickness surrounded by two interference filters of 5 periods. To fabricate this structure, the applied current density during electrochemical etching was periodically varied in time (Chi Do et al. 2011). Figure 3.7b presents the measured reflectance spectrum of the microcavity. A transmission resonance can be observed at 643 nm of wavelength between two high reflectivity bands, which represent the bandgap. The line width of the pass band is around 20 nm, assigned to the strong dependence of the reflectance with the inhomogeneities in the layers.

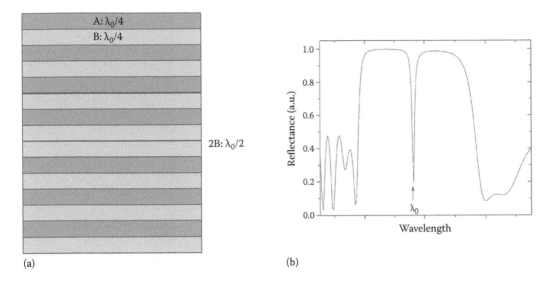

(a)   (b)

FIGURE 3.6   (a) Squematic representation of a resonant microcavity. (b) Simulated reflectance spectrum of the resonant microcavity where the refractive index of A and B layer were set to 2.2 and 1.4, respectively.

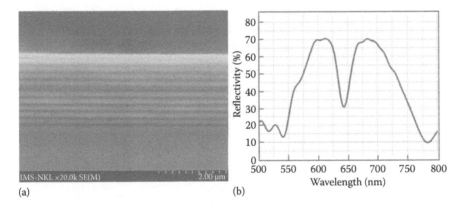

(a)   (b)

FIGURE 3.7   (a) Field-emission scanning electron microscope image of microcavity cross-section with λ/2 wavelength thickness spacer for a centered wavelength of 650 nm. (b) Experimental reflectance spectrum of the PSi microcavity. (From Chi Do T. et al., *Adv. Nat. Sci.: Nanosci. Nanotechnol.* 2, 035001, 2011. The Vietnam Academy of Science and Technology (VAST). CC BY-NC-SA. Copyright 2011: IOP Publishing. With permission.)

3.3.2 APPLICATIONS

These 1D structures have been suggested for several applications. Due to the possibility to control the refractive index in depth, PSi-based waveguides have been proposed (Araki et al. 1996a,b; Loni et al. 1996). In order to fabricate PSi-based waveguides different strategies have been used. The most common method is formed by a three-layer structure where a core layer of different porosity is embedded between two cladding layers of higher porosity. Using this kind of structure, Loni et al. (1996) achieved light guidance for waves with 1.28 μm of wavelength in as-prepared samples and for 0.63 μm in oxidized samples, given that oxidation prevents the absorption in the visible range. Araki et al. (1996a,b) proposed an alternative structure where a core PSi layer is between a bottom cladding PSi layer of lower refractive index and a top thin layer of evaporated Al. By surface excitation, light guidance was observed with polarization dependence. These waveguides are important in order to achieve an all silicon integrated optical circuit where the information is guided in the form of photons (Lazarouk et al. 1998).

Another interesting application is the fabrication of light-emitting devices. The photoluminescence of a PSi layer can be adjusted in the visible range by controlling the fabrication process (Canham 1990). However, the broad spectrum can be inappropriate where high monochromatic emission is required. The photoluminescence spectrum can be narrowed with the use of a microcavity (Pavesi 1997). In this case, the reflectance dip in the reflectance spectrum corresponds to a transmittance peak for light coming from the spacer. Consequently, only the photoluminescence of this narrow wavelength range will escape. In addition, the position of the photoluminescence can be adjusted by controlling the porosity of the spacer layer (Chan and Fauchet 1999). This effect can be implemented in PSi-based light emitting diodes (LED) where electrical injection is performed through silicon/PSi heterojunction. In this way, electroluminescence PSi-based devices can be obtained (Araki et al. 1996a,b; Pavesi et al. 1996). These devices have some advantages with respect to conventional LEDs such as spectral purity of the emission with a narrow emission band, improved directionally in the emission, or an improved yield of the device. However, considering practical applications of PSi LEDs, they were not continuously operable due to charging, the long-term stability of the PSi LED is poor, and the passivation of the PSi is generally difficult due to the huge surface area of the porous structure.

In the field of nonlinear optics, PSi-1D photonic crystals also can be used for different applications. Due to the narrow emission line and light confinement effects in the planar optical resonator, nonlinear optical applications can be exploited including second and third harmonic generation (Dolgova et al. 2001; Martemyanov et al. 2004) and optical bistability (Pham et al. 2011). As an example, Figure 3.8a shows the linear spectra of the s-polarized radiation reflected from a PSi-based microcavity for different incident angle while Figure 3.8b shows the second-harmonic

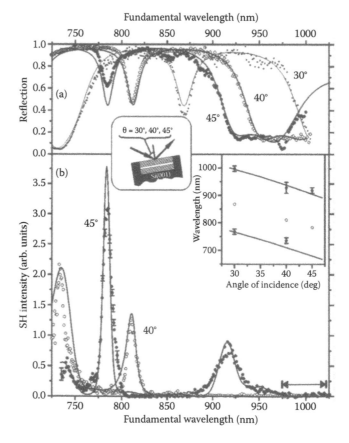

FIGURE 3.8   (a) The linear reflection spectra of a PSi microcavity measured for various angles of incidence. (b) The second harmonic generation spectra for angles of 45° (filled circles) and 40° (open circles). Inset: The angular dependence of positions of the second harmonic generation peaks at the microcavity mode (open circles), at the bandgap edges (filled circles), and the calculated angular dependences of the bandgap edges (lines). (From Dolgova T.V. et al., *Appl. Phys. Lett.* 81(15), 2725, 2002. Copyright 2002: AIP Publishing LLC. With permission.)

intensity spectra. It can be observed that the second-harmonic intensity is strongly enhanced at 785 nm of wavelength for an angle of 45° and at 810 nm of wavelength for an angle of 40°. Due to these wavelengths, the fundamental field is in resonance (Dolgova et al. 2002).

Another possible application of these structures is their use as filtered photodetectors. Optical PSi-filters can enhance the spectral performance of silicon photodiodes. PSi layers can modulate the spectral photoresponse of silicon substrates (Krüger et al. 1997). The use of a PSi Bragg reflector can enhance this photoresponse (Torres-Costa et al. 2007). Figure 3.9a shows a cross-section of a PSi-based photodetector, which consists of two Bragg reflectors, each of them composed of five periods. The ITO contact deposited on the top is also visible. The reflectance and responsivity of the devices are shown in Figure 3.9b. It can be observed that the responsivity of the substrate can be modulated by the transmittance of the PSi filter; as for high reflectance wavelengths, the responsivity is low and vice versa.

One of the main applications of this PSi-1D photonic crystal is their use as optical sensors. In the visible and IR range, the effective refractive index of a PSi layer can be defined as a homogeneous mixture of silicon and air. If pores are filled by another substance, its effective refractive index is increased, resulting in a redshift of the reflectance spectrum (Snow et al., 1999; De Stefano et al. 2003; Jalkanen et al. 2009). Figure 3.10 shows the experimental reflectance spectrum of a

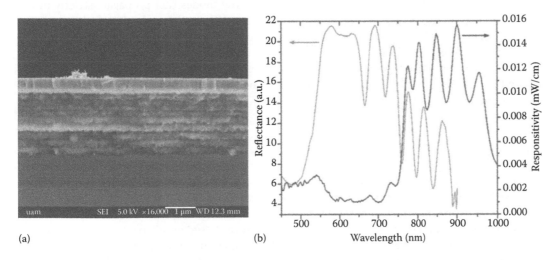

(a)   (b)

FIGURE 3.9   (a) Cross-section scanning electron microscopy image of a PSi filtered photodetector. (b) Experimental reflectance and responsivity of the PSi filtered photodetecting element. (From Torres-Costa V. et al., *Mat. Sci. Eng. C* 27(5–8), 954, 2007. Copyright 2007: Elsevier. With permission.)

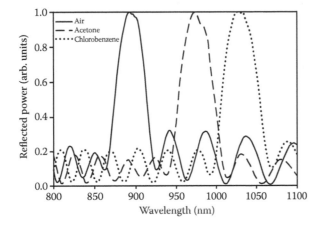

FIGURE 3.10   Experimental normalized reflectance spectrum of a PSi-based Bragg mirror after exposure to acetone and clorobenzene. (From Snow P.A. et al., *J. Appl. Phys.* 86(4), 1781, 1999. Copyright 1999: AIP Publishing LLC. With permission.)

PSi-based Bragg reflector after exposure to different substances. A red shift can be observed in the stop-band of the 1D structure after the exposition to acetone and chlorobenzene. As formed, the reflectance peak is centered at 890 nm of wavelength while after exposure of the structure to acetone and chlorobenzene, the stop-band is centered at 975 nm and 1025 nm of wavelength, respectively. Furthermore, due to the high chemical activity of PSi, biosensors based on PSi multilayers also can be obtained (DeLouise et al. 2005; Ouyang et al. 2005).

3.4 2D PHOTONIC CRYSTALS

2D photonic crystals can be described as a periodic structure along two of its axes (x-y) and homogeneous along its third axis (z). For these structures, one can describe the propagation within the x-y plane in terms of a band structure. For certain circumstances, a photonic bandgap can be opened. Inside this gap, no extended states are permitted and incident light is reflected. Unlike 1D photonic crystals, in these structures light cannot propagate in any direction within this plane.

Another characteristic of these 2D photonic crystals is the decoupling of the electromagnetic field into two scalar fields, one for each polarization. Any modes that propagate strictly in the x-y plane will be invariant under reflection in this plane. That allows classifying the modes as transversal electric (TE) where **H** and **E** are normal and parallel to the x-y plane, respectively, and transversal magnetic (TM) where **H** is parallel to the x-y plane and **E** is normal to the x-y plane. Both polarizations can exhibit different behavior for each structure. In general, TM bandgaps are favored in a lattice of isolated high-ε region and TE bandgaps are favored in a connected lattice (Joannopoulos et al. 2008).

3.4.1 BULK 2D PHOTONIC CRYSTALS

2D photonic crystals are much easier to fabricate than 3D structures. Several methods have been proposed in order to fabricate 2D-photonic crystals based on PSi. In the present section, we will review some of the most common methods.

One of the main methods to obtain photonic structures based on PSi contemplates the use of proton beam writing combined with the electrochemical etching method (Teo et al. 2006). This method is based on the selective formation of PSi due to defect regions caused by the ion beam. When an ion beam is focused on a silicon wafer surface, the crystal lattice is damaged, producing defect regions, which reduces the localized hole density and hole current (Svensson et al. 1991; Breese et al. 2006). The defect distribution inside the silicon structure depends on the energy and fluency of the ion beam (Ziegler et al. 2010). The irradiated regions can inhibit the PSi formation during the electrochemical etching. Then, PSi can be removed by immersion in basic solutions and ordered silicon structures can be obtained (Azimi et al. 2012). Using this method, Teo et al. (2004) fabricated arrays of high aspect-ratio silicon pillars. Following the same method, Dang et al. (2012) studied the fabrication of these kinds of structures with different radius and lattice constants, by changing the main parameters in the fabrication such as energy and fluency of the ion beam or the applied current density during the electrochemical process, and their optical properties as photonic crystals. Figure 3.11 shows different arrays of these silicon pillars where the ion beam was focused to different sizes in both directions allowing changing the period. By controlling the applied current density and the etching time during the sequent electrochemical etching, the ratio between the radius and the lattice parameter can be adjusted.

This kind of structure shows different complete bandgaps for TM polarization. In Figure 3.12a, the PBS of a 2D square lattice of silicon pillars with a radius of $r = 0.25a$, a being the lattice parameter, is presented. Two different complete bandgaps for TM polarization can be observed; one between the first band and second band, and other between the third band and the fourth band. In Figure 3.12b, the map gap for TM polarization is shown in the function of the radius of silicon pillars, where the radius is given in functions of the lattice parameter. It can be observed that for radius lower than $0.08a$, there is no complete bandgap. When increasing the radius of the silicon pillar, a complete bandgap for TM polarization is opened between the first and the second band.

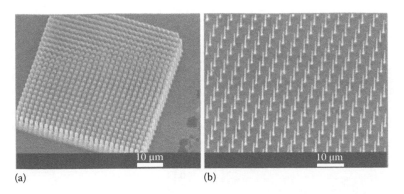

(a)   (b)

FIGURE 3.11 Scanning electron microscopy images of silicon pillars in square lattice. (a) Period of 2 μm with a large *r/a* ratio, where *r* is the radius of the pillars and *a* is the lattice period. (b) Period of 4 μm with small *r/a* ratio. (From Dang Z. et al., *Nanoscale Res. Lett.* 7, 416, 2012. Published by Springer as open access.)

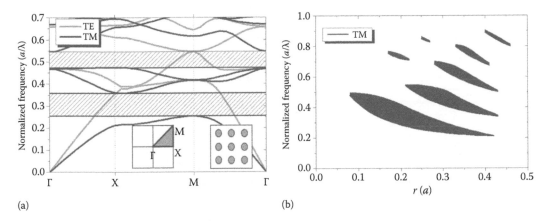

(a)   (b)

FIGURE 3.12 (a) Photonic band structure of square lattice of silicon pillars for a ratio *r/a* = 0.285. Bands are plotted along high-symmetry lines of the Brillouin zone, where frequency is given in normalized units (*a/λ*). The representation of the 2D Brillouin zone and the squematic of 2D structure used for the calculations are shown in the inset of PBS. (b) Gap map of TM gaps in function of the radius of the pillars given in units of the lattice parameter *a*.

When the radius is higher than 0.15*a*, new complete gaps appear for higher ranges of normalized frequency. For radius bigger than 0.45*a*, there is not a gap.

On the other hand, complete bandgaps for TE polarization can be obtained with complementary structures such as square meshes of dielectric veins. Recio-Sánchez et al. (2012a) developed a novel fabrication process to obtain PSi-based 2D photonic crystal with this ordered pattern. The fabrication process consists of a 1 keV Argon ion bombardment through a Cu grid with this pattern. The Cu grid is placed on the top of the PSi layer. After the bombardment, the Cu grid is removed and the PSi ordered structure is obtained. Figure 3.13a shows a PSi-based structure of square veins of 2 μm wide, leaving air holes of 5 μm × 5 μm.

The PBS computed for this 2D structure for a PSi layer of 40% porosity ($\varepsilon = 4.84$) is shown in Figure 3.13b. A complete bandgap for TE polarization between 0.35 and 0.375 of normalized frequency can be clearly observed. In addition, there are several partial gaps in the main high-symmetry directions for both TE and TM polarization. The position and shape of the gaps of these structures can be tuned by changing either the pattern of the dielectric veins or their dielectric constant, that is, the porosity of the PSi layer. Figure 3.14a shows the TE complete gap map in function of the width of the dielectric veins, given this width in function of the lattice parameter *a*, being the dielectric material crystalline silicon ($\varepsilon = 11.56$). Figure 3.14b presents the TE complete gap map for all possible PSi porosity values, for a structure based on a square mesh of dielectric veins with a width of 0.285*a*. The porosity of the PSi layer and its dielectric constant were related through the Bruggeman model (Bruggeman 1935). It can be noted that the width of

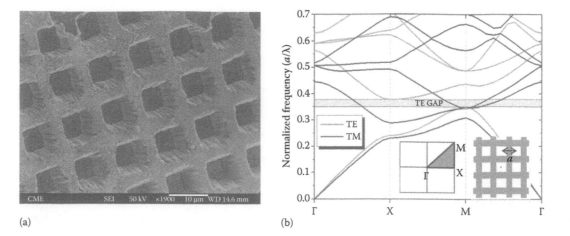

(a)   (b)

FIGURE 3.13 (a) Scanning electron microscope image of the PSi-based photonic structure. (b) 2D computed PBS for the square mesh of dielectric veins with a width of 0.285a with a dielectric material being a PSi layer of 40% porosity (ε = 4.84). (From Recio-Sánchez G. et al., *J. Nanotec.* 2012, 106170. Published by Creative Commons Attribution License as open access.)

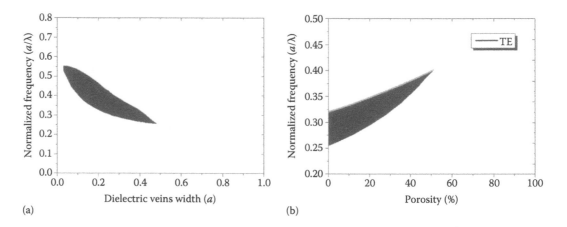

(a)   (b)

FIGURE 3.14 (a) Complete TE map gap as a function of the dielectric veins width being the dielectric constant ε = 11.56. (b) Complete TE map gap as a function of the PSi layer porosity. (From Recio-Sánchez G. et al., *J. Nanotec.* 2012, 106170. Published by Creative Commons Attribution License as open access.)

the dielectric veins as well as the porosity of the PSi layer have a strong influence on the size and frequency of the photonic bandgaps.

In order to obtain 2D photonic crystals with complete bandgaps for both polarizations, other kinds of structures have to be fabricated such as a hexagonal lattice of air columns. To fabricate those structures based on PSi, the most common is the use of macroporous silicon. A detailed description of the method can be found elsewhere (Lehmann and Föll 1990; Birner et al. 1998). Briefly, macropores with a micrometer pore radius can be formed by electrochemical etching in a hydrofluoridic acid solution in a doped n-type silicon wafer while the backside of the wafer is illuminated, or in a high resistivity p-type silicon wafer (Lehmann 1993, 1995). In this process, the pores will grow in a random pattern. However, if ordered pits are generated in a silicon surface, pore formation will initiate at these pits. These pits can be formed by different techniques including standard lithography. With this method, 2D-photonic structures studied before such as square lattice of silicon pillars or square lattice of connected dielectric veins also can be obtained (Birner et al. 1998; Belloti et al. 2002).

Using this technique, combining macroporous silicon with standard lithography, Grüning et al. (1996) fabricated a two-dimensional photonic crystal with a complete bandgap centered at

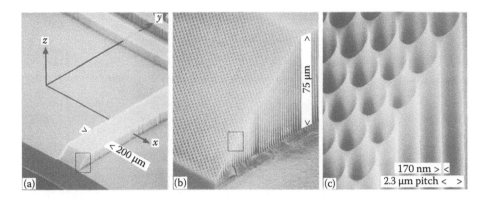

FIGURE 3.15 Scanning electron microscope of a patterned and micromachined layer of macroporous silicon forming a two-dimensional triangular lattice. (a) The 200 μm wide and 75 μm high bars of PSi are produced by micromechanically etching PSi. (b) A tenfold of the inset in (a) is shown. The edges of the bars were formed by micromechanically etching the layer. (c) A tenfold magnification of the inset in (b) is shown. (From Gruning U. et al., *Appl. Phys. Lett.* 68(6), 747, 1996. Copyright 1996: AIP Publishing LLC. With permission.)

infrared wavelength. They generated a triangular lattice with a lattice constant of 2.3 μm, and a pore diameter of 2.13 μm and 75 μm of depth, as can be seen in Figure 3.15.

PBS for both TE and TM polarization were computed (Figure 3.16) showing a completed bandgap centered at 4.9 μm of wavelength. Transmission spectra in the two main high-symmetry directions Γ-M and Γ-K (corresponding to the x and y directions in Figure 3.15a) were measured. The transmission spectra agreed with the theoretical calculations, showing a sharp drop in the transmission in the band regions for both polarizations and directions measured.

The width and position of the bandgap can be tuned by changing the main characteristic of the photonic structure such as lattice constants and pore radius. These parameters can be adjusted by determining the experimental parameters. In this way, one can design the photonic crystal to obtain the bandgap over the IR range (Birner et al. 2000; Schilling et al. 2001a; Wehrspohn and Schilling 2003).

3.4.2 DEFECTS IN 2D PHOTONIC CRYSTALS

By incorporating defects in 2D photonic crystals, a single localized mode or a set of closely spaced modes, which have frequencies inside the bandgaps, can be obtained. These defects can be created using the different techniques reviewed before. Using proton beam writing combined with the electrochemical etching method, defects can be created by changing the ordered proton beam writing (Dang et al. 2013). In the same way, point or line defects can be introduced into 2D macroporous silicon photonic crystals by designing a suitable mask for the lithography process (Müller et al. 2000). Figure 3.17 shows different kinds of defects based on this method. In Figure 3.17a, a single point defect can be observed. In this case, a missed hole is obtained in a 2D hexagonal lattice. Figure 3.17b shows a bent linear defect inside the same kind of 2D triangular lattice.

By introducing a line defect into a 2D photonic crystal, waveguides can be obtained (Leonard et al. 2000). In order to demonstrate waveguiding through these line defects, Schilling et al. (2001c) incorporated a 27-μm long line defect along the Γ-K direction into a triangular 2D photonic crystal with an r/a ratio of 0.43 ($r = 0.64$ μm) (see Figure 3.18c). The line defect creates new states inside the bandgap. Figure 3.18a shows the measured transmission through the line defect. Pronounced Fabry–Perot resonances can be observed due to the multiple reflections at the waveguide facets. Comparing the experimental spectrum with the FTDT-transmission calculation (Figure 3.18b) revealed very good agreement. In addition, the measured and calculated resonance indicated small losses inside the sample.

Besides line defects, also point defects consisting only in one missing point that breaks the symmetry are of special interest. They also create photonic states inside the bandgaps of perfect

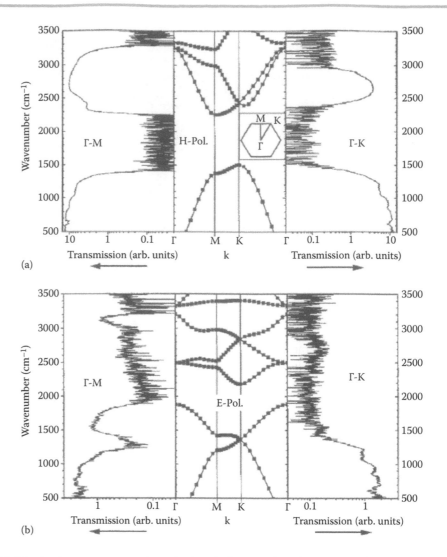

FIGURE 3.16   In the center part, the calculated photonic band structure of the two-dimensional triangular lattice shown in Figure 4.8 is plotted, the inset showing the first Brillouin zone. On the left-(right)-hand side, the transmission spectra of the sample for the Γ-M (Γ-K) direction are shown between 500 and 3500 cm⁻¹, (a) for TM polarization and (b) for TE polarization. (From Gruning U. et al., *Appl. Phys. Lett.* 68(6), 747, 1996. Copyright 1996: AIP Publishing LLC. With permission.)

photonic crystals. The light fields belonging to these point defect states are confined in a small volume resulting in very high energy densities (Joannopoulos et al. 2008). These point defects can be described as microcavities surrounded by perfect reflecting walls. Then, resonance peaks with very high quality factors are expected in the transmission spectra.

A point defect in a PSi-based 2D photonic crystal can be created by removing one pore in a perfect 2D photonic crystal using macroporous silicon (Müller et al. 2000). In order to study resonance peaks of these defects, Kramper et al. (2001) included a point defect between two line defects serving as waveguides. Figure 3.19a and b show the resulted structure, which consists of an array of air cylinders with a lattice constant of 1.5 μm and depth of 100 μm into which two line defects and a point defect have been incorporated. While the central part of the sample acts as an optical cavity around the point defects, the line defects serve as waveguides for coupling the resonator. Figure 3.19c shows the experimental and calculated transmission spectrum of the perfect 2D photonic crystal. A clear bandgap can be observed between 3.4 μm and 5.8 μm, which agrees with the theoretical calculation. In the inset of Figure 3.19c, the calculated resonances peaks from the point defects are shown.

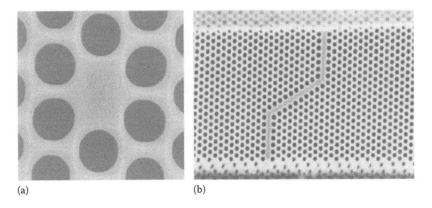

(a)   (b)

FIGURE 3.17   Scanning electron microscopy image of (a) the region around a missing etch pit after electrochemical pore growth and subsequent pore widening by oxidation/etching steps. The distance between the pores is 1.5 μm, pore diameters are 1.15 μm and (b) another structure with a bent waveguide. The width of this very well defined PSi bar is 30 μm. (From Müller F. et al., *J. Porous Mater.* 7, 201, 2000. Copyright 2000: Springer Science and Business Media. With permission.)

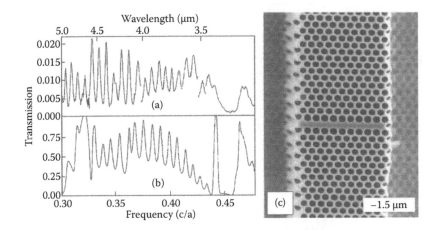

FIGURE 3.18   (a) Measured and (b) calculated H-polarized transmission spectrum of a 27-μm long waveguide directed along Γ-K covering the spectral range of H-bandgap of the surrounding perfect photonic crystal. The small stopgap at frequency of 0.45 c/a is caused by the anticrossing of two even waveguides modes. (c) Scanning electron microscopy image of the line defect. (From Schilling J. et al., *J. Opt. A: Pure Appl. Opt.* 3, S121, 2001. Copyright 2001: IOP Publishing. With permission.)

Measuring transmission between these defects demanded an optical source with very narrow line width. In Figure 3.20, the measured resonances of the point defect are shown. The resonance peaks are centered at 3.616 μm and 3.843 μm, with a Q value of 640 and 190, respectively. Wavelengths at which resonance peaks are shown do not totally agree with the theoretically predicted positions by FDTD calculations due to slightly different dynamics of the pore formation close to defects (Müller et al. 2000).

3.4.3 FINITE 2D PHOTONIC CRYSTALS (SLABS)

Since 2D photonic crystals cannot provide light confinement in the z direction (given that their dielectric structure is only periodic in the x-y plane), a way to avoid out-of-plane losses is the use of photonic crystal slabs. These are two-dimensional periodic structures with a finite thickness that use index guiding to confine light in the third dimension (Johnson et al. 1999; Chow et al.

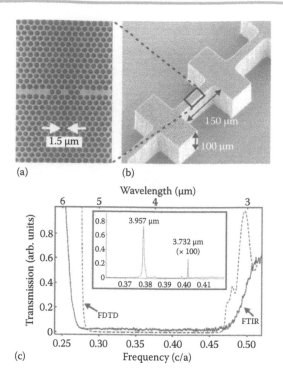

(a)   (b)

(c)

FIGURE 3.19 (a) A top view of a zoom into the region of the crystal containing the microresonator. (b) An overview of the photonic crystal substrate. (c) The solid curve shows the bandgap spectrum of the substrate in (b) measured by FTIR. The dashed curve displays the outcome of numerical simulations (FTDT) for a crystal without defects. When defects are introduced, two resonances appear as shown in the inset. (From Kramper P. et al., *Phys. Rev. B*, 64, 233102, 2001. Copyright 2001: the American Physical Society. With permission.)

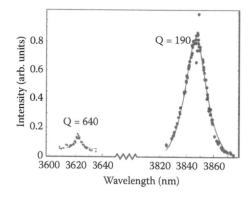

FIGURE 3.20 Two of high-Q resonances centered at wavelengths 3.621 and 3.843 μm. The norentzian profiles were fitted to the experimental data and yield quality factors of 640 and 190. (From Kramper P. et al., *Phys. Rev. B*, 64, 233102, 2001. Copyright 2001: the American Physical Society. With permission.)

2000). The finite thickness of the photonic crystal slab leads to major differences compared to the ideal 2D photonic crystal. First, the modes cannot decouple anymore into TE and TM polarization. However, since the *x-y* plane where the dielectric constant is periodic is a mirror plane of the photonic structure, the modes can be labeled "even" and "odd," depending if the magnetic fields are even (or odd) with respect to the *x-y* plane of symmetry. Second, not all the modes can be guided inside the photonic crystal slab. Most of them are either guided in the cladding or scattered out of the slab. This continuum of states is usually called light of cone.

A photonic crystal slab can consist of a thin 2D photonic crystal (core) surrounded by two layers of lower effective dielectric constant. These structures provide an index guiding by total reflection in the direction normal to the *x-y* plane, similar to the guiding in a planar wave-guide. In this case, the typical photonic crystal slab is the so-called "air-bridge," where the core structure is surrounded by air. These structures can be fabricated from the selective formation of PSi by electrochemical etching after a proton beam with different energies and fluencies is focused on a silicon wafer. From this method, freestanding photonic slabs can be obtained. Recio-Sánchez et al. (2012b) fabricated freestanding slabs of square lattice of cylindrical air holes in a silicon matrix based on this technique (Figure 3.21). To fabricate this structure, first a 250-KeV proton beam was focused and scanned in both directions to define the square grid. In order to obtain a freestanding structure, a high-energy proton beam of 1 MeV was used to define the supports.

Furthermore, these authors demonstrated the flexibility of this technique for tuning the frequency ranges and sizes of the photonic bandgaps. Figure 3.22 shows the computed PBS for the photonic slab presented in Figure 3.21 where the radius *r/a* and *h/a* (*r* being the radius of the air holes, *a* is the lattice parameter, and *h* is the height of the slabs) were set to 0.38 and 0.4,

FIGURE 3.21   Scanning electron microscopy image of the freestanding Si-based photonic crystal slab. The slab is a square mesh of cylindrical air holes in an Si matrix. (From Recio-Sánchez G. et al., *Nanoescale. Res. Lett.* 7, 449, 2012. Published by Springer as open access.)

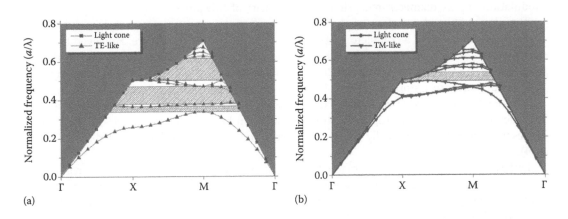

FIGURE 3.22   Photonic band structure corresponding to a photonic slab of square lattice of cylindrical air holes in an Si matrix. The ratios *r/a* and *h/a* were set to 0.38 and 0.4, respectively, ε = 11.56 being the dielectric constant of Si. (a) Slab band with even symmetry with respect to the *z*-plane (TE-like). (b) Slab bands with odd symmetry with respect to the *z*-plane (TM-like). (From Recio-Sánchez G. et al., *Nanoescale. Res. Lett.* 7, 449, 2012. Published by Springer as open access.)

respectively. Different bandgaps can be observed below the light cone for both kinds of symmetries. The frequency range and size of the bandgaps can be determined by changing the ratios r/a and h/a. They can be controlled by changing the main parameters in the fabrication process such as the proton beam line spacing, which determines the lattice parameter, the proton fluency used, or the applied current density during the electrochemical etching.

Another typical silicon-based photonic crystal slab consists of a thin Si slab surrounded by two layers of lower effective dielectric constant such as Si oxide (Jamois et al. 2002). These structures can be fabricated by macroporous silicon methods reviewed before and are easier to integrate in a photonic chip. However, the low index contrast between the core and the clouding makes the mode profiles quite extended. Another possibility to obtain this kind of structure based on macroporous silicon consists of fabricating the clouding of the core by porosity modulation (Jamois et al. 2003).

3.5 3D PHOTONIC CRYSTALS

The fabrication of 3D photonic crystals with a complete bandgap is not an easy task because they require a complex 3D connectivity and strict alignment requirements in order to be able to control the propagation of light in the three spatial dimensions. Theoretically, the first structure that was discovered that shows a 3D complete photonic bandgap was an arrangement of photonic atoms in a diamond structure. This structure was fabricated by the group of Yablonovitch et al. (1991) at microwave frequencies. However, it is quite complex to miniaturize this structure to work at lower frequency. Several methods have been proposed to obtain these structures based on macroporous silicon including the use of <111> n-type silicon substrate where the pores grow in <111> crystalline directions (Jäger et al. 2000) and by combining photochemical macropore etching in silicon and subsequent drilling of two pores set with a focused ion beam (Chelnokov et al. 2000; Wang et al. 2003).

Another possibility to create 3D photonic crystals based on macroporous silicon is based on the modulation of pore diameter with pore depth in a 2D periodic structure. Schilling et al. (2001b) fabricated this kind of 3D photonic crystal based on macroporous silicon with modulated pore diameter. These structures were fabricated using a lithographic prestructuring process. The initiation spots for the pores were defined by photolithographic process on n-type silicon wafers, presenting a triangular pattern. A subsequent alkaline etching forms pits where pores start to grow. The macropores were obtained by a photochemical etching. To achieve periodic variation of the pore diameter with pore depth, the illumination intensity was varied periodically while the pores grew into the silicon substrate. The resulted structure can be observed in Figure 3.23. The macropores were arranged in a triangular lattice with a lattice parameter of 1.5 μm and the modulation of pore diameter in depth had a periodicity of 1.69 μm.

The photonic band structure of the 3D photonic crystal was computed and it is shown in Figure 3.24. There is not a complete 3D photonic bandgap for this structure. However, there are several gaps for wave vectors in the Γ-K-M plane corresponding to waves traveling in the x-y plane, which have been well studied in 2D photonic structures (Birner et al. 1998; Schilling et al. 2001a). Furthermore, a stop-band exists in the Γ-A direction, which corresponds to waves traveling parallel to the pore axis (z axis). The transmission spectrum along this direction (Figure 3.24 left side) was measured. A drop in the transmission can be observed around 1500 cm^{-1}, which agrees well with the theoretically predicted bandgap in that direction (gray shaded bar in Figure 3.24).

Different structures based on macroporous silicon have been proposed as alternative 3D photonic crystals with complete bandgaps. Schilling et al. (2005) fabricated a 3D structure of orthorhombic symmetry with the primitive lattice vector $\mathbf{a}$ = (111), $\mathbf{b}$ = (1/2 $\sqrt{3}$/2 1/2), and $\mathbf{c}$ = (100). To fabricate this structure, they used a combination of a photoelectrochemical macropores etching process and focused-ion-beam drilling. To do that, a 2D hexagonal structure was fabricated by a prelithography pattern of silicon wafer and the sequent electrochemical etching. This pattern forms a 2D photonic crystal (Birner et al. 1998; Schilling et al. 2001a). Then, the structure is cleaved and a trench is milled out of the porous structure applying a dual beam focused-ion-beam. The resulted structure is shown in Figure 3.25.

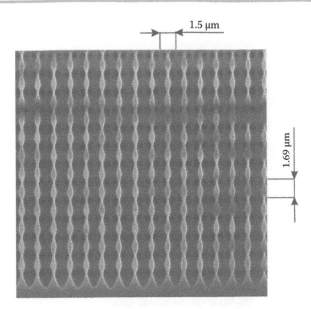

FIGURE 3.23   Scanning electron microscopy image showing a longitudinal section of the modulated pore structure. (From Schilling J. et al., *Appl. Phys. Lett.* 78(9), 1180, 2001, Copyright 2001: AIP Publishing LLC. With permission.)

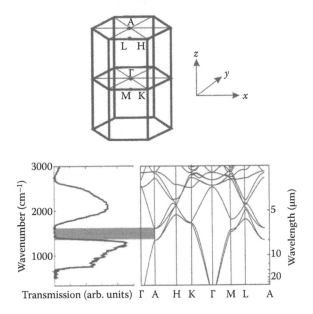

FIGURE 3.24   (a) Hexagonal Brillouin zone. (b) Transmission along Γ-A. (c) Calculated 3D photonic band structure. The gray bar indicates the bandgap in the Γ-A direction and the corresponding spectral region of low transmission. (From Schilling J. et al., *Appl. Phys. Lett.* 78(9), 1180, 2001, Copyright 2001: AIP Publishing LLC. With permission.)

This structure was first studied by Hillerbrand et al. (2003). Figure 3.26 shows the first Brillouin zone (Figure 3.26a) and the computed photonic band structure (Figure 3.26b) where the ratio between the radius of the pores (r) and the lattice constant (a) of both pore sets, r/a, was set as 0.38. A complete bandgap at low normalized frequencies between the second and the third bands can be observed. Consequently, they should be insensitive to modest disorder. To check the optical behavior of this structure, the optical reflections were measured along the z direction, which agreed well with the predicted position of the bandgap.

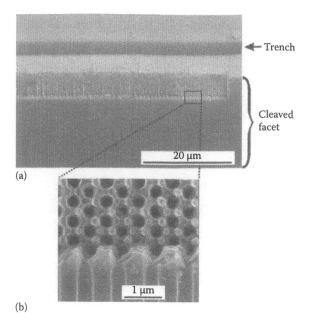

(a)

(b)

FIGURE 3.25 Scanning electron microscope images of the 3D photonic crystal. (a) Titled overview. In the front, the perpendicular running etched pores are visible. In their upper 5 µm, the orthogonal pore set was drilled with the focused ion beam. Due to the tilt, the etched trench is visible too. It separates the ridge containing the drilled structure with the rest of the sample. (b) Detailed view of the drilled region showing the alignment of the etched pores and drilled pores. (From Schilling J. et al., *Appl. Phys. Lett.* 86, 011101, 2005, Copyright 2005: AIP Publishing LLC. With permission.)

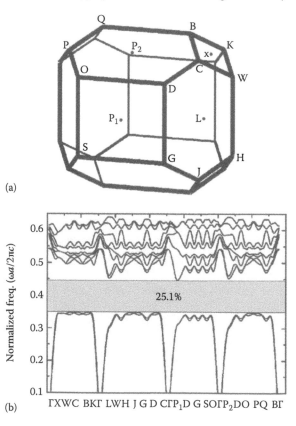

(a)

(b)

FIGURE 3.26 (a) First Brillouin zone for the orthorhombic structure. (b) Photonic band structure for the orthorhombic structure with a ratio $r/a = 0.38$ for both pore sets. (From Schilling J. et al., *Appl. Phys. Lett.* 86, 011101, 2005. Copyright 2005: AIP Publishing LLC. With permission.)

Other kinds of 3D photonic crystals based on Si with complete bandgaps have been proposed. Noda et al. (1999) published an Si-based 3D woodpile fabricated by wafer bonding and selective etching. Dang et al. (2013) also studied the fabrication of this kind of structure based on the selective formation of PSi after a proton beam is focused with different fluency and energy. Additionally, other kinds of 3D photonic crystals with complete bandgaps can be obtained by the use of Si and SiO_2 opals. Under certain conditions, colloids self-organize into periodic structures with cubic symmetry (Vos et al. 1996). Besides its simple fabrication process, this structure does not have a complete structure. However, the so-called inverse opal, which is the inverse structure, has a 3D complete bandgap (Blanco et al. 2000).

3.6 APPLICATIONS OF 2D AND 3D PHOTONIC CRYSTALS

There are many proposed applications related to PSi-based 2D and 3D photonic crystals. Since they can control the light propagation inside them, they have received a lot of interest in a wide variety of fields. Large kinds of theoretical applications have been proposed using photonic crystals, which can be performed with PSi-based photonic crystals. One of the main applications of these structures is to achieve light guiding. By incorporating line defects in a perfect 2D and 3D photonic crystal, it is possible to form waveguides that permit transmission of light only along these defects (Jamois et al. 2003). In the same way, point defects can generate novel applications including efficient light emitter devices (Makarova et al. 2006). By incorporating both kinds of defects in a perfect 3D photonic crystal, it is possible to obtain an all-optical chip, which can be the base of future photonic technology.

Other kinds of applications related to PSi-based photonic crystals include the fabrication of more efficient optical fibers, filters, suppression of unwanted spontaneous emissions (Yablonovitch 1987), and the use of slabs to improve the efficiency of solar cells (Demésy and John 2012).

In spite of all these applications, the main use of PSi-based photonic crystals is to obtain high quality sensors (Vega et al. 2013). Mixing the optical properties of photonic crystals with the structural properties of PSi such as their high surface area and surface physiochemical reactivity, a wide variety of high sensitive sensors can be achieved (Pacholski 2013). In addition, the possibility to turn PSi into a biocompatible material opens the way to obtain sensitive biosensors from a PSi-based photonic crystal (Lee and Fauchet 2007).

3.7 OUTLOOK

In this chapter, we have overviewed the main progress on PSi-based photonic crystals. It has been demonstrated that PSi is a very suitable material for development of photonic crystals due to its interesting optical properties and its simple fabrication process, which is totally compatible with the current silicon-based technology. A review about different techniques to obtain photonic crystals in one, two, and three dimensions has been shown. One-dimensional photonic crystals based on PSi can be easily fabricated by simply changing the applied current density during electrochemical etching. In this way, Bragg reflectors, rugate filters, or resonant microcavities can be obtained. These one-dimensional crystals have been demonstrated as very efficient structures for different applications including waveguides, light emitting devices, tunable photodetectors, nonlinear optics, and optical sensors and biosensors. However, considering commercial applications, a passivation of the layers is needed, which is not a simple task due to the high surface area and chemical reactivity of PSi layers. In addition, the inhomogeneities in the layer and roughness surface should be improved for an appropriate use of these structures.

On the other hand, the fabrication of two-dimensional PSi-based photonic crystals by different techniques such as the common method using macroporous silicon has been reviewed. Other methods include the selective formation of PSi during the electrochemical etching when a proton beam is focused on the silicon wafer, and the direct ion bombardment on a PSi surface through a metal grid. All these methods allow the optimal integration of two-dimensional photonic crystals with a bandgap in the infrared range in the current technology. In addition, they permit the incorporation of waveguides and microresonators, which would be essential for an all-optical device. Three-dimensional PSi-based photonic crystals can also be fabricated. However, a complex three-dimensional connectivity and strict alignment are required in order to control the propagation of light in the three spatial dimensions. The modulation of pore diameter with pore depth in two-dimensional periodic structures or a combination of photochemical macropores etching and focused ion beam drilling are some of the most common methods used. The transmission measurements on these bulk and finite structures have shown a good agreement with the theoretical predictions in order to fabricate optical devices for use in the infrared wavelength range.

Since all the results reviewed in this chapter make PSi a very promising candidate for the development of novel photonic devices, some challenges have to be improved. In spite of PSi having been thoroughly studied during the last two decades, more progress in the fabrication processes must be done in order to have a more precise control of the pore structures. The different reviewed methods to fabricate two- and three-dimensional photonic crystals need further studies in order to obtain homogeneous structures that perfectly match with the predicted calculations. In addition, the miniaturization of these structures to operate in the visible range is a novel but difficult challenge because it requires structuring the devices in a nanometric scale. Furthermore, while three-dimensional photonic crystals can control the light in three spatial directions, a real three-dimensional guiding of photons has proven to be quite challenging. The lack of systematic design principles, as well as the fabrication of a large area and free-defect three-dimensional photonic crystals, makes the development of an all-PSi-photonic device, which can control guiding the light in three dimensions, yet to be demonstrated. However, photonic crystals based on PSi also can be used in different applications. The possibility of becoming PSi into a biocompatible material has allowed the development of high sensitive biosensors from two-dimensional PSi-based photonic crystals, even other kinds of biomedical applications including encoded microcarriers.

Thus, it is expected that in the near future the optimization of PSi-based photonic crystals fabrication processes will give the opportunity for the development of all-silicon based industry compatible photonic devices, which can revolutionize the next decades.

REFERENCES

Araki, M., Koyama, H., and Koshida, N., 1996a. Fabrication and fundamental properties of an edge-emitting device with step-index porous silicon waveguide. *Applied Physics Letters,* 68, 2999.

Araki, M., Koyama, H., and Koshida, N., 1996b. Controled electroluminescence spectra of porous silicon diodes with a vertical optical cavity. *Applied Physics Letters,* 69, 2956.

Azimi, S., Breese, B.H., Dang, Z.Y., Yan, Y., Ow, Y.S., and Bettiol, A.A., 2012. Fabrication of complex curved three-dimensional silicon microstructures using ion irradiation. *Journal of Micromechanics and Microengineering,* 22, 015015.

Belloti, P., Dal Negro, L., Gaburro, Z., and Pavesi, L., 2002. P-type macroporous silicon for two-dimensional photonic crystals. *Journal of Applied Physics,* 92(12), 6966–6972.

Berger, M., Dieker, C., Thonissen, M. et al., 1994. Porosity superlattice a new class of Si heterostructures. *Journal of Physics D; Applied Physics,* 27, 1333–1336.

Berger, M., Thönissen, M., Arens-Ficher, R. et al., 1995. Investigation and design of optical properties of porosity superlattice. *Thin Solid Films,* 255, 313–316.

Berger, M., Arens-Fischer, R., Thönissen, M. et al., 1997. Dielectric filters made of PS: Advanced performance by oxidation and new layer structures. *Thin Solid Films,* 297, 237–239.

Birner, A., Grüning, U., Ottow, S. et al., 1998. Macroporous silicon: A two-dimensional photonic bandgap material suitable for the near-infrared spectral range. *Physica Status Solidi (a),* 165, 111–117.

Birner, A., Li, A.P., Müller, F. et al., 2000. Transmission of a microcavity structure in a two-dimensional photonic crystal based on macroporous silicon. *Materials Science in Semiconductor Processing,* 3, 487.

Blanco, A., Chomski, E., Grabtchak, S. et al., 2000. Large-scale synthesis of a silicon photonic crystal with a complete three-dimensional bandgap near 1.5 micrometers. *Nature,* 405, 437.

Bovard, B., 1993. Rugate filter theory: An overview. *Applied Optics,* 32(28), 5427–5442.

Breese, M.B.H., Champeaux, F.J.T., Teo, E.J., Bettiol, A.A., and Blackwood, D.J., 2006. Hole transport through proton-irradiated p-type silicon wafers during electrochemical anodization. *Physical Review B,* 73, 035428.

Bruggeman, D., 1935. Berechnung verchiendener physikalischer konstanten von heterogenen substanzen. *Annalen der Physik,* 416(7), 636–664.

Busch, K., Lölkes, S., Wehrpohn, R., and Föl, H., 2006. *Photonic Crystal.* Weinhein: Wiley-VCH.

Canham, L., 1990. Silicon quantum wire array fabrication by electrochemical and chemical dissolution of wafers. *Applied Physics Letters,* 57(10), 1046–1048.

Chan, S. and Fauchet, P.M., 1999. Tuneable, narrow and directional luminescence from porous silicon light emitting devices. *Applied Physics Letters,* 75(2), 274.

Chan, C., Lu, Q., and Ho, K., 1995. Orden-N spectral method for electromagnetic waves. *Physical Review B,* 51, 16635–16642.

Chelnokov, A., Wang, K., Rowson, S., Garoche, P., and Lourtioz, J.M., 2000. Near-infrared Yablonovite-like photonic crystals by focused-ion-beam etching of macroporous silicon. *Applied Physics Letters,* 77(19), 2943.

Chi Do, T., Bui, H., Van Nguyen, T. et al., 2011. A microcavity based on porous silicon multilayer. *Advances in Natural Sciences: Nanoscience and Nanotechnology,* 2, 035001.

Chongjun, J., Bai, Q., Miao, Y., and Ruhu, Q., 1997. Two-dimensional photonic band structure in the chiral medium- transfer matrix method. *Optics Communication,* 142, 179–183.

Chow, E., Lin, S.Y., Johnson, S.G. et al., 2000. Three-dimensional control of light in a two-dimensional photonic crystal slab. *Nature,* 407, 983–986.

Cunin, F., Schmedake, T.A., Link, J.R. et al., 2002. Biomolecular screening with encoded porous-silicon photonic crystals. *Nature Materials,* 1, 39–41.

Dang, Z., Breese, M.B.H., Recio-Sánchez, G. et al., 2012. Silicon-based photonic crystals fabricated using proton beam writing combined with electrochemical etching method. *Nanoscale Research Letters,* 7, 416.

Dang, Z., Banas, A., Azim, S. et al., 2013. Silicon and porous silicon mid-infrared photonic crystals. *Applied Physics A,* 112, 517–523.

DeLouise, L., Kou, P., and Miller, B., 2005. Cross-correlation of optical microcaviy biosensor response with immobilized enzyme activity. Insights into biosensor sensivity. *Analytical Chemistry,* 77, 3222–3230.

Demésy, G. and John, S., .2012. Solar energy trapping with modulated silicon nanowire photonic crystals. *Journal of Applied Physics,* 112, 074326.

De Stefano, L., Rendina, I., Moretti, L., and Rossi, A.M., 2003. Optical sensing of flammable substances using porous silicon microcavities. *Materials Science and Engineering: B,* 100, 271–273.

Dobson, D., 1999. An efficient method for band structure calculations in 2D photonic crystals. *Journal Computational Physics,* 149, 363–376.

Dolgova, T.V., Maidykovskii, A.I., Martemyanov, M.G. et al., 2001. Giant second harmonic generation in microcavities based on porous silicon photonic crystals. *JETP Letters,* 73(1), 8–12.

Dolgova, T.V., Maidykovski, A.I., Martemyanov, M.G. et al., 2002. Giant microcavity enhancemnt of second-harmonic generation in all-silicon photonic crystals. *Applied Physics Letters,* 81(15), 2725–2727.

Elson, J. and Tran, P., 1996. Coupled-mode calculation with the R-matrix propagator for the dispersion of surface waves on truncated photonic crystal. *Physical Review B,* 54, 1711–1715.

Grüning, U., Lehmann, V., Ottow, S., and Busch, K., 1996. Macroporous silicon with a complete two-dimensional photonic band gap centered at 5 μm. *Applied Physic Letters,* 68(6), 747–749.

Hillerbrand, R., Senz, S., Hergert, W., and Gösele, U., 2003. Macroporous-silicon-based three-dimensional photonic crystal with a large complete band gap. *Journal of Applied Physics,* 94, 2758.

Hirschman, K., Tsybeskov, L., Duttagupta, S., and Fauchet, P., 1996. Silicon-based visible light-emitting devices integrated into microelectronic circuits. *Nature,* 384(6607), 338–341.

Ho, K., Chan, C., and Soukoulis, C., 1990. Existence of a photonic gap in periodic dielectric structures. *Physical Review Letters,* 65, 3152–3155.

Jäger, C., Finkenberger, B., Jäger, W., Chistophersen, M., Carstensen, J., and Föll, H., 2000. Transmission electron microscopy investigation of formation of macropores in n- and p-Si(100)/(111). *Materials Science and Engineering: B,* 69–70, 199–204.

Jalkanen, T., Torres-Costa, V., Salonen, J. et al., 2009. Optical gas sensing properties of thermally hydrocarbonized porous silicon Bragg reflectors. *Optics Express,* 17(7), 5446.

Jamois, C., Wehrspohn, B., Schilling, J., Müller, F., Hillerbrand, R., and Herget, W., 2002. Silicon-based photonic crystal slabs: Two concepts. *IEEE Journal of Quantum Electronics,* 38(7), 805–810.

Jamois, C., Wehrspohn, R.B., Andreani, L.C., Hermann, C., Hess, O., and Gösele, U., 2003. Silicon-based two-dimensional photonic crystal waveguides. *Photonics and Nanostructures-Fundamentals and Applications,* 1, 1–13.

Joannopoulos, J., Villeneuve, P., and Fan, S., 1997. Photonic crystals: Putting a new twist on light. *Nature,* 368, 143–149.

Joannopoulos, J., Johnson, S., Winn, J., and Meade, R., 2008. *Photonic Crystals: Molding the Flow of Light.* Princeton, NJ: Princeton University Press.

John, S., 1987. Strong localization of photons in certain disordered dielectric superlattices. *Physical Review Letters,* 58, 2486–2489.

Johnson, S. and Joannopoulos, J., 2001. Block-iterative frequency-domain methods fro Maxwell's equations in a planewave basis. *Optics Express,* 8(3), 173–190.

Johnson, S., Fan, S., Villeneuve, P.R., Joannopoulos, J.D., and Kolodziejski, L.A., 1999. Guided modes in photonic crystals slab. *Physical Review B,* 60, 5751.

Koshida, N. and Koyama, H., 1992. Visible electroluminescence from porous silicon. *Applied Physic Letters,* 60(3), 347–349.

Kramper, P., Birner, A., Agio, M. et al., 2001. Direct spectroscopy of a deep two-dimensional photonic crystal microresonator. *Physical Review B,* 64, 233102.

Krüger, M., Marso, M., Berger, M.G. et al., 1997. Color sensitive photodetector based on porous silicon superlattices. *Thin Solid Films,* 297(1–2), 241.

Lazarouk, S., Jaguiro, P., and Borisenko, V., 1998. Integrated optoelectronic Unit Based on porous silicon. *Physica Status Solidi (a),* 165(1), 87–90.

Lee, M. and Fauchet, P.M., 2007. Two-dimesnional silicon photonic crystal based biosensing platform for protein detection. *Optics Express,* 15(8), 4530–4535.

Lehmann, V., 1993. The physics of macropore formation in low doped n-type silicon. *Journal of the Electrochmical Society,* 140(10), 2836–2843.

Lehmann, V., 1995. The physics of macroporous silicon formation. *Thin Solid Films,* 255(1–2), 1–4.

Lehmann, V. and Föll, H., 1990. Formation mechanism and properties of electrochemically etched trenches in n-type silicon. *Journal of the Electrochemical Society,* 137, 653.

Leonard, S., van Driel, H.M., Birner, A., Gösele, U., and Villeneuve, P.R., 2000. Single-mode transmission in two-dimensional macroporous silicon photonic crystal waveguide. *Optics Letters,* 25, 1550.

Loni, A., Canham, L.T., Berger, M.G. et al., 1996. Porous silicon multilayer optical waveguides. *Thin Solid Films,* 276(1–2), 143–146.

Lorenzo, E., Oton, C.J., Capuj, N.E. et al., 2005. Porous silicon-based rugate filters. *Applied Optics,* 44(26), 5415–5422.

Macleod, H.A., 1969. *Thin-Film Optical Filters.* Adam Hilger Ltd, London.

Makarova, M., Vuckovic, J., Sanda, H., and Nishi, Y., 2006. Two-Dimensional Porous Silicon Photonic Crystal Light Emitters, in *Conference on Lasers and Electro-Optics/Quantum Electronics and Laser Science Conference and Photonic Applications Systems Technologies, Technical Digest. Optical Society of America,* paper CFI5.

Martemyanov, G., Kim, E.M., Dolgova, T.V., Fedyanin, A.A., Aktsipetov, O.A., and Marowsky, G., 2004. Third-harmonic generation in silicon photonic crystals and microcavities. *Physical Review B,* 70, 073311.

Mazzoleni, C. and Pavesi, L., 1995. Applications to optical components of dieletric porous silicon multilayers. *Applied Physics Letters,* 67(20), 2983–2985.

Meade, S., Yoon, M., Ahn, K., and Sailor, M., 2004. Porous silicon photonic crystals as encoded microcarriers. *Advanced Materials,* 16(20), 1811–1814.

Müller, F., Birner, A., Gösele, U., Lehmann, V., Ottow, S., and Föll, H., 2000. Structuring of macroporous silicon for applications as photonic crystals. *Journal of Porous Materials,* 7, 201–204.

Noda, S., Yamamoto, N., Imada, M., Kobayashi, H., and Makoto, O., 1999. Alignment and stacking of semiconductor photonic bandgaps by wafer-fusion. *Journal of Lightwave Technology,* 17, 1948.

Ouyang, H., Christophersen, M., Viard, R., Miller, B.L., and Fauchet, P.M., 2005. Macroporous silicon microcavities for macromolecule detection. *Advanced Functional Materials,* 15, 1851–1859.

Pacholski, C., 2013. Photonic crystal sensor based on porous silicon. *Sensor,* 13, 4694–4713.

Pavesi, L., 1997. Porous silicon dielectric multilayers and microcavities. *La Rivista Del Nuovo Cimento,* 20(10), 1–76.

Pavesi, L. and Mulloni, V., 1999. All porous silicon microcavities: Growth and physic. *Journal of Luminescence,* 80, 43–52.

Pavesi, L., Guardini, R., and Mazzoleni, C., 1996. Porous silicon resonant cavity light emitting diodes. *Solid State Communications,* 1051, 97.

Pendry, J. and Mackinnon, A., 1992. Calculation of photon dispersion relations. *Physical Review Letters,* 69, 2772–2775.

Pham, A., Qiao, H., Guan, B., Gooding, J.J., and Resse, P.J., 2011. Optical bistability in mesoporous silicon microcavity resonators. *Journal of Applied Physics,* 109(9), 093103.

Recio-Sánchez, G., Torres-Costa, V., Manso-Silván, M., and Martín-Palma, R., 2012a. Nanostructured porous silicon photonic crystal for applications in the infrared. *Journal of Nanotecnology,* 2012, 106170.

Recio-Sánchez, G., Dang, Z., Breese, M.B.H., Torres-Costa, V., and Martín-Palma, R.J., 2012b. Highly flexible method for the fabrication of photonic crsytal slabs based on the selective formation of porous silicon. *Nanoscale Research Letter,* 7, 449.

Sailor, W., Mueller, F., and Villeneuve, P., 1998. Augmented-plane-wave method for photonic band-gap materials. *Physical Review B,* 57, 8819–8822.

Sakoda, K., 2005. *Optical Properties of Photonic Crystals.* Berlin: Springer.

Sakoda, K. and Shiroma, H., 1997. Numerical method for localized defect modes in photonic lattices. *Physical Review B,* 56, 4830–4835.

Salem, M., Sailor, M., Sakka, T., and Ogata, Y., 2007. Electrochemical preparation of a rugate filter in silicon and its deviation from the ideal structure. *Journal of Applied Physics,* 101, 063503.

Schilling, J., Birner, A., Müller, F. et al., 2001a. Optical characterization of 2D macroporous silicon photonic crystals with badngaps around 3.5 and 1.3 µm. *Optical Materials,* 17, 7–10.

Schilling, J., Müller, F., Matthias, S., Wehrspohn, R.B., Gösele, U., and Busch, K., 2001b. Three-dimensional photonic crystals based on macroporous silicon with modulated pore diameter. *Applied Physics Letters,* 78(9), 1180–1182.

Schilling, J., Wehrspohn, R.B., Birner, A. et al., 2001c. A model system for two-dimensional and three-dimensional photonic crystals: Macroporous silicon. *Journal of Optics A: Pure and Applied Optics,* 3, S121–S132.

Schilling, J., White, J., Scherer, A., Stupian, G., Hillebrand, R., and Gösele, U., 2005. Three-dimensional macroporous silicon photonic crystal with large photonic band gap. *Applied Physics Letters,* 86, 011101.

Snow, P.A., Squire, E.K., Russell, P.J., and Canham, L.T., 1999. Vapor sensing using the optical properties of porous silicon Bragg mirrors. *Journal of Applied Physics,* 86(4), 1781–1784.

Soüzer, H. and Haus, J., 1992. Photonic bands: Convergence problems with plane-wave method. *Physical Review B,* 45, 13962–13972.

Svensson, B., Mohadjeri, B., Hallén, A., Svensson, J.H., and Corbett, J.W., 1991. Divacancy acceptor level in ion-irradiated silicon. *Physical review B,* 43, 2292.

Teo, E., Breese, M.B.H., Tavenier, E.P. et al., 2004. Three-dimensional microfabrication in bulk silicon using high-energy protons. *Applied Physics Letters,* 84(16), 3203–3204.

Teo, E, Breese, M.B.H., Bettiol, A.A. et al., 2006. Multicolor photoluminescence from porous silicon using focused high energy helium ions. *Advanced Materials,* 18(1), 396–402.

Theib, W., 1996. The dielectric function of porous silicon: How to obtain it and how to use it. *Thin Solid Films,* 276(1-2), 7–12.

Theib, W., 1997. Optical properties of porous silicon. *Surface Science Report,* 29(3–4), 91–192.

Torres-Costa, V., Martín-Palma, R., and Martínez-Duart, J., 2007. All silicon color-sensitive photodetector in the visible. *Materials Science & Engineering C,* 27(5–8), 954–956.

Vega, D., Reina, J., and Rodriguez, A., 2013. Macroporous silicon photonic crystals for gas sensing. In *Electron Devices (CDE), 2013 Spanish Conference on IEEE.* 143–146.

Vos, W.L., Sprik, R., van Blaaderen, A., Imhof, A., Lagendijk, A., and Wegdam, G.H., 1996. Strong effects of photonic band structures on the diffraction of colloidal crystals. *Physical Review B* 53, 16231.

Wang, K., Chelnokov, A., Rowson, S., and Lourtioz, J.-M., 2003. Extremely high-aspect-ratio patterns in macroporous substrate by focused-ion-beam etching: The realization of three-dimensional lattices. *Applied Physics A: Material Science & Processing,* 76, 1013–1016.

Ward, A. and Pendry, J., 2000. A program for calculating photonic band structures, Green's function and transmission/reflection coefficients using a non-orthogonal FDTD method. *Computer Physics Communications,* 128, 590–621.

Wehrspohn, R. and Schilling, J., 2003. A model system for photonic crystals: Macroporous silicon. *Physica Status Solidi (a),* 197(3), 673–687.

Yablonovitch, E., 1987. Inhibited spontaneous emission in solid-state physics and electronics. *Physical Review Letters,* 58, 2059–2062.

Yablonovitch, E., 2001. Photonic crystals: Semiconductor of light. *Scientific American,* 285(6), 47–55.

Yablonovitch, E., Gmitter, T., and Leung, K., 1991. *Physical Review Letter,* 67, 2295.

Ziegler, J., Ziegler, M., and Biersack, J., 2010. SRIM—The sttoping and range of ions in matter. *Nuclear Instruments and Methods in Physics Research Section B,* 268, 1818–1823.

Passive and Active Optical Components for Optoelectronics Based on Porous Silicon

Raghvendra S. Dubey

CONTENTS

4.1	Introduction		78
4.2	Optical Filters		79
	4.2.1	Distributed Bragg Reflector Filters	79
	4.2.2	Fabry–Perot Interference Filters	83
	4.2.3	Long Wave Pass Filters	85
4.3	Microlens		86
4.4	Antireflective Coating		89
4.5	Diffraction Optical Elements		93
4.6	Optical Waveguides		95
4.7	Integrated Optoelectronics		99
Acknowledgments			104
References			104

4

4.1 INTRODUCTION

Interconnects has become a primary bottleneck in integrated circuit design. Due to continuous scaling of CMOS technology, in the coming future it will be more challenging to meet the optimal requirements such as delay, power, bandwidth, noise, and so on with the conventional interconnects. In deep submicrometer VLSI technologies, interconnect plays a vital role. However, the interconnection delays have put the limitations on the operating speed of microelectronics devices. One promising candidate to satisfy these performance objectives is optical interconnects, which is the main motivation to look for silicon photonics. Even in the communication market, photonics has seen a big development in the transmission of more information at very high speed rate. Optical devices are widely demanded in telecommunications to facilitate optical interconnects and for electrical-optical and optical-electrical conversions. A technology that has merged the photonics and silicon microelectronics components is known as silicon microphotonics. The keys of success of the silicon industry are availability of single material silicon, availability of a natural oxide of silicon, existence of single leading CMOS process technology, possibility of integration of huge devices, and availability of existing silicon technology. The passive components required for the networks communication are waveguides, power splitters/combiners, modulators/switches, isolators, multiplexers/demultiplexers, and so on, whereas active components are the laser diodes, amplifiers, LEDs, and so on.

The main disadvantages associated with these systems are the difficulty in aligning and their bulky nature. In fiber optics communication, optical fiber and lenses are used but again the difficulty in aligning and integration of these systems is still unsolved. However, in integrated optics, the optical parameters of the material can be easily controlled. The source and the detector can be integrated on a single chip. Therefore, in optoelectronics technology there is need of new materials for integration. The optical interconnects have the ability to transfer large bandwidth signals over long distances while minimizing power requirements, which outweighs the complications of integrating electro-optical technologies. The example of silicon photonic integrated circuits is the hybrid integration of active components and silica-based planer light wave circuits, which provides a means for photonic component integration within a chip. In this integration, the passive components are realized by using silica waveguides while active components are hybridized within the silica. By using this approach, various photonic components have been integrated such as multiwavelength light source, optical wavelength selector, wavelength converters, all optical time division multiplexers, and so on (Kato and Tohmori 2000). A full integrated optical system based on silicon oxynitride waveguides, silicon photodetectors, and CMOS trans-impedance amplifiers has also been realized (Hilleringmann and Goser 1995). The first all silicon-integrated opto-coupler has been demonstrated whose fabrication using ion implantation into SiO_2 is compatible with standard silicon technology (Rebohle et al. 2001). Silicon continues to play a key role in future nano electronic and photonic devices due to its availability of well-established technology. A full isolation by porous oxidized silicon process (FIPOS) was first reported where the porous silicon (PSi) layers were used as passivation in integrated circuits (Watanabe and Sakai 1971). Afterward, the silicon-on-insulator (SOI) in integrated circuits technology, the silicon on sapphire technology (SOS), and silicidation of PSi were reported (Takai and Itoh 1986). A low temperature photoluminescence was observed reported through electrochemically etched silicon wafers (Imai 1981; Pickering et al. 1984). Room temperature photoluminescence through PSi has boosted up the research on PSi for its potential application in silicon-based integrated optoelectronics (Canham 1990). The possibility of stimulated emission in nanostructured silicon has been demonstrated despite the severe competition with fast non-radiative processes (Pavesi et al. 2000; Khriachtchev et al. 2001; Luterova et al. 2002).

The progress in semiconductor materials for optoelectronic applications has raised the hope of silicon-based optical interconnects for faster response and high-speed communications. PSi offers major potential for integrated optoelectronics technology and can accept the new challenges to fabricate silicon-based photonic devices and retain advantages of silicon technology.

4.2 OPTICAL FILTERS

Optical filters are the primary need of several optoelectronic devices. Such filters are the devices that have provision to pass or stop the light of different wavelengths or a range of wavelengths. In simple words, an optical filter is nothing more than an optical component with a wavelength dependent reflection/transmission. In general, these filters work on the mechanism of absorption/diffraction interference and can be designed for a fixed or tunable wavelength range as per the requirement. PSi material is demanded since its discovery in 1956 and it is demanded material due to its wide application in solar cells, chemical and biosensor and advanced photonic components such as microcavity and so forth. The fabrication of PSi is not only easy but also this technique provides tailoring of PSi layer properties as per application need. Among several applications of PSi, it is fascinating material for the fabrication of optical filters which has the ability to transmit/reflect the light of a specific wavelength. In general, optical filters are classified as stop-band, pass-band, and a combination of both. A stop-band filter prohibits the propagation of light with a range of wavelength of interest whereas a pass-band filter allows the propagation of a band of light. The combination of both filters can pass as well as stop a specific wavelength range of light. PSi-based filters have the ability to tune the refractive index profile of its layer by controlling porosity. Moreover, the porosity can be attained during the fabrication by optimization of its process parameters. Another possibility of tuning of the refractive index is by applying multiple current densities or changing time, which enables filter preparation based on multilayers of PSi. The unique property of electrochemical etching process is that the change in anodization current does not affect the properties of the previous PSi layer that has already been formed. The electrochemical etching of silicon is a dissolution process that needs positive charge carriers to continue it. Once already-formed PSi is depleted of charge carriers, the layer is inert to further dissolution. Therefore, any change in anodization condition only affects the porosity and pore size of the layer forming below the previously prepared layer. In this way, a multilayer structure of PSi is fabricated by tuning of applied current density as a function of time.

4.2.1 DISTRIBUTED BRAGG REFLECTOR FILTERS

A simplest optical filter is distributed Bragg reflector (DBR), which is also known as one-dimensional photonic crystal (1DPC). Such a multilayer structure consists of two alternating layers of distinct refractive index and periodic in one direction as shown in Figure 4.1a. 1DPC possesses a gap known as photonic band within which light propagation is completely prohibited for a specified band of frequency/wavelength as depicted in Figure 4.1b as a gray region. Within the photonic band 100% reflection can be attained as shown in Figure 4.1c. Due to the wave nature of light and interference caused by the periodic structure, there exists a bandgap for light in photonic crystals.

A complete photonic bandgap is a range of frequencies in which there are no propagating solutions of Maxwell's equations for any wave vector surrounded by propagating states above and below the gap. A bandgap of the photonic crystals is created with appropriate selection of three

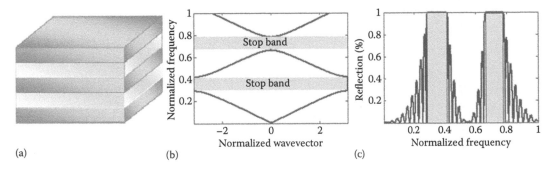

(a)   (b)   (c)

FIGURE 4.1   Distributed Bragg reflector showing (a) periodicity of their layers in one-direction, (b) dispersion relation, and (c) reflection spectra of DBR, respectively.

parameters of lattice topology, spatial period, and dielectric constants of the constituent materials in which propagation of electromagnetic waves is prohibited. The widening and shrinking of the photonic bandgap shows great dependence on lattice constant, dielectric contrast, thickness of individual layers, number of bilayers, angle of incidence, and so on (Dubey and Gautam 2007).

Using the electrochemical etching method, the multilayer structure of PSi can be easily prepared by the periodic variation of the anodization parameters, which leads to periodic variations of the refractive index of each layer. In order to form such multilayer structures of PSi, the modulation of different current density is needed for a specific duration of time.

A step-by-step preparation of the DBR filter is shown in Figure 4.2. In order to allow the hydrofluoric acid concentration to equilibrate throughout the porous matrix, the current can be kept to zero for a short regeneration period during the application applied to each current density. The magnitude and duration of applied current density are responsible for the thickness and porosity of the etched layer. When applied current density is modulated with respect to the time (see left of Figure 4.2), PSi layers of different porosity are formed (see right of Figure 4.2). A short regeneration period introduced between each current pulse is helpful to prevent the formation of undesirable porosity gradients. PSi has a great potential in photonics applications due to its tunability of refractive index by controlling the porosity of its layer during formation. Hence, the bandwidth of the stop/pass band can be tuned with refractive index contrast of two constituent layers.

Following the Bragg condition, one can fabricate an optical filter by anodizing a silicon wafer. It means the thickness of each layer must be necessarily quarter wavelength ($d_1 = \lambda_c/4n_1$ and $d_2 = \lambda_c/4n_2$). Here, λ_c is the center wavelength of the high reflection band. Mathematically, the structure of a Bragg reflector can be expressed as HLHLHLHLHL.........HL = (HL)m. Here, H and L represent high and low refractive index layers and m is the number of stacks of high and low refractive index layers. Using anodic etching of silicon wafer, a multilayer structure of PSi low and high refractive index layers was fabricated. The starting wafer type was boron-doped silicon substrate with 0.015-ohm cm resistivity (Kordas et al. 2004). The etching was performed with an electrolyte of 11.7 M HF and 10.3 M C_2H_5OH. Alternate layers of low and high refractive index were fabricated by

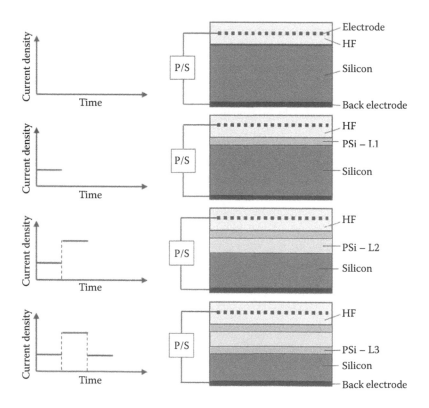

FIGURE 4.2 An illustration of step-by-step preparation of multilayer PSi structure/filter. (From Weiss M., Tunable porous silicon photonic band gap structure: Mirrors for optical interconnect and optical switching, PhD Thesis, University of Rochester, 2005. With permission.)

periodically changing the anodization time and current density. Using an electrochemical polishing process, the prepared multilayer structure was detached from the substrate; as a result freestanding stacks of low and high refractive layers were observed using FESEM, which is shown in Figure 4.3.

It was observed that as the number of periods is increased, a wide nontransparent band appears above 1500 nm in the case of 40 layers and shifts gradually toward the shorter wavelength down to 1080 nm for the case of 100 layers. With optimized process parameters, they could fabricate a multilayer structure that acts as a reflector in near IR (800–900 nm and above 1100 nm) range. Depending on the pore sizes, porous structure is divided into different categories: micro, meso, macro, and nanoporous. Therefore, the property of mesoporous filters can be easily tailored by tuning refractive index via porosity at visible and IR spectral ranges. Therefore, mesoporous silicon optical filters have their applicability not only to visible or near IR but also mid and far IR filter applications where cooling of the filter is done to cryogenic temperatures as in the case of astronomical applications. Mesoporous filters are made of a single material of silicon; therefore, they are mechanically and thermally more stable due to which there is no possibility of thermal or mechanical stresses between the individual layers, whereas conventional interference filters do have issues such as these because they are composed of layers with dissimilar mechanical properties and thermal expansion coefficients.

To fabricate mesoporous filters, boron-doped silicon wafers (resistivity from 1 to 100 mΩ·cm) is preferred, while phosphorous-doped silicon wafers can also be used if thickness and mechanically stability are not a major concern. A silicon wafer of (100) orientation is commonly used; however, wafers of other orientations would be used for anisotropy etching. Figure 4.4a depicts cross-sectional surface morphology of far IR mesoporous silicon filters. Gray color layers are high porosity mesoporous silicon layers (low refractive index) and whiter layers are low porosity (high refractive index). According to wavelength range of interest, the thickness of the mesoporous silicon structure, for example, will be 2 μm for visible-range filters and 200–300 μm for far IR filters. While PSi superlatives can be formed on both p- and p^+-doped silicon wafers, p^+-doped wafers are preferred if high refractive index contrast or thick multilayer structures are to be fabricated. Mesoporous silicon superlattices can be fabricated on n- and n^+-doped wafers as well, although not as thick and mechanically stable as those made on p^+-doped wafers (25-μm-thick mesoporous multilayer structure etched on n$^+$-doped wafer is the thickest etched thus far). With tuning of porosity, a direct filter can be fabricated but high porosity layers make the filter fragile. This problem would be serious with the drying of the mesoporous silicon sample when the capillary force in pores disturbs the mechanical integrity. In addition to this, environmental instability of mesoporous silicon structure is foremost a problem, which originates from the very high surface area of mesoporous silicon (200–250 m^2/cm^3). Because of this, a blue shift of filter wavelength, destruction of quarter wavelength thickness layers, and loss of reflectivity are observed, which degrades the performance of filters. Similarly, degradation in transmission is also reported through infrared transmission filters in such circumstances (Kochergin and Foll 2009).

Environmental stability of far IR filters is improved by using a magnetron sputtered silicon layer onto the mesoporous silicon structure, so that the thermal expansion coefficient of a sputtered silicon

FIGURE 4.3 FESEM image of freestanding PSi bilayers of 20 alternate layers. (From Kordas K. et al., *Opt. Mater.* 25, 25725, 2004. Copyright 2004: Elsevier. With permission.)

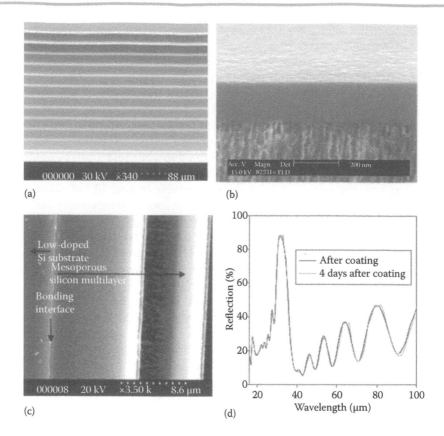

FIGURE 4.4 (a) SEM images of far IR mesoporous silicon filter, (b) silicon-sputtered mesoporous silicon filter, (c) mesoporous silicon filter close to the bonding interface, and (d) reflection spectrum through a mesoporous silicon filter. (From Kochergin V. and Foll H., *Porous Semiconductors: Optical Properties and Applications.* Springer-Verlag, London, 2009. Copyright 2009: Springer. With permission.)

layer would be the same as that of the mesoporous silicon multilayer and silicon substrate. Figure 4.4b shows a SEM image of a silicon-sputtered mesoporous silicon filter. Reflection analysis has confirmed the environmental stability of such a filter as shown in Figure 4.4d. No change in reflectivity is observed from the silicon-sputtered mesoporous silicon filter even after aging three days.

Another problem with mesoporous silicon IR filters is degradation of transmission due to free carrier absorption in silicon substrate beneath the prepared mesoporous layer. For mid IR region, transmission can be maintained at cryogenic temperatures but if we are talking about room temperature operation, it needs removal of the mesoporous silicon multilayer from the unetched silicon substrate. Further, this removal causes problems such as high porosity rendering free-standing membranes fragile. To overcome this problem, a two-layer silicon wafer composed of a highly doped Si layer with the suitable thickness for the mesoporous layer bonded to a low-doping density handle wafer is used. This mechanism maintains high carrier concentration needed for mesoporous multilayer formation on silicon transparent throughout for far IR range. Figure 4.4c depicts an SEM image of a portion of the mesoporous silicon filter around the bonding interface. FTIR measurement at room temperature is done to verify the transparency of the mesoporous silicon multilayer and transmission is observed in the far IR range.

DBR filters are fabricated by alternating the etching current between a high and a low value, respectively. However, rugate filters are prepared by modulating current gradually, which processes a smooth index profile. These filters also have a forbidden band similar to DBR filters but slightly narrower than the bandwidth of a quarter-wave-based DBR filter. The advantage of rugate filters is that higher harmonics can be suppressed. The multilayer structure of PSi was produced by electrochemical etching of boron-doped silicon wafer of (100) orientation. Figure 4.5a shows the cross-sectional view of prepared rugate filters, which is composed of 30 periods with the stopbands at 1400, 1610, 1852, 2129, and 2449 nm.

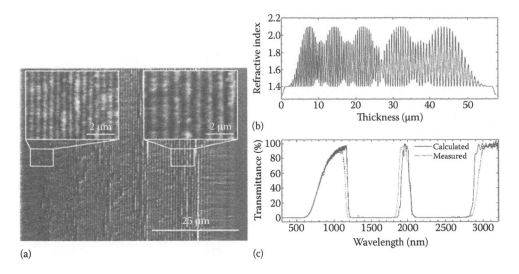

FIGURE 4.5 (a) SEM image of 30 periods rugate filters, (b) refractive index profile as a function of depth, and (c) transmission spectra. (From Ishikura N. et al., *Opt. Mater.* 31, 102, 2008. Copyright 2008: Elsevier. With permission.)

Figure 4.5b depicts the refractive index profile of prepared pass-band filter as a function of physical depth. The filter consists of five rugate structures (30 periods) with the stop-bands at 1300, 1495, 1719, 2250, and 2588 nm. Figure 4.5b shows transmission behavior and compared with calculated. A pass-band centered at wavelength 1950 nm can be observed between two stop-bands, left and right sides. The pass-band position, width, and number can be tailored simply by properly controlling the stop-band positions of rugate structures constructing the filter (Ishikura et al. 2008).

4.2.2 FABRY–PEROT INTERFERENCE FILTERS

A Fabry–Perot interference filter is another class of filters that has a narrow pass-band employed interferometer. A Fabry–Perot interference filter can be easily prepared by employing two DBRs separated by a spacer of half wave thickness. The purpose of inserting a spacer between DBRs is to break the periodicity of the dielectric layer, which yields a pass-band. In such a device, multiple beam interference in the spacer layer makes high transmission over a narrow band. In a similar way, microcavities are formed by introducing an active material between two DBR mirrors and this arrangement produces an enhancement of spontaneous emission at the energy of the cavity mode.

Using the anodization technique, a Fabry–Perot interference filter was fabricated and SEM image is shown in Figure 4.6a,b. It is comprised of 12 pairs of low and high refractive index layers with a spacer of half wavelength thickness. Figure 4.6c shows reflectivity of Fabry–Perot interference filters where maximum transmittance can be observed within the center of the stop-band wavelength range. Reflectance plotted at different center wavelengths shows the tuning of peak transmittance within the whole visible and IR ranges (Mangaiyarkarasi et al. 2006).

PSi is a versatile material to host itself as passive and active components in advanced optoelectronic and photonic devices. Optical filters have wide application such as solar blind non-line-of-sight communications, electrical spark imaging, and photolithography, chemical, and biological analyses. The wavelength ranges of these filters can vary for ultraviolet ($\lambda < 400$ nm), deep ultraviolet ($\lambda < 300$ nm), or even far ultraviolet ($\lambda < 200$ nm) ranges. There is a technological challenge to fabricating such sophisticated filters for the deep and far UV range.

Macroporous silicon UV filters have advantages over the conventional interference filters, such as omni-directionality. This enables high and wide rejection range, truly short-pass (down to far UV) transmission, which makes this filter for its promising application. The fabrication of such filters needs freestanding membranes of PSi so that transmittance can be attained at the UV wavelength range. Figure 4.7 shows top and cross-sectional images of a cleaved, freestanding

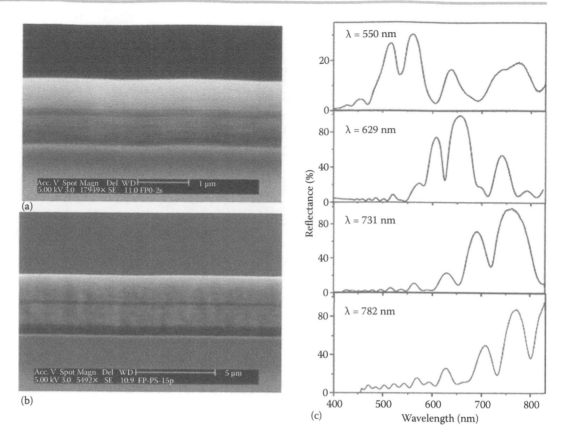

FIGURE 4.6 (a) SEM image of Fabry–Perot interference filter without thick PSi layer, (b) SEM image of F-P filter formed at the top of a 2-μm thick PSi layer, and (c) reflectance spectrum at wavelength 550, 629, 731, and 782 nm. Current densities used for the low and high porosity layers are 50 and 85 mA/cm² and the same high current density (85 mA/cm²) is applied to form a central layer. (From Dharmalingam et al., *Proc. SPIE.* 6125, 61250X, 2006. Copyright 2006: SPIE. With permission.)

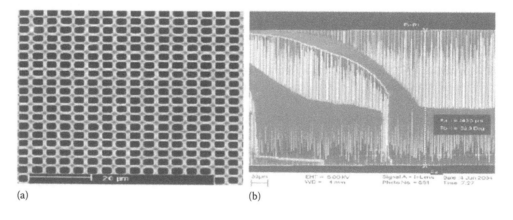

FIGURE 4.7 (a) SEM images of UV macroporous silicon membrane top view and (b) cross-sectional view. (From Kochergin V. and Foll H., *Porous Semiconductors: Optical Properties and Applications.* Springer-Verlag London Limited, UK, 2009. Copyright 2009: Springer. With permission.)

macroporous silicon membrane. A thin Si_3N_4 protective layer was deposited to avoid the exposure of the wall to the chemical during the etching process (Kochergin and Foll 2009).

To fabricate a UV macroporous filter, a low pressure chemical vapor deposition was used for the UV coating of PSi pore walls. Figure 4.8a depicts the SEM image of a single-layer silicon nitride coated macroporous silicon membrane, while Figure 4.8b shows an image of the five layers of coated $SiO_2/Si_3N_4/SiO_2/Si_3N_4/SiO_2$ macroporous silicon membrane (Kochergin and Foll 2009).

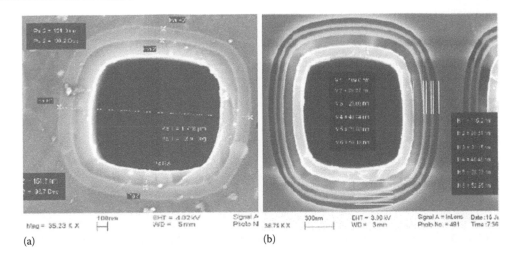

(a)   (b)

FIGURE 4.8 (a) SEM images of macroporous silicon membrane coated with single-layer silicon nitride and (b) five layers of $SiO_2/Si_3N_4/SiO_2/Si_3N_4/SiO_2$. (From Kochergin V. and Foll H., *Porous Semiconductors: Optical Properties and Applications.* Springer-Verlag London Limited, UK, 2009. Copyright 2009: Springer. With permission.)

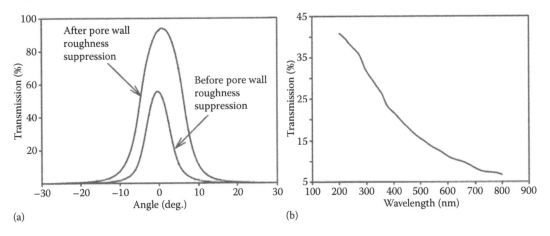

(a)   (b)

FIGURE 4.9 (a) Near field as a function of angle of incidence and (b) far field transmission spectrum in accordance to wavelength. (From Kochergin V. and Foll H., *Porous Semiconductors: Optical Properties and Applications.* Springer-Verlag London Limited, UK, 2009. Copyright 2009: Springer. With permission.)

The roughness of the pore walls is an important factor, which can be tailored to get required transmission levels. Figures 4.9a and b depict the tuning of transmission of light in near and far-fields, which reveals the far field transmission up to 35% in deep UV (300 nm and below) and near field transmission up to 90%. Though the near field transmission was close to the practical value, improvement of far-field transmission through macroporous silicon filters is further possible.

4.2.3 LONG WAVE PASS FILTERS

Long wave pass filters based on PSi have been investigated due to their significant application. This filter can support long wave passes in similar ways as diamond particle based filters do. Depending on pore-to-pore distance of macropore arrays, it is possible to make scattering of light. Long wave pass filter can scatter the light of wavelength less than edge wavelength and transmit above to it. The applications of such filters are demanded in astronomy, Fourier transform spectroscopy, and so on. There are two variants of these filters: a long wave pass or short

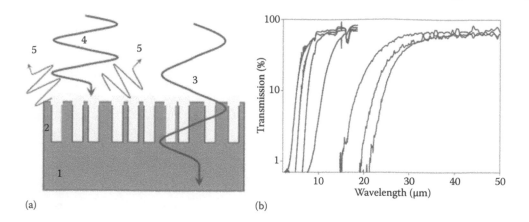

FIGURE 4.10   (a) Principle of operation of macroporous silicon long wave pass filter and (b) transmission spectra of various long wave pass filters. (From Kochergin V. and Foll H., *Porous Semiconductors: Optical Properties and Applications.* Springer-Verlag London Limited, UK, 2009. Copyright 2009: Springer. With permission.)

wave pass filter, which works on the concept of absorption and interference of light. The working mechanism is that it filters when the rejection of light is due to absorption (in case of absorption-based filter); however, it filters when rejection of light is due to reflection from a multilayer structure (in the case of interference filter) (Kochergin and Foll 2009).

The graphically working mechanism of the long wave pass filter is shown in Figure 4.10a where the light of wavelength 3 is enough for long pass and transmission from filter 1, macroporous layer 2 is an effective transparent medium, light 4 with wavelengths lower than the average distance between pores, which gives strong scattering 5 and, hence, it is forbidden. Figure 4.10b shows transmission spectra of various filters, which confirm high transmission in infrared regions of wavelength. The transmission of long wave pass filters was found to be affected through various process parameters during the fabrication of macroporous arrays using a *p*-type silicon wafer. The optical and mechanical qualities can be improved by optimizing current density and temperature with proper selection of silicon doping and electrolyte concentration.

4.3 MICROLENS

Microlens is a type of lens that has a diameter less than 1 mm and smaller than 10 μm. Microlenses can be designed for desired optical properties and such small lenses are comprised of a single or multiple lenses in one, two, and three dimensions. Normally, there are two surfaces such as one plane and one spherical convex surface is enclosed as a single element to form a microlens. A single microlens is used to couple the light into the optical fibers, whereas microlens arrays are used to enhance the incident light collection efficiency of CCD (charge-coupled device) arrays. Microlens arrays are used in various applications such as in digital projectors, digital camera, photodiodes, mobile phone cameras, 3D imaging and displays, and photovoltaic cells.

Novel optical oxidized PSi (OPS) microlenses were produced using selectively oxidized PSi/ fully isolated by porous oxidized silicon technology and silicon bulk micromachining. Before preparing the oxidized PSi microlenses, various silicon wafers were processed and studied the morphology of oxidized PSi-silicon interface for their use as OPS microlens surface. The etching of the mask Si_3N_4 layer constantly was performed during the anodization process to form the lens. By measuring the diameter of the OPS lens, the focal length could be calculated. The prepared OPS microlens was integrated and merged with the previously prepared optical network based on v-grooves, optical fiber, electrical network substrate based on FIPOS/SOPS, and flip chip bonded LED/PD (Ha et al. 2004). The focal length as a function of anodization time, open area window, and silicon nitride layer thickness was studied. Saturation in focal length was observed as a function of an Si_3N_4 open window area. By using oxidized PSi microlens and prereported result, a new optoelectronic multichip module configuration silicon substrate was suggested.

A fabrication approach of PSi with gradient refractive index (GRIN) has been reported, which has been used for the preparation of a planar microlens on silicon on insulator (SOI) substrate. There are two electrochemical etching to synthesize PSi, one is single-tank and the other is double-tank approach. The double-tank electrochemical method was adopted for the fabrication of PSi for planar GRIN microlens. The proposed electrochemical etching method was found to be a suitable one in order to prepare PSi with their effective optical thickness (EOT) distribution, which has shown good approximation equivalent to that of the GRIN lens as shown in FTIR spectra, Figure 4.11a. A high reflection was noticed in the center, which was due to the lower porous structure of silicon. It is also important to note that the effective optical thickness depends on the position from the center of the microlens.

Figure 4.11b depicts the measured distribution of the EOT of a GRIN sample with 10 mm diameter. A quadratic relationship between EOT and position was expected for the GRIN lens. A reduced value of EOT was attributed to an increased porosity of prepared porous structure. Such fabricated planar GRIN lenses in SOI substrate was claimed to be appropriate for rapid prototyping of GRIN lenses. The fabricated PSi-based GRIN microlens by this method has given a better distribution of porosity all over the layer, which gives the effective optical thickness. There was low porosity at the center with maximum effective optical thickness (Zhong et al. 2013).

Many optical devices were fabricated based on multilayer PSi by varying the refractive indices. Variation in the refractive indices changes the porosity of the PSi, which is a very important parameter for optical property. The variation of the refractive index can be done either along the optical axis or perpendicular to the optical axis. Using porous silicon, various planar GRIN lenses were designed and fabricated. For such lenses, the refractive index was having a quadratic position dependency parameter perpendicular to an optical axis. The fabricated PSi GRIN lenses have an advantage over conventional GRIN devices such as these are compatible with silicon technology with applicability in optoelectronics. These lenses were optically transparent in near infrared spectral region and covered the optical communication window. Positive GRIN lens and negative GRIN lens were designed and fabricated. The uniform distribution of porosity in the layer could be achieved for fabricating the GRIN lens in PSi. The positive GRIN lens was fabricated by producing optical thickness greater at the center and decreased toward the edge and the negative GRIN lens was fabricated by producing the smallest optical thickness at the center and increased toward the edge. By varying the etching current density, the porosity and layer thickness can be attained on the silicon wafer. The flat (stainless steel) disk electrode was replaced by a ring-shaped anode, which touches the silicon wafer during the fabrication of the PSi layer in the electrochemical cell, which has the optical thickness required for positive GRIN lens. Point type anode for electrical contact with the silicon wafer where only the center point touches the silicon wafer was used for the fabrication of the PSi layer with low refractive index, which gives the optical thickness needed for the negative GRIN layer. GRIN device arrays were fabricated by modifying the anode. Positive and negative lens arrays were fabricated using multiple rings and

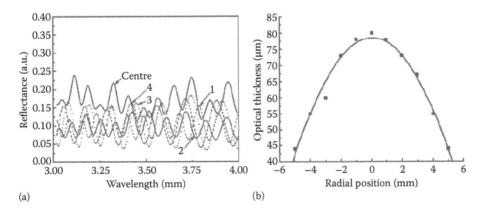

(a)   (b)

FIGURE 4.11   (a) FTIR spectra and (b) optical thickness as a function of radial distance of GRIN lens. (From Zhong F.-R. et al., *Optoelectron. Lett.* 9(2), 0105–0107, 2013. Copyright 2013: Springer. With permission.)

multiple points as anodes in the electrochemical cell and these lenses were tested using Newton's rings. The interference pattern obtained by interference between light reflected from the PSi layer (lens) and the silicon wafer is called Newton rings. The interference fringes are the function of radial distance from the center of the lens. These inference fringe changes are due to change in the optical thickness. The relationship between optical thickness and the radius was determined by inference fringes. The fabricated positive lens shows the less dense rings at the center and dense rings are more toward the edge and vice versa for the negative lens (Ilyas and Gal 2006a).

The same group has reported the fabrication of a planar GRIN microlens using PSi. The fabrication of the planar GRIN microlens was done using an electrochemical cell with a ring type anode material, which was in contact with a silicon wafer. The uniform porosity and layer thickness all over the layer was obtained using this type of anode. The two point contacts were formed by ring electrode at the edges of the slab and electrolyte. The optical thickness was determined by Newton's rings. Lateral distribution of porosity was seen in PSi GRIN lenses. The transparency of such GRIN lens was seen at the center due to low porosity and the increased transparency toward the edges in the visible region. The focal length of the lens was varied during fabrication by changing the thickness of the PSi layer. This produced a planar microlens that can be used in silicon-based devices due to the transparency obtained in the near infrared region (Ilyas and Gal 2006b).

Figure 4.12a shows an interferogram of a GRIN lens fabricated into a 2.4-cm silicon wafer. Newton's rings can be seen, which comprises consecutive dark/bright rings representing a variation of one wavelength in the optical thickness. Figure 4.12b depicts relative optical thickness for a 10-mm diameter GRIN lens, in which the solid line represents the quadratic position dependence that was expected. The electrochemical etching of silicon using a simple ring electrode could produce an optical thickness distribution to a very good approximation equivalent to that of a GRIN lens. The dashed line shows the calculated optical thickness of the PSi layer by assuming a ring electrode configuration. Such fabricated GRIN lenses would be useful for light coupling efficiency silicon sensors, detectors, and waveguides.

Concave mirrors are used in many applications in the field on MEMS and optoelectronics as those mirror surfaces are capable of trapping the light as a spot. The fabrication of a concave mirror using ion-irradiation and electrochemical anodization has been investigated. The fabricated concave mirror was point and line focused. The annuli were spin coated on a silicon wafer, which was patterned by UV lithography. In this process, ion irradiation was used for converting the irradiated annuli into an HF resistant mask with PSi formed within a small area of aperture. The photoresist were removed and the wafer was electrochemically etched depending on annuli dimensions after the irradiation step. The final electropolishing was carried out to detach the thick PSi and formed the concave silicon surface at the end of anodization.

Cylindrical concave lines and spherical concave mirrors were fabricated using this process. The above-discussed fabrication process is illustrated in Figure 4.13. Figure 4.13a through c depict the fabrication of concave mirrors and concave cylinders. Figure 4.13d defines the parameters (A)

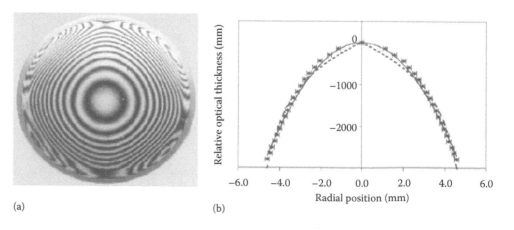

(a) (b)

FIGURE 4.12 (a) Interferogram of PSi GRIN lens showing Newton's rings, and (b) relative optical thickness of a GRIN lens in accordance to radial distance. (From Ilyas S. and Gal M., *Appl. Phy. Lett.* 89, 211123 (2006). Copyright 2006: AIP. With permission.)

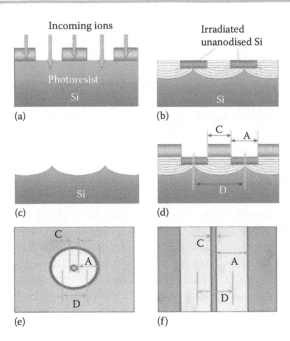

FIGURE 4.13   Fabrication process of concave mirrors and concave cylinders. (a, b, c) Schematic process flow of concave mirrors and concave cylinders fabrication. (d) Defines respective parameters. (e, f) The photoresist patterns used for fabricating the concave mirrors and cylinders, respectively. (From Sheng O. Y. et al., *Opt. Express.* 18(14), 14511–14518, 2010. Copyright 2010: OSA. With permission.)

the irradiated annulus/lines, (C) the diameter of the central aperture, and (D) final diameter of the mirror/cylinder. Figure 4.13e and f show the photoresist patterns used for fabricating the concave mirrors and cylinders, respectively. A multilayer Bragg reflector based on concave PSi was fabricated and helps to focus the wavelength and reflect the light selectively. The Bragg reflector designed based on concave PSi was able to reflect blue, red, and green wavelength light and was able to focus on a spot. The Bragg reflector designed for visible wavelength can be extended to infrared wavelength range by changing the thickness of each layer (Sheng et al. 2010).

The microlenses were fabricated by reflow (melting) photoresist pillars on the silicon substrate. Then dry etching was done to get lens shape. The anodiazation process was carried out for producing nanoporous silicon after the microlenses fabrication. The fabrication of a nonplanar surface was performed whose property depends on the electric field distribution and the anodization exposure. The bulk micromachining technique was used for creating the channels with sloped walls to characterize the effects of a nonplanar surface on the anodization process. P-type and n-type silicon substrates were used and the anodization was carried out in a single tank PFTE cell. Two different morphologies of nanoporous silicon were formed by p-type silicon substrate. The n-type sample anodization was carried out in an avalanche breakdown for the mechanism of hole generation. The thickness and porosity was more at the bottom edges of the sloped channel walls. The nanoporous silicon layer prepared on p-type has given better antireflection property and used as antireflection coating on microlens application (Sheng et al. 2010).

4.4 ANTIREFLECTIVE COATING

The antireflective coating was constructed by Lord Rayleigh (John Strutt) in the nineteenth century. The antireflective coating was produced by Fraunhofer in 1817. Fraunhofer observed reduction of reflection due to etching of surface carried out in the atmosphere of sulfur and nitric acid vapors. Antireflective coatings on solar panels were made for better transmission and glare reduction. The Fresnel equation gives the basic mathematical model of reflection coating and transmission of light in the medium can be tailored via refractive index of the material which quantifies the speed of light in the current medium with respect to that in vacuum. For single layer coating, the Fresnel equation considers both reflection and refraction for basic mathematical model by taking two assumptions that the reflected waves should have constant intensity and one wave is reflected per interface. Taking this into account, the optical interactions such as scattering, absorption, and so on are negligible.

Figure 4.14a illustrates that there is no reflection if destructive interference is between light reflected from the coating substrate and the air coating interfaces. Therefore, the refractive index of coating layer n_c in case of an ideal homogeneous antireflection coating should have the following two conditions: $n_c = (n_a n_s)^{1/2}$, where n_a and n_s are the refractive indices of the air and substrate, respectively, and $d = \lambda/4n_c$, where d is the thickness of the coating, and λ is the wavelength of the incident light (Raut et al. 2011; Yao and He 2014). For multilayer coating as shown in Figure 4.14b, the basic mathematical equation is somewhat different. Antireflection can be achieved by adjusting the reflective index and the thickness of each layer, and the minimum sum of reflection can be obtained. In recent years, different profiles for gradient reflective index layers have been investigated for broadband and omnidirectional antireflection coating such as linear, parabolic, cubic, Gaussian, quintic, exponential, exponential-sine, and Klopfenstein. Assembling a multilayer hetero-structure, which has suited the gradient-reflective index profile, is important to obtain an antireflection property within a wide range of wavelengths.

PSi is used as an antireflection coating material because its properties can be tuned by refractive index, which is controllable during the fabrication process. The refractive index of porous/density graded silicon depends on the volume fraction because the PSi is a sponge-like material marked with nano voids. The refractive index can be decreased by decreasing the volume fraction of silicon, thereby increasing the proportion of air. The PSi layer as an antireflection coating was fabricated by the electrochemical etching and chemical stain etching method. The PSi layers with different thickness were prepared by changing the anodization charge in electrochemical etching. The best cells based on antireflection coating of PSi could produce 13.3% (prepared electrochemical etching) and 12.6% (prepared chemical etching) (Lipinski et al. 2003).

Antireflection surfaces based on subwavelength have also been recently investigated. Lithography techniques were used to get subwavelength structure of PSi n. The growth of vertically aligned silicon nanowires by silver-induced electroless etching of silicon has decreased the reflection loss by ~2% (Srivastava et al. 2010). This silicon nanowire surface acts as a tunable antireflection layer, which is controlled by nanowire length. SEM images of vertically aligned silicon nanowires and schematic presentation of subwavelength structures and multiple coatings of different refractive indexs are shown in Figure 4.15.

The reflectance of antireflection coating of PSi was compared with other coating materials such as SiO_2 and ZnO/TiO_2 as shown in Figure 4.16a. The texturing of PSi surfaces could decrease the reflection of light and increase the light trapping centers at optimum level as comparison to SiO_2 and ZnO/TiO_2 antireflection coating layers.

For the PSi formation a P-type silicon wafer was used as a substrate and SiO_2 on silicon wafer was obtained by a thermal oxidation method. The TiO_2 layer was fabricated using RF sputtering and followed by a ZnO layer using DC sputtering. ZnO/TiO_2 on Si wafer was formed using a sputtering technique. Figure 4.16b depicts the current–voltage characteristics of solar cells based on

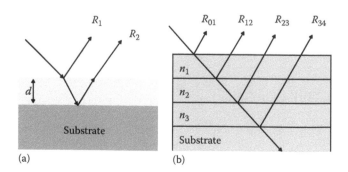

FIGURE 4.14   (a) Light propagation through single-layer coating and (b) multilayer coating on substrate. (From Lin Y. et al., *Prog. Mater. Sci.* 61, 94–143, 2014. Copyright 2014: Elsevier. With permission.)

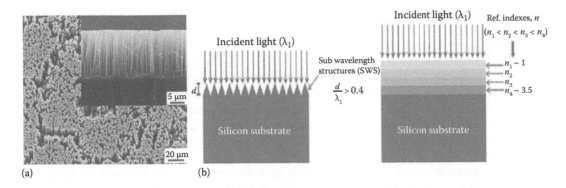

FIGURE 4.15 (a) SEM images of vertically aligned silicon nanowires arrays and (b) schematic diagram of sub-wavelength structures and multiple coatings of different refractive index layers. (From Srivastava S. K. et al., *Sol. Energy Mater. Sol. Cells.* 94, 1506–1511, 2010. Copyright 2010: Elsevier. With permission.)

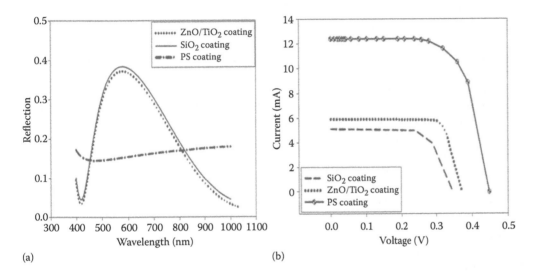

FIGURE 4.16 (a) Reflectance spectra of SiO_2 coating, ZnO/TiO_2 and PSi coating and (b) current–voltage characteristics of solar cells based SiO_2 coating, ZnO/TiO_2 coating and PSi coating. (From Aziz W. J. et al., *Optik.* 122(16), 1462–1465, 2011. Copyright 2011: Elsevier. With permission.)

various materials coating. The high roughness of the PSi layer could improve the photoconversion efficiency of solar cells due to formation of light trapping centers. The photovoltaic conversion efficiencies of 11.23% were achieved by antireflection coating of PSi as compared with SiO_2 (3.34%) and ZnO/TiO_2 (4.41%) (Aziz et al. 2011).

Further, improvement in the properties of antireflection coating has been investigated by using rare earth metals such as cerium and lanthanum on a PSi layer, which maintains the instability of the hydro-silicon bond on the top surface which undergoes oxidation and reduces the life span. A PSi layer was produced by the electrochemical anodization method and deposition of lanthanum on it was done using a spontaneous chemical method; however, cerium deposition on PSi was done using the electrochemical method. The change in the morphological properties showed the possibility to enhance photoluminescence intensity by forming light trapping centers. The rare earth metal treated PSi layers decreased the reflection loss and thus could be considered an antireflection coating. The effective lifetime of the antireflection coating was also increased using rare earth metal on a PSi layer (Atyaoui et al. 2013).

The n-type Si wafer was etched by electrochemical process to fabricate the porous silicon structure. The porous surface formed on the front polished side had discrete pores, whereas the PSi surface formed on the unpolished backside had small pores. The high degree of roughness in the porous structure was used as an antireflection coating because the surface texture was found to reduce the light reflection. This parameter is important for enhancing the photoconversion process for solar cell devices (Ramizy et al. 2011).

The influence of thermal treatment on the optical properties of lithium metalized PSi has been reported (Haddadi et al. 2015). A PSi layer was produced by electrochemical anodization using a p-type silicon wafer. Further, lithium bromide (metallic salt) was used as a source of lithium ions, which were deposited on the PSi layer by an immersion plating technique. PSi coated with Li ion was thermally treated at different temperatures from 100°C to 800°C. After thermal treatment, the good deposition of Li ions on the PSi and also a change in the chemical composition were observed. As compared to untreated PSi, the treated one has exhibited a reflection loss in the whole wavelength range; however, it was more dominant for Li/PSi annealed at 200°C. The observed light absorption improvement was due to the increased optical path obtained by the Li/PSi layer and promising for antireflection application.

A vertical double cell electrochemical etching system was used on each silicon wafer and single PSi layer was fabricated by applying a constant current density during etching while a multilayer structure was fabricated by varying the current density. The homogeneity in porosity and thickness of single and multilayer PSi samples were measured using an ellipsometer at different spots of the sample. By correlating the thickness and porosity measurement, it was found that the homogeneities in the porosity and thickness of layers were dependent on applied current density. Electrolyte aging was also studied and observed that the porosity was decreased while thickness was increased on aging. However, the interface width between PSi-silicon was decreased with electrolyte aging for a week and then it was stabilized. Multilayer PSi structures were found more homogeneous than mono PSi layers, which revealed that the multilayer structure used for antireflection coating was less sensitive to local variations in current density (Selj et al. 2011).

A screen-printed multicrystalline silicon solar cell with PSi antireflection coating was presented and observed in reduction of reflectance about 4.7% in the wavelength range from 400–100 nm with enhanced cell efficiency of 13.2%. The employed PSi layer was fabricated on the front surface of the cell using electrochemical etching. As a result, a mesoporous sponge-like structure for light trapping and light diffusing was formed by KOH etching. The fabricated porous structure could be used as antireflection as charge density was observed to be increased accordingly and the porosity increased, which ultimately reduced the index of refraction. Comparison of PSi antireflection coating formed on alkaline textured mc-Si surface with conventional SiN$_x$ coating was performed and found that PSi coating is promising as it was uniform on the whole area of the mc-silicon wafer and was a good light diffuser for broader wavelength. The fabricated PSi layer was integrated in solar cells and observed an enhanced cell efficiency up to 13.27% as compared to SiN$_x$ coating based solar cell which could produce cell efficiency 11.43% (Kwon et al. 2007).

An improvement in antireflection coating of PSi was investigated by coating an additional layer of ZnO on it (Salman et al. 2012). The PSi layer was fabricated using a photoelectrochemical etching method on an n-type silicon wafer and then a ZnO film was deposited on the prepared PSi layer by the radio frequency sputtering method. The fabricated porous structure consisted of nanosilicon crystals and nanopores that were responsible for the reduction in the refractive index because of high porosity and, hence, loss in reflection was observed. In comparison to single layer PSi coating, ZnO/PS layers based antireflection coating was dominant, which decreased reflection and increased the light trapping from 400 to 1000 nm. The low reflective ZnO/PS was integrated in a solar cell and observed enhanced cell efficiency up to 18.15%.

The PSi layer is known for an ultra-efficient antireflection coating material; however, a graded layer with varying expanded band-gap offers increased absorption in visible spectrum regions. A p-type crystalline silicon wafer of <100> orientation was etched using electrolyte solution of HF:H$_2$O:C$_2$H$_5$OH in the volume fraction of 1:1:2 and etching was done at 40 mA/cm^2 for 5 min. The prepared porous structure consisted of isolated silicon pillars with steeper sidewalls. A lower reflectance was observed because of light scattering and trapping of weakly absorbed photons. Within a nanoporous structure, there would be a total internal reflection, which furnished the coupling of light and altered the direction and, hence, the optical path length could be enhanced (Dubey and Gautam 2009).

4.5 DIFFRACTION OPTICAL ELEMENTS

An optical component consisting of microstructured surfaces is known as the diffractive optical element (DOE). Light beams can be reshaped by placing DOEs in front of light beams, which changes the light pattern. The structure of the element is important for reshaping the light beam because it changes the light pattern through grating diffraction. The diffractive concept is used for many applications such as beam splitter (spot arrays), diffractive lenses, diffractive diffusers, corrector plates, beam shapers, diffractive line generator, off-axis illumination, and so on.

DOEs are used in imaging and antireflection coating due to their unique properties. Arranging PSi layers alternatively with DOEs gives the importance of using these layers for waveguides and Bragg mirrors in sensing applications. The chromatic response of a PSilayer designed with DOE changes by multiplication of reflection of oscillating thin film was studied. In chemical sensing applications, PSi and porous polymer were used as templates due to responses of different analytes and due to the absorption of analytes in the porous structure. Figure 4.17a shows a fiber optical sensor illustration using PSi diffraction grating. The PSi diffraction grating with a fiber optical sensor was placed closely parallel to illuminating and reading fibers with a detector and a beam collimating lens (Golub et al. 2010).

The higher response of spectral selectivity of the sensor with PSi diffractively structured was observed compared to the sensor with uniform PSi layer. Other setups consist of the fiber optical sensor with PSi diffractive lens acting as a dispersive sensing element and a focusing lens is depicted in Figure 4.17b. An increase in spectral selectivity was obtained compared to a uniform PSi layer structure. The PSi DOE sensor can also be coupled with the linear fiber array or a light detector array for measuring the diffraction efficiency in the place of a single reading fiber and illuminating fibers. PSi DOE sensors are designed in such a way that they act as simple low-resolution spectrometers. Such sensors were operated using vapors of the analyte molecules, which is penetrated inside the PSi sponge-like structure by displacing air and condensed to a liquid phase. Due to this, change in the refractive index and absorption of the liquid in the pores was observed. The response of DOEs with the PSi layers was determined, which was due to diffraction efficiency changes with wavelengths for the case of water condensate. The sensor's spectral was found to be dependent on diffraction efficiency in accordance to the wavelength. To make diffraction gratings and diffractive optical elements based on PSi electrochemical etching along with interferometric lithography/ion-implantation/laser micromachining or photolithography can be used. By employing an electrochemical setup, uniform PSi layers were fabricated. The produced structure consisted of pores with nanocrystalline silicon rods.

Figure 4.18 depicts the formed PSi layers with uniform thickness and its applicability for diffractive elements due to created surface profile. Afterward, preparation of DOEs of PSi was done using a photolithography technique. The PSi grating sensor was experimentally verified and showed an increase in the spectral selectivity of PSi grating sensor compared to uniform PSi sensor due to the light reflected off the wavelength arrived at the fixed reading fiber position (Golub et al. 2010).

Diffraction-based biosensors can be operated by fixed wavelength and the detection angle, which gives the diffraction efficiency variation because chemical and biological species were

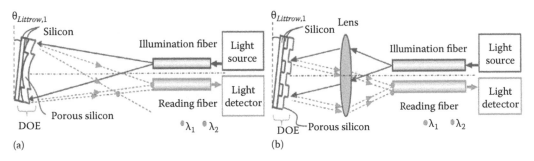

FIGURE 4.17   (a) Fiber optical sensors using PSi diffraction grating with collimating lens, and (b) using PSi diffractive lens which works as a dispersive sensing element as well as a focusing lens. (From Golub M. et al., *Appl. Opt.* 49(8), 1341–1349, 2010. Copyright 2010: OSA. With permission.)

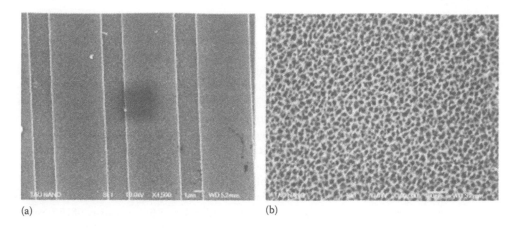

(a)   (b)

FIGURE 4.18   (a) SEM image top view magnification ×4500, and (b) top view magnification ×100; 000 of porous silicon binary diffraction grating. (From Golub M. et al., *Appl. Opt.* 49(8), 1341–1349, 2010. Copyright 2010: OSA. With permission.)

present on the diffraction grating. PSi diffraction based sensors with porous diffraction grating were fabricated at low cost and sensing at high efficiency. PSi substrate was prepared through electrochemical etching. Silicon grating stamps were fabricated by standard contact lithography and reactive ion-etching techniques (Ryckman et al. 2010). PSi-diffraction based biosensors can produce the grating height ranges from 10 to 120 nm. PSi grating undergoes partial oxidation for producing silica surface required to attach biochemical species.

The PSi-diffraction based biosensor (PSi-DBB) was exposed to water vapor during the first investigation. Then 3-aminopropyltriethoxysilane was used to study the molecule attachment and sensing of PSi-DBB. A diffraction technique was developed in such a complicated geometry that the interaction of light matter could be improved by confinement of light. The grating material was optimized and the PSi was incorporated, which has shown a decrement in refractive index and an increment in the response of the analyte by changing the relative diffraction efficiency. Figure 4.19 shows a biosensor based on PSi diffraction and the result of a diffraction experiment performed at 647 nm line 67° angle. This type of PSi–DBB was developed with low cost and wavelength measurement performed by bulky equipment was resolved.

Diffraction gratings surface can be fabricated on the porous silicon or silicon by various techniques such as an optical lithography or optical recording which produces submicrometer periodic length of the gratings however, E-beam lithography or holography is promising one technique to achieve and adjust the line density. For biochemical sensing application the specimen (in the form of suspended cells or molecules) is allowed to penetrate the pores which changes the optical properties of the porous silicon stack and gives significant change of the in-coupling angles of the grating coupled PSi/oxidized PSi waveguide. In experimental work, photoresist was spin coated on silicon wafer. The submicron period gratings for coupling purposes by holographic

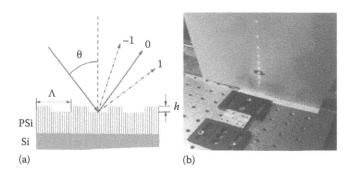

(a)   (b)

FIGURE 4.19   (a) PSi diffraction based biosensor, and (b) visible diffraction from a porous grating with a 60-nm grating height. (From Ryckman J. D. et al., *Appl. Phys. Lett.* 96, 171103, 2010. Copyright 2010: AIP. With permission.)

lithography were obtained after inserting the photoresist in holographic equipment. The light-assisted anodization of PSi was done using holographic images. Then the mask was formed by ion-implantation and subsequent anodization of the silicon surface was performed (Nagy et al. 2005). The photoresist was removed by a wet chemical method and thermal annealing steps of the implanted samples were done. The electrochemical etching of the implanted silicon substrate was carried out. After this step, the diffraction grating formed at the porous crystalline interface. Then PSi layers were removed by etching for AFM microscopy. The use of holographic exposure facilitates the adjustment of the periodic length of gratings in the submicron range. AFM study of one- and two-dimensional gratings confirmed that the higher current density anodization process could give larger amplitude of the grating with increases in surface roughness.

The novel simple PSi structure based on the grating optical sensor was fabricated in such a way that high reflectivity PSi stacks were added between the substrate and grating to improve the diffraction efficiency of the top surface grating, which can improve detection sensitivity (Mo et al. 2014). A high reflectivity PSi multilayer was prepared by altering the electrochemical etching followed by dry photo-etching and reactive ion etching methods for the fabrication of PSi grating diffractive devices on the silicon substrate. The performed numerical simulations showed the high reflectivity PSi stacks could not prevent the loss of energy radiated in the silicon substrate. This has also improved the internal surface area for sensing by immobilizing the molecules attachment. APTES (amino propyltriethoxysilane) was used for the sample sensing measurement in PSi based on the multilayer dielectric grating. The top surface graded showed an improvement in the diffraction efficiency. The biosensing sensitivity was enhanced due to the use of PSi high-reflectivity stacks between the substrate and surface grating. The loss of radiation energy to the substrate was prevented by PSi stacks; hence, energy of light was highly concentrated (reflected much more) on the surface grating. The biomolecules were immobilized to the grating; before and after diffraction, the intensity was maximal at grating. The developed PSi based on the multilayer dielectric grating optical sensor could give high sensitivity—10 times more sensitive compared to those PSi diffraction grating biosensors without a high-reflectivity quarter-wave stack.

4.6 OPTICAL WAVEGUIDES

Optical waveguides are demanded in optical communication for global interconnects, for example, to serve as data bus and clock distribution networks. Optical waveguides are the natural replacement over the metallic waveguides, which are having losses at optical frequencies. Optical waveguides based on PSi came into existence in the 1990s. The infiltration of the liquids in the pores demonstrated a change in the intensity of guided light and this has enabled the application of PSi as a sensor in the form of a waveguide. PSi optical waveguides allow the micromachining of photonic integrated circuits as it can be exploited to confine, manipulate, and guide the photons. Optical waveguides based on PSi multilayers have attracted more interest due to unique guiding mechanisms in comparison to conventional waveguides based on total internal reflection.

PSi is used in many optoelectronics applications by controlling the refractive index of the PSi. One more parameter for the application of PSi for optoelectronics applications is that if PSi has a large internal surface area, then the surface combination rates will be increased which can reduce the free-carrier lifetime. The nonlinear property of the PSi waveguide has to be measured to know the free carrier lifetime measurement.

PSi slab waveguides were prepared by electrochemical etching of silicon wafers. A three-layered structure was formed by an etching process out of which a thick core layer was sandwiched between the two low refractive index cladding layers. The thickness of each layer was optimized based on the refractive index in order to ensure the single mode operation of fabricated slab waveguide. A channel waveguide was patterned after the three-layer slab waveguide fabrication by scanning laser lithography. Finally, a prepared waveguide sample was kept in an oven to get a thin uniform oxide layer on the etched PSi surface. Figure 4.20 shows the schematic of a PSi waveguide (see Figure 4.20a) and SEM image of prepared PSi cladding layer (see Figure 4.20b).

Figure 4.21 depicts the SEM images of silicon on an insulator (SOI) waveguide and PSi waveguides. The mode contours shown in Figure 4.21a were superimposed on the micrographs and

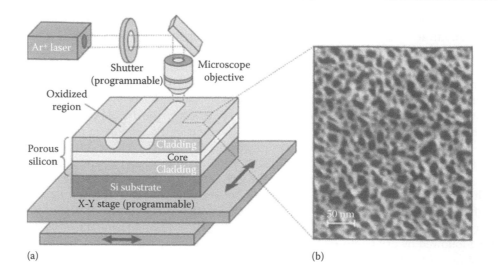

(a)

(b)

FIGURE 4.20 (a) Illustration of channel waveguide patterning using scanning laser lithography system, and (b) SEM image of PSi cladding layer of high porosity. (From Paveen A. et al., *Opt. Express,* 17(5), 3396, 2009. Copyright 2009: OSA. With permission.)

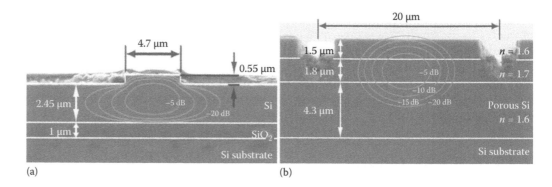

(a)

(b)

FIGURE 4.21 (a) Cross-sectional SEM images of silicon on insulator waveguide, and (b) PSi waveguide. (From Paveen A. et al., *Opt. Express,* 17(5), 3396, 2009. Copyright 2009: OSA. With permission.)

calculated by knowing the refractive index and SOI ridge waveguide dimensions. Figure 4.21b shows the 2D Gaussian mode shape of a PSi waveguide and the Gaussian mode shape was measured by far-field diffraction angles of light emerging from a PSi waveguide.. The nonlinear property of a PSi waveguide was compared with a crystalline silicon waveguide. The PSi waveguide was comprised of 70–80% air and 20–30% silicon. The carrier-based nonlinearities in a PSi waveguide were observed to be stronger and faster compared to the crystalline silicon waveguide (Apiratikul et al. 2009). This property makes the PSi waveguide application an optical switch or fast electro-optic modulators.

The fabrication of a PSi waveguide using an electrochemical process and a photolithography technique which needs masking is not good due to low resistance of polymer, which has been reported as a result of application of acid in the electrochemical process. The alternative technique used to overcome this problem is direct laser writing for patterning on the PSi surface for channel waveguide. The PSi slab waveguide was fabricated by electrochemical etching of crystalline silicon and the Bragg reflected waveguide was patterned on PSi upper layer using laser ablation/oxidation. The fabricated waveguide was oxidized by thermal treatment. Finally, the waveguide edge was cleaved parallel to grating to allow measurement of a Bragg grating waveguide (BGW) transmission spectrum by end fire fiber coupling (Rea et al. 2008).

Figure 4.22a–c depict the stepwise process of Bragg grating waveguide fabrication with core and cladding refractive index 1.65 and 1.52, respectively. Figure 4.22d shows the regular air trenches etched by laser light in the PSi surface and Figure 4.22e shows a single-element end,

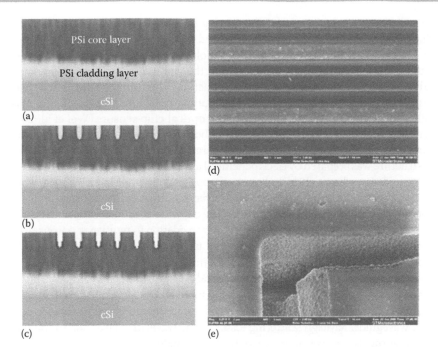

FIGURE 4.22   (a–c) SEM images of fabrication step of Bragg grating waveguide, (d) fabricated top view image of Bragg grating waveguide structure, and (e) SEM image of the single element end showing regular wall. (From Ilaria R. et al., *J. Phys. Condens. Matter*. 20, 365203, 2008. Copyright 2008: IOP. With permission.)

which is found to be as regular walls with a porosity gradient along the vertical direction. This work claimed the optical losses using Bragg grating fabricated by a laser ablation fabrication method. Such prepared PSi slab waveguide with Bragg grating can be used for chemical and biological sensing with easy and low cost fabrication.

PSi-based waveguides are used for many sensor-based applications. The PSi waveguide with grating coupler is an advanced component for the integrated optical chip biosensor. PSi waveguide was fabricated by an electrochemical etching method on a silicon wafer. The low porosity waveguide was fabricated based on the parameter variation in the electrochemical etching and then the waveguide was thermally oxidized. The photoresist grating was formed on the waveguide using a lithography technique. Prism coupler is used for the measurement of reflectance properties of the PSi waveguide sensor when DNA oligos molecule is exposed to a light source.

Figure 4.23a depicts a SEM image of a typical photoresist grating on a PSi waveguide. The opened pores without any disturbances between the photoresist grating lines can be observed, which supports the infiltration of biomolecules into the waveguide structure of PSi. Figure 4.23b

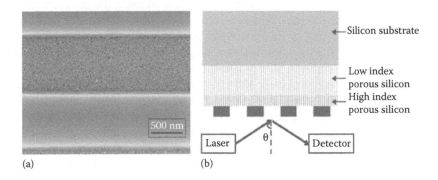

FIGURE 4.23   (a) SEM image of photo resist grating on the PSi waveguide, and (b) configuration measurement for PSi waveguide with grating coupled. (From Wei X. and Kang C., *Proc. SPIE*. 7167, 71670C, 2009. Copyright 2009: SPIE. With permission.)

shows a measurement arrangement for the grating coupled PSi waveguides, where laser light is made incident onto the waveguide and the reflected light intensity is fetched to a photodetector. The prism coupler waveguide showed that the molecule attachment was very poor where the grating coupler showed that a large surface area in the waveguide made the biomolecules immobilized and the guided mode interaction was satisfactory. The study of reflectance spectra has the resonance near 38°, which is due to the field distribution in the upper PSi layer. This type of device has suitability for the application in the integrated optical chip sensors (Wei et al. 2009).

For fabricating a PSi multilayer structure, two methods were preferred; that is, by altering the anodization parameter and by changing dopant concentration (Loni et al. 1996). Multilayer PSi optical waveguide structure was prepared by electrochemical anodization. This layered structure consisted of a porous waveguide layer, lower cladding layer, and upper more porous layer. The waveguide layer was sandwiched between an upper porous layer and a lower cladding layer. A pattering layer was prepared using silicon nitride instead of photoresist to avoid the chemical reaction between PSi and photoresist. Further, as-prepared planar samples were converted by oxidation to multilayer porous oxides in ambient air conditions. The strip waveguide was formed using oxidized samples. The oxidized samples were etched by reactive ion etching. The property of the optical waveguide was measured by coupled light either in infrared laser or in visible laser inside and outside the cleaved samples. The PSi refractive index depended on the porosity, which was also measured by Bruggeman effective medium approximation. The propagation losses were observed to be high in the multilayer waveguide compared to a conventional waveguide. Rayleigh scattering losses in the waveguide layer due to the porous nature were negligible at the used wavelength. The optical waveguide based on multilayer PSi structure was well worked in the infrared region; however, oxidation of a multilayer structure could be useful for extending the wavelength range for the visible region.

In another work, oxidized PSi optical waveguides were demonstrated with boron-doped silicon wafer used as the starting material (Maiello et al. 1997). The porous waveguide optical property was observed to be improved by thermal oxidation. The optical fiber was used as a probe for measuring the scattering losses on the surface of the waveguide. The coupled light was sent through the waveguide and output light was controlled by placing a rutile prism. Low propagation loss was found in the visible range wavelength. Near field and refractive index measurements were shown with the less refractive index layer sandwiched between the oxidized PSi core and the bulk silicon. The low refractive region in the waveguide was responsible for guiding performance of the structure.

The development of a new optical biosensor by using an anti-resonant reflecting optical waveguide (ARROW) based on PSi material as a transducer was demonstrated. The antiresonance condition of the planar porous silicon based ARROW was adjusted before the protein attachment for transverse electric (TE) and transverse magnetic (TM) mode polarizations. ARROW on PSi was prepared by the electrochemical anodization method. By adjusting the electrochemical anodization parameter, an upper PSi layer (core layer) with pores open, which allows the biomolecule, was easily obtained. The cladding layer was obtained with a smaller pore diameter, which confirmed that the layer was impermeable to biomolecules. The oxidation of waveguide was carried out to get the transparent porous waveguide in the visible range after anodization. The porosities and refractive indices for each layer were calculated using the Bruggemann model. The reflectivity coefficient at the interface between core and the first cladding for ARROW structure was measured (Hiraoui et al. 2012). The reflectivity was observed to be increased according to an increase in core thickness. The thickness was functionalized before and after protein attachment. The thickness and reflectivity were also measured in TE mode and TM mode of polarization. Near-field measurement was done before and after protein attachment. The light was guiding in the core layer after protein attachment. The prepared structure was suitable to realize label-free optical biosensors. The overlapping between the optical field and the molecule attachment in the porous silica surface was the main advantage of this type of waveguide.

Similarly, the PSi waveguide was fabricated and functionalized for infrared wavelength range. The oxidation of PSi has extended the wavelength up to the visible range. The multilayered and graded refractive index structures were optimized by changing the current density and were utilized for fabricating optical waveguides. The optical waveguide structure can be used for optical biosensor application (Shokrollahi and Zare 2013). In other reported work, biomolecule detection was studied for functionalized PSi membrane waveguide (Rong et al. 2008). After formation the PSi film was removed from the silicon substrate and during the detachment of the porous film, the diameter of the pore was increased in the PSi film at the bottom and further it was oxidized. Waveguide was prepared where formvar polymer film was on the PS membrane. The PSi membrane was placed in such a way that large opened pores were faced toward the air interface for facilitating molecule infiltration. The DNA hybridization detection was done and observed that when complementary DNA was exposed to the probe DNA, the two strands of DNA were bound together. The refractive index of the PSi was increased during hybridization, which has changed the dispersion of the waveguide mode. The change in the resonance angle of the waveguide was measured by a prism coupler. The change in the refractive index of the PS membrane and polymer layer thickness gives change in resonance angle of the substrate and the guided mode. The biomolecule in the nanoporous waveguide and the binding across in the waveguide shows the high sensitivity of the PSi membrane-based waveguide. Many biosensing applications are attracted toward this kind of PSi membrane waveguide.

4.7 INTEGRATED OPTOELECTRONICS

Silicon-based optoelectronics is a colossal innovation that is picking up development. This innovation expects to influence the expense adequacy and renowned performance of silicon-based electronic circuits by integrating photonic components on digital and wireless silicon integrated circuits. The extensive capability of silicon-based optoelectronic integrated circuits (OEICs) is starting to be acknowledged and this range holds a guarantee for an extensive variety of utilization from high-capacity on-chip optical interconnects to optical communications receivers and transmitters. PSi is a sort of synthetic element silicon that has exhibited nanoporous gaps in its microstructure, rendering a high surface-to-volume ratio and due to its inherent electronic and transport characteristics, this material is suitable for development of photonic and sensing devices. The traditionally used method to fabricate PSi is by the anodic electrochemical etching method and property of obtained PSi depends on the process parameters/conditions.

Due to the excellent property of PSi such as its photoluminescence, electroluminescence, wave guiding, and so on, this material has a wide scope of applications in integrated optoelectronic technology due to its high-speed optical interconnection. An optoelectronic unit was developed with an aluminum PSi LED connected to a photodetector by an aluminum waveguide. The purpose of using PSi was to produce strong visible light under optical or electrical excitation. It was demonstrated that the internal light excitation was more responsible for higher photoresponse instead of external light excitation. An integrated optoelectronic system was developed that was comprised of two aluminum PSi Schottky junctions and within that an alumina layer was sandwiched (Samsonov 1978). The one end was acting as an LED and the other end was acting as a photodetector. A niobium mask was placed on the surface of the aluminum electrodes, which acted as a mask for selective anodization of aluminium as well as a reflector, which could confine the light spreading in the anodic alumina layer. Around a 20-mm gap was maintained between the LED and the photodetector. The PSi surface is protected by an anodic aluminium oxide (alumina) in order to protect it from atmospheric oxygen and it also plays a major role in the device; it acts as an optical waveguide in transmitting the light emitted by one of the Schottky junctions. Optimal light guiding can be ensured when a refractive index of PSi is less than that of alumina. The designed and developed technology of the silicon integrated optoelectronic unit was tested and found promising for high-speed optical interconnections in integrated circuits. The use of PSi-based LEDs has given hope of integration development of new electronic and optoelectronic devices.

MEMS is a technology of miniaturizing electromechanical elements using a microfabrication technique. Micromechanical photothermal spectroscopy integration with suitable sorbent material can be fabricated in the chip-scale device with IR-cooled detector. MEMS photothermal spectroscopy helps in detecting the trace gases using optical MEMS coated functionalized sorbent materials. MEMS spectrometer for high-resolution IR spectroscopy of gases should contain a tunable infrared source, a biomaterial microbride, interferometric displacement readout, and functionalized polymer sorbent material. This setup is helpful for the detection of vapor phase analyte. The microstructure absorbs radiation and heated at infrared wavelengths corresponding to rotational or vibrational molecular resonances of the sorbent material or sorbed analyte. Due to heating, biomaterial microbridges get bent, which can be optically read out. The presence of trace analyte vapors, which were sorbed on the polymer material, gives the changes in the photothermal spectrometer. The filter helps for measurement of microphotothermal spectrometer signals for high resolution.

The fabrication of the chip scale tunable filter based on MEMS Fabry–Perot etalon using PSi distributed Bragg reflector mirrors was demonstrated (Kozak et al. 2014). The PSi on silicon wafer was fabricated using an electrochemical etching method. The porosity of the layer was maintained by optimizing the etching current density and other parameters of electrochemical etching. By using this method, alternate layers were formed with different refractive indices to get Bragg reflectors as shown in Figure 4.24a. Further, PSi-based Fabry–Perot interferometer was fabricated with mid- and long-wave infrared regions, which are depicted in Figure 4.24b. MEMS combined with this fabricated PSi-based Fabry–Perot interferometer could work well as a good sensor.

MEMS devices needed to be encapsulated to protect them from external environment and low package pressure maintenance. Therefore, getter has been demanded to maintain the low package pressure of MEMS devices. The getter designed and fabricated in such a way that it should have high porosity to absorb gases. In this context, PSi is used as getter for MEMS devices due to its high porosity. Due to large surface area with more porosity, PSi helps for reacting with gases. This PSi getter with cavity helps to reduce the air damping and parasitic capacitance, which makes better performance for resonators. Two types of MEMS devices were reported where one device was with getter and the other without getter. In the first device with getter, a cavity was produced by electrochemical etching of the silicon substrate. However, a thick PSi layer to act as getter is produced by electrochemical etching. The encapsulated MEMS device with cavity and PSi as getter was fabricated (Mohammad et al. 2011). The fabricated MEMS device with encapsulation was tested for a resonant profile with a network analyzer. The MEMS device with cavity and PSi has shown higher quality factor than the MEMS device with only cavity. The MEMS device was also tested for pressure inside the device and the MEMS device with cavity and PSi getter could show low pressure than the MEMS device with cavity only. The PSi reacts with oxygen gases to form the oxide layer that will trap the other gas molecules; therefore, the low pressure was maintained in the cavity. The encapsulated MEMS device with cavity and PSi getter was observed to be protected from the external environment without any other material or other masking process.

PSi is a material whose optical properties can be tuned during its formation and it involves a quantum confinement effect in the nanometer range. Due to which, this material can be an efficient one for light emission at room temperature. Using PSi, light emitting diodes (LEDs) were fabricated and integrated with a standard industrial process. A yellow-orange luminescence was observed by the naked eye at room temperature.

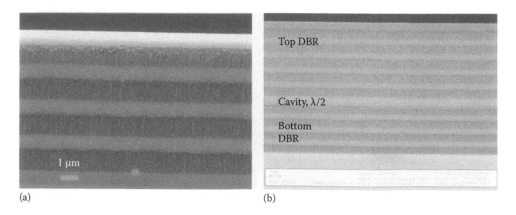

(a)   (b)

FIGURE 4.24   (a) SEM images of fabricated Bragg reflector, and (b) cavity prepared in between two Bragg reflectors. (From Kozak D. A. et al., *ECS Trans.* 64(1), 197–203, 2014. Copyright 2014: ECS. With permission.)

Figure 4.25a depicts typical I-V characteristics of LED based on PSi. A limiting on forward current was noticed that was due to high series resistance in LED. The cause of this was attributed to the resistance of the poly contact lines, the crystalline substrate, and the polysilicon layer. Different anodization current waveforms were applied and some uncommon characteristics were observed, which are shown in Figure 4.25b. An LED fabricated at 75 mA/cm² (constant current density) showed an emission peak at 700 nm; however, a peak at 800 nm was observed for the same magnitude of current density (nonconstant). This is because the application of constant current density could give higher porosity whereas it was low for the case of application of nonconstant current density (Barillaro et al. 2003).

On-chip PSi micro-thruster was reported for the robotic platform (Churaman et al. 2013). A robotic platform needs actuators for storing and for faster energy release. The high storage and faster energy release actuators were designed and implementation of thrusters used in space applications. MEMS technology was used for designing microthrusters with low cost fabrication. PSi has the ability to generate thrust. Therefore, PSi was fabricated and integrated with a confinement chamber, quantify thrust and impulse as a function of confinement. The PSi micro-thruster was produced by a two-wafer process. The device chip was produced using low stress, low-pressure chemical vapor deposition method. The nitride remained wafer was patterned using a lithographic technique and the nitride region was etched by reactive ion etching for opening the windows for exposed silicon. An electronic initiator was patterned using lithography. These initiators were connected to band pads. On the top of the electronic initiator, a spin on thermoplastic material was patterned, which acts as an electrical insulator. The exposed silicon regions were etched by galvanic etching on the chip. The cap chip was fabricated on the silicon wafer. Two chips were fabricated and assembled. For evaluating the force and impulse of the PSi, a setup was developed using a Kistler force sensor. PSi acted as a solid propellant for analysis of the single microthruster output. The silicon porosity was maintained and a cap chip was used for the confinement of the reaction products in the microthruster performance level using a force sensor. This type of tested microthruster can be used for the robotic applications for jumping locomotion.

A different approach has been reported to study the influence of localized PSi regions and the obtained results were found promising for the integration of active and passive devices on various substrates.

In this work, planar inductors were integrated on the PSi substrates. The dotted lined area in Figure 4.26a is a localized PSi region while Figure 4.26b depicts the SEM image of PSi integrated with an inductor on it. A complete experimental study was explored and concluded that the use of a hybrid substrate is a promising one for the integration of active and passive devices which

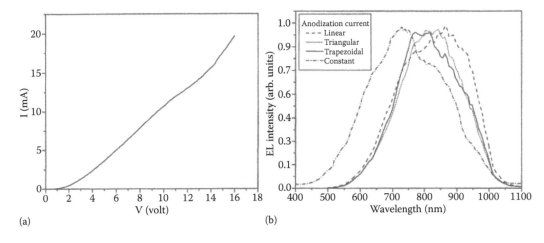

(a)   (b)

FIGURE 4.25   (a) I-V characteristics of PSi LED device, and (b) electroluminescence spectra of various LEDs fabricated with different current waveforms. (From Barillaro G. et al., *Mater. Sci. Eng. B* 101, 266–269, 2003. Copyright 2003: Elsevier. With permission.)

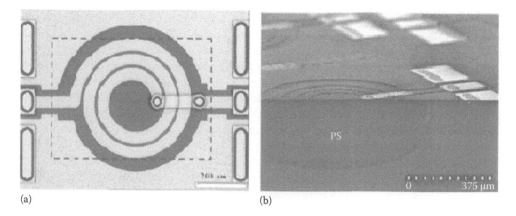

(a)   (b)

FIGURE 4.26 (a) Top view of an inductor with PSi region defined by the dashed line, and (b) SEM image of an inductor integrated on PSi. (From Capelle M. et al., *Nanoscale Res. Lett.* 7(523), 1–8, 2012. Copyright 2012: Springer. With permission.)

could be due to increased performance of a passive device and further it can be improved using oxidized PSi (Capelle et al. 2012).

We know that the PSi is a good electrical conductivity material with very low thermal conductivity, due to which it can be used as a buffer layer in micromachining technology. This material has shown excellent performance as microsensing devices and intensive attention has been focused toward compatibility of PSi and anodic bonding technologies. Due to sponge-like nanostructures (specific surfaces up to 500 m²/cm³), this material effectively interacts with several chemicals and biological molecules and acts as a transducer. When this material is exposed to chemical or biological substances, optical properties such as refractive index, photoluminescence, and electrical conductivity is altered, which helps the device to sense the target material. The key concept behind this sensing phenomenon is that when it is exposed to chemicals such as hydrocarbon, acetone, and so on, as well as biological species, the refractive index of the PSi will change. Under the test of vapor or liquid phase, due to capillary condensation in nanometric pores, it changes the refractive index of PSi microcavity and hence a red shift in reflectance spectrum occurs. For testing of biological species, refractive index of the PSi surface causes variation due to formation of covalent bonding. The development of two types of sensors based on PSi is reported where for one sensor, the analysis chamber was dogged into the crystalline silicon, while for the second sensor, the analysis chamber was directly into the glass (De et al. 2006).

Figure 4.27 shows fabricated sensing based on PSi. For the first type of device fabrication, after electrochemical process and drying, packing and drilling was done to apply gas or liquid

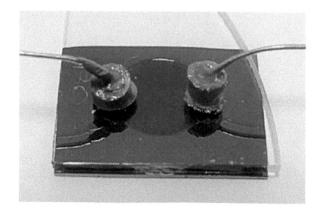

FIGURE 4.27 Integrated PSi based optical sensor. (From De Stefano L. et al., *Sens. Actuators B* 114, 625–630, 2006. Copyright 2006: Elsevier. With permission.)

substances, which are under test. The glass and silicon chip was anodically bonded and then stainless steel capillary tubes were glued to the holes to make simple in–out operations.

PSi is demanded in the integration of several devices such as chemical and biosensors, waveguides, MEMS, and so on. PSi is used as a chemical and biological sensing element due to its morphology and physical properties. The use of PSi in optical sensing is based on the photoluminescence or reflectance properties, but normally PSi has a disadvantage that it cannot be used as a transducer material for optical sensing because the interaction of transducer material with analyte is not specific when exposed to a gaseous or liquid substance, which is based on change in photonic properties. To overcome this problem, there is a limit to modify the chemical or physical of the PSi hydrogenated surface to enhance the sensor selectivity for specific biochemical interactions. This can be done by fabricating the sensor using as-etched PSi which has very reactive Si-H terminated surface. The optical biosensing using a PSi-based photonic device was designed and fabricated. The PSi monolayer, which acts as a Fabry–Perot interferometer, is a basic transducer in the biosensing. This device has high sensitivity to change in refractive index, and several trails were made in the interaction of biomolecules such as DNA single strand, glutamine binding protein from *Escherichia coli*. The biochemical sensor based on 1D photonic bandgap structures has both reduced group velocity of light within the sample, which increases the interaction of analytes and hence sensitivity. In this type of sensor, measurement is based on the reflection peak. PSi optical microcavity was used to study the molecular bending of the GlnBP and glutamine with higher sensitivity with respect to the interferometric characterization. The other type of optical sensing is used, that is, PBG structure with Bragg mirror. The optical sensor with PBG structure distributed Bragg mirror was designed and simulated. The Bragg low porosity layer was produced by the etching method. To perform a biosensing experiment, a post-etching was done to increase the size of the pores and to eliminate the superficial nano residue induced after the etching. In this way, infiltration of biomolecular probes into the pores can be well performed. However, this process eliminates the native bonds (Si–H), which are considered for functionalization treatment. The refractive index of the prepared PSi layer was estimated based on the response of the sensor using optical fiber connected to the tungsten lamp as a light source and an optical analyzer. After functionalization, PSi chip with various temperatures, FTIR measurement was performed (Rendina et al. 2007). The link between the inorganic and organic phases was found to be stable up to 250°C. In addition to it, Si–C and C–N–C bonds were thermodynamic strength, which indicated that a passivated chip could work even after the anodic bonding process.

A fabricated hybrid PSi-glass is shown in Figure 4.28a. After incubation of the labeled bioprobes bonded to the chip, fluorescence macroscopic measurement was done by illuminating the chip spotted with labeled proteins. It can be seen in Figure 4.28b that the image is highly fluorescent and homogeneous all over the surface. Figure 4.28c depicts a scratch on the chip surface, which indicates that the fluorescent probe can penetrate into the inner pores of the chip.

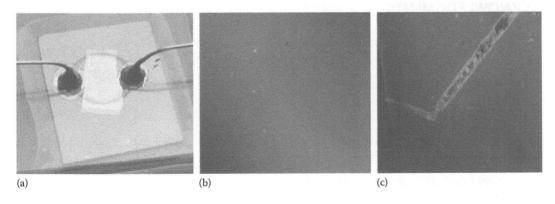

(a)   (b)   (c)

FIGURE 4.28   Pictures of PSi hybrid glass-chip: (a) an element of a lab-on-chip based optical transducer, (b) fluorescent image of labeled proteins, and (c) particular of a scratch on the chip surface. (From Ivo R. et al., *Physica E.* 38, 188–192, 2007. Copyright 2007: Elsevier. With permission.)

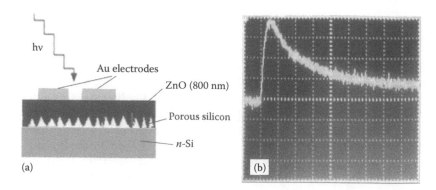

FIGURE 4.29   (a) Schematic diagram of proposed ZnO/PSi detector and (b) photo response curve of fabricated ZnO UV detector based on PSi. (From Thjeel H. A. et al., *Adv. Mater. Phys. Chem.* 1, 70–77, 2011. Copyright 2011: Scientific Research Publishing. With permission.)

ZnO nanofilms were grown on the PSi to fabricate a UV detector. This detector could be able to decrease the response time from a few seconds to a few hundred microseconds with enhanced photoresponsivity of the ZnO film deposited on the PSi layer by polyamide nylon (Thjeel et al. 2011). If we compare ZnO nanofilms/PSi-based UV detectors with silicon ultraviolet photodetectors, it overcomes a major limitation of low quantum efficiency in the deep UV range due to the passivation layer and age reduction of the Si photodiode. Figure 4.29a depicts a schematic diagram of a proposed UV detector. For the fabrication of a UV photoconductive detector, PSi structure was prepared with photochemical etching and then a ZnO thin film was deposited on the PSi using a thermal chemical spray pyrolysis technique.

The response time of the ZnO UV detector on the PSi layer was measured with a nitrogen laser and the trace of the output pulse on the digital oscilloscope of 200 MHz band width is shown in Figure 4.29b. It is observed that the signal with a rise time (10–90%) was of the order of 180 µs and the fall time (1–1/e) was about 750 µs. The slow decay time was attributed to the slow escape of holes from the tarps. Generally, traps in a wide band semiconductor are extremely deep; therefore, a broad peak centered around 540 nm can be observed.

In summary, PSi is a versatile material which has broad applications in electronics, optoelectronics, biosensors, gas sensors, batteries, medicinal application, and so on. Silicon-based technology is fully capable of integrating electrical and optical components on a single chip. Using PSi, it is possible to link silicon technology with optoelectronic devices. By adding optical functionality to the well-established silicon-manufacturing infrastructure, an enhanced performance of both existing microelectronic and developing photonic devices can be achieved. PSi is an evergreen material for a new generation of devices.

ACKNOWLEDGMENTS

The author expresses thanks to S. Saravanan and Krishna Teja (doctoral fellows) and M.Y. Thanuja (assistant professor) for their help in the drafting of this chapter.

REFERENCES

Apiratikul, P., Rossi, A.M., and Murphy, E. (2009). Nonlinearities in porous silicon optical waveguides at 1550 nm. *Opt. Express*, **17**(5), 3396.

Atyaoui, M., Dimassi, W., Atyaoui, A., Jalel, R., and Ezzaouia, H. (2013). Improvement in photovoltaic properties of silicon layer with rare earth (Ce, La) as antireflection coatings. *J. Lumin.* **141**, 1–5.

Aziz, J., Ramizy, A., Ibrahim, K., Hassan, Z., and Khalid, O. (2011). The effect of anti-reflection coating of porous silicon on solar cells efficiency. *Optik.* **122**(16), 1462–1465.

Barillaro, G., Diligenti, A., Piotto, M., Allegrini, M., Fuso, F., and Pardi, L. (2003). Non-constant anodization current effects on spectra of porous silicon LEDs. *Mater. Sci. Eng. B* **101**, 266–269.

Canham, L.T. (1990). Silicon quantum wire array fabrication by electrochemical and chemical dissolution of wafers. *Appl. Phys. Lett.* **57**, 1046.

Capelle, M., Billoue, J., Poveda, P., and Gautier, G. (2012). RF performances of inductors integrated on localized p⁺-type porous silicon regions. *Nanoscale Res. Lett.* **7**(523), 1–8.

Churaman, A., Morris, J., Currano, J., and Bergbreiter, S. (2013). On-chip porous silicon microthruster for robotic platforms. In: *Proceedings of 17th International Conference on Solid State Sensors, Actuators and Micro Systems (Transducers and Eurosensors XXVII)*, June 16–20, Barcelona, pp. 1599–1602.

De Stefano, L., Maleck, K., Rossi, M., Rotiroti, L., Della, G., Moretti, L. and Rendina, I. (2006). Integrated silicon-glass opto-chemical sensors for lab-on-chip applications. *Sens. Actuators B* **114**, 625–630.

Dubey, R. and Gautam, D. (2007). Photonic bandgap analysis in one-dimensional porous silicon photonic crystal by transfer matrix method. *Optoelectron. Adv. Mater.–Rap. Commun.* **1**(9), 436–441.

Dubey, R. and Gautam, D. (2009). Synthesis and characterization of nanocrystalline porous silicon layer for solar cells applications. *J. Optoelectron. Biomed. Mater.* **1**(1), 8–14.

Golub, A., Hutter, T., and Ruschin, S. (2010). Diffractive optical elements with porous silicon layers. *Appl. Opt.* **49**(8), 1341–1349.

Ha, L., Kim, H., Yeo, K., and Kwon, S. (2004). An oxidized porous silicon (OPS) microlens implemented on thick OPS Membrane for a silicon-based optoelectronic-multichip module (OE-MCM). *IEEE Photon. Technol. Lett.* **16**(6) 1519–1521.

Haddadi, I., Slema, B., Amor, B., Bousbih, R., Bardaoui, A., Dimassi, W., and Hatem, E. (2015). Effect of rapid thermal treatment on optical properties of porous silicon surface doped lithium. *J. Lumin.* **160**, 176–180.

Hilleringmann, U. and Goser, K. (1995). Optoelectronic system integration on silicon: Waveguides, photo detectors and VLSI CMOS circuits on one chip. *IEEE Trans. Quantum Electron. Dev.* **42**, 841–846.

Hiraoui, M., Haji, L., Guendouz, M., Lorrain, N., Moadhen, A., and Oueslati, M. (2012). Towards a biosensor based on anti resonant reflecting optical waveguide fabricated from porous silicon. *Biosens. Bioelectron.* **36**(1), 212–216.

Ilyas, S. and Gal, M. (2006a). Single and multi-array GRIN lenses from porous silicon. In: *Proceedings of IEEE Conference on Optoelectronic and Microelectronic Materials and Devices, COMMAD*, Dec. 6–8, Perth, WA, pp. 245–247.

Ilyas, S. and Gal, M. (2006b). Gradient refractive index planar microlens in Si using porous silicon. *Appl. Phys. Lett.* **89**, 211123.

Imai, K. (1981). A new dielectric isolation method using porous silicon. *Solid-State Electron.* **24**, 159.

Ishikura, N., Fujii, M., Nishida, K. et al. (2008). Broadband rugate filters based on porous silicon. *Opt. Mater.* **31**, 102–105.

James, D., Milne, S., Keating, J. et al. (2006). Nano-Porous Silicon antireflection coatings for microlens application. In: *Proceedings of International Conference on Nanoscience and nanotechnology*, July 3–7, Brisbane, Queensland, pp. 431–434.

Kato, K. and Tohmori, Y. (2000). PLC hybrid integration technology and its application to photonic components. *IEEE J. Sel. Top. Quantum Electron.* **6**(1), 4–13.

Khriachtchev, L., Rasanen, M., Novikov, S., and Sinkkonen, J. (2001). Optical gain in Si/SiO$_2$ lattice: Experimental evidence with nanosecond pulses. *Appl. Phys. Lett.* **79**, 1249.

Kochergin, V. and Foll, H. (ed.) (2009). *Porous Semiconductors: Optical Properties and Applications*. Springer-Verlag, London.

Kordas, K., Beke, S., Edit, A., Uusimaki, A., and Leppavuori, S. (2004). Optical properties of porous silicon. Part II: Fabrication and investigation of multilayer structures. *Opt. Mater.* **25**, 257–260.

Kozak, A., Stievater, H., Pruessner, W., Kerry, N., and Rabinovich, S. (2014). Porous silicon MEMS infrared filters for micromechanical photo thermal spectroscopy. *ECS Trans.* **64**(1), 197–203.

Kwon, H., Lee, H., and Ju, K. (2007). Screen-printed multicrystalline silicon solar cells with porous silicon antireflective layer formed by electrochemical etching. *J. Appl. Phys.* **101**, 104515.

Lipinski, M., Bastide, S., Panek, P., and Levy, C. (2003). Porous silicon antireflection coating by electrochemical and chemical methods for silicon solar cells manufacturing. *Phys. Stat. Sol. (a)* **197**(2), 512–517.

Loni, A., Canham, L., Berger, G. et al. (1996). Porous silicon multilayer optical waveguides. *Thin Solid Films* **276**(1–2), 143–146.

Luterova, K., Pelant, I., Mikulskas, I. et al. (2002). Stimulated emission in blue-emitting Si- implanted SiO$_2$ films? *J. Appl. Phys. Lett.* **91**, 2896.

Maiello, G., La, S., Ferrari, A. et al. (1997). Light guiding in oxidized porous silicon optical waveguides. *Thin Solid Films* **297**, 311–313.

Mangaiyarkarasi, D., Breese, H., Ow, S., Kambiz, V., and Daniel, B. (2006). Porous-silicon-based Bragg reflectors and Fabry–Perot interference filters for photonic applications. *Proc. SPIE* 6125, 61250X.

Ha, M.-L., Kim, J.-H., Yeo, S.-K., and Kwon, Y.-S. (2004). An oxidized porous silicon (OPS) microlens implemented on thick OPS membrane for a silicon-based optoelectronic-multichip module (OE-MCM). *IEEE Photon. Technol. Lett.* **16**(6), 1519–1521.

Mo, J., Liu, Y., Liu, C., and Jia, Z. (2014). Porous silicon based on multilayerdielectric-grating optical sensors with enhanced biosensing. *Phys. Stat. Sol. (a)* **211**(7), 1651–1654.

Mohammad, W., Wilson, C., and Kaajakari, V. (2011). Introducing porous silicon as a getter using the self aligned maskless process to enhance the quality factor of packaged MEMS resonators. In: *Proceedings of Joint Conference of the IEEE International Frequency Control and the European Frequency and Time Forum (FCS)*, May 2–5, San Francisco, pp. 1–4.

Nagy, N., Volk, J., Hámori, A., and Bársony, I. (2005). Sub micrometer period silicon diffraction gratings by porous etching. *Phys. Stat. Sol. (a)* **202**(8), 1639–1643.

Pavesi, L., Dal, L., Mazzoleni, C., Franzo, G., and Priolo, F. (2000). Optical gain in silicon nanocrystals. *Nature* **408**, 440–444.

Pickering, C., Beale, J., Robbins, J., Pearson, J., and Greef, R. (1984). Optical studies of the structure of porous silicon films formed in p-type degenerate and non-degenerate silicon. *J. Phys. C: Solid State Phys.* **17**(35), 6535.

Ramizy, A., Hassan, Z., Omara, K., Al-Dourib, Y., and Mahdi, M.A. (2011). New optical features to enhance solar cell performance based on porous silicon surfaces. *Appl. Surf. Sci.* **257**, 6112–6117.

Raut, H., Anand, V., Nair, S., and Ramakrishna, S. (2011). Anti-reflective coatings: A critical, in-depth review. *Energy Environment Sci.* **4**, 3779–3804.

Rea, I., Marino, A., Iodice, M., Coppola, G., Rendina, I., and de Stefano, L. (2008). A porous silicon Bragg grating waveguide by direct laser writing. *J. Phys. Condens. Matter.* **20**, 365203.

Rebohle, L., Von, J., Borchert, D., Frob, H., Helm, M., and Skoupa, W. (2001). Efficient blue light emission from silicon: The first integrated Si-based optocoupler. *Electrochem. Solid State Lett.* **4**(7), G57–G60.

Rendina, I., Rea, I., Rotiroti, L., and de Stefano, L. (2007). Porous silicon-based optical biosensors and biochips. *Physica E.* **38**, 188–192.

Rong, G., Ryckman, D., Mernaugh, L., and Weiss, S.M. (2008). Label-free porous silicon membrane waveguide for DNA sensing. *Appl. Phys. Lett.* **93**, 161109.

Ryckman, D., Liscidini, M., Sipe, E., and Weiss, S.M. (2010). Porous silicon structures for low-cost diffraction-based biosensing. *Appl. Phys. Lett.* **96**, 171103.

Salman, J.A., Khalid, O., and Hassan, Z. (2012). Effective conversion efficiency enhancement of solar cell using ZnO/PS antireflection coating layers. *Sol. Energy.* **86**, 541–547.

Samsonov, G.V. (ed.) (1978). *Physico-Chemical Properties of Oxides.* Metallurgija, Moscow USSR (in Russian).

Selj, H., Marstein, S., Thogersen, A., and Foss, E. (2011). Porous silicon multilayer antireflection coating for solar cells; Process considerations. *Phys. Stat. Sol. (c)* **8**(6), 1860–1864.

Sheng, O.Y., Breese, H., and Azimi, S. (2010). Fabrication of concave silicon micro-mirrors. *Opt. Express.* **18**(14), 14511–14518.

Shokrollahi, A. and Zare, M. (2013). Fabricating optical waveguide based on porous silicon structures. *Optik.* **124**, 855–858.

Srivastava, S., Dinesh, K.D, Singh, K., Kar, M., Vikram, K., and Husain, M. (2010). Excellent antireflection properties of vertical silicon nanowire arrays. *Sol. Energy Mater. Sol. Cells.* **94**, 1506–1511.

Takai, H. and Itoh, T. (1986). Porous silicon layers and its oxide for the silicon-on-insulator structure. *J. Appl. Phys.* **60**, 222.

Thjeel, A., Suhail, M., Naji, N., Al-Zaidi, G., Muhammed, S., and Naum, A. (2011). Fabrication and characteristics of fast photo response ZnO/porous silicon UV photoconductive detector. *Adv. Mater. Phys. Chem.* **1**, 70–77.

Watanabe, Y. and Sakai, T. (1971). Application of a thick anode film to semiconductor devices. *Rev. Electron. Commun. Lab.* **19**, 899.

Wei, X., Kang, C., and Weiss, M. (2009). Porous silicon waveguide with integrated grating coupler for DNA sensing. *Proc. SPIE* **7167**, 71670C.

Yao, L. and He, J. (2014). Recent progress in antireflection and self-cleaning technology from surface engineering to functional surfaces. *Prog. Mater. Sci.* **61**, 94–143.

Zhong, R., Lu, Y., Jia, H., and Tian, M. (2013). Microlens fabricated in silicon on insulator using porous silicon. *Optoelectron. Lett.* **9**(2), 105–107.

Electronics

Electrical Isolation Applications of PSi in Microelectronics

Gael Gautier, Jérôme Billoué, and Samuel Menard

CONTENTS

5.1	Introduction	110
5.2	Application to Active Device Isolation	110
5.3	Application to RF Passive Devices	113
	5.3.1 General Considerations on Device Processing	113
	5.3.2 Interconnect Structures	113
	5.3.3 Inductors	116
	5.3.4 RF Functions	119
	5.3.5 Hybrid Substrates for Active and Passive Devices	119
5.4	Electrical Properties of PSi	120
	5.4.1 Dielectric Permittivity	121
	5.4.2 DC Conductivity	122
	5.4.3 AC Conductivity	122
5.5	Conclusion	123
References		123

5.1 INTRODUCTION

Historically, electronic was the first discipline that sought to exploit the properties of porous silicon (PSi) in the 1980s. Then, PSi was used after oxidation to isolate bipolar devices. Radio frequency (RF) devices also took advantage of the isolating properties of PSi. Indeed, highly resistive substrates are required to reduce eddy currents and capacitive couplings. Moreover, the insulating properties of PSi, added to the ability to locate these areas in different resistivity wafers, make this material potentially interesting in terms of development of monolithic insulator/semiconductor substrates. This chapter begins with the presentation of many implementation processes of PSi peripheries in active devices and the use of this material as a silicon on insulator (SOI) substrate. Then, a large overview of RF microelectronic applications is presented. Finally, an ultimate section will be dedicated to direct current (DC) and alternative current (AC) electrical properties of PSi. The performance improvement brought by PSi namely originates from its dielectric behavior.

5.2 APPLICATION TO ACTIVE DEVICE ISOLATION

SOI is a specific way in chip processing. In this case, thick bulk silicon wafers (generally between 0.5 and 1 mm thick) are replaced by wafers that have three layers: a thin surface layer of silicon where the transistors are formed, an underlying layer of insulating material, and a substrate (silicon wafer). The insulating layer is usually made of silicon dioxide.

Among the most promising techniques for producing SOI substrates suitable for fabrication of high-performance devices are those based on the oxidation of PSi. PSi has a unique set of material properties and its oxidation is now well mastered (Earwaker et al. 1991). In addition, PSi anodization technology on large area wafers seems to be mature enough to emerge in the industry (Boehringer et al. 2012; Desplobain et al. 2014). The use of PSi or oxidized PSi leads to many SOI fabrication techniques for localized isolation or full wafer isolation.

The first type of isolation structure is called insulation by porous oxidized silicon (IPOS). This technology was suggested by Pogge and Poponiak (1975). Then, it was mainly developed by Watanabe et al. (1975). This bipolar IC technology involved oxidized PSi islands to isolate transistors from each other. A so-called "n-type process," using n-type epitaxial layers on p-type substrates, was first demonstrated. A silicon nitride layer was used to localize the anodization. Afterward, the oxidation was performed. Figure 5.1 shows a schematic cross-section of the final NPN device and the associated process flow. Isolation voltages around 210 V were reached. A "p-type" process was also performed. In this case, p^+-type wells were performed before the anodization. Nakajima and Kato (1977) reported isolation voltages between 130 and 150 V. Unfortunately, the leakage currents are one order higher than for conventional PN junctions.

A second technology called full insulation porous oxidized silicon (FIPOS) was also developed in the 1980s. FIPOS was first proposed by Imai and Unno (1984). This technology is close to SOI. It was mainly performed for complementary metal oxide semiconductor (CMOS) devices for digital applications. The main differences with conventional SOI process are the use of a proton implantation and oxidized PSi. Specifically, a Si_xN_y layer is firstly deposited on a p-type substrate. Then, a photolithography step allows localizing boron-doped $p+$ wells. After photoresist removing, the whole surface of the wafer is implanted with hydrogen protons. A low temperature annealing in the range of 400–500°C leads to a type inversion near the silicon surface except in the $p+$ wells. Afterward, the anodization process is performed to convert p regions into PSi. Finally, an annealing at high temperature (700°C) is performed in order to annihilate the n-type doping by releasing H^+ and to oxidize PSi.

Recently, the same method was employed using fluorine implantation. Actually, implanted fluorine has demonstrated a donor effect in silicon upon annealing at low temperature (600°C). This doping is reversible as the fluorine out-diffuses during higher temperature annealing (1100°C). NMOS transistors performed in active regions with thickness less than 200 nm completely surrounded by oxidized PSi were characterized showing the electronic integrity of the active area (Veeramachaneni et al. 2011).

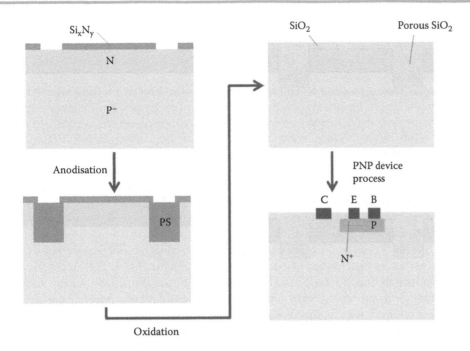

FIGURE 5.1 Typical "*n* type" IPOS process for bipolar NPN transistor integration. (From Watanabe Y. et al., *J. Electrochem. Soc.* 122, 1351, 1975. Copyright 1975: The Electrochemical Society. With permission.)

Many other uses of PSi involving epitaxial layers for substrate isolation were proposed. Nevertheless, to our knowledge, these processes have not been performed and have remained concepts (Watanabe and Sakai 1972; Pogge and Poponiak 1975; Bean and Runyan 1977; Frye and Leamy 1983).

The most promising approaches that have been developed are those based on structures such as *p/p+/p* or *n/n+/n* (Holmstrom and Chi 1983; Barla et al. 1986; Bomchil et al. 1988; Tsao et al. 1991), where PSi is only formed in a thin (1–3 μm) heavily doped buried layer accessible through an upper layer (Figure 5.2). If we consider the latter structure, in a lightly doped *n*-type silicon substrate, a heavily doped n layer is created by ionic implantation. Then, an epitaxial growth of a lightly doped n layer is performed (Figure 5.2a and b). Generally, the optimum concentration is around 10^{19} cm^{-3}, leading to a homogeneous porous layer that subsequent oxidation transforms into an oxide equivalent to silicon dioxide throughout the whole thickness of the porous layer. It

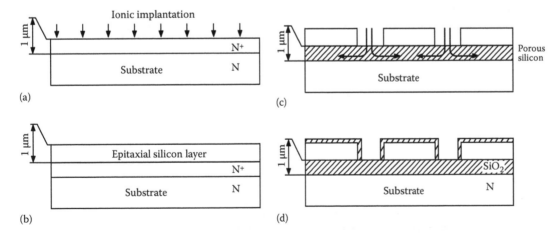

FIGURE 5.2 Successive technological steps leading to SOI structure by oxidation of a buried PSi layer. In a lightly doped n-type silicon substrate a heavily doped (antimony) n layer is created by ionic implantation (a) followed by the epitaxial growth of a lightly doped n layer (b). The anodization is then performed through the openings (c) and a subsequent oxidation transforms PSi into oxide (d). (From Bomchil G. et al., *Microelectronic Eng.* 8, 293, 1988. Copyright 1988: Elsevier Science. With permission.)

was also shown that the thickness of the epitaxial layer could be reduced to 100 nm. Then, circuits were successfully fabricated using these PSi structures in submicron technologies (Thomas et al. 1989). Konaka and co-workers (1982) also reported the formation of epitaxial Si islands on PSi and a subsequent oxidation to isolate active regions.

Many studies were also done on epitaxial silicon growth directly on full sheet PSi layers for SOI applications. The techniques covered a wide variety of processes. For instance, Vescan et al. (1988) and co-workers deposited silicon by low-pressure vapor-phase epitaxy (LPVPE) on PSi at much lower temperatures than in conventional chemical vapor deposition (CVD) epitaxy. The epitaxy was performed at 823°C and 0.03 mbar using $SiCl_2H_2$. In addition, the authors reported that the original microstructure of PSi was only slightly modified during the epitaxial process. Full oxidation of the buried PSi layer was subsequently possible. Oules et al. (1992) mentioned rapid thermal processing-CVD reactor using silane diluted in hydrogen at a temperature of 830°C and a total pressure of about 2 Torr during a processing time in the order of a minute for a final thickness of 800 nm.

Zheng et al. (1998) proposed a more complex stack involving a double-layer PSi system with a spongy microstructure layer on top of a dendritic microstructure layer. This structure was fabricated on a moderately doped p-type Si wafer using a two-step anodization process, and was used as the substrate for Si molecular beam epitaxy (MBE). The isolation between Si epi layer and the substrate was evaluated, with a breakdown voltage higher than 500 V.

Finally, Sanda et al. (2005) developed a new device layer transfer technology with porous layer splitting. Then, they showed that CMOS FETs on more than 300-nm thick epitaxial films could be successfully fabricated on epitaxial layers with different thicknesses over PSi. A fabricated active layer was also successfully transferred on a flexible plastic substrate without altering the performances. A similar approach for transfer substrates was developed since 1990 by Canon Inc. in Japan by Yonehara et al. (1994) and Sato et al. (1995). ELTRAN® (epitaxial layer transfer) was the first SOI industrial process to provide SOI wafer that involves PSi (Yonehara and Sakaguchi 2001). The method involves silicon anodization, wafer bonding, and etching-back PSi in conjunction with hydrogen annealing. The newest technique consists of splitting within the PSi layer by means of water jet, which allows the seed wafer to be reused several times in order to reduce manufacturing costs. A wide range of SOI/buried oxide thickness can be achieved using this technique from 10 nm to 2–3 μm with thickness uniformity of less than 5%.

In the field of AC switch periphery isolation, some structures were proposed and tested. Generally, deep p+ wells are anodized to produce locally PSi. These wells can be performed by diffusion (Menard et al. 2012) (see Figure 5.3) or by aluminum thermo-migration (Gautier et al. 2006). At the opposite, localized PSi can be employed to perform, after dopant diffusion, deep isolation wells. Many dopant sources have been tried successfully to dope macroporous silicon: BBr3 (Déhu et al. 1995), H_3BO_3, H_3PO_4 (Astrova et al. 1999), or $AlCl_3$ (Amato et al. 1997).

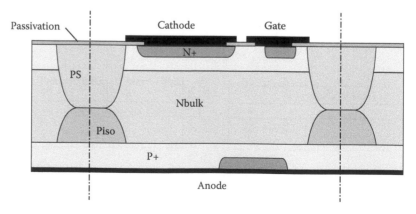

FIGURE 5.3   TRIAC periphery using PSi as a junction termination. PS and Piso refer to porous silicon and through wafer p-well, respectively. (From Menard S. et al., *Nanoscale Res. Lett.* 7, 566, 2012. Copyright 2012 Springer under Creative Commons Attribution License. With permission.)

5.3 APPLICATION TO RF PASSIVE DEVICES

Many parasitic capacitances limit the performances of RF devices at high frequencies but the main contribution is due to the substrate. Consequently, the permittivity of the underlying material must be reduced. Knowing that the dielectric permittivity of PSi is greatly lower than for Si bulk, the capacitive couplings must be greatly reduced employing this material under devices. Moreover, the circulation of currents in the metal coils generates a magnetic field. According to Lenz's law, the variation of B with time induces reverse currents (called Eddy currents) in the lines or in inductor windings. The Eddy current flow in the substrate region is responsible for a magnetic field image generation. Then, a magnetic coupling with metal leads to increased losses. Thanks to the high resistivity of PSi, leakage currents are highly decreased.

In this part, we show how PSi could be an efficient candidate for high performance RF devices such as line interconnects, inductors, or more complex structures such as filters. Moreover, the unique capacity of PSi to be localized in small areas can be used to perform hybrid substrates on which mixed active/passive circuits can be integrated.

5.3.1 GENERAL CONSIDERATIONS ON DEVICE PROCESSING

In this section, we give a description of steps that are susceptible to be involved in the fabrication of RF circuits, such as coplanar waveguides (CPW) or inductors with a particular emphasis on the issues related to PSi processes.

The first phase is generally dedicated to the anodization localization on restricted areas of the substrate. The simplest way that can be employed to localize the etching is the use of specific wafer holders with the appropriate seal geometry. However, in the case of low dimension apertures (<1 cm^2), hard masks must be employed. Many HF-resistant materials are reported in the literature (see Section 5.3.5 dedicated to hybrid Si/PSi substrates). Nevertheless, most of the time, this mask must be removed before metallization. Consequently, resists and polymers are preferable. Finally, PSi possesses a high specific surface. Therefore, it is more sensitive and reactive than bulk silicon and drying, annealing, or etching can be delicate steps that can lead to the modification or destruction of the PSi structure (Ayvazyan 1999; Manotas et al. 2001). Therefore, the PSi stress appears here as a crucial issue for PSi substrates processing. This phenomenon can be significantly reduced if the etched area or the PSi layer thickness is lowered (Capelle et al. 2014d).

Then, PSi must be stabilized by annealing at low temperature (below 500°C) in O_2 or N_2 ambient. Thereafter, if an oxidation must be achieved, an annealing is performed at higher temperature.

Generally, a thin SiO$_2$ cap layer is deposited on PSi to seal the pores and to ensure a complete DC isolation between the substrate and the device. The deposition must be performed at low temperature, generally by plasma enhanced chemical vapor deposition (PECVD). A low dielectric material, generally BCB (benzocyclobutene, $\varepsilon_r = 2.56$), can be also used.

Many metal layers can be employed to perform the strips of inductors, contacts, or connections. The most common materials used in microelectronics are copper and aluminum. Lines must be thick (several microns) in order to reduce ohmic losses. A wide range of deposition techniques are available such as evaporation (mainly for Al), sputtering, or (electro)chemical plating (mainly for copper). In the case of copper, a thin adhesion layer is required (Ti, for example).

5.3.2 INTERCONNECT STRUCTURES

Each RF system includes transmission line interconnects between components. Conventional interconnect structures such as Microstrips (MS) or CPW, realized in standard industrial processes on silicon substrates, typically suffer from poor quality factors in the RF and millimeter-wave ranges. The use of PSi could be a solution to reduce the losses in the substrate in such structures. From these simple devices, it is possible to quantify the benefits of PSi. In the case of MS, the ground is located on the backside of the chip. Then, the device performances are more sensitive to the substrate characteristics. Nevertheless, the electrical measurements and the extraction of PSi electrical properties from S-parameters are easier on a coplanar

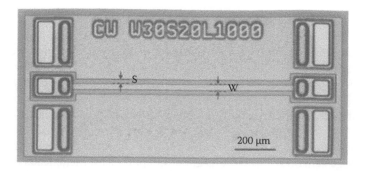

FIGURE 5.4 Typical CPW integrated on PSi. S and W are respectively the distance line to the ground and the line width. (From Capelle M. et al., *International Journal of Microwave and Wireless Technologies* 6, 39, 2014. Copyright 2014: Cambridge University Press and the European Microwave Association. With permission.)

device. This is the reason why most of the studies that have been reported in the literature are focused on CPW. CPW design consists of three strips, a conductive line surrounded by two ground lines and generally separated from the substrate with a thin oxide layer (see Figure 5.4) (Capelle et al. 2014). This device is easier to use for test purposes using ground signal ground (GSG) probes.

These structures are sensitive to capacitive effects and eddy current losses in the Si substrate. So, its electrical characteristics become crucial for monolithic RF systems if high performances have to be reached. To investigate how efficiently PSi provides RF isolation, figures-of-merit (FoM) like power losses (PL), insertion losses (IL), or line attenuation (α) are required. They are extracted from S-parameters measurements. IL can only be computed if the return loss (S_{11}) of the line equals to zero. Therefore, in various samples of PSi substrates, each line geometry should be designed in order to adapt the characteristic impedance close to 50 ohms (Peterson and Drayton 2001). Consequently, α is mainly used for CPW characterization and can be extracted from [S] matrices of two waveguides with the same cross-section but different lengths. The total attenuation at high frequencies can be decomposed into three contributions: the conductor loss (α_C), originating from heat dissipation and reduction of the effective section due to skin effect, the substrate losses (α_S), depending on the electrical properties of the substrate (ε_{eff}, $\tan\delta$, or $\sigma_{substrate}$), and radiation losses (α_R), associated with the electromagnetic field radiated out of the device. α_R is completely negligible versus α_C and α_S. Therefore, the measured power loss differences are mainly due to the material electrical properties. PSi CPW electrical performances measured by various workers are summarized in Table 5.1.

Several authors clearly demonstrated the effect of the PSi thickness on RF performances (Welty et al. 1998; You et al. 2003; Contopanagos et al. 2008; Issa et al. 2011). For example, You and co-workers (2003) showed that, at 10 GHz, α is about 10.4 dB/mm. They worked on a CPW grown on top of a 0.01 Ω·cm bulk silicon. This value is falling down to about 0.7 dB/mm and 0.25 dB/mm with 20-μm and 70-μm thick PSi layers, respectively. This is still quite lossy, but the improvement over bulk silicon is significant. Contopanagos et al. (2008) came to the same conclusion estimating the power losses of copper CPW on various PSi thicknesses (see Figure 5.5).

Nam and Kwon (1998) also reported large improvements on the line electrical performances if the PSi is fully oxidized. At 10 GHz, α is about 0.3 dB/mm on 20-μm thick oxidized porous silicon (OPS) layer. Similar observations resulted from the Park and co-workers study (Park and Lee 2003).

Peterson and Drayton (2001) concluded that the device must be isolated from the PSi. On a 26-μm thick PSi layer, the attenuation of CPW with or without PECVD SiO_2 capping (0.48 μm) is 0.68 dB/mm or 0.83 dB/mm, respectively, at 10 GHz. Finally, Itotia and Drayton (2002) pointed out the influence of the porosity on the substrate losses. For a given PSi thickness, they demonstrated that high porosities are efficient for RF isolation.

TABLE 5.1   RF Electrical Characteristics of Various Coplanar Waveguides (CPW) on *p*-Type OPS or Mesoporous Silicon

Reference	Si Substrate Resistivity ($\Omega \cdot$cm)	Porosity (%)	PSi Thickness (μm)	Post Anodization Processes	Line Geometry (W/L/G) (μm)	Losses	F (GHz)
Nam and Kwon 1998	8–10	–	20	350°C, 30 min, dry O_2 1060°C, 3 min, wet O_2	100/2000/30	$\alpha = 0.3$ dB/mm	4
Welty et al. 1998	1–3	–	Bulk 1 7 15	300°C, 1 h, dry O_2	5/1000/2.5	$\alpha = 7.2$ dB/mm $\alpha = 4.2$ dB/mm $\alpha = 1.95$ dB/mm $\alpha = 1.6$ dB/mm	10
Peterson et al. 2001	14–21	56	26	350°C, 30 min, dry O_2 350°C, 30 min, dry O_2 + SiO_2	94/15000/53	$\alpha = 0.83$ dB/mm $\alpha = 0.68$ dB/mm	10 10
Itotia and Drayton 2002	10–25	51 85	34 20	350°C, 30 min, dry O_2	47/–/12	$\alpha = 0.27$ dB/mm $\alpha = 0.18$ dB/mm	10
You et al. 2003	0.01	–	Bulk 20 70	–	–	$\alpha = 10.4$ dB/mm $\alpha = 0.7$ dB/mm $\alpha = 0.25$ dB/mm	10
Park and Lee 2003	0.8–1.2	70–85	200	500°C, 1 h, wet O_2 1050°C, 2 min, dry O_2	15/30/2000	IL = 0.4 dB/mm	20
Contopanagos et al. 2008	1–10	–	25	300°C, 3 h, dry N_2 420°C, 1 h, dry N_2	80/35/5000	PL = 80% PL = 30%	10
Issa et al. 2011	0.005	–	150	300°C, 2 h, dry O_2 420°C, 1 h, dry O_2	180/200/20 180/2000/20 20/200/180 20/2000/180	$\alpha = 1$ dB/mm $\alpha = 1.2$ dB/mm	10
Capelle et al. 2014a	0.02	40–50	20 50 160	300°C, 1 h, dry N_2	70/500/20	$\alpha = 0.7$ dB/mm $\alpha = 0.34$ dB/mm $\alpha = 0.23$ dB/mm	20

Note: All the CPW are made of gold or aluminum. F is the loss measurement frequency and t_{PSi} is the PSi thickness.

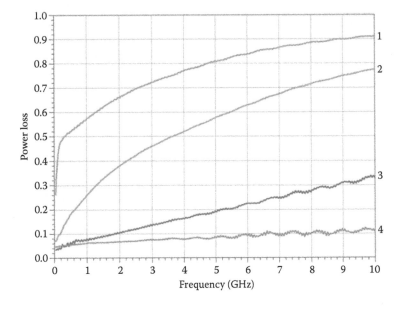

FIGURE 5.5   Measured normalized power loss of the CPW on three dies and a reference. (1) Bare Si die; (2) die with a 25-μm-thick PSi; (3) die with a 50-μm-thick porous Si; (4) reference minimum loss line (alumina substrate). (From Contopanagos H. et al., *Solid State Electron.* 52, 1730, 2008. Copyright 2008: Elsevier Ltd. With permission.)

5.3.3 INDUCTORS

Another substrate-sensitive device in systems on chip (SoC) is the planar inductor. Its main relevant frequency-dependent parameter is the quality factor Q, obtained when one port is shorted. Q can easily be derived from the admittance matrix [Y] by applying Equation 5.1, as defined below.

(5.1)
$$Q = -\frac{\mathrm{Im}\{Y_{11}\}}{\mathrm{Re}\{Y_{11}\}}$$

By using the transformations summarized in Pozar (2011), [Y] can be deduced from the S parameters. In case of network measurements, an additional three-step de-embedding procedure is applied in general to the raw results in order to remove the electrical effect of the feeding lines (Cho and Burk 1991).

Q-factor is a sensitive indicator of resistive and capacitive losses of the substrate, causing a degradation of the inductor electrical properties. Energy losses may result from capacitive couplings occurring between the metal lines and the substrate. In addition, the circulation of an alternative current in the inductor metal winding generates a varying magnetic field. Reverse currents, called eddy currents, are then induced in the substrate. They lead to Joule losses and are responsible for proximity effects in the inductor metal coil (Huo et al. 2006).

The self-resonance frequency (f_r) is also a typical parameter able to quantify substrate losses. f_r occurs when the inductive reactance of the device is equal to the parasitic capacitive reactance coupling the coil and the substrate.

The geometry of a spiral planar inductor can be defined by the strip width (W), the spacing between adjacent turns (S), the internal radius (R_{int}), the number of turns (N_t), the spacing to the surrounding coplanar ground plane (S_g), and the metal thickness (t) (see Figure 5.6).

Table 5.2 summarizes the electrical performances measured for spiral inductors performed on various PSi layers.

Many authors showed the improvement of inductor electrical characteristics using PSi layer (Yu et al. 2000; Chong et al. 2005; Contopanagos and Nassiopoulou 2007). The reduction of Si substrate parasitic effect results in higher Q-factor and resonant frequency f_r. Moreover, the coupling with the underlying silicon bulk also limits the device's performances (Royet et al. 2003). By increasing the PSi thickness, inductance and quality factor curves are shifted to high frequencies, clearly indicating the improvement of f_r, a parameter representative of inductive device application. Moreover, the increase of the quality factor maximum is also a significant demonstration of

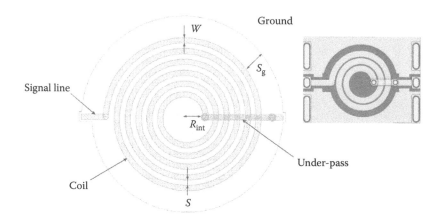

FIGURE 5.6 Picture and schematic of a coplanar spiral inductor. The typical geometrical parameters of a coplanar spiral inductor are the strip width (W), the spacing between adjacent turns (S), the internal radius (R_{int}), the number of turns (N_t), the spacing to the surrounding coplanar ground plane (S_g), and the metal thickness (t). In this configuration, an underpass is required to take the contacts in the center of the coil.

TABLE 5.2   Electrical Performances (Maximum Q-Factor and Resonance Frequency) of Coplanar Spiral Inductors on *p*-Type Porous Silicon

Reference	R_{Si} (Ω·cm)	t_{PS} (μm)	R_{int}	W	S	N_t	S_g	t_{metal}	L (nH)	Qmax at f (GHz)	f_r (GHz)
Nam and Kwon 1997	5–7	25	–	5	5	4.5	–	2.3 (Au)	6.3	13.3 (4.6)	13.8
Yu et al. 2000	5–10	Bulk	–	25	25	1.5	–	6 (Au)	1.3	3.4 (3)	>10
		1							1.3	3.9 (3)	
		2							1.3	4.8 (3)	
Kim et al. 2001	0.07	54	–	100	10	2	–	4 (Al)	7.6	3.5 (0.57)	2.95
		109								6 (1.29)	3.6
		200								14 (1.74)	3.7
Royet et al. 2003	0.015	150	150	100	50	3.5	–	3 (Cu)	11.4	6 (0.59)	–
		300								16 (1)	
Chong et al. 2005	0.01-0.05	Bulk	60	12	4	5.5	60	3 (TiAl)	4.52	4.2 (1.9)	5.4
		200	60	12	4	5.5	60		4.52	11.4 (4.86)	13.4
			60	6	4	5.5	–		4.27	10.2 (9)	18.4
			60	24	4	5.5	–		5.4	9.6 (3.1)	8.1
			60	12	4	5.5	–		0.89	16.5 (15.9)	>20
			60	12	4	5.5	–		28.52	8.7 (1.6)	3.4
			30	12	4	5.5	–		3.09	11 (8)	17.7
			120	12	4	5.5	–		7.91	11.9 (4)	8.6
Contopanagos and Nassiopoulou 2007	–	50	–	–	–	2	–	1 (Al)	3.7	15 (3.9)	4
								1 (Cu)		32 (3.9)	
Capelle et al. 2011	0.0015	100	80	30	10	1.5	50	1 (Al)	1	6 (4.5)	–
Billoué et al. 2011	30–50	Bulk	75	30	10	2.5	50	3 (Cu)	14	19.92 (3)	14
		5								21.2 (3.2)	14.2
		50								25.5 (4.2)	14.4
		100								28 (4.7)	15.2
Gautier et al. 2012	0.02	Bulk	180	30	10	5.5	50	3 (Cu)	20	7.7 (0.2)	1.2
		20								9 (0.3)	2.2
		50								10.7 (0.4)	2.7
		65								11 (0.47)	2.9
		80								11.5 (0.5)	3
Sarafis et al. 2013b	1–10	200	–	20	10	2.5	–	1.3 (Al)	3.3	10(5)	10

Note: All the dimensions are in μm. The inductance value *L* is determined for low frequencies ($f \rightarrow 0$). In the case of Capelle et al. (2011), Si had *n*-type. In the case of Nam and Kwon (1997), PSi was oxidized.

substrate loss reductions (Kim et al. 2001; Billoué et al. 2011; Gautier et al. 2012). This phenomenon is especially well pronounced for high-doped wafer (see Figure 5.7).

If mesoporous or macroporous silicon can be suitable for RF isolation applications, some studies have proven that morphologies that are more complex can also be used. Indeed, Capelle et al. (2011) presented an original structure that combines the mesoporous Si smooth surface and the macroporous silicon advantages. Indeed, this material is less stressed mechanically than mesoporous Si. Then, higher thicknesses could be reached, even for porosities of around 40%. The maximum Q-factor measured for integrated inductors built on mesoporous/macroporous silicon bilayers are drastically higher than those deposited on low-doped *n*-type silicon. These results have been confirmed for all the inductor designs, and an improvement from 50% up to 216% was observed regarding inductors' characteristics. Li and co-workers (2007b) proposed a variant that involves a post-CMOS selectively grown porous silicon (SGPS) technique in order to improve the Q-factor of integrated inductors. The devices were fabricated in a standard RF CMOS process, and PSi layers are selectively grown after processing from the backside of the silicon wafer. For a 2.1 nH inductor, a 105% increase (from 9.5 to 19.4) in Q-factor peak is achieved when only 1 μm of silicon remains under the device. Using this technique, PSi can also be localized only under passive elements (Li et al. 2007a) (see Figure 5.8).

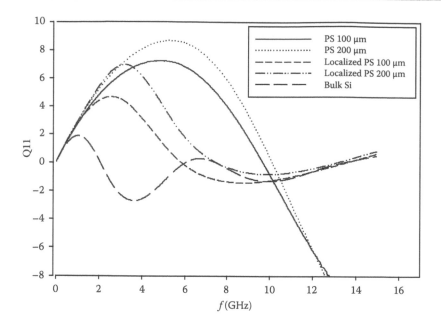

FIGURE 5.7   Evolution of inductors Q-factors (Q11) with the frequency (*f*) for different PSi layer thicknesses (100, 200 μm) in the case of full-sheet porous substrates and localized PSi (*W* = 10 μm, N_t = 5.5, R_{int} = 30 μm).

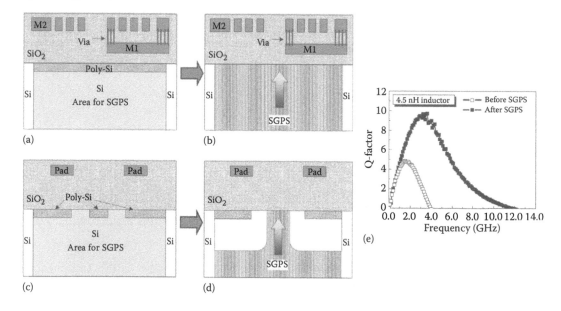

FIGURE 5.8   Schematic illustrations for RF integrated inductor and isolation structure with SGPS technique. The cross-section of inductor (a) before and (b) after the SGPS technique and the crosstalk isolation structure (c) before and (d) after the formation of SGPS trench (e). Extracted Q-factors values of a 4.5 nH integrated inductor as a function of frequency before and after SGPS technique. (From Li C. et al., *Solid. State Electron.* 51, 989, 2007. Copyright 2007: Elsevier Ltd. With permission.)

In the case of localized PSi regions, the effects of the surrounding bulk silicon were also clearly demonstrated by Capelle et al. (2012) for highly doped *p*-type silicon. The authors suggested a simplified electric model of an inductor integrated on localized PSi region. It makes them possible to illustrate all the parasitic elements brought by the hybrid substrate. Thus, they showed that highly doped silicon regions around and under localized PSi are responsible of the substrate loss increase, regarding full sheet PSi substrate. However, whatever the thickness of the PSi layer they demonstrated that the RF device performances, which are clearly reduced, are still acceptable compared to bulk silicon wafers (see Figure 5.7).

Finally, simulations can be performed, generally using finite element method, in order to quantify the losses and to design new inductor geometries (Contopanagos et al. 2007; Gautier et al. 2008). In particular, using ANSYS's 3D full-wave electromagnetic field simulator HFSS, Zhou et al. (2014) showed that PSi could significantly improve the performances of backside silicon-embedded inductors. Indeed, when a normal CMOS substrate resistivity of 10–30 Ω·cm is used, these power inductors are subject to significant capacitive effects. With the PS layer, the quality factor peak of a 1-mm² inductor is expected to increase from 6.2 to over 11 for a PS layer thickness of 40 μm. In parallel, the operating frequency is supposed to rise from around 70 MHz to over 200 MHz.

5.3.4 RF FUNCTIONS

Interconnects, inductors, and capacitors are the basic passive devices for developing integrated wireless communication systems. These elements also can be added to antennas whose performances can be improved by the use of PSi. Therefore, RF functions like filters have been investigated with PSi layer interface to improve electrical performances (Itotia and Drayton 2004).

Fang et al. (2005) showed performances of planar integrated low-pass filters made of two inductors and a metal-insulator-metal (MIM) on 30-μm thick OPS technology. The –3 dB S_{21} bandwidth was 2.925 GHz and the mid-band IL was 0.874 dB at 500 MHz.

Billoué et al. (2011) estimated the influence of different mesoporous silicon layer thicknesses on notch filter characteristics. IL is significantly reduced from 1.05 dB for bulk silicon (30–50 Ω·cm) to 0.4 dB for a 100-μm PSi layer. In addition, the filter bandwidth was improved by 250%.

Chong and Xie (2005) demonstrated also the drastic reduction of capacitive couplings under bond pads using PSi layers. In their work, they pointed out that f_r was raised up to 56.2 GHz by using a 200-μm thick PSi layer. Without any PSi, the measured f_r value was only 8.2 GHz on a p^+ substrate.

Itotia and Drayton (2005) investigated radiating element integrated on PSi substrate. An aperture coupled patch antenna, designed to resonate at 15 GHz, has been integrated on a 23-μm thick PSi layer. The measured return losses of the antenna showed a resonant frequency at 14.7 GHz with a bandwidth of 3 GHz. This demonstration showed that PSi could enable the excitation of radiating elements on silicon wafers. In other words, PSi can play the role of the feed substrate.

5.3.5 HYBRID SUBSTRATES FOR ACTIVE AND PASSIVE DEVICES

Developing RF monolithic microwave integrated circuits (MMIC) requires integrating miniature microwave functions on low-resistivity silicon substrate employing CMOS technology. However, passive devices suffer from high dielectric losses due to insufficient isolation of this substrate. Consequently, hybrid substrates mixing both PSi isolating regions and surrounding semiconductor material, where active devices can be integrated, is of great interest. Some authors have successfully demonstrated the feasibility of such substrates in the field of sensors (Barillaro et al. 2007) or optoelectronics devices (Hirschman et al. 1996), for instance. This target can be reached using HF-resistant material in order to localize the PSi formation during anodization such as fluoro-polymers (Defforge et al. 2012), photoresists (Hourdakis and Nassiopoulou 2014), nonstoichiometric silicon nitride (Kim et al. 2003a; Celigueta et al. 2005), Au/Cr layer (Lammel and Renaud 2000), or polysilicon on top of an SiO_2 layer (Kaltsas and Nassiopoulou 1997; Kouassi et al. 2012).

Capelle et al. (2014b) showed that hybrid PSi/silicon substrates for monolithic integration of RF common mode filters with electrostatic discharge (ESD) protection diodes can be performed. Moreover, their results showed that active devices are fully functional and the RF function performances were raised regarding to p^+-type silicon. Indeed, the cutoff frequency was increased by 8.8 GHz with the formation of an 80-μm thick PSi well below the inductors and the bump pads (Figure 5.9).

Moreover, the same group presented the integration of an electromagnetic interference (EMI) filter with ESD protection diodes (Capelle et al. 2014c). Measuring the S_{21} parameter, they showed

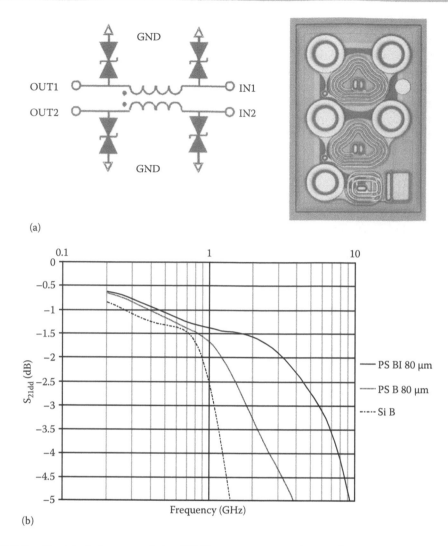

(a)

(b)

FIGURE 5.9   Electrical diagram of the ECMF component and optical microscope picture of the component integrated on a Si/PSi hybrid substrate (a). Evolution of the ECMF differential mode attenuation (S_{21dd}) versus frequency. The effect of the PSi regions geometries was investigated for a fixed PSi thickness (80 μm) (b). PS B refers to porous silicon below the bump pads and PS BI to porous silicon below the bump pads and inductors. (From Capelle M. et al., *Appl. Phys. Lett.* 104, 072104, 2014. Copyright 2014: AIP Publishing LCC. With permission.)

that the rejection frequency at 1 GHz was improved by 36 dB regarding to p^+-type silicon. All these performances clearly showed the potentialities of PSi for developing new RF devices for next communication standards (see Figure 5.10).

5.4 ELECTRICAL PROPERTIES OF PSi

This part is dedicated to the electrical properties of PSi. Two intrinsic properties of isolating materials must be considered: the dielectric permittivity and the electrical conductivity. These parameters must be lowered for isolation applications. In addition, in the case of RF devices, the losses due to the substrate can be correlated with the displacement of charges into the substrate and therefore depend on its dielectric and conduction properties at operation frequencies, generally around some GHz. In the following sections, we will show how, by varying PSi morphologies, we can modulate the dielectric permittivity and the electrical conductivity in DC and AC regimes.

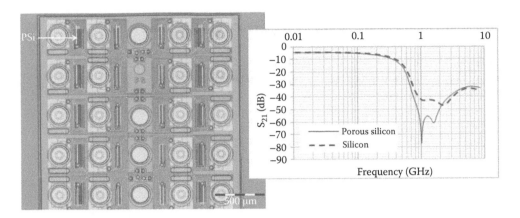

FIGURE 5.10 Optical microscope picture of passive EMI filters with active electrostatic discharge protections integrated on a Si/PSi hybrid substrate. The inductors are integrated on the PSi close to the diodes fabricated on a p+ silicon region. The evolution of the differential mode attenuation (S_{21}) with the frequency is also represented. (From Capelle M. et al., *Proceedings of International Conference on Porous Semiconductors–Science and Technology*, PSST-2012, March 9–14, 2014, p. 128. With permission.)

5.4.1 DIELECTRIC PERMITTIVITY

Dielectric permittivity is generally extracted from metal/PSi/c-Si/metal capacitance measurements applying reverse biases at high frequencies (above 10 kHz) (Adam et al. 1995). Effective dielectric permittivity (ε_{eff}) measurements from coplanar transmission lines also could be an interesting solution to calculate the intrinsic dielectric constant of PSi at higher frequencies (Sarafis et al. 2013a; Sarafis and Nassiopoulou 2014). In this case, high thicknesses must be employed to reduce the influence of the bulk under the PSi layer. Table 5.3 summarizes some results extracted from the literature. The measurements were performed on microporous or mesoporous silicon generally made from *p*-type substrates.

The dielectric properties of PSi are mainly governed by the porosity. Indeed, low dielectric constants can be measured for high porosity layers. Two types of approximation are generally used to model the behavior of ε_{PSi} as a function of the porosity. The first one is Vegard's law, which considerers PSi as a homogeneous mixture of silicon and air. Then, for a given porosity (P), PSi dielectric constant (ε_{PSi}) can be described by the following equation:

$$(5.2) \qquad \varepsilon_{PSi} = \varepsilon_{Si} + P(\varepsilon_{air} - \varepsilon_{Si})$$

where ε_{air} and ε_{Si} are the air and silicon relative dielectric constant equal to 1 and 11.7, respectively.

This equation successfully described the results of Zimin and Komarov (1996). In this case, porosities from 30 to 68% were considered. Kim et al. (2003b) also used this approximation to model the evolution of ε_{PSi} as a function of the porosity.

Cox (1997) summarized many results in their work and it seems that for porosities higher than 60%, Vegard's law overvalues ε_{PSi}. In this case, the Bruggeman approximation could be more accurate (Wang and Pan 2008). If we consider pores as spheres, the following model can be deduced (Badoz et al. 1993):

$$(5.3) \qquad P\frac{\varepsilon_{air} - \varepsilon_{PS}}{\varepsilon_{air} + 2\varepsilon_{PS}} + (1 - P)\frac{\varepsilon_{Si} - \varepsilon_{PS}}{\varepsilon_{Si} + 2\varepsilon_{PS}} = 0$$

Astrova and Tolmachev (2000) also applied this model in a 3-phase system (Si/air/SiO$_2$) in order to evaluate the oxidation influence on ε_{PS} value. Indeed, the impact of ambient air aging or high temperature thermal oxidation must also be considered. In the case of high porosity layers (>70%), the effective dielectric constant is slightly impacted by the oxidation fraction. For moderate porosity layers, between 30% and 50%, the improvement is more pronounced (Pan et al. 2005).

TABLE 5.3   Measured Dielectric Constants from Various PSi Layers Reported in the Literature

Reference	Silicon Type, Orientation, Resistivity ($\Omega \cdot cm$)	Electrolyte	Duration (min)	J (mA/cm^2)	Porosity (%)	PS Thickness (μm)	ε_{PS}
Badoz et al. 1993	p, (100), 1	HF(50%):EtOH	–	–	75	20 to 80	2.25 ± 0.25
			–	–	85		1.75 ± 0.25
Adam et al. 1995	p, (100), 5.1–6.9	HF(50%):EtOH	–	–	45	3	5.5
			–	–	60		3.5
			–	–	68		3
			–	–	78		1.7
Ben-Chorin et al. 1995	p, (100), 5	HF(50%):EtOH (1:1)	–	30	75	–	2
Balagurov et al. 2001	p, (100), 10	HF(50%):EtOH	–	–	75	6	3
Axelrod et al. 2002	p, (100), 10–30	HF(50%):EtOH (1:2)	19	35	72	30	2.6
Adamyan et al. 2007	p, (100), 5	HF(50%):EtOH (1:2)	–	–	57	0.5	5.64
	p, (111), 10				33	0.7	8.2
Adam et al. 1995	p, (100), 0.015	HF(50%):EtOH	–	–	60	9	5.5
Peng et al. 1996	p, (100), 5–10	HF(50%):EtOH (1:1)	–	20	80	< 10	3.2
Zimin and Komarov 1996	p, n, (100), 0.01 to 7.5	HF(50%):EtOH (1:1)	10–60	50–60	30 to 68	55 to 190	8.6 to 4.2
Kim et al. 2003b	p, (100), 0.01	HF(50%)	–	50	24	50	9
		HF(50%):EtOH (2:5)	–	100	78		3
Sarafis et al. 2013a	p, (–), 0.002	HF(50%):EtOH (2:3)	200	20	84	140	2.2
	p, (–), 0.005		180	20	70	168	3.7
Menard et al. 2014	p, (111), 6–12	HF(50%):H$_2$O:Acetic acid (4.63:1.45:2.14)	20	50	66	35	4.5 ± 0.5
	p, (111), 0.08–0.12		10		47	22	5.2 ± 1
	p, (111), 0.01–0.015		10		36	21	5.9 ± 0.2

5.4.2 DC CONDUCTIVITY

Many studies of electrical transport in PSi deal with high porosity (60–80%) luminescent silicon. With such material, it has been found that DC electrical conductivity (σ_{DC}) is very low. The measured σ_{DC} generally varies between 10^{-14} and 10^{-8} S/cm (Fejfar et al. 1995; Lee et al. 1996). Most of the time, σ_{DC} is extracted from I-V measurements. For a complete review on the electrical properties extraction and conductivity models, one can see Menard et al. (2012). In the case of mesoporous silicon, with low porosities (30–50%), values from 10^{-4} to 10^{-7} S/cm have been reported (Lubianiker et al. 1996). This value is drastically reduced if an annealing is performed (Balagurov et al. 2000).

Whatever the morphology, the electrical behavior is strongly modulated by the porosity (Bouaïcha et al. 2006; Khardani et al. 2006). For instance, one can see the evolution of the PSi resistivity as a function of the porosity for various morphologies obtained from three different p-type wafer resistivities (Figure 5.11) (Menard et al. 2014).

Moreover, the electrical conductivity of mesoporous silicon is thermally activated with activation energies (E_a) generally between and 30 meV and 80 meV for temperatures higher than 300 K (Fejfar et al. 1995; Diligenti et al. 1996; Khirouni et al. 1997; Pan et al. 2003; Islam et al. 2009). In the case of microporous silicon, E_a can reach 1.1 eV (Lubianiker and Balberg 1997).

5.4.3 AC CONDUCTIVITY

The AC conductivity (σ_{AC}) increases with the frequency. In addition, this behavior is similar for low or high porosity layers. For frequencies above 10 kHz, σ_{AC} is proportional to the frequency (Ben-Chorin et al. 1995). This is a consequence of hopping transport of charge carriers. In this case, the electrons perform a random walk into a fractal Si network in the presence of the electric field. Moreover, some surface effects also can be visible for frequencies below 1 kHz (Parkhutik 1996). In

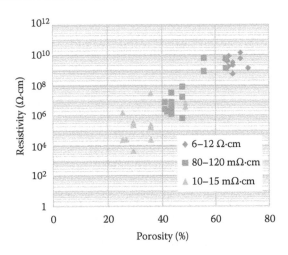

FIGURE 5.11 Resistivity (ρ_{ps}) versus porosity at room temperature for different PSi morphologies obtained from *p*-type wafers with resistivities from 6–10 to 0.010–0.015 Ω·cm. (From Menard S. et al., *Proceedings of International Conference on Porous Semiconductors–Science and Technology,* PSST-2012, March 9–14, 2014, p. 799. With permission.)

RF regimes, beyond 1 GHz, the use of transmission lines (CPW) can lead to loss tangent (tan δ) calculations. Then, PSi dielectric constant or conductivity can be deduced. From CPW on 50-μm thick PSi layers, Kim et al. (2003b) measured a loss tangent varying from 3×10^{-4} to 10^{-2} S/cm at 10 GHz and porosities included in the interval (24–78%). For the same PSi layer thickness, Contopanagos and co-workers (2008) have also shown an evolution of the silicon (6.85 Ω·cm)/mesoporous silicon tanδ ratio from 200 to 20. The frequency was varied from 1 to 10 GHz. Sarafis et al. (2013a) used the same kind of measurements in order to extract AC conductivities of mesoporous silicon layers with porosities between 70 and 80%. tanδ was estimated to be around 5×10^{-4} S/cm at 10 GHz. Finally, Capelle (2013) performed RF measurements on aluminum microstrips (MS). Between 12 and 15 GHz, the resistivity is around 1 kΩ·cm for 40/50% porosity mesoporous silicon, with a high thickness of 550 μm. These two results from Capelle et al. (2013) and Sarafis et al. (2013a) seem to be in good accordance taking into account the porosity differences.

5.5 CONCLUSION

In this chapter, we have shown the main applications of electrical isolating PSi in microelectronic devices. In fact, the electrical resistivity of PSi, described in the last part of this chapter, is greatly higher than bulk material. It could be modulated over 7 decades varying the porosity from 20 to 70% at room temperature. Moreover, the dielectric permittivity of PSi is also greatly lower than the Si one. Values close to 4 can be reached for porosities between 60 and 70%. Considering these intrinsic PSi performances and knowing that this material can also be easily oxidized, this material can be used for the realization of SOI substrates or in active device periphery. Second, PSi layers can be used as RF substrates. Indeed, numerous studies showed very promising results in terms of loss reduction or quality factors improvement of RF devices using such a solution. Finally, PSi still has the particularity, upon all "classical substrates," such as glass or high resistivity silicon, to provide localized isolation regions. Consequently, high-performance mixed integrated circuits, involving active devices as well as passive components, can be considered.

REFERENCES

Adam, M., Horvath, Z.J., Barsony, I., Szolgyemy, L., Vazsonyi, E., and Tuyen, V.V. (1995). Investigation of electrical properties of Au/porous Si/Si structures. *Thin Solid Films* **255**, 266–268.
Adamyan, A., Adamian, Z., and Aroutiounian, V. (2007). Capacitance method for determination of basic parameters of porous silicon. *Phys. E* **38**, 164–167.

Amato, G., Boarino, L., Brunetto, N., and Turnaturi, M. (1997). Deep "cold" junctions by porous silicon impregnation. *Thin Solid Films* **297**(1), 321–324.

Astrova, E.V. and Tolmachev, V.A. (2000). Effective refractive index and composition of oxidized porous silicon films. *Mater. Sci. Eng. B* **69**, 142–148.

Astrova, E., Voronkov, V., Grekhov, I., Nashchekin, A., and Tkachencko, A. (1999). Deep diffusion doping of macroporous silicon. *Tech. Phys. Lett., Springer* **25**, 958–961.

Axelrod, E., Givant, A., Shappir, J., Feldman, Y., and Sa'ar, A. (2002). Dielectric relaxation and porosity determination of porous silicon. *J. Non-Cryst. Solids* **305**, 235–242.

Ayvazyan, G.E. (1999). Anisotropic warpage of wafers with anodized porous silicon layers. *Phys. Stat. Sol. (a)* **175**(2), R7–R8.

Badoz, P.A., Bansahel, D., Bomchil, G., Ferrieu, F., Halimaoui, A., Perret, P., Regolini, J.L., Sagnes, I., and Vincent, G. (1993). Characterisation of porous silicon: Structural, optical and electrical properties. *Proc. Mat. Res. Soc.* **283**, 97–108.

Balagurov, L.A., Yarkin, D.G., and Petrova, E.A. (2000). Electronic transport in porous silicon of low porosity made on a p^+ substrate. *Mater. Sci. Eng. B* **69–70**, 127–131.

Balagurov, L.A., Bayliss, S.C., Kasatochkin, V.S., Petrova, E.A., Unal, B., and Yarkin, D.G. (2001). Transport of carriers in metal/porous silicon/c-Si device structures based on oxidized porous silicon. *J. Appl. Phys.* **90**, 4543–4548.

Barillaro, G., Bruschi, P., Pieri, F., and Strambini, L.M. (2007). CMOS-compatible fabrication of porous silicon gas sensors and their readout electronics on the same chip. *Phys. Stat. Sol. (a)* **204**, 1423–1428.

Barla, K., Bomchil, G., Herino, R., Monroy, A., and Gris, Y. (1986). Characteristics of SOI CMOS circuits made in N/N+/N oxidised porous silicon structures. *Electron. Lett.* **22**(24), 1291–1293.

Bean, K.E. and Runyan, W.R. (1977). Dielectric isolation: Comprehensive, current and future. *J. Electrochem. Soc.* **124**, 5C–12C.

Ben-Chorin, M., Möller, F., Koch, F., Schirmacher, W., and Eberhard, M. (1995). Hopping transport on a fractal: AC conductivity of porous silicon. *Phys. Rev. B* **51**, 2199–2213.

Billoué, J., Gautier, G., and Ventura, L. (2011). Integration of RF inductors and filters on mesoporous silicon isolation layers. *Phys. Stat. Sol. (a)* **208**(6), 1449–1452.

Boehringer, M., Artmann, H., and Witt, K. (2012). Porous silicon in a semiconductor manufacturing environment. *J. Microelectromech. Syst.* **21**(6), 1375–1381.

Bomchil, G., Halimaoui, A., and Herino, R. (1988). Porous silicon: The material and its applications to SOI technologies. *Microelectron. Eng.* **8**(3), 293–310.

Bouaïcha, M., Khardani, M., and Bessaïs, B. (2006). Correlation of electrical conductivity and photoluminescence in nanoporous silicon. *Mater. Sci. Eng.* **C26**, 486–489.

Capelle, M. (2013). Intégration monolithique de composants bipolaires et de circuits mixtes Si/Si poreux. PhD thesis. Université François Rabelais de Tours, Tours, France.

Capelle, M., Billoué, J., Poveda, P., and Gautier, G. (2011). N-type porous silicon substrates for integrated RF inductors. *IEEE Trans. Electron. Dev.* **58**(11), 4111–4114.

Capelle, M., Billoué, J., Poveda, P., and Gautier, G. (2012). RF performances of inductors integrated on localised p-type porous silicon regions. *Nanoscale Res. Lett.* **7**, 523–531.

Capelle, M., Billoué, J., Poveda, P., and Gautier, G. (2014a). Study of porous silicon substrates for the monolithic integration of radiofrequency circuits. *Int. J. Microwave Wireless Technol.* **6**, 39–43.

Capelle, M., Billoué, J., Concord, J., Poveda, P., and Gautier, G. (2014b). Monolithic integration of common noise filter with ESD protection on silicon/porous silicon hybrid substrate. *Appl. Phys. Lett.* **104**, 072104.

Capelle, M., Billoué, J., Poveda, P., and Gautier, G. (2014c). Monolithic integration of EMI filters with ESD protection on silicon/porous silicon hybrid substrates. In: *Proceedings of 9th International Conference on Porous Semiconductors–Science and Technology*, March 9–14, Alicante, Spain, pp. 128–129.

Capelle, M., Billoué, J., Poveda, P., and Gautier, G. (2014d). Evaluation of the strain in mesoporous silicon substrates for RF circuit integration. In: *Proceedings of 9th International Conference on Porous Semiconductors–Science and Technology*, March 9–14, Alicante, Spain, pp. 350–351.

Celigueta, I., Arana, S., Gracia, F.J., and Castaiio, E. (2005). Selective formation of porous silicon using silicon nitride and SU-8 masks for electroluminescence applications. In: *Proceedings of the IEEE Spanish Conference on Electron Devices*, Feb. 2–4, Tarragona, Spain, pp. 331–334.

Cho, H. and Burk, D. (1991). A three step method for the de-embedding of high frequency s-parameter measurements. *IEEE Trans. Electron. Dev.* **38**, 1371–1375.

Chong, K. and Xie, Y. (2005). Low capacitance and high isolation bond pas for high-frequency RFICs. *IEEE Electron Dev. Lett.* **26**, 746–748.

Chong, K., Xie, Y., Yu, K., Huang, D., and Chang, F. (2005). High performance inductors integrated on porous silicon. *IEEE Electron Dev. Lett.* **26**, 93–95.

Contopanagos, H. and Nassiopoulou, A.G. (2007). Integrated inductors on porous silicon. *Phys. Status Solidi (a)* **204**(5), 1454–1458.

Contopanagos, H., Zacharatos, F., and Nassiopoulou, A. G. (2008). RF characterization and isolation properties of mesoporous Si by on-chip coplanar waveguide measurements. *Solid State Electron.* **52**, 1730–1734.

Cox, T.I. (1997). Porous silicon layer capacitance. In: L. Canham (Ed.) *Properties of Porous Silicon.* INSPEC, London, pp. 185–191.

Defforge, T., Capelle, M., Tran Van, F., and Gautier, G. (2012). Plasma deposited fluoropolymer film mask for local porous silicon formation. *Nanoscale Res. Lett.* **7**, 344–350.

Déhu, P., Senes, A., and Miserey, F. (1995). P wells made of porous silicon for power devices: Determination of the formation steps. *Thin Solid Films* **255**, 321–324.

Desplobain, S., Ventura, L., Defforge, T., and Gautier, G. (2014). Lateral homogeneity of porous silicon film on silicon wafer. In: *Proceedings of International Conference on Porous Semiconductors–Science and Technology*, March 9–14, Alicante-Benidorm, Spain, pp. 404–405.

Diligenti, A., Nannini, A., Pennelli, G., Pellegrini, V., Fuso, F., and Allegrini, M. (1996). Current transport in free-standing porous silicon. *Appl. Phys. Lett.* **68**, 687–689.

Earwaker, L.G., Briggs, M.C., Nasir, M.I., Farr, J.P.G., and Keen, J. M. (1991). Analysis of porous silicon silicon-on-insulator materials. *Nucl. Instrum. Meth. Phys. Res., Sect. B* **56**, 855–859.

Fang, J., Liu, Z.W., Chen, Z.M., Liu, L.T., and Li, Z.J. (2005). Realization of an integrated planar LC low-pass filter with modified surface micromachining technology. In: *Proceedings of IEEE Conference on Electron Devices and Solid-State Circuits*, Dec. 19–21, Hong Kong, pp. 729–732.

Fejfar, A., Pelant, I., Sipeck, E., Kocka, J., Juska, G., Matsumoto, T., and Kanemitsu, Y. (1995). Transport study of self-supporting porous silicon. *App. Phys. Lett.* **66**, 1098–1100.

Frye, R.C. and Leamy, H.J. (1983). Method of forming dielectrically isolated silicon semiconductor materials utilizing porous silicon formation. US Patent 4380865.

Gautier, G., Ventura, L., Jérisian, R., Kouassi, S., Leborgne, C., Morillon, B., and Roy, M. (2006). Deep trench etching combining aluminum thermo-migration and electrochemical silicon dissolution. *Appl. Phys. Lett.* **88**, 212501.

Gautier, G., Leduc, P., Semai, J., and Ventura, L. (2008). Thick microporous silicon isolation layers for integrated RF inductors. *Phys. Stat. Sol. (c)* **5**(12), 3667–3670.

Gautier, G., Capelle, M., Billoué, J., Defforge, T., Leduc, P., and Poveda, P. (2012). Porous silicon: Application to RF microelectronic devices. In: *Proceedings of WOCSDICE/EXMATEC conference*, May 28–June 1, Island of Porquerolles, France, pp. 1–4.

Hirschman, K., Tsybeskov, L., Duttagupta, S., and Fauchet, P. (1996). Silicon-based visible light-emitting devices integrated into microelectronic circuits. *Nature* **384**, 338–341.

Holmstrom, R.P. and Chi, J.Y. (1983). Complete dielectric isolation by highly selective and self-stopping formation of oxidized porous silicon. *Appl. Phys. Lett.* **42**(4), 386–388.

Hourdakis, E. and Nassiopoulou, A.G. (2014). Single photoresist masking for local porous Si formation. *J. Micromech. Microeng.* **24**(11), 117002.

Huo, X., Chan, P.C.H., Chen, K.J., and Luong, H.C. (2006). A physical model for on-chip spiral inductors with accurate substrate modeling. *IEEE Trans. Electron Dev.* **53**, 2942–2949.

Imai, K. and Unno, H. (1984). FIPOS (full isolation by porous oxidized silicon) technology and its application to LSIs. *IEEE Trans. Electron Dev.* **ED-31**, 297–302.

Islam, M.N., Ram, S.K., and Kumar, S. (2009). Mott and Efros Shklovskii hopping conductions in porous silicon nanostructures. *Phys. E* **41**, 1025–1028.

Issa, H., Ferrari, P., Hourdakis, E., and Nassiopoulou, A. (2011). On-chip high-performance millimeter-wave transmission lines on locally grown porous silicon areas electron devices. *IEEE Transactions on, IEEE* **58**, 3720–3724.

Itotia, I.K. and Drayton, R.F. (2002). Porosity effects on coplanar waveguide porous silicon interconnect. In: *Proceedings of IEEE Meeting of Microwave Theory and Techniques Society, MTT-S*, Jun. 2–7, Seattle, WA, pp. 681–684.

Itotia, I.K. and Drayton, R.F. (2004). Loss reduction methods for planar circuit designs on lossy substrates. In: *Proceedings of IEEE APS International Symposium*, Jun. 20–25, Minneapolis, pp. 1447–1450.

Itotia, I.K. and Drayton, R.F. (2005). Aperture coupled patch antenna chip performance on lossy silicon substrates. In: *Proceedings of IEEE APS International Symposium*, July 3–10, Washington, DC, pp. 377–380.

Kaltsas, G. and Nassiopoulou, A. (1997). Bulk silicon micromachining using porous silicon sacrificial layers. *Microelectron. Eng.* **35**, 397–400.

Khardani, M., Bouaïcha, M., Dimassi, W., Zribi, M., Aouida, S., and Bessaïs, B. (2006). Electrical conductivity of free-standing mesoporous silicon thin films. *Thin Solid Films* **194**, 243–245.

Khirouni, K., Bourgouin, J.C., Borgi, K., Maaref, H., Deresmes, D., and Stievenard, D. (1997). DC current-voltage characteristics and admittance spectroscopy of an Al porous Si barrier. In: *Proceedings of MRS Fall Meeting*, December 2–6, 1996, Boston, pp. 619–623.

Kim, H.S., Zheng, D., Becker, A.J., and Xie, Y.H. (2001). Spiral inductors on Si p/p+ substrates with resonant frequency of 20 GHz. *IEEE Electron Dev. Lett.* **22**, 275–277.

Kim, H., Chong, K., and Xie, Y. (2003a). Study of the cross-sectional profile in selective formation of porous silicon. *Appl. Phys. Lett.* **83,** 2710–271.

Kim, H.S., Xie, Y.H., Devicentis, M., Itoh, T., and Jenkins, K.A. (2003b). Unoxidized porous Si as an isolation material for mixed-signal integrated circuit applications. *J. Appl. Phys.* **93**, 4226–4231.

Konaka, S., Tabe, M., and Sakai, T. (1982). A new silicon-on-insulator structure using a silicon molecular beam epitaxial growth on porous silicon. *App. Phys. Lett.* **41**(1), 86–88.

Kouassi, S., Gautier, G., Thery, J., Desplobain, S., Borella, M., Ventura, L., and Laurent, J.-Y. (2012). Proton exchange membrane micro fuel cells on 3D porous silicon gas diffusion layers. *J. Power Sources* **216**, 15–21.

Lammel, G. and Renaud, P. (2000). Free-standing, mobile 3D porous silicon microstructures. *Sens. Actuators A* **85**, 356–360.

Lee, W.H., Choochon, L., and Jang, J. (1996). Quantum size effects on the conductivity in porous silicon. *J. Non-Cryst. Solids* **198–200**, 911–914.

Li, C., Liao, H., Yang, L., and Huang, R. (2007a). High-performance integrated inductor and effective cross-talk isolation using post-CMOS selective grown porous silicon (SGPS) technique for RFIC applications. *Solid-State Electron.* **51**(6), 989–994.

Li, C., Liao, H., Wang, C., Yin, J., Huang, R., and Wang, Y. (2007b). High-Q integrated inductor using post-CMOS selectively grown porous silicon (SGPS) technique for RFIC applications. *IEEE Electron. Dev. Lett.* **28**, 763–767.

Lubianiker, Y. and Balberg, I. (1997). Two Meyer-Nedel rules in porous silicon. *Phys. Rev. Lett.* **78**, 2433–2436.

Lubianiker, Y., Balberg, I., Partee, J., and Shinar, J. (1996). Porous silicon as a near-ideal disordered semiconductor. *J. Non-Cryst. Solids* **198**, 949–952.

Manotas, S., Agulló-Rueda, F., Moreno, J.D., Ben-Hander, F., and Martınez-Duart, J.M. (2001). Lattice-mismatch induced-stress in porous silicon films. *Thin Solid Films* **401**(1), 306–309.

Menard, S., Fèvre, A., Valente, D., Billoué, J., and Gautier, G. (2012). Non oxidized porous silicon based power AC switch peripheries. *Nanoscale Res. Lett.* **7**, 566–576.

Menard, S., Fèvre, A., Capelle, M., Defforge, T., Billloué, J., and Gautier, G. (2014). Dielectric behavior of porous silicon grown from *p*-type substrates. In: *Proceedings of International Conference on Porous Semiconductors–Science and Technology, PSST-2012*, March 9–14, Alicante-Benidorm, Spain, pp. 799–800.

Nakajima, S. and Kato, K. (1977). An isolation technique for high speed bipolar integrated circuits. *Rev. Electr. Commun. Lab.* **25**, 1039–1051.

Nam, C.M. and Kwon, Y.S. (1997). High-performance planar inductor on thick oxidized porous silicon (OPS) substrate. *IEEE Microwave Guided Wave Lett.* **7**(8), 236–238.

Nam, C., and Kwon, Y. (1998). Coplanar waveguides on silicon substrate with thick oxidized porous silicon (OPS) layer. *IEEE Microwave Guided Wave Lett.* **8**(11), 369–371.

Oules, C., Halimaoui, A., Regolini, J.L., Perio, A., and Bomchil, G. (1992). Silicon on insulator structures obtained by epitaxial growth of silicon over porous silicon. *J. Electrochem. Soc.* **139**, 3595–3599.

Pan, L.K., Huang, H.T., and Sun, C.Q. (2003). Dielectric relaxation and transition of porous silicon. *J. Appl. Phys.* **94**, 2695–2699.

Pan, L.K., Chang, Q.S., and Li, C.M. (2005). Estimating the extent of surface oxidation by measuring the porosity dependent dielectrics of oxygenated porous silicon. *App. Surf. Sc.* **240**, 19–23.

Park, J.Y. and Lee, J.H. (2003). Characterization of 10 μm thick porous silicon dioxide obtained by complex oxidation process for RF application. *Mater. Chem. Phys.* **82**(1), 134–139.

Parkhutik, V.P. (1996). Residual electrolyte as a factor influencing the electrical properties of porous silicon. *Thin Solid Films* **276**, 195–199.

Peng, C., Hirschman, K.D., and Fauchet, P.M. (1996). Carrier transport in porous silicon light-emitting. *J. Appl. Phys.* **80**, 295–300.

Peterson, R.L. and Drayton, R.F. (2001). Dielectric properties of oxidized porous silicon in a low resistivity substrate. In: *Proceedings of IEEE Conference MTT-S*, May 20–24, Phoenix, AZ, pp. 767–700.

Pogge, H.B. and Poponiak, M.R. (1975). Method of fabricating semiconductor device embodying dielectric isolation. US Patent 3919060.

Pozar, D. M. (2011). *Microwave Engineering.* 4th ed. John Wiley & Sons, New York.

Royet, A.S., Cuchet, R., Pellissier, D., and Ancey, P. (2003) On the investigation of spiral inductors processed on Si substrates with thick porous Si layers. In: *Proceedings of the 33rd Conference on European Solid-State Device Research*, Sept. 16–18, Estoril, Portugal, pp. 111–114.

Sanda, H., McVittie, J., Koto, M., Yamagata, K., Yonehara, T., and Nishi, Y. (2005). Fabrication and characterization of CMOSFETs on porous silicon for novel device layer transfer. In: *Proceedings of the IEEE Electron Devices Meeting*, Dec. 5, Washington, DC, pp. 679–682.

Sarafis, P. and Nassiopoulou, A. (2014). Dielectric properties of porous silicon for use as a substrate for the on-chip integration of millimeter-wave devices in the frequency range 140 to 210 GHz. *Nanoscale Res. Lett.* **9**, 418–426.

Sarafis, P., Hourdakis, E., and Nassiopoulou, A.G. (2013a). Dielectric permittivity of porous Si for use as substrate material in Si-integrated RF devices. *IEEE Trans. Electron Dev.* **60**(4), 1436–1443.

Sarafis, P., Hourdakis, E., Nassiopoulou, A.G., Roda Neve, C., Ben Ali, K., and Raskin, J.P. (2013b). Advanced Si-based substrates for RF passive integration: Comparison between local porous Si layer technology and trap-rich high resistivity Si. *Solid-State Electron.* **87**, 27–33.

Sato, N., Sakaguchi, K., Yamagata, K., Fujiyama, Y., and Yonehara, T. (1995). Epitaxial growth on porous Si for a new bond and etchback Silicon-on-Insulator. *J. Electrochem. Soc.* **142**(9), 3116–3122.

Thomas, N.J., Davis, J.R., Keen, J.M. et al. (1989). High-performance thin-film silicon-on-insulator CMOS transistors in porous anodized silicon. *IEEE Electron Dev. Lett.* **10**(3), 129–131.

Tsao, S., Guilinger, T.R., Kelly, M.J., Kaushik, V.S., and Datye, A.K. (1991). Porous silicon formation in N–/N+/N– doped structures. *J. Electrochem. Soc.* **138**(6), 1739–1743.

Veeramachaneni, B., Winans, J.D., Hu, S., Kawamura, D., Fauchet, P.M., Witt, K., and Hirschman, K.D. (2011). A novel technique for localized formation of SOI active regions. *Phys. Stat. Sol. (c)* **8**(6), 1865–1868.

Vescan, L., Bomchil, G., Halimaoui, A., and Perio, A. (1988). Low-pressure vapor-phase of silicon on porous silicon. *Materials Lett.* **7**, 94–98.

Wang, M. and Pan, N. (2008). Predictions of effective physical properties of complex multiphase materials. *Mater. Sci. Eng. R* **63**, 1–30.

Watanabe, Y. and Sakai, T. (1972). Semiconductor device and method of producing the same. US Patent 3640806.

Watanabe, Y., Arita, Y., Yokohama, T., and Igarashi, Y. (1975). Formation and properties of porous silicon and its application. *J. Electrochem. Soc.* **122**, 1351–1355.

Welty, R.J., Park, S.H., Asbek, P.M., Dancil, K.S., and Sailor, M.J. (1998). Porous silicon technology for RF integrated circuit applications. In: *Proceedings of IEEE Topical Meeting on Silicon Monolithic Integrated Circuits in RF Systems*, Sept. 18, Ann Arbor, MI, pp. 160–163.

Yonehara, T. and Sakaguchi, K. (2001). EltranR: Novel SOI wafer technology. *JSAP Intern.* **4**, 10–16.

Yonehara, T., Sakaguchi, K., and Sato, N. (1994). Epitaxial layer transfer by bond and etch back of porous Si. *App. Phys. Lett.* **64**(16), 2108–2110.

You, S.Z., Long, Y.F., Xu, Y.S. et al. (2003). Fabrication and characterization of thick porous silicon layers for rf circuits. *Sens. Actuators A* **108**, 117–120.

Yu, M., Chan, Y., Laih, L., and Hong, J. (2000). Improved microwave performance of spiral inductors on Si Substrates by chemically anodizing a porous silicon layer. *Microwave Opt. Technol. Lett.* **26**, 232–234.

Zheng, D.W., Cui, Q., Huang, Y.P. et al. (1998). A low temperature Silicon-on-Insulator fabrication process using Si MBE on double-layer porous silicon. *J. Electrochem. Soc.* **145**, 1668–1672.

Zhou, J., Whu, R., Billoué, J., and Gautier, G. (2014). Backside silicon-embedded inductor using porous silicon layer for substrate effect suppression. In: *Proceedings of IEEE Conference on Electron Device and Solid State Circuits*, June 18–20, Chengdu, China.

Zimin, S.P. and Komarov, E.P. (1996). Capacitance of structures with a thick layer of porous silicon. *Tech. Phys. Lett.* **22**, 808–809.

Surface Micromachining (Sacrificial Layer) and Its Applications in Electronic Devices

Alexey Ivanov and Ulrich Mescheder

6

CONTENTS

6.1   Introduction   130

6.2   Porous Silicon for CMOS-Compatible Processing in Microelectronics
        Production Line   130

6.3   Freestanding Mechanical Elements   131

        6.3.1   Cantilevers   131

        6.3.2   Microphones   133

        6.3.3   Thermal Emitters   135

        6.3.4   Free Elements   135

6.4   Cavities and Wafer-Through Holes   136

6.5   Layer-Transfer Process   137

6.6   Electrical Isolation   139

References   140

6.1 INTRODUCTION

Sacrificial layers are used in microelectronics and microsystems technology to form freestanding functional elements, which are isolated thermally, electrically, or mechanically from substrate. Surface micromachining (SMM) is an approach to fabricate microelectromechanical systems (MEMS). SMM is compatible with CMOS technology, and thus allows the monolithic integration of microelectronics and MEMS. In SMM, typically silicon oxide, photoresist, or aluminum are used as a sacrificial layer to form void space between substrate and freestanding functional structures on top of it (Mescheder 2004; Korvink and Paul 2006; Zhang 2013). The fabrication is usually performed on the top surface of a silicon substrate and provides only thin separation of functional elements from the substrate. Another approach to define freestanding functional elements is bulk micromachining, where fabrication processes from the backside of a substrate and full-wafer-depth processes are used. Typically, in bulk micromachining the substrate is thinned down from the backside after formation of functional elements on top of the substrate, for example, with anisotropic etch process, to free the functional elements. Whereas for SMM sticking of the functional elements can occur after the sacrificial layer etching step (during rinsing due to capillary forces), bulk micromachining is not CMOS compatible, especially when using KOH for anisotropic etching of the substrate from the backside of the wafer. Additionally, the need to conduct processes on the wafer backside requires extra process steps and increases production costs. Another approach to define freestanding structures is SOI-technology, where the buried oxide can be used as a sacrificial layer to disconnect the device layer from the handle wafer or can be used as a very defined etch stop when etching from the backside, thus resulting in freestanding structures with well defined thickness (i.e., thickness of device layer).

Introduction of porous silicon (PSi) as a sacrificial layer with superior properties, such as high reactivity, allowed fabricating devices in surface micromachining without depth limitations (French et al. 1997). However, integration of sacrificial PSi in a fabrication process requires consideration of some specific issues concerning formation of porous volume, its drying and removal. Special care during electrochemical formation of PSi is required if this porous layer is formed under already existing structures. Because of hydrogen evolution during divalent dissolution of silicon in pore formation regime, it might be advisable to slow down the process in order to protect fragile functional structures from damage. Additionally, in order to fully undercut the functional elements, good process control is necessary. Proper selection of a localization method or etch stop technique, for example, with current focusing (Zeitschel et al. 1999), is necessary. During removal of PSi in diluted KOH solution or similar etchant, it is also necessary to keep the etch rate slow to avoid damage of the functional elements by hydrogen bubbles. There are also alternative processes for PSi removal, such as removal of oxidized PSi in HF, which helps to avoid the problem of hydrogen bubbles. More details on general fabrication aspects are covered in *Porous Silicon: Formation and Properties*.

First proposals on application of PSi sacrificial layer for fabrication of freestanding poly-silicon or silicon nitride cantilevers, beams, bridges, and membranes have been suggested in the late 1980s and early 1990s (Tu 1988; Guilinger et al. 1991; Steiner et al. 1993). Since then, many applications have been demonstrated. Most of them are already described, such as pressure sensors (Chapter 6 in *Porous Silicon: Biomedical and Sensor Applications*), and gyroscopes and accelerometers (Chapter 7 in *Porous Silicon: Biomedical and Sensor Applications*). Additionally, sacrificial porous layer is used for thermal isolation of sensing elements on hotplates in gas sensors (Chapter 11 in *Porous Silicon: Biomedical and Sensor Applications*), and in bolometers and gas flow sensors (Chapter 8 in *Porous Silicon: Biomedical and Sensor Applications*). In this chapter, further applications of sacrificial PSi in micromechanical/microelectronic devices, such as microphones, probe cantilevers, and free elements, are discussed. Applications of sacrificial PSi for electrical isolation of radio-frequency microcomponents and silicon-on-nothing technology are also covered in this chapter.

6.2 POROUS SILICON FOR CMOS-COMPATIBLE PROCESSING IN MICROELECTRONICS PRODUCTION LINE

In order to integrate a new process in microelectronics, it is necessary to keep it CMOS-compatible. On one hand, this means compatibility with the fabrication infrastructure, where introduction of a new process means investment in new equipment. On the other hand, the process itself must meet strict requirements concerning operation conditions and chemicals used, in order to avoid degradation or even damaging of sensitive microelectronic circuits. They are as follows:

- When using PSi as a sacrificial layer, as a preparation step for anodization (in case of electrochemical pore formation), doping regions have to be formed. These regions are needed to provide electric contact to wafer backside or to define etch stops during the anodization process. These high-temperature processes can be performed in the beginning together with n-well and p-well CMOS processes required for fabrication of electronic circuits.
- It is known that silicon is susceptible to structural rearrangement when annealed in a deoxidizing atmosphere, such as a hydrogen ambient, at temperatures exceeding 900°C (Habuka et al. 1995; Ogino et al. 1997). Structural rearrangement makes the porous structure more dense and, hence, more difficult to remove (Bell and Wise 1998). In case of oxidizing atmosphere, PSi is quickly oxidized, which can induce internal stress and cracking of the porous layer with functional elements on it. Therefore, formation of PSi should be performed after all high-temperature processes, such as annealing, diffusion, or oxidation.
- During anodization, electronic circuits on the substrate have to be correspondingly sealed to avoid contact with hydrofluoric acid. If dielectrics have already been deposited on the substrate, they can work as such protection.
- In order to make the anodization process completely metal free, silicon (consumable) electrodes can be used instead of platinum electrodes in the anodization bath.
- Removal of sacrificial porous layer should be done in the last steps because micromechanical components, which are being released, are typically very fragile and require support by the sacrificial layer until the last fabrication steps. Since at the last steps the electronic components are already encapsulated, not CMOS-compatible diluted KOH can be applied. If CMOS-compatible etchant is desired, TMAH or stain etching in hydrofluoric and nitric acids in water can be used instead (Vitanov et al. 2009).
- Safety and good reproducibility of a process are also of big concern when integrating it into microelectronic mass production. Introduction of fully automatic anodization tools helps to minimize risks for workers and provides good process control (Boehringer et al. 2012).

6.3 FREESTANDING MECHANICAL ELEMENTS

6.3.1 CANTILEVERS

Typical electromechanical applications of cantilevers fabricated with sacrificial PSi, such as hotplates and accelerometers, have been described in another book (Chapters 7 and 11 in *Porous Silicon: Biomedical and Sensor Applications*). In this section, applications that are more exotic are described.

Bell and Wise have used silicon cantilevers fabricated with sacrificial PSi as neural probes (Bell and Wise 1998). In their CMOS-compatible fabrication process, etch stop was defined with a lightly doped p-type silicon region in a heavily doped n-type substrate. This way, 500 μm wide and 14.3 mm long probes have been fabricated.

Another application of cantilevers was demonstrated by Kim et al. (2005). In production of integrated circuits, an important step is testing of the devices for failure. In order to connect integrated circuits to the test equipment, micro-probes are used. Integrated circuits nowadays are getting smaller, with more dense inputs and outputs, and to provide a reliable connection becomes challenging. Kim et al. (2005) have fabricated a micro-probe card with cantilevers, using PSi as a sacrificial layer. The fabrication process is shown in Figure 6.1. In this process, lateral formation of PSi in the 20-μm thick buried n^+ silicon layer was done in 43 m% aqueous HF solution for 25 min at room temperature by applying a current density of 10 mA/cm^2 (Figure 6.1f). In order to bend the cantilevers to the top, an annealing process was performed (Figure 6.1g). Fabricated cantilevers are shown in Figure 6.2.

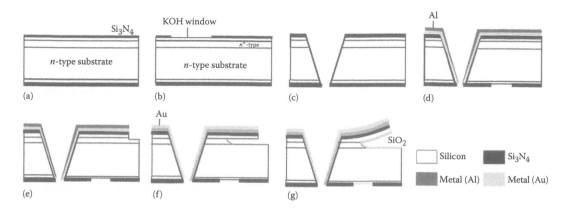

FIGURE 6.1   Fabrication of micro-probe card with sacrificial porous silicon. (a) Starting wafer ($n/n^+/n/n^+$ structure) and Si_3N_4 deposition. (b) KOH open window formation with RIE process (Si_3N_4 dry etching). (c) KOH process and Si_3N_4 redeposition. (d) Au electroplasting process and Al evaporation process. (e) RIE process (back-side Si_3N_4 etching). (f) Anodic reaction process. (g) Annealing process (silicon microprobe card fabrication). (From Kim Y.-M. et al., *ETRI J.* 27, 433, 2005. Copyright 2005: the Electronics and Telecommunications Research Institute. With permission.)

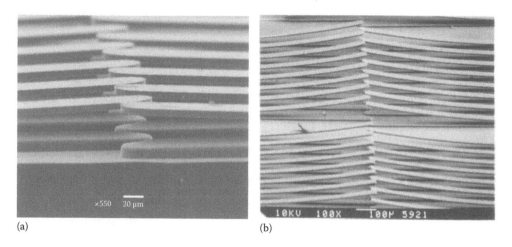

FIGURE 6.2   SEM micrographs of micro-probe card with sacrificial PSi: (a) before annealing, (b) after annealing. (From Kim Y.-M. et al., *ETRI J.* 27, 433, 2005. Copyright 2005: the Electronics and Telecommunications Research Institute. With permission.)

Teva et al. (2010) have applied a metal assisted etching process to generate a sacrificial porous layer and fabricated boron-silica-glass cantilevers this way (Figure 6.3a). The process of metal assisted etching is described in detail in Chapter 10 in *Porous Silicon: Formation and Properties*. In the fabrication process of Teva et al. (2010), in the beginning, a catalytic layer of metal with a thickness of 4.5 nm is deposited on a lightly doped *p*-type or *n*-type silicon substrate (substrate thickness 375 μm) with e-beam deposition and structured with a lift-off process. Different metals have been tested: Au, Ag, Ni, Cr, and Pt. As electrolyte, a solution composed of hydrogen peroxide (H_2O_2, 30%), hydrofluoric acid (HF, 40%), and ethanol in 1:1:1 proportion by volume was used. Preliminary tests with the metals showed that Cr was etched away in the solution before the pore formation process started. With Ni and Pt, no pore formation was observed. The highest etch rate was observed for gold; therefore, this metal was used for fabrication of the cantilevers. The process of metal assisted etching allowed forming a porous layer all through the wafer. This way, a PSi layer was formed from the backside up to the nondoped polysilicon layer on the frontside (Figure 6.3a-2). Then, boron-silica-glass and nickel were deposited and structured on the front side to form cantilevers (Figure 6.3a-3). In the last step, the porous volume and the front side polysilicon layer were removed in an aqueous KOH solution. With metal assisted etching, macropores with relatively thick walls are typically generated; therefore, the porous layer fabricated

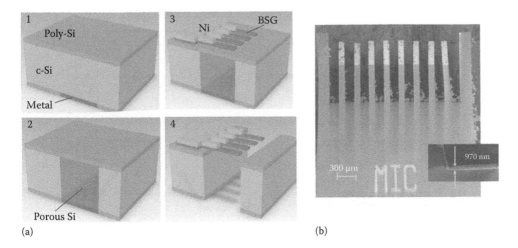

FIGURE 6.3 Boron-silica-glass cantilevers fabricated with metal assisted etching: (a) fabrication process, where 1—substrate with non-doped polysilicon on both sides and structured metal catalyst layer on the backside, 2—after metal assisted etching, 3—after formation of the cantilevers on the front side, 4—final structure after removal of the porous volume; (b) SEM micrograph of the fabricated structures. (From Teva J. et al., *J. Micromech. Microeng.* 20, 015034, 2010. Copyright 2010: IOP Publishing. With permission.)

this way is much less reactive than the micro/mesoporous silicon fabricated electrochemically. That is why high concentrated KOH of 7 M at high temperature of 60°C was used for removal of the formed PSi. With these conditions, the removal of the porous layer and of the polysilicon took 15 min (Figure 6.3a-4). Fabricated cantilevers of length 400 µm and width 100 µm had thickness of 970 nm (Figure 6.3b).

6.3.2 MICROPHONES

Sacrificial layers out of PSi can be used to form thin membranes for pressure sensors and microphones, where the porous layer defines the gap between the membrane and the substrate. Application of PSi for pressure sensors is covered in detail in Chapter 6 in *Porous Silicon: Biomedical and Sensor Applications.* Microphones, although in principle similar to pressure sensors, represent a separate class of MEMS devices, as they need so-called acoustic holes for dynamic response. Thus, they are covered separately in this section.

Different to cantilevers, beams, or bridge structures, membranes are continuous thin layers that cover the underlying sacrificial layer completely. Therefore, access to the sacrificial porous layer for its formation and removal needs special processes. If a membrane is formed with anodization from the front side of a substrate, access of electrolyte should be provided to the region beneath the membrane layer. This can be done either with access openings in the membrane layer (Lee 1995) or by performing anodization in two steps, with formation of an upper layer of PSi with low porosity (thus more dense), which withstands a PSi removal step, and a bottom layer with higher porosity. After PSi removal, the membrane is sealed with, for example, polymer, epitaxial silicon, or poly-silicon. In comparison to microphones fabricated with conventional sacrificial materials, such as photoresist (Ganji 2011) or phosphorous-silicate glass (Li et al. 2001), a much larger air gap can be achieved with PSi as sacrificial layer.

Few approaches to fabrication of microphones with sacrificial PSi have been published. Kronast et al. (2001) have fabricated a single-chip capacitive microphone. Double-side processing was applied (Figure 6.4) as follows: First, a surface p^+ doping of both sides of the low-doped p-type silicon wafer was done. Then, a layer of 0.5-µm thick PSi was formed in the front side in a solution of 1:1 of 50 m% HF and ethanol through a low-stress LPCVD silicon nitride (Figure 6.4a). After removal of the nitride layer, 0.8-µm thick SiO_2 was sputtered on the front side. The oxide layer, together with the PSi layer, formed the sacrificial layer defining an air gap of 1.3 µm between the

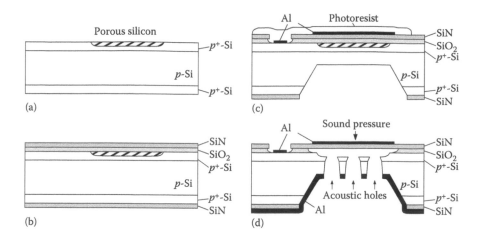

FIGURE 6.4 Fabrication of capacitive microphone with silicon nitride diaphragm: (a) silicon substrate with p^+-Si surface doping and PSi; (b) deposition of SiO_2 and SiN; (c) deposition of aluminum for the electrode and the contact wires and thinning of the wafer in KOH; and (d) final structure with etched acoustic holes and removed PSi and oxide in the airgap. (From Kronast W. et al., *Sens. Actuators A* 87, 188, 2001. Copyright 2001: Elsevier. With permission.)

membrane and the counter electrode. The membrane of the microphone (diaphragm) was formed by LPCVD-deposition of silicon-rich low-stress silicon nitride on the oxide layer (Figure 6.4b). After that, a contact via the front side p^+ layer, which is to be the counter electrode, was etched. Finally, an upper electrode on silicon nitride and interconnections were formed with aluminum and protected with photoresist during further steps. Then, the wafer under the membrane was thinned from the backside in 40% KOH until rest thickness of 20 µm (Figure 6.4c), and acoustic holes were etched with RIE until the front side oxide layer. Finally, the PSi and the oxide layers were removed in 1% KOH and HF, respectively (Figure 6.4d). Microphones of different dimensions and with round and square shaped electrodes have been produced this way (Figure 6.5). The fabricated microphones showed frequency response beyond 20 kHz and achieved sensitivity in the range of mV/Pa.

Ning et al. (2004) have used a similar approach with front side anodization and backside KOH etching. In contrast to the previous example, the PSi layer was defined here with the shallow p^+-doping, and later oxidized. The backside KOH etching to thin the backplate and to form acoustic holes was performed using a heavy boron doping etch stop technique. The fabricated microphone showed sensitivity in the range from −55 dB to −45 dB for the frequency range of 500 Hz to 25 kHz. The cut-off frequency was higher than 20 kHz.

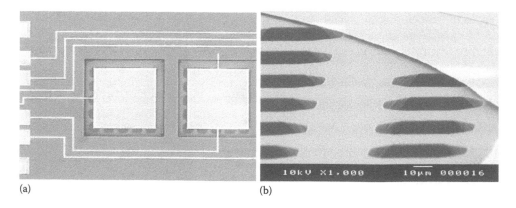

FIGURE 6.5 Single-chip capacitive microphones: (a) photograph of two square diaphragms of side length 0.5 mm, the acoustic holes are visible through the transparent silicon nitride film and (b) SEM micrograph showing enlarged view of the acoustic holes through the partly removed diaphragm. (Reprinted with kind permission of W. Kronast.)

6.3.3 THERMAL EMITTERS

One of the applications of cavities fabricated with sacrificial PSi is for thermal isolation of temperature-sensitive functional elements from a substrate. Application of this technique for fabrication of bolometers and gas sensors is discussed in Chapters 8 and 11 in *Porous Silicon: Biomedical and Sensor Applications*. Thermal isolation is also needed for thermal emitters to effectively emit the generated heat out of the device. Dobrzański and Piotrowski (1998) fabricated a thermal emitter with sacrificial PSi technology this way, where a polysilicon emitter was placed over a silicon nitride membrane. The bottom surface of the silicon nitride membrane was sputtered with tungsten working as a reflector. Experiments showed that a high-resistance polysilicon emitter operated at constant voltage exhibited unstable operation due to a strong dependence of its resistance on temperature. Reduction of the sheet resistance of the polysilicon emitter to 10 $\Omega/\square$ improved the issue. The fabricated emitter with active area of 0.17 mm^2, operated in Kr/Xe mixture, showed a conversion efficiency of about 0.3 for input power of 0.093 W/mm^2. The maximum input power for stable operation was determined to be 0.1 W/mm^2, with the corresponding maximum integral luminance of about 10 mW/(mm^{2}srd).

6.3.4 FREE ELEMENTS

Sacrificial PSi, especially in combination with 3D localization techniques, such as hydrogen implantation (read Chapters 3 and 12 in *Porous Silicon: Formation and Properties*), also can be used to fabricate free structures, that is, elements that are completely detached from a substrate. Rajta et al. (2009) fabricated a 3D micro-turbine with high aspect ratio (Figure 6.6a) in this way. The 10-μm deep 3D structure was formed by proton beam writing with high energy proton beam of 1.6 MeV (for the rotor) and 2 MeV (for the housing) to create high resistivity regions acting as etch stop. In order to provide release of the structures, a 37-μm deep layer of PSi was anodized. Due to thin silicon walls remaining under the underetched rotor after PSi removal, an additional well-controlled silicon etch step in poly-silicon etchant (HF:HNO$_3$:CH$_3$COOH = 3:2:5) was used. Due to widening of the implanted regions in a depth of about 3 μm at both sides around the projected range and minimum lateral distance between two neighboring elements of 2.5 μm required for the current to flow during anodization (Figure 6.6b), the authors could reach minimum distance between the shaft and the bearing of the rotor of 5 μm. With this gap, wobbling of the rotor around the shaft could not be avoided. As an alternative process, the authors suggested fabricating the rotor and the housing in separate substrates, and then assembling the microturbines, although such a process can hardly provide high throughput and needs further improvement.

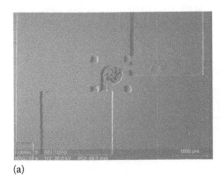
(a)

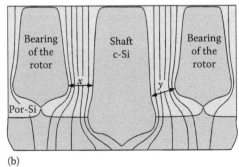
(b)

FIGURE 6.6 Micro-turbine fabricated with sacrificial PSi and proton beam writing: (a) SEM micrograph, (b) schematic view of the geometry showing widening of the implanted regions in depth, current lines during anodization and the limiting distances *x* and *y* defining the minimal achievable distance between the rotor and the shaft. (From Rajta I. et al., *Nucl. Instrum. Meth. Phys. Res. B* 267, 2292, 2009. Copyright 2009: Elsevier. With permission.)

6.4   CAVITIES AND WAFER-THROUGH HOLES

One of the most obvious applications of sacrificial PSi is for structuring, that is, formation of cavities of various shapes in silicon substrates. In this case, PSi does not play any special role and is just a by-product of the etch process. Various applications of the silicon anodization process in general and of the microporous silicon formation in particular have been proposed. In this section, some of them, mostly related to optoelectronics and microelectronics, are discussed.

Lang et al. (1994) proposed using sacrificial PSi for fabrication of flow channels (Figure 6.7). In order to get small undercutting of the mask and steep walls with an angle of 45°, heavily doped *p*-type or *n*-type substrates with resistivity of 0.2 Ω·cm and 0.01 Ω·cm, respectively, and a metal mask were applied.

Deep grooves are also needed to position optical fibers in silicon integrated optical circuits. Joubert et al. (2000) formed 75-μm deep pseudo-V shaped grooves for this purpose by anodization of *n*⁺-type silicon through a Cr-Au masking layer. The pseudo-V shapes resulted from crystallographic orientation dependent anodization rate at the chosen process conditions. With this technique, arrays of optical fibers were positioned with accuracy better than 1 μm (Figure 6.8). This precision allowed reducing the losses of the optical connections with 8-μm fibers to less than 0.5 dB at the wavelengths of 1.3 μm and 1.55 μm used in telecommunication.

An important question of using the anodization process for structuring is to control and predict etch shapes. Etch front propagation during pore generation for simple mask openings and backside electrical contacts were studied by Ivanov et al. (2011). Mescheder and Kovacs (2006) developed a process to control etch form development for real 3D cavities in silicon substrates (Figure 6.9), which can be used to form microoptical or fluidic structures (Ivanov et al. 2009).

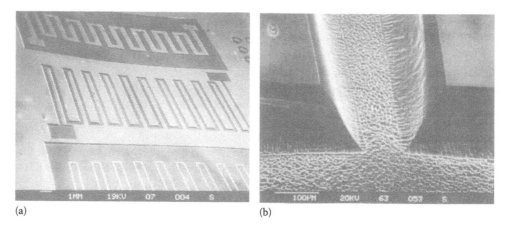

(a)   (b)

FIGURE 6.7   SEM micrographs of a flow channel fabricated with sacrificial PSi: (a) overview, (b) magnified. (From Lang W. et al., *Sens. Actuators A* 43, 239, 1994. Copyright 1994: Elsevier. With permission.)

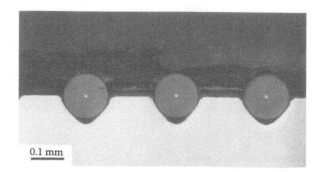

0.1 mm

FIGURE 6.8   Optical micrograph of three single mode optical fibers positioned in grooves fabricated with sacrificial PSi technology. (From Joubert P. et al., *J. Porous Mater.* 7, 227, 2000. Copyright 2000: Springer Science and Business Media. With permission.)

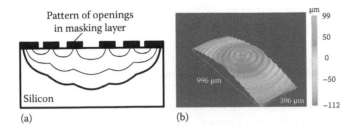

FIGURE 6.9 Three-dimensional shape control by two-dimensional complex patterns of openings in a protective masking layer: (a) principle of the technique, cross-section of silicon substrate. (Idea from Mescheder U. and Kovacs A., DE Patent 10,234,547 (2006).) (b) Example of a Fresnel-like structure obtained with the technique, profiler measurement. (From Ivanov A. and Mescheder U., *Adv. Mater. Res.* 325, 666, 2011. Copyright 2011: WILEY-VCH Verlag. With permission.)

FIGURE 6.10 Wafer-through holes formed with metal assisted etching. (From Teva J. et al., *J. Micromech. Microeng.* 20, 015034, 2010. Copyright 2010: IOP Publishing. With permission.)

The metal assisted process applied to fabricate cantilevers by Teva et al. (2010) shown in Section 6.3.1, was also used by the same authors to form wafer-through holes. Resulting structures are shown in Figure 6.10. Such openings can be used in microfluidic, microoptical, or microelectronic hybrid devices as contact windows for fluidic, optical, or electronic access.

6.5 LAYER-TRANSFER PROCESS

During production of electronic devices, it is difficult to handle thin substrates. However, use of thick substrates will increase material costs. One solution to this problem is application of thick handling wafer (donor wafer) with thin device layer on it, which is separated from the handling wafer in the last fabrication step, and the wafer is reused to fabricate the next electronic device layer.

Sacrificial PSi suits very well as an intermediate separation layer. The fabrication process is based on epitaxial growth of monocrystalline silicon on a PSi layer and PSi sintering (Chapter 12 in *Porous Silicon: Formation and Properties*). All necessary electronic devices and structures or further layers are formed on this epitaxially grown layer. In the end, the fabricated device layer is separated from the donor wafer mechanically with a fine water jet or with a PSi wet etch process.

Layer transfer process based on PSi was applied in ELTRAN® process for fabrication of SOI wafers (Ichikawa et al. 1995; Sato et al. 1995; Sakaguchi and Yonehara 2000) (see Chapter 5 in this book) and nowadays is used for fabrication of thin photovoltaic cells (Brendel et al. 1998, 2003; Fave et al. 2004; Solanki et al. 2004) (for more details, see Chapter 10 in this book).

Burghartz et al. (Zimmermann et al. 2008; Burghartz et al. 2010; Burghartz 2011) used a PSi-based layer transfer process to release ultra-thin Si chips from the wafer by preprocessing the wafer as shown in Figure 6.11a: A two-step anodization was used to form a fine porous layer at the surface and a course porous layer deeper in the substrate. For preparing the later release, a thermal treatment (1100°C in hydrogen) was used to sinter the porous layers. This process is similar to the advanced porous silicon membrane (APSM) Bosch process to form a sealed cavity for absolute pressure sensors described in Armbruster et al. (2003). Due to sintering, the fine porous layer transforms into a micro/nano cavity-rich single-crystalline silicon layer, while the coarse porous layer converts to an approximately 200-nm thick cavity below the recrystallized film. The micro/nano cavity-rich single-crystalline silicon layer acted as a seed of a silicon epitaxial layer. The epitaxial layer defined the thickness of the chip released after IC-device fabrication. The released thin flexible CMOS-chip is shown in Figure 6.11b.

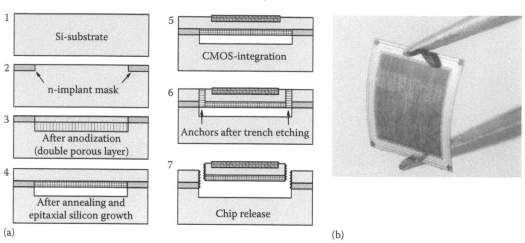

FIGURE 6.11   Chipfilm™ layer transfer process to separate thin chips from the substrate: (a) process flow, where 1—starting *p*-type substrate (heavily boron doped at the wafer surface), 2—with n^+ etch stop regions, 3—after anodization (PSi double layer), 4—after sintering in hydrogen and growth of device-quality silicon epitaxial layer, 5—after CMOS integration above the buried cavity, 6—after trench etching at the chip edges down into the buried cavity (formation of anchors), 7—release of the chip. (Adapted with kind permission of Prof. J.N. Burghartz from Zimmermann M. et al., In: *Proceedings of Annual Workshop on Semiconductor Advances for Future Electronics and Sensors 2008 (SAFE2008)*, November 27, Veldhoven, Netherlands, pp. 531–534.) (b) Released thin flexible CMOS-chip. (From Burghartz J.N. et al., *Solid-State Electron.* 54, 818, 2010. Copyright 2010: Elsevier. With permission.)

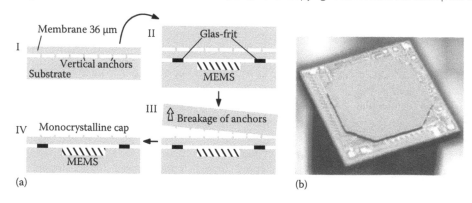

FIGURE 6.12   Layer transfer with APSM for encapsulation of MEMS: (a) process steps in wafer cross-section, where I—separation of a monocrystalline membrane in donor wafer with APSM process, II—bonding of the membrane to the MEMS wafer with glass-frit, III—detaching the substrate of the donor wafer from the membrane, IV—encapsulated MEMS device; (b) photograph of a pressure sensor encapsulated with monocrystalline silicon membrane. (From Prümm A. et al., *Sens. Actuators A* 188, 507, 2012. Copyright 2012: Elsevier. With permission.)

Prümm et al. (2012) have also used the APSM process as a layer transfer process for encapsulation of MEMS devices (Figure 6.12). In contrast to the previous example, anchors were formed during the sintering process within the PSi buried volume. This way, thin silicon membrane attached to the substrate with vertical anchors was fabricated (Figure 6.12a-I). After bonding of the membrane to a MEMS wafer with glass-frit (Figure 6.12a-II), the substrate of the donor wafer was mechanically detached from the membrane with controlled breaking occurring at the anchors (Figure 6.12a-III). The result was the MEMS device encapsulated with monocrystalline silicon membrane (Figure 6.12a-IV). The authors demonstrated this technique by encapsulation of absolute pressure sensors, where the formed cap enclosed a reference pressure (Figure 6.12b).

6.6   ELECTRICAL ISOLATION

All the way through the history of microelectronic technology, increasing the density of electronic components by miniaturization and working at higher frequencies for cheaper and faster integrated circuits remains the trend. By reducing the distance between single semiconductor devices and increasing operating frequencies, researchers always faced the problem of parasitic currents resulting in low reliability and increased energy consumption. Various solutions to solve this problem have been developed, such as introduction of SOI technology in the early 1980s, where reduction of parasitic currents was achieved by placing electronic components on the buried oxide layer (Kononchuk and Nguyen 2014). Although silicon dioxide can provide good electrical isolation, it comes to its limits nowadays, and novel approaches, such as the one to use air (or vacuum) as isolator, as in silicon-on-nothing technology (SON) (Mizushima et al. 2000; Sato et al. 2001; Hsu and Lee 2006; Kilchytska et al. 2008), are getting more attention. Demand for superior electrical isolation is especially acute for high radio frequency (RF) semiconductor devices. Sacrificial PSi process is an effective solution to form electrical RF components insulated with large air gaps from the substrate.

Ding et al. (Ding et al. 2001, 2003; Liu et al. 2003) used a sacrificial PSi layer to fabricate planar coils for CMOS-compatible RF circuits (Figure 6.13a). PSi of thickness of 30 µm was formed in 10 min in a mixture of hydrofluoric acid (40%) and absolute ethanol in a volume ratio of 3:1 at current density 40 mA/cm². A silicon dioxide layer of 200 nm and silicon nitride layer of 200 nm were used as a mask during anodization process and removed after formation of PSi by immersing the substrates in 10% HF for 2 h without attacking the porous layer. The two-layer aluminum RF structure was deposited on 1-µm thick SiO_2 layer with silicon nitride as an intermediate isolation layer. Finally, PSi was removed through the openings in SiO_2/Si_3N_4 layers in 5% TMAH solution with 1.6% dissolved silicon and 0.6% $(NH_4)_2S_2O_8$. As reported by Yan et al. (2001), the chosen etch solution has zero etch rate of aluminum. The fabricated planar coil is shown in Figure 6.13b.

It is worth noting that electrical isolation with PSi can be also done without its removal, either by oxidation of PSi to form a porous oxide layer (Imai 1981; Holmstrom and Chi 1983; Tsao 1987; Bomchil et al. 1989; Oules et al. 1992; Wu 2001) or by using native PSi (Capelle et al. 2011,

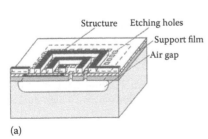

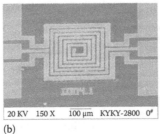

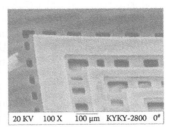

(a)

(b)

FIGURE 6.13   Planar coil for RF-circuits with air gap for electrical isolation formed with sacrificial porous silicon: (a) schematic view; (b) SEM micrographs of the fabricated structure—overview and detail of the coil lines and etching holes. (From Liu Z. et al., *Sens. Actuators A* 108, 112, 2003. Copyright 2003: Elsevier. With permission.)

2012, 2014). However, these applications lie beyond the frame of this chapter and are discussed in Chapter 5 of this book and Chapter 11 in *Porous Silicon: Biomedical and Sensor Applications*.

REFERENCES

Armbruster, S., Schafer, F., Lammel, G. et al. (2003). A novel micromachining process for the fabrication of monocrystalline Si-membranes using porous silicon. In: *Proceedings of 12th International Conference on Solid-State Sensors, Actuators and Microsystems, TRANSDUCERS' 2003*, June 8–12, Boston, pp. 246–249.

Bell, T. and Wise, K. (1998). A dissolved wafer process using a porous silicon sacrificial layer and a lightly-doped bulk silicon etch-stop. In: *Proceedings of the Eleventh Annual International Workshop on Micro Electro Mechanical Systems, MEMS 98*, January 25–29, Heidelberg, pp. 251–256.

Boehringer, M., Artmann, H., and Witt, K. (2012). Porous silicon in a semiconductor manufacturing environment. *Microelectromech. Syst. J.* **21**, 1375–1381.

Bomchil, G., Halimaoui, A., and Herino, R. (1989). Porous silicon: The material and its applications in silicon-on-insulator technologies. *Appl. Surf. Sci.* **41**, 604–613.

Brendel, R., Artmann, H., Oelting, S., Frey, W., Werner, J., and Queisser, H. (1998). Monocrystalline Si waffles for thin solar cells fabricated by the novel perforated-silicon process. *Appl. Phys. A: Mater. Sci. Process.* **67**, 151–154.

Brendel, R., Feldrapp, K., Horbelt, R., and Auer, R. (2003). 15.4%-efficient and 25 μm-thin crystalline Si solar cell from layer transfer using porous silicon. *Phys. Stat. Sol. (a)* **197**, 497–501.

Burghartz, J.N. (Ed.) (2011). *Ultra-Thin Chip Technology and Applications*, Springer, Heidelberg.

Burghartz, J.N., Appel, W., Harendt, C., Rempp, H., Richter, H., and Zimmermann, M. (2010). Ultra-thin chip technology and applications, a new paradigm in silicon technology. *Solid-State Electron.* **54**, 818–829.

Capelle, M., Billoue, J., Poveda, P., and Gautier, G. (2011). N-type porous silicon substrates for integrated RF inductors. *Electron. Dev., IEEE Trans.* **58**, 4111–4114.

Capelle, M., Billoué, J., Poveda, P., and Gautier, G. (2012). RF performances of inductors integrated on localized p⁺-type porous silicon regions. *Nanoscale Res. Lett.* **7**, 1–8.

Capelle, M., Billoué, J., Concord, J., Poveda, P., and Gautier, G. (2014). Monolithic integration of common mode filters with electrostatic discharge protection on silicon/porous silicon hybrid substrate. *Appl. Phys. Lett.* **104**, 072104.

Ding, Y., Liu, Z., Cong, P., Liu, L., and Li, Z. (2001). Preparation and etching of porous silicon as a sacrificial layer used in RF-MEMS devices. In: *Proceedings of 6th International Conference on Solid-State and Integrated-Circuit Technology, IEEE*, October 22–25, Shanghai, pp. 816–819.

Ding, Y., Liu, Z., Liu, L., and Li, Z. (2003). A surface micromachining process for suspended RF-MEMS applications using porous silicon. *Microsyst. Technol.* **9**, 470–473.

Dobrzański, L. and Piotrowski, J. (1998). Micromachined silicon thermopile and thermal radiators using porous silicon technology. *IEE Proc.: Optoelectron.* **145**(5), 307–311.

Fave, A., Quoizola, S., Kraiem, J., Kaminski, A., Lemiti, M., and Laugier, A. (2004). Comparative study of LPE and VPE silicon thin film on porous sacrificial layer. *Thin Solid Films* **451**, 308–311.

French, P., Gennissen, P., and Sarro, P. (1997). Epi-micromachining. *Microelectron. J.* **28**, 449–464.

Ganji, B.A. (2011). Design and fabrication of a novel MEMS silicon microphone. In: Basu P.S. (Ed.) *Crystalline Silicon—Properties and Uses*. InTech, Rijeka, pp. 313–328.

Guilinger, T.R., Kelly, M.J., Martin, Jr. S.B., Stevenson, J.O., and Tsao, S.S. (1991). Porous silicon formation and etching process for use in silicon micromachining. US Patent 4,995,954.

Habuka, H., Tsunoda, H., Mayusumi, M., Tate, N., and Katayama, M. (1995). Roughness of silicon surface heated in hydrogen ambient. *J. Electrochem. Soc.* **142**, 3092–3098.

Holmstrom, R. and Chi, J. (1983). Complete dielectric isolation by highly selective and self-stopping formation of oxidized porous silicon. *Appl. Phys. Lett.* **42**, 386–388.

Hsu, S.T. and Lee, J.-J. (2006). Fabrication of silicon-on-nothing (SON) MOSFET fabrication using selective etching of Si1-xGex layer. US Patent 7,015,147.

Ichikawa, T., Yonehara, T., Sakamoto, M. et al. (1995). Method for producing semiconductor articles. US Patent 5,466,631.

Imai, K. (1981). A new dielectric isolation method using porous silicon. *Solid-State Electron.* **24**, 159–164.

Ivanov, A. and Mescheder, U. (2011). Silicon electrochemical etching for 3D microforms with high quality surfaces. *Adv. Mater. Res.* **325**, 666–671.

Ivanov, A., Kovacs, A., Mescheder, U., Kuhn, S., and Burr, A. (2009). Optimisation of surface quality of 3D silicon master forms for injection molding of optical micro elements. In: *Proceedings of Mikrosystemtechnik-Kongress 2009*, October 12–14, Berlin. VDE-Verl., Berlin and Offenbach, pp. 726–729.

Ivanov, A., Kovacs, A., and Mescheder, U. (2011). High quality 3D shapes by silicon anodization. *Phys. Stat. Sol. A* **208**, 1383–1388.

Joubert, P., Guendouz, M., Pedrono, N., and Charrier, J. (2000). Porous silicon micromachining to position optical fibres in silicon integrated optical circuits. *J. Porous Mater.* **7**, 227–231.

Kilchytska, V., Flandre, D., and Raskin, J.-P. (2008). Silicon-on-nothing MOSFETs: An efficient solution for parasitic substrate coupling suppression in SOI devices. *Appl. Surf. Sci.* **254**, 6168–6173.

Kim, Y.-M., Yoon, H.-C., and Lee, J.-H. (2005). Silicon micro-probe card using porous silicon micromachining technology. *ETRI J.* **27**, 433–438.

Kononchuk, O. and Nguyen, B.-Y. (Eds.) (2014). *Silicon-On-Insulator (SOI) Technology—Manufacture and Applications.* Elsevier, Amsterdam.

Korvink, J.G. and Paul, O. (Eds.) (2006). *MEMS: A Practical Guide to Design, Analysis, and Applications.* W. Andrew Pub., Norwich, NY.

Kronast, W., Müller, B., Siedel, W., and Stoffel, A. (2001). Single-chip condenser microphone using porous silicon as sacrificial layer for the air gap. *Sens. Actuators A* **87**, 188–193.

Lang, W., Steiner, P., Richter, A., Marusczyk, K., Weimann, G., and Sandmaier, H. (1994). Application of porous silicon as a sacrificial layer. *Sens. Actuators A* **43**, 239–242.

Lee, J.H. (1995). Method for fabricating a semiconductor device using a porous silicon region. US Patent 5,445,991.

Li, X., Lin, R., Kek, H., Miao, J., and Zou, Q. (2001). Sensitivity-improved silicon condenser microphone with a novel single deeply corrugated diaphragm. *Sens. Actuators A* **92**, 257–262.

Liu, Z., Ding, Y., Liu, L., and Li, Z. (2003). Fabrication planar coil on oxide membrane hollowed with porous silicon as sacrificial layer. *Sens. Actuators A* **108**, 112–116.

Mescheder, U. (2004). *Mikrosystemtechnik: Konzepte und Anwendungen* (German Edition). Vieweg+Teubner Verlag, Springer, Wiesbaden.

Mescheder, U. and Kovacs, A. (2006). Verfahren zur bildung einer ausnehmung in der oberfläche eines werkstücks, insbesondere zur herstellung von mikroformen. Patent DE 10,234,547.

Mizushima, I., Sato, T., Taniguchi, S., and Tsunashima, Y. (2000). Empty-space-in-silicon technique for fabricating a silicon-on-nothing structure. *Appl. Phys. Lett.* **77**, 3290–3292.

Ning, J., Liu, Z., Liu, H., and Ge, Y. (2004). A silicon capacitive microphone based on oxidized porous silicon sacrificial technology. In: *Proceedings of 7th International Conference on Solid-State and Integrated Circuits Technology 2004, IEEE.* October 18–21, Beijing, pp. 1872–1875.

Ogino, T., Hibino, H., and Homma, Y. (1997). Step arrangement design and nanostructure self-organization on Si surfaces. *Appl. Surf. Sci.* **117**, 642–651.

Oules, C., Halimaoui, A., Regolini, J., Perio, A., and Bomchil, G. (1992). Silicon on insulator structures obtained by epitaxial growth of silicon over porous silicon. *J. Electrochem. Soc.* **139**, 3595–3599.

Prümm, A., Kraft, K.-H., Gottschling, P. et al. (2012). Monocrystalline thin-film waferlevel encapsulation of microsystems using porous silicon. *Sens. Actuators A* **188**, 507–512.

Rajta, I., Szilasi, S., Fürjes, P., Fekete, Z., and Dücső, C. (2009). Si micro-turbine by proton beam writing and porous silicon micromachining. *Nucl. Instrum. Methods Phys. Res., Sect. B* **267**, 2292–2295.

Sakaguchi, K. and Yonehara, T. (2000). SOI wafers based on epitaxial technology. *Solid State Technol.* **43**(6), 88–92.

Sato, N., Sakaguchi, K., Yamagata, K., Fujiyama, Y., and Yonehara, T. (1995). Epitaxial growth on porous Si for a new bond and etchback silicon-on-insulator. *J. Electrochem. Soc.* **142**, 3116–3122.

Sato, T., Nii, H., Hatano, M. et al. (2001). SON (silicon on nothing) MOSFET using ESS (empty space in silicon) technique for SOC applications. In: *Technical Digest. International Electron Devices Meeting, 2001. IEDM'01. IEEE*, December 2–5, Washington, pp. 37.1.1–37.1.4.

Solanki, C., Bilyalov, R., Poortmans, J., Nijs, J., and Mertens, R. (2004). Porous silicon layer transfer processes for solar cells. *Sol. Energy Mater. Sol. Cells* **83**, 101–113.

Steiner, P., Richter, A., and Lang, W. (1993). Using porous silicon as a sacrificial layer. *J. Micromech. Microeng.* **3**, 32–36.

Teva, J., Davis, Z.J., and Hansen, O. (2010). Electroless porous silicon formation applied to fabrication of boron–silica–glass cantilevers. *J. Micromech. Microeng.* **20**(015034), 1–11.

Tsao, S.S. (1987). Porous silicon techniques for SOI structures. *Circuits Dev. Mag., IEEE* **3**, 3–7.

Tu, X.-Z. (1988). Fabrication of silicon microstructures based on selective formation and etching of porous silicon. *J. Electrochem. Soc.* **135**, 2105–2107.

Vitanov, P., Goranova, E., Stavrov, V., Ivanov, P., and Singh, P. (2009). Fabrication of buried contact silicon solar cells using porous silicon. *Sol. Energy Mater. Sol. Cells* **93**, 297–300.

Wu, S.-L. (2001). Fipos method of forming SOI CMOS structure. US Patent 6,331,456.

Yan, G., Chan, P.C., Hsing, I. et al. (2001). An improved TMAH Si-etching solution without attacking exposed aluminum. *Sens. Actuators A* **89**, 135–141.

Zeitschel, A., Friedberger, A., Welser, W., and Müller, G. (1999). Breaking the isotropy of porous silicon formation by means of current focusing. *Sens. Actuators A* **74**, 113–117.

Zhang, D. (Ed.) (2013). *Advanced Mechatronics and MEMS Devices.* Springer, New York.

Zimmermann, M., Appel, W., Burghartz, J.N., Ferwana, S., and Harendt, C. (2008). Ultra-thin chip fabrication and assembly process. In: *Proceedings of Annual Workshop on Semiconductor Advances for Future Electronics and Sensors 2008 (SAFE2008)*, November 27, Veldhoven, Netherlands, pp. 531–534.

Porous Silicon as Substrate for Epitaxial Films Growth

Eugene Chubenko, Sergey Redko, Alexey Dolgiy,
Hanna Bandarenka, and Vitaly Bondarenko

7

CONTENTS

7.1 Introduction 144

7.2 Homoepitaxy on PSi 144

7.3 Heteroepitaxy on PSi 148

    7.3.1 Heteroepitaxial Films of II-VI and IV-VI Compound Semiconductors 148

    7.3.2 Heteroepitaxial Films of III-V Compound Semiconductors 152

        7.3.2.1 GaAs 152

        7.3.2.2 InSb 153

        7.3.2.3 GaN 153

    7.3.3 Germanium and Silicon-Germanium Alloys 157

    7.3.4 Diamond Films 157

    7.3.5 Other Materials 159

7.4 Summary 159

Acknowledgments 159

References 159

7.1 INTRODUCTION

Porous silicon (PSi) remains crystalline even at high porosity (p ~90%) (Unagami and Seki 1978; Duttagupta and Fauchet 1997), allowing an epitaxial growth of films of various materials. Mechanical properties of PSi strongly depend on the porosity and can be controlled by the anodization regimes (Duttagupta and Fauchet 1997) to provide PSi buffer layers with specified characteristics for the epitaxial growth. The possibility of the Si homoepitaxy on PSi to form an intercomponent isolation in ICs was patented as early as 1972 by the Nippon Telephone and Telegraph Public Corporation (Watanabe and Sakai 1972). This application dominated over a long period and was studied intensively for the fabrication of the silicon-on-insulator (SOI) structures (Bondarenko et al. 2005). Based on the Si epitaxy on PSi, Canon Inc. has developed the ELTRAN technology for the commercial production of high-quality 300-mm SOI wafers (Yonehara 2002). From the mid 1980s, PSi applicability for the heteroepitaxial growth of various films at the silicon substrate has been studied. These studies were initiated by Luryi and Suhir (1986). They theoretically predicted a possible approach for the growth of dislocation-free lattice-mismatched heteroepitaxial layers on a patterned substrate, such as PSi. After that, many efforts to validate the Luryi/Suhir theoretical model and to use PSi for the heteroepitaxy have been performed (Xie and Bean 1990a; Bondarenko et al. 1994, 1996; Zubia and Hersee 1999; Levchenko et al. 1999; Novikov et al. 2003; Christiansen et al. 2006; Huangfu et al. 2013; Ye and Yu 2014). Recently, research of Ge and SiGe epitaxy (Kim et al. 2007; Blanchard et al. 2011; Aouassa et al. 2012; Gouder et al. 2014a,b) also demonstrated that PSi is indeed a strong contender for specific "pseudo-substrate" applications (Dariani 2014).

Chapter 7 is a comprehensive review of research works on homo- and heteroepitaxy on PSi.

7.2 HOMOEPITAXY ON PSi

The homoepitaxial growth of single-crystal Si films on the PSi layer is mainly intended for the formation of SOI structures (Konaka et al. 1982; Yonehara et al. 1994) and thin-film solar cells (Beaucarne et al. 2006). PSi is used (Figure 7.1) at that as a starting layer to form the buried oxide (BOX) layer for the SOI technology, a buried Bragg reflector layer for solar cells, or a sacrificial layer to fabricate a free high-quality Si epitaxial film. In the latter case, the epitaxial Si film is detached from the substrate and transferred to another one that is used for the formation of SOI

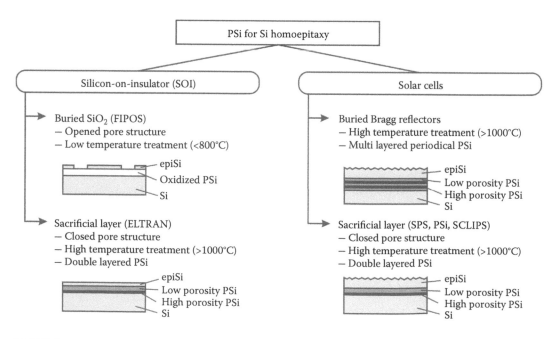

FIGURE 7.1   Main applications of epitaxial Si films grown on PSi.

wafers and thin film solar cells. Either of the two methods imposes peculiar requirements on the characteristics of the PSi layer and epitaxial single-crystal Si film.

For the BOX fabrication, it is important to keep PSi porous structure unchanged during the epitaxial process (Konaka et al. 1982). Otherwise, PSi loses its reactivity associated with the large specific surface area and cannot be transformed into the SiO_2 layer selectively because its oxidation rate is comparable with the oxidation rate of bulk Si. However, a high-temperature treatment of PSi changes the material properties driven by the reduction of the excess free energy due to broken atomic bonds at the inner surfaces of the porous structure. The surface mobility of the Si atoms on the inner surface of the PSi layer leads to a sintering process at temperatures above 850°C. The sintering transforms the initial pores into spherical voids with a diameter of 50 nm to 1 μm depending on the morphology and the porosity of the starting PSi layer. Therefore, a low-temperature (600–800°C) epitaxial technique, such as MBE (Oules et al. 1992), is of advantage.

When PSi is used as a sacrificial layer, the temperature of the epitaxial process can exceed 1000°C. Moreover, before the epitaxy, the PSi sintering procedure at 1000–1200°C is made on purpose, which results in the PSi restructuring (Labunov et al. 1986, 1987). During restructuring, PSi forms a top single-crystal layer with the closed surface, which serves as a seed layer for the high-temperature epitaxy as well as a buried layer, which is mechanically weak and therefore allows a later detachment of the epitaxial film.

Low defect density in the epitaxial films is of great importance for the SOI technology. For the solar cells, the defect density can be an order of magnitude higher although it is also of no less importance as it directly affects their efficiency. For the solar cells, high surface roughness (up to several μm) is permissible for antireflection purposes.

Requirements on the epitaxial film thickness are various as well. The epitaxial film thickness can be from 50 nm to 2 μm for SOI but it can amount to several tens of microns (20–50 μm typically) for solar cells.

Unagami and Seki (1978) pioneered in the single-crystal silicon homoepitaxial growth on the PSi layer. Further researches have been directed to the development of optimal conditions for the formation of the PSi layers and epitaxial silicon films to realize the FIPOS technology (Imai 1981) (see Chapter 5 of this book). To prevent the PSi sintering, the MBE (Konaka et al. 1982; Beale et al. 1985; Lin et al. 1986), plasma-induced (Takai and Itoh 1983) and low-pressure CVD (Vescan et al. 1988; Oules et al. 1989, 1992), and electron beam evaporation (Ito et al. 1990; Yasumatsu et al. 1991) techniques have been used. Typical temperatures of the epitaxy were 750–850°C. To prevent Si atoms from the migration on the PSi inner surface and thus to avoid the PSi sintering, a pretreatment in the H_2/HCl gas (Bomchil et al. 1989) or in the H_2O_2 based solution (Vescan et al. 1988) as well as the thermal pre-oxidation (Bomchil et al. 1989) or chemical pre-oxidation (Lin et al. 1986) can be used directly before the epitaxial step. In most cases, meso-PSi layers for the Si homoepitaxial growth are formed in the p^+-type silicon wafers. In spite of the PSi layer presence, epitaxial films always succeed to the substrate crystallographic orientation (Unagami and Seki 1978). The porous structure has a pronounced effect on the generation of stacking faults, which are the major defects in the epitaxial layers (Labunov et al. 1983; Vescan et al. 1988; Oules et al. 1989; Yasumatsu et al. 1991). When the PSi porosity is less than 46–50%, the stacking fault density in the epitaxial films grown on PSi is comparable with the stacking fault density in the epitaxial film grown on bulk silicon (10 cm^{-2}). This value increases considerably with the PSi porosity because minimum surface density of nucleation centers is necessary to obtain good epitaxial films bridging over the spaces (pores) of the structure. Increased strains in high porosity PSi also contribute to the defect density (Labunov et al. 1984). Minimum obtainable thickness of the epitaxial film grown on the PSi layer with open pores and pre-oxidized walls is 8–10 nm and depends on porosity increasing considerably at the porosity above 50%, preventing ultra-thin SOI manufacturing (Yasumatsu et al. 1991). It should be noted that since the PSi volume increases considerably at the oxidation, as early as in the work by Takai and Itoh (1983) a double PSi layer consisted of the thin (200 nm) low-porous layer as a seed layer for the high-quality epitaxy and the thicker (1.8 μm) high-porous layer for the following oxidation was proposed to prevent silicon wafer from bending after the oxidation.

The indirect correlation between the substrate conductivity type and value and the defect density in the epitaxial film has been revealed. The stacking faults density in the epitaxial layers grown on the substrates of high conductivity (~0.01 Ohm·cm) was no more than 10 cm^{-2}, but that in the epitaxial layers grown on the substrates of low (8–15 Ohm·cm) conductivity was 10^7–10^8 cm^{-2}. This associated with the difference in the structure of PSi formed in the substrates of various conductivities (Oules et al. 1992; Liu et al. 2003a). The high stacking fault density (10^5–10^8 cm^{-2}) in the epitaxial layer is typical for the use of unsintered single and even double layered PSi (Jin et al. 2000). It can be significantly reduced by controlling both the porous structure and the thermal treatment before the epitaxial growth.

The process of the silicon epitaxial growth on the surface of PSi with open pore channels has been theoretically investigated by the numerical simulation method (Novikov et al. 1998; Novikov 1999). The epitaxial film has been shown to grow due to the formation of the overhang layer and to depend strongly on the deposition rate of Si atoms onto the surface. At too high growth rate, pores have no time to cure resulting in the increase of the epilayer roughness.

The development of the ELTRAN technology (developed at Canon) (Yonehara et al. 1994) and layer transfer technology for crystalline thin-film silicon solar cells (developed at Sony) (Tayanaka and Matsushita 1996; Tayanaka et al. 1998), either of the two using the layer transfer approach, call for another way of looking at the PSi layer formation. The PSi oxidation is not needed for these technologies, so PSi can be treated in the temperature conditions wherein its sintering can take place. Moreover, for the successful layer transfer procedure, PSi should offer appropriate mechanical properties, that is, it should be resistant enough to withstand thermal and mechanical loads, and at the same time, it should be fragile enough to provide the epitaxial layer detachment from the substrate without the damage of the thin epitaxial film. In the ELTRAN technology, the starting ~12-μm thick and 15% porous uniform single meso-PSi layer is pre-oxidized at 400°C in dry oxygen to stabilize its structure and keep its high specific surface area and chemical reactive properties (Herino et al. 1984). When the low-temperature SiO_2 layer was removed in diluted HF from the outer surface, the 1-μm thick silicon epitaxial layer by the LPCVD at 900°C in the SiH_2Cl_2, and H_2, at 80 Torr grew (Yonehara et al. 1994). The density of stacking fault defects in the epitaxial film was 10^3–10^4 cm^{-2}. The stacking fault density has been significantly reduced to $3.5 \cdot 10^2$ cm^{-2} by the prebaking in H_2 atmosphere at temperature up to 1150°C, resulting in the pore close up at the PSi surface in accordance with the mechanism studied by Labunov et al. (1986, 1987) or by the special pre-injection procedure whereby a small additional amount of silicon was provided from the gas phase during the hydrogen prebaking so that the remaining pores in the PSi surface closed up. These operations provided the smooth surface for the epitaxial growth to be made and the temperature of the epitaxy to be increased to 1000–1050°C (Sato et al. 1995, 1996). High quality of the epitaxial layers grown on the PSi layers at the temperatures above 1000°C has been demonstrated even in a number of early works on the epitaxial growth on the PSi layers (Unagami and Seki 1978; Labunov et al. 1983, 1984).

To facilitate the detachment of the epitaxial layer from the seed wafer, the stacked double PSi system consisted of the 15-μm thick low-porous top layer and the 0.5-μm high-porous layer underneath, so-called "zipper" layer (Yonehara and Sakaguchi 2001; Yonehara 2002). The detachment of the epitaxial layer occurs along the high-porous layer of smaller mechanical strength.

For the thin-film silicon solar cells, a layer transfer process similar to ELTRAN named the sintered PSi (SPS) process has been proposed (Tayanaka et al. 1998). However, a porous three-layer system in which the layer of high porosity is sandwiched between two layers of low porosity permits high quality epitaxial growth and subsequent detachment of the epitaxial layer is used. Since no stabilization of the PSi by the thermal oxidation is used, the surface and the volume of the porous multilayer reorganize during the epitaxy: in the low-porous layer, the surface closes (as for ELTRAN) to form a planar surface for the epitaxial growth and voids form in the volume. The high-porous layer transforms into weak Si bridges, enabling transfer of epitaxial silicon film to low-cost silicon substrate (Tayanaka and Matsushita 1996). The process effectively works at the 20% porosity of the low-porous layer and 50% porosity of the high-porous layer and at the sintering temperature of 1100°C (Müller and Brendel 2000).

Subsequently several similar processes have been developed for the solar cells using the silicon epitaxial films grown on the sacrificial PSi layers. These are the PSI (porous Si) developed at the ZAE Bayern (Brendel 1997; Brendel et al. 2001), solar cells by liquid phase epitaxy over PSi (SCLIPS) process developed at Canon (Nishida et al. 2001), ELTRAN-like process developed at the University of Stuttgart (Bergmann et al. 2002) and at IMEC (Solanki et al. 2004), as well as some related processes (see Radhakrishnan et al. 2014 and references therein and Chapter 10 of this book). In the majority of these processes, the double PSi layers are used, although the single 10–15 μm thick PSi layers are employed as well (Krinke et al. 2001; Nishida et al. 2001). What all the processes have in common is that no stabilization of the PSi structure by the thermal pre-oxidation is used and the surface and the volume of the PSi reorganize during annealing and epitaxy. Recently Karim et al. (2014) studied strains in the double PSi structure depending on the thickness of the surface low-porous PSi layer. The increase of the thickness of this layer and thermal annealing time has been shown to reduce strains in the low-porous PSi layer, resulting in the silicon epitaxial films with lower density of stacking faults. Main stresses in the double PSi structure are generated in the lower high-porous PSi layer. During the annealing step, voids are formed in this layer, resulting in the stress relaxation. The increase of the stresses in the homoepitaxial films grown on the PSi with the porosity above 40% confined by higher lattice deformation and defect density was shown by Lamedica et al. (2002).

Usually for the homoepitaxial silicon growth on the PSi layers, planar growth substrates are used. But the PSI process starts from the randomly pyramid textured growth substrates formed by anisotropic chemical etching in the alkaline solution (Brendel et al. 2001; Krinke et al. 2001; Cai et al. 2010) to fabricate a textured epitaxial Si film using PSi for the layer transfer. This allows reflectivity of solar cells to be reduced. The silicon CVD deposition smoothes the pyramid texture, and the top side is almost planar. Therefore, usually the surface of the homoepitaxial film grown on the PSi structure is textured with random inverted pyramids (Matte et al. 2013). Stain etching process in the HF/HNO_3 solution has also been proposed instead of the electrochemical anodization to form PSi for the PSI process (Terheiden et al. 2011).

Silicon epitaxial films for the solar cells have been mainly grown by the atmospheric pressure chemical vapor deposition (APCVD) or low pressure chemical vapor deposition (LPCVD) at temperatures of 1050–1100°C, though ion assisted deposition (IAD) (Krinke et al. 2001) and liquid phase-epitaxy (LPE) (Nishida et al. 2001) have been used as well. No essential difference in the Hall mobility and resistivity of epitaxial films grown by the CVD at atmospheric and low pressures has been found (Fave et al. 2004).

In solar cells, PSi can also be used as a Bragg reflector, that is, as a buried layer on the surface of which a silicon epitaxial film of the active solar cell region is grown. This approach, studied both theoretically (Zettner et al. 1998) and practically (Bilyalov et al. 2002) and developed intensively at IMEC from the point of view of the silicon epitaxial growth, is not different from the methods discussed previously. However, since the PSi layer is a functional part of the completed device, the change of the PSi optical properties, in particular refractive index, due to the sintering during the high-temperature epitaxy should be taken into account (Müller and Brendel 2000). The use of low-temperature deposition methods such as low energy plasma enhanced chemical vapor deposition (LEPECVD) is possible (Bilyalov et al. 2002) but unjustified. Therefore, as long as the PSi surface rearranging cannot be fully avoided, the calculation of optical characteristics of reflectors should take into account this factor. Duerinckx et al. (2006) discussed the solar cells with reflectors based on reorganized PSi. In addition to Bragg buried reflective PSi mirrors consisting of alternate PSi layers of uniform thickness and various porosity (Duerinckx et al. 2006), it is possible to form chirped buried reflective PSi mirrors consisting of alternate PSi layers with the increased thickness of each subsequent layer (Kuzma-Filipek et al. 2008) to broaden its reflectivity band. The number of PSi layers can reach 37 (Kuzma-Filipek et al. 2012). The porosities of low-porous and high-porous PS layers usually are 20–25% and 40–55% correspondingly. Total thickness of the porous silicon layer can amount to 4–5 μm. The low-porous PSi layer, which served as the seed layer during the epitaxial growth, is always located at the top of PSi structure. The PSi baking is performed in hydrogen at 1130–1150°C for a longer time than usual up to 1 h to ensure full reorganization of the PSi layer stack (Van Hoeymissen et al. 2011). Then the single-crystal silicon epitaxial layer is deposited by the APCVD method. In the work by Kuzma-Filipek et al. (2007) an advantage of APCVD over LPCVD for the silicon epitaxial growth on PSi was shown. At LPCVD, full closure of the porous surface layer may not occur, resulting in lower quality silicon epitaxial layer. The defect density in the silicon film grown by APCVD does not exceed 10^3 cm^{-2} (Kuzma-Filipek et al. 2007).

Note that low-grade starting silicon wafers can be used for the technology of thin-film silicon solar cells. Therefore, when multilayered PSi stacks are used for the thin-film silicon solar cells, the PSi layers act as getter (Kuzma-Filipek et al. 2009; Radhakrishnan et al. 2012) to adsorb effectively atoms of various elements such as Ag, Cl, Cu, Ge, Fe, Ni, and S and therefore to reduce their concentration in the epitaxial film by the order of magnitude (Kuzma-Filipek et al. 2009), increasing the efficiency of the device operation.

7.3 HETEROEPITAXY ON PSi

Silicon is a dominant semiconductor material of modern microelectronics. However, many other semiconductors exceed silicon in electronic and optical properties. However, nowadays, three-dimensional crystals and consequently wafers of these semiconductors with diameters approximate to silicon wafers (up to 300 mm) are not produced on a commercial scale. Possible solution of that problem is a growing of high-quality heteroepitaxial films of other semiconductors on large silicon substrates. The realization of this idea enables heteroepitaxial structures to be processed on the high-performance processing lines of modern microelectronic production and making electronic and optoelectronic devices and circuits showing the parameters that are not achieved by silicon or material of the heteroepitaxial film separately. The growth of heteroepitaxial films on silicon is a complicated task due to the difference in the crystal lattice parameters (lattice constant a) and thermomechanical properties (coefficient of thermal expansion CTE, α, in first place). This induces stress and initiates defects in heteroepitaxial structures (Figure 7.2). The difference in a results in the mismatch strain and the difference in α leads to the thermal stress. The mismatch strain and thermal stress cause the elastic macro-deformation that at the critical value results in the plastic deformation and crystal defect initiation. In turn, defects cause local strains and elastic macro-deformation over again.

One of the most efficient solutions for the high-quality heteroepitaxial film growth on silicon is the use of buffer layers formed at the silicon surface before the heteroepitaxy to compensate the lattice mismatch and difference in CTE (Dariani 2014). The possibility to control PSi mechanical parameters by porosity and thickness variation and the PSi compatibility with silicon technology make it a versatile material of buffer layers for the heteroepitaxy on silicon.

7.3.1 HETEROEPITAXIAL FILMS OF II-VI AND IV-VI COMPOUND SEMICONDUCTORS

Compound II-VI and IV-VI semiconductors have the direct bandgap structure that has traditionally determined their main application area as optoelectronics. Such materials can be used to fabricate LED and photodetectors of visible and near IR ranges.

For the first time, the heteroepitaxial growth of compound II-VI and IV-VI semiconductors on the PSi buffer layer was demonstrated by Bondarenko et al. (1994) and Levchenko et al. (1999). They grew heteroepitaxial PbS films by the MBE method at 365°C using an undoped stoichiometric PbS as a source. PbS films formed on the buffer PSi were found to inherit crystallographic

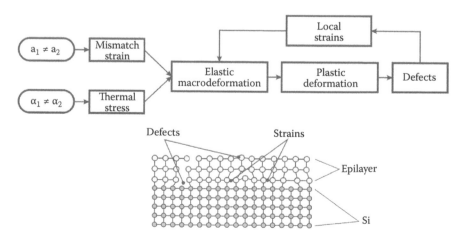

FIGURE 7.2   Deformations, strains, and stresses in epitaxial heterostructures.

orientation of the initial Si wafers. On the other hand, PbS films directly grown on monocrystal-line Si had the polycrystalline structure, which presented as a system of disoriented crystallites and irregular cracks. Heterostructures PbS/PSi/Si (100) also tended to cracking due to the absence of the effective dislocation sliding system in the epitaxial layer of (100) orientation, whereas PbS/PSi/Si (111) heterostructures were characterized by a high crystal perfection even at the 3.5-μm thickness of PbS films. An increase of PbS film thickness led to a narrowing FWHM of PbS peak on a double-crystal rocking curve of the X-ray diffraction maximum (Figure 7.3).

The heteroepitaxial PbS/PSi/Si heterostructures were used to fabricate IR-photodetectors (Yakovtseva et al. 2000). Current-voltage characteristics of the PbS/PSi/Si photodetectors demonstrated classical Schottky-barrier behavior at 77 K (Figure 7.4).

The rectification coefficient approximated 80 at a voltage of 0.5 V. At low voltages (Figure 7.4, inset), a shift of current-voltage characteristics occurred due to the thermal noise of environment at 300 K indicating a high photosensitivity of the photodiodes. Photosensitivity was observed in the wide shift range asymmetric with respect to zero value being characteristic of classical Schottky-barrier photodiodes. The resistance area product R_0A was calculated to evaluate the

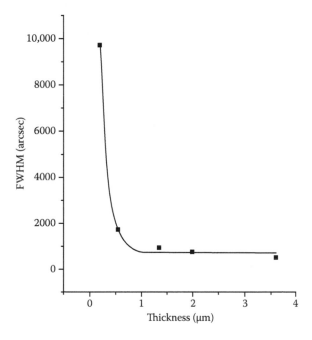

FIGURE 7.3 FWHM of X-ray diffraction reflex (111) versus the thickness of PbS layer grown on PS. (From Levchenko V.I. et al., *Thin Solid Films* 348, 131, 1999. Copyright 1999: Elsevier. With permission.)

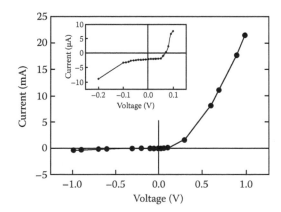

FIGURE 7.4 Low-temperate current-voltage characteristics of a Schottky-barrier photodiode. (From Yakovtseva V. et al., 182, 195, 2000. Copyright 2000: Wiley-VCH. With permission.)

photodiode sensitivity. Figure 7.5 shows R_0A versus bias voltage derived by the numerical differentiation of the experimental current-voltage characteristic and a similar experimental dependence of signal-to-noise ratio. These dependences correlate well as seen from Figure 7.5. The value $R_0A = 80$ Ohm·cm^2 at $V = 0$ is in line with the best Schottky-barrier photodiodes fabricated in multi-layer Pb/Pb$_{1-x}$Sn$_x$Se/CaF$_2$/Si structures (John et al. 1996). The best signal-to-noise ratio was equal to 500, which is very close to that for PbS films grown on CaF$_2$.

The beneficial influence of the buffer PSi layer was also demonstrated for PbTe films (Belyakov et al. 1997; Zimin et al. 1999). Deposited PbTe films had the ordered crystalline structure and orientation (200) independent on the Si substrate orientation. The PbTe crystallites grew along this direction in case of a weak bond with the wafer since a minimum of a total free surface energy of PbTe corresponds to its [100] direction. The PbTe films deposited on the micro-PSi partially consisted of amorphous phase was observed to have slightly worse quality of the crystalline structure. However, FWHM of the PbTe (100) peak on the X-ray spectrum did not change.

Chang and Lee (2000a,b, 2001) studied the epitaxial growth of ZnSe films on the buffer PSi layers formed on the Si wafers of *n*-, *p*-types. The ZnSe films were deposited in the CVD reactor at 450°C temperature of the substrate at the nucleation stage and at 650°C at the stationary growth stage. The ZnSe monocrystalline films of (111)-orientation formation was possible only on the buffer PSi layers fabricated at the accurately controlled regimes (see Table 7.1). Any deviation from these regimes led to the growth of the ZnSe polycrystalline films. The ZnSe/PSi/Si structures formed at the above-mentioned optimal regimes demonstrated PL in the visible range and accompanied by maximums at 440 and 480 nm of wavelength, which corresponds to monocrystalline ZnSe. The ZnSe/PSi/Si based PIN structures photosensitive to the short wavelengths were produced (Chang and Lee 2000b). Recently, Levchenko et al. (2015) deposited high-quality heteroepitaxial ZnSe films on PSi by MBE method at 400°C using an undoped stoichiometric ZnSe as a source (Figure 7.6).

The epitaxial growth of high-quality ZnO films on the PSi buffer layers for the development of UV- and blue-range optoelectronic devices was also attempted (Kim et al. 2012a,b,c). The deposition of the ZnO layer was realized by the PA-MBE method at 750°C temperature of the substrate. It should be noticed that the temperature of the initial ZnO seed layer deposition varied in the range of 150–550°C. Kim et al. (2012a) reported a dependence of crystallinity of the epitaxial ZnO films on the thickness of the ZnO seed layer. Its thickness increase from 50 to 350 nm led

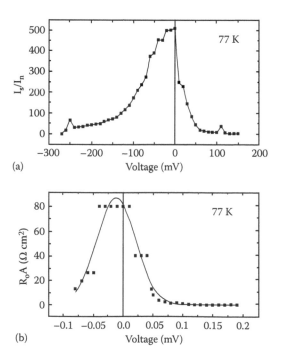

(a)

(b)

FIGURE 7.5 Low-temperature (a) signal-to-noise ratio and (b) product versus bias voltage. (From Yakovtseva V. et al., *Phys. Stat. Sol. (a)* 182, 195, 2000. Copyright 2000: Wiley-VCH. With permission.)

TABLE 7.1   A^{II}B^{VI} and A^{IV}B^{VI} Heteroepitaxy on PSi

Semi-conductor	Silicon (Type, Orientation, ρ, $\Omega \cdot$cm)	Deposition Method	Porous Silicon				Results	Application	Ref.
			UIPAC Class	d, μm	P, %	Regime (Solution, Parts, j, mA/cm², Time)			
PbS	n^+, (100), 0.01	MBE	Meso	2–5	20–40	HF:(CH$_3$)$_2$CHOH = 1:1; 10–30; 120 s	Monocrystalline layer with irregular cracks. Substrate orientation inherits.	Near IR photodetectors	Bondarenko et al. 1994; Levchenko et al. 1999
PbS	n^+, (111), 0.01	MBE	Meso	2–5	20–40	HF:(CH$_3$)$_2$CHOH = 1:1; 10–30; 120 s	Monocrystalline layer without cracks. Substrate orientation inherits.	Near IR photodetectors	Bondarenko et al. 1994; Levchenko et al. 1999
PbTe	p^+, (111), 0.03	CVD	Meso	37–62	20–24	HF (48%); 10; 30–60 min	Films consisting of grains 20–60 μm size, with (200) orientation	Near IR photodetectors	Zimin et al. 1999
PbTe	p, (100), 10	CVD	Meso	1–3	22–25	HF:(CH$_3$)$_2$CHOH = 1:1; 30; 1–10 min	Films consisting of grains 20–60 μm size, with (200) orientation	Near IR photodetectors	Zimin et al. 1999
PbTe	n, (100), 4.5	CVD	meso	20–67	5–7	HF (48%); 10; 20–60 min	Films consisting of grains 20–60 μm size, with (200) orientation	Near IR photodetectors	Zimin et al. 1999
PbTe	p^+, (111), 0.03	CVD	Micro	2–27	19–22	HF (48%); 5; 10–60 min	Films consisting of grains 20–60 μm size, with (200) orientation	Near IR photodetectors	Zimin et al. 1999
ZnSe	p, (111), 3	CVD	Meso	–	–	HF:C$_2$H$_5$OH = 3:2; 30; —	Monocrystalline film	Blue region photodetectors	Chang and Lee 2000a,b, 2001
ZnSe	n, (111), 3	CVD	Meso	–	–	HF:C$_2$H$_5$OH = 5:5; 40	Monocrystalline film	Blue region photodetectors	Chang and Lee 2000a,b, 2001
ZnSe	n, (111), (100), 0.01	MBE	Meso	–	–	HF:H$_2$O =1:3	Monocrystalline film	–	Levchenko et al. 2015
ZnO	p, (111), 1-10	PA-MBE	Meso	–	–	HF:C$_2$H$_5$OH = 1:1; 10; 30 min	Monocrystalline film	ZnO based LED, photodetectors and piezoelectric devices	Kim et al. 2012a,b,c

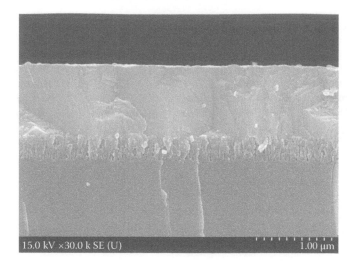

FIGURE 7.6   SEM image of cross-section of ZnSe/PSi/Si heteroepitaxial structure fabricated by MBE method at 400°C.

to narrowing FWHM of the ZnO (0002) peak. Such results are similar to those of the epitaxial growth of the PbS films on PSi (Levchenko et al. 1999). Simultaneously with the thickening of the seed ZnO layer that was in direct contact with PSi, its structure changed. At the small thickness, the structure of the heteroepitaxial ZnO film inherits the structure of PSi and can be termed as pillar or rod-like. Stresses and defects concentrate in the seed layer. At the higher thickness, the structure of the ZnO seed layer is more similar to that of monocrystalline ZnO. Kim et al. (2012b) marked an importance of temperature of the ZnO seed layer formation as well. The optimal temperature for ZnO is 350°C. Such temperature provides the fabrication of this semiconductor layer, which has minimal possible FWHM of diffraction peaks. The ZnO films on PSi formed by Kim et al. (2012a,b,c) were monocrystalline in nature and demonstrated an intense band-to-band luminescence in the UV region with a maximum at 380 nm (3.3 eV).

Data on the heteroepitaxial growth of compound $A^{II}B^{VI}$ and $A^{IV}B^{VI}$ semiconductors are tabulated in Table 7.1.

7.3.2 HETEROEPITAXIAL FILMS OF III-V COMPOUND SEMICONDUCTORS

7.3.2.1 GaAs

The monolithic integration of optical devices based on GaAs and Si electronic circuits is one of the major goals of epitaxial growth of GaAs on Si. The quality of epitaxial GaAs layer should be comparable to the bulk GaAs. However, Si and GaAs have 4% lattice mismatch, a large difference in expansion coefficients and the polar/nonpolar heterointerface between them. In spite of these problems, several techniques have been developed, which allow depositing single crystalline GaAs films with dislocation densities of less than 10^7 cm^2 on the silicon substrates. These techniques include the two-step growth technique, the incorporation of a strained layer superlattice, the thermal annealing cycles, the use of misoriented Si substrates, and the growth on patterned Si substrates.

Lin et al. (1987) pioneered in the GaAs epitaxial growth on the PSi buffer layer by the MBE method to produce GaAs/PSi heterojunction diodes. The thin intermediate GaAs layer was first deposited on the PSi layer at low temperature and the thick GaAs layer was grown at higher temperature. Both microtwins and stacking faults were shown to originate from the GaAs/PSi interface. Nevertheless, the GaAs/PSi heterojunction shows good rectification characteristics with a reasonably low leakage current. One of the reasons of high imperfection at the interface is GaAs penetration into the porous layer that can be avoided by the deposition of the additional Si sublayer on PSi (Mii et al. 1988; Kovyazina et al. 1994; Hasegawa et al. 2000). The gas pressure at MBE also influences the quality of the growing GaAs film (Kang et al. 1992, 1996). When the RHEED patterns were studied *in situ* at every growth stage, it was found that only at the end of the process the structure became the As-stabilized single-domain GaAs epilayer. The quality of the GaAs film grown on PSi is higher than that grown on single-crystal silicon and tensile stress is half (Kang et al. 1992).

When the MOCVD method was used for the GaAs deposition on PSi, the heterogeneity in the Ga and As concentrations through the film thickness appeared (Kovyazina et al. 1994, 1995). Practically only the thin interfacial defect-free layer (30–50 nm) partially penetrated into the pores and containing Ga and As in the stoichiometric ratio was observed. The formation of polycrystal grains and defects began in any distance from the interface and was likely caused by the external stress of the film. With MOCVD and MBE, the crystal perfection of GaAs heteroepitaxial films grown on PSi has improved with the thickness. MOCVD GaAs epitaxial films became crystalline only at thicknesses higher than 1.8 μm. It is worthy of note that the polycrystalline GaAs films grew on single-crystal silicon at the same regimes. The use of MOCVD allows thick (5–7 μm) crack-free GaAs epitaxial films to be grown on PSi. The overall quality and internal tension of the structures were shown to depend on the PSi porosity (Kovyazina et al. 1995).

The use of the chemical beam epitaxy (CBE) method for the GaAs deposition on PSi allows reducing the deposition temperature and thermal load on the heteroepitaxial structure as compared with MOCVD (Saravanan et al. 2002). The chemical-mechanical polishing (CMP) of the surface of the GaAs heteroepitaxial film allowed decreasing the roughness and internal stress and improved PL characteristics.

7.3.2.2 InSb

InSb is a narrow-band semiconductor compound with the high electron mobility attractive for the application for infrared imaging systems, free space communications, magneto-resistive sensors, and high-speed photodetectors. Heteroepitaxial InSb films should be grown on the substrates of high resistivity to prevent current leakage. The heteroepitaxy of InSb on PSi was first performed by Farag et al. (2008) using the liquid phase epitaxy (LPE). InSb films were deposited from the cooling saturated InSb metallic melt at a temperature of 350–280°C. As a result, the InSb films of high crystallinity were grown. The films had (100)-orientation different from that of the initial Si wafers. The heterostructures InSb/PSi/Si demonstrated a strong rectifying behavior under an applied potential.

7.3.2.3 GaN

GaN is a wide-band-gap semiconductor with high electron mobility, large value of breakdown electric field, and excellent thermochemical characteristics used in UV and blue range optoelectronics and high electron mobility transistors. Heteroepitaxial GaN growth on foreign substrate is a common approach because of limited availability of bulk GaN wafers. To produce high-quality epitaxial films, it is necessary to have substrates with lattice parameters, chemical and physical characteristics correlated with those of the film grown. Most modern GaN-based devices employ sapphire Al_2O_3 or 6H-SiC substrates. However, advantages of Si wafers attract developing alternative approaches using heteroepitaxy of GaN on Si with different buffer layers, in particular, AlN, AlOx, and PSi.

The use of the buffer PSi layer between the GaN film and the Si substrate partially compensates the misfit between the top face of the Si substrate and the bottom face of the GaN film. Additionally, any microcracking of the PSi layer decreases the stress transmitted to the GaN layer (Tsao et al. 1999). However, perfect epitaxial films should be oriented. The film orientation is defined by the substrate crystal structure. The greater the difference in the crystal structure and symmetry between the film and substrate, the more strained and imperfect is the film grown. However, even in the case when the crystal structure and symmetry of the film and substrate are closely related, but the substrate surface is nonuniform and contains various defects, the quality of the film is degraded strongly. Therefore, the PSi layer, despite the ability of stress relaxation, distinctly decreases orientation properties of the substrate surface. Note that the epitaxial film formation refers to the first-order phase transitions. Nevertheless, the phase transition process is a stochastic process. Therefore, the pores at the substrate surface just increase the nucleation randomness thereby degrading the layer orientation. That is why Matoussi et al. (2001) failed to grow by MOCVD device-quality GaN films directly on the PSi surface.

To reduce the nucleation randomness and improve the growing layer orientation, the surface of the buffer PSi layer should be made smooth. Chaaben et al. (2004) used a porous double-layer system with a low surface porosity and a high porosity in the depth. It was shown that continuous GaN epitaxial layer grown over this structure consisted of variously oriented large crystals of size 2–3 μm. The low-temperature photoluminescence was studied to reveal excitonic transition energies below the DBE line at 3.44 and 3.42 eV related to structural and surface defects in polycrystalline GaN. Matoussi et al. (2008, 2010) determined that at high annealing temperatures, the high-porous PSi layer can crack, resulting in the peeling of the entire on-PSi structure. The decrease of the PSi porosity from 60 to 40% reduces crack probability and has a beneficial effect on the surface uniformity of the GaN layer after thermal annealing at temperatures up to 1050°C (Chaaben et al. 2006).

To provide the nucleation of the hexagonal GaN material, the surface of the PSi layer can be coated with AlN (Boufaden et al. 2003; Chaaben et al. 2004, 2006). The increase in the AlN layer thickness has beneficial effect on the GaN film quality and reduction of the number of crystal-lattice orientations of GaN grains. Since PSi sintering occurs at temperatures above 850°C for the high-temperature AlN and GaN epitaxy, the PSi structure should be stabilized by pretreatment and the surface of the PSi should be prebaked to close up the pores. The buffer layers of AlN on PSi formed at the optimal regimes provided further deposition of the heteroepitaxial GaN films with better alignment due to lattice mismatch between GaN and Si reduced to 2.5% (Chaaben et al. 2006).

The formation of the GaN epitaxial film at the PSi surface can be made in the epitaxial lateral overgrowth (ELO) or NHELO (nanoheteroELO) regimes. In this case, GaN begins to grow laterally from grains already formed and finally single clusters join together to form a continuous film (Beaumont et al. 2001). This technique allows obtaining less stressed high-quality GaN layers with a small number of defects. Such layers showed advanced photo-emissivity. Liang et al. (2003) formed the PSi layer by RIE of Si through the anodic alumina template instead of the Si anodization. Then the 400-nm thick GaN films were grown by the MBE method (Figure 7.7).

The optical studies show fivefold luminescence enhancement and a small shift of the luminescence peak at 3.4 eV for these films as compared with films grown on the single-crystal silicon. The authors associate this effect with the stress reducing in the GaN/PSi film. A similar technique was used by Zang et al. (2006). The pattern of the anodic alumina template was

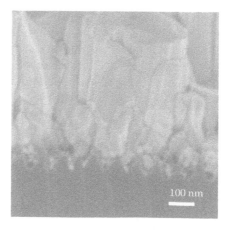

FIGURE 7.7 Oblique cross-sectional scanning electron microscope view of the nanoheteroepitaxial GaN on Si interface. The Si nanopores contain numerous small GaN particles formed in the early stages of growth. (From Liang J. et al., *Appl. Phys. Lett.* 83, 1752, 2003. Copyright 2003: AIP Publishing. With permission.)

extended to the silicon wafer using CF_4-based inductively coupled plasma (ICP) to form the 160–180 nm thick PSi layer. Before the GaN epitaxial growth, the 3-nm thick AlN film was deposited on the PSi layer by the MOCVD method to prevent a direct reaction of Ga and Si and then buffer 70-nm thick $Al_xGa_{1-x}N$ layer was formed. The 1.5–2 μm thick GaN film grown consisted of a conjoined pyramids with angle between faces of 62°. The study by the Raman spectroscopy showed a considerable decrease of the internal stress in the heteroepitaxial film grown on the patterned substrate as compared with the film deposited on the planar silicon surface. A considerable luminescence enhancement and small shift of the luminescence peak at 3.4 eV was established as well.

Ishikawa et al. (2008) grew perfect single-crystal GaN films by the MOCVD method on the 20–100 μm thick PSi layer formed by the conventional anodization. Two types of structures were studied. For the first type, 50-nm thick AlN was deposited before the 1-μm thick GaN deposition. For the second type, before the 200-nm thick n-GaN growth, the 200-nm thick AlN layer was deposited followed by the deposition of 20 pairs of n-GaN/n-AlN (20/5 nm) layers. SEM studies of the structures formed revealed that the PSi morphology changed by the action of high temperature of the GaN epitaxy, namely, the nanostructure changed to the large enough foam-like structure with 100-nm sized cavities (Figure 7.8). At the same time, the intermediate buffer layer covered completely the PSi layer, preventing Ga ELO directly on the PSi layer. Despite the change of structure, the PSi layer was equal to the lattice matching task. Some cavities were at the surface of the porous layer, resulting in the hexagonal defect formation in the GaN layer. The XRD analysis of a 1-μm thick GaN film grown on PSi with a 50-nm thick AlN layer showed that diffraction peaks from the c-face symmetric reflections were clearly observed and no other peaks except for the diffraction resulting from the c-face were observed. The Raman spectroscopy revealed the stress reduction, and the structures with intermediate n-GaN/n-AlN layers were less stressed at that (Figure 7.9). The improvement of photoluminescence was observed as well.

Finally, the use of the PSi buffer layers allows growing the epitaxial GaN films at the surface of the silicon substrates. The single-crystal structure of the GaN film can be obtained by varying the technological parameters. The additional buffer AlN layer at the PSi layer surface prevents ELO on the PSi layer and improves the perfection of the GaN film.

Data on heteroepitaxial growth of compound III-V semiconductors are tabulated in Table 7.2.

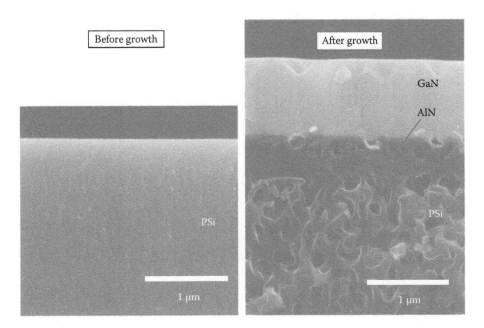

FIGURE 7.8 Cross-sectional SEM images of PSi substrate and 1-μm-thick GaN film on PSi substrate. (From Ishikawa H. et al., *J. Cryst. Growth* 310, 4900, 2008. Copyright 2008: Elsevier. With permission.)

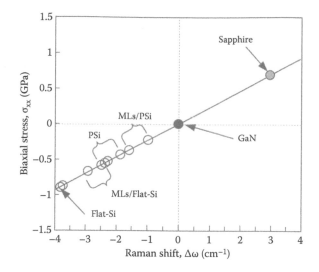

FIGURE 7.9 Biaxial stresses of GaN films on various substrates estimated from the Raman spectra. (From Ishikawa H. et al., *J. Cryst. Growth* 310, 4900, 2008. Copyright 2008: Elsevier. With permission.)

TABLE 7.2 A^{III}B^{V} Heteroepitaxy on PSi

| Semi-conductor | Silicon (Type, Orientation, ρ, Ω·cm) | Deposition Method | Porous Silicon | | | | Results | Application | Ref. |
			UIPAC Class	d, μm	P, %	Regime (Solution, Parts, j, mA/cm², Time)			
AlN	p, (111)	MOVPE	Meso	2	40	–	Polycrystalline films	Buffer for GaN epitaxy	Chaaben et al. 2006
GaAs	p, (100), 0.02	MBE	–	–	–	HF (30%)	Very good crystallinity	GaAs/PSi heterojunction diodes	Lin et al. 1987; Mii et al. 1988
GaAs	p, (100), 0.02	MOCVD	–	–	30	HF (12%)	Better than on buck Si	–	Kovyazina et al. 1994, 1995
GaAs	p, (100), 0.001	MBE	–	20	–	HF:H$_2$O:C$_2$H$_5$OH = 1:1:2; 20	Good crystalline quality	–	Hasegawa et al. 2000
GaAs	p, (100), 1.72–2.53	MBE	–	–	–	HF (48%); 30–100; 10 min	Better than on bulk Si	Optoelectronics	Kang et al. 1992, 1996
GaAs	p, (100), 0.01–0.02	CBE	Micro	12	–	HF:C$_2$H$_5$OH = 12:1	Biaxial tensile stress reduced, good PL intensity	Optoelectronics	Saravanan et al. 2002
GaN	p, (100), 1–3	MOVPE	–	1–3	–	Vapor etching	Polycristalline film	–	Matoussi et al. 2001 2010
GaN	p, (111)	MOVPE	Meso	2	40	–	Polycristalline film	–	Chaaben et al. 2006
GaN	p, (111), 0.005–0.018	MOCVD	Meso	20–100	–	–	Perfect single-crystal film	High-power electronics	Ishikawa et al. 2008
InSb	p, (111), 20	LPE	Meso	–	–	HF:C$_2$H$_5$OH = 1:1; 10; 30 min	Monocrystalline film with (100) orientation	InSb based devices	Farag et al. 2008

7.3.3 GERMANIUM AND SILICON-GERMANIUM ALLOYS

PSi presenting suitable substrate for growth of the heteroepitaxial layers with decreased stresses and deformations has also applied to form films of the Si_xGe_{1-x} alloy and Ge. Heteroepitaxial films of these materials can act as pseudo- or virtual substrates to each other. However, the direct formation of the Si_xGe_{1-x} and Ge films on Si substrates is difficult to realize because of the great lattice mismatch of Si and Ge, which is evaluated as 4.18%.

The first attempts to fabricate the SiGe heteroepitaxial film by the MBE method were made by Xie and Bean (1990a,b). The study of the $Ge_{0.16}Si_{0.84}$ films grown on the 800 nm buffer PSi layer did not show significant improvement of their quality. It was explained as follows. The stress field in the Si_xGe_{1-x} film does not have the exponential decaying behavior because of the nature of the interconnected growth area in the PSi substrate. However, later works showed that PSi layers formed in the same substrates have beneficial effect on the stress relaxation in Si_xGe_{1-x} and Ge films. Aouassa et al. (2012) and Gouder et al. (2014a,b) used the UHV-MBE deposition of Ge to fabricate Si_xGe_{1-x} films. They deposited Ge on the buffer PSi layers of the thickness about 200 nm at 400–700°C. The epitaxial deposition of Ge atoms on PSi, which is not accompanied by the Si deposition, at the temperatures preventing fast PSi sintering allowed Ge penetration into pores. In this way, Si_xGe_{1-x} structures, which had Ge contents of 50 and 96%, were formed. These structures were characterized by the presence of the thin layer (40–150 nm) of relaxed or slightly stressed Ge with the flat surface (RMS ≈ 3–19 nm) and the top layer free of dislocations on the surface. The method of the partial low-temperature oxidation at 500°C of PSi to fabricate Si_xGe_{1-x} films on PSi was proposed by Kim et al. (2007). A tensile-strained $Si_{0.77}Ge_{0.23}$ layer grown on such substrates by MBE at 500°C was free from dislocations and microcracks.

7.3.4 DIAMOND FILMS

Extremely high thermal conductivity, chemical resistance, unique hardness, and low friction coefficient determine an interest to the study of synthesis and features of different diamond materials. In addition, diamond is characterized by the record electron and hole mobility. Diamond films can be used in high-performance ICs, thermal- and radiation-hardened electronics, acoustic delay lines, electron multipliers, solar-blind detectors, electrochemical electrodes, and sensors able to operate in an aggressive environment, and many other unique devices. Interest to heteroepitaxial diamond films on Si and particularly PSi is caused by a combination of promising features of diamond and high availability of Si in microelectronics.

Spitzl et al. (1994) were the first who presented the work on heteroepitaxial growth of diamond films on PSi by MPCVD. Just before diamond deposition, PSi was cleaned in a CH_4/H_2 mixture at voltage of 200 V to increase a number of the nucleation centers of diamond nanocrystals. Diamond films were found to consist of smaller grains and to be less stressed than those on monocrystalline Si. Later, Spitzl et al. (1995) and Raiko et al. (1996) showed that morphology of diamond films depends on PSi morphology; however, the mechanism of the dependence was not described in the above-mentioned papers. Liu et al. (1995) performed heteroepitaxial growth of diamond film on PSi by the method of chemical vapor deposition with a hot filament (HFCVD). The deposition of diamond films followed previous ultrasonic treatment of the sample with a diamond abrasive. The diamond films were characterized by good crystallinity. The authors noticed that diamond films tended to grow on PSi through the phase of SiC.

Authors of the above-mentioned papers connected decrease of diamond film grains with a higher level of roughness of PSi surface in contrast to monocrystalline Si wafer. However, Iyer and Srinivas (1997) controverted this supposition underlining that diamond film growth in those works was accompanied by the pretreatment of the PSi surface. They experimentally showed that avoiding pretreatment leads to growth of diamond film of similar structure on both PSi and monocrystalline Si. Very slow growth of diamond films on PSi was also noticed. Nevertheless, Iyer and Srinivas (1997) presented no explanation of such growth kinetics. Results of researchers from the United States (Khan et al. 1998) and later several groups almost completely proved the conclusions of Iyer and Srinivas (1997): using non-treated PSi leads to long and weak heteroepitaxial growth of diamond films by MPCVD. They coated the PSi surface with diamond nanopowder to improve uniformity of the epitaxial films. It caused a decrease of cluster formation and an increase of adhesion of the formed diamond films. Two-step diamond deposition on the PSi by HFCVD with the growing of the preliminary seed film on first step also allows forming dense diamond films (Liao et al. 2000). The further results of a comprehensive study of diamond HFCVD deposition on PSi (Baranauskas et al. 1999, 2000, 2001) stated that inappropriate parameters of PSi can lead to destruction of the porous layer during the epitaxial process; diamond film forms in two stages: nucleation on the tops of Si crystallites of PSi and following coalescence and growth of new grains on the film deposited on the first stage; diamond film tends to inherit a porous PSi structure.

The first proposed application of diamond films on PSi was protection of the more fragile PSi layer underneath (Fernandes et al. 1999; Wang et al. 2000). Growth of the PL stability of PSi covered with diamond film by MPCVD accompanied by pretreatment of the PSi surface as well as mechanical hardness of the sample surface increase were noted.

Low work function and high electric strength of diamond have driven researchers to study a possibility of its application as cold emission cathodes. Two papers are known that have reported results on the study of diamond films on PSi for the above-mentioned application. Chen et al. (2004) formed diamond films by MP CVD while Arora et al. (2008) used HF CVD for this purpose. Decrease of switching voltage for electron emission in contrast to diamond films grown on monocrystalline Si was found in both works. What is more, this potential was five times less at MP CVD epitaxy than that at HF CVD.

Several papers have been published on the application of diamond films on PSi for fabrication of effective electrochemical electrodes (Ferreira et al. 2005, Chen et al. 2006, Miranda et al. 2010). Chemical inertness of diamond provides a wide working voltage range. At the same time, developed surface of PSi improves specific parameters and response at CV-analysis.

Summarizing we should note that heteroepitaxy of diamond films on PSi has still not been widely spread because carrying out this process aiming to protect PSi is not profitable. Furthermore, diamond heteroepitaxy on PSi for the sake of diamond film fabrication is beside the purpose and has no advantages over alternative methods, which have been recently developed and are technologically worse. Moreover, nowadays many materials demonstrating most of the attractive diamond features can be produced (diamond-like films, Si carbide, etc.).

TABLE 7.3   Deposition of Miscellaneous Materials on PSi by Nonepitaxial Methods

Material	Deposition Technique	Results	Ref.
BST $Ba_xSr_{1-x}TiO_3$	Pulsed laser deposition (PLD)	Polycrystalline films	Liu et al. 2004
ITO $(In_2O_3)_x-(SnO_2)_{1-x}$	RF-sputtering, sol-gel spin coating	Slightly stressed polycrystalline films or crystalline film	Ghosh et al. 2002; Daoudi et al. 2003
CdS	Electrochemical deposition	100 nm grains	Zhang et al. 2001
CdSe	Electrochemical deposition	Oriented smooth polycrystalline film	Chubenko et al. 2009
GaAs	Electrochemical deposition	Polycrystalline films	Lajnef et al. 2010
GaN	RF-sputtering, Ga_2O_3 nitridization	Porous oriented film, 0.5–2 crystals gathered in clusters	Dong et al. 2006; Samsudin et al. 2014
MnSb	Physical vapor deposition (PVD)	Net-like film resembling PSi surface morphology	Dai et al. 2007
SiC	APCVD, RF-sputtering	Polycrystalline film, porous film	Severino et al. 2006; Naderi and Hashim 2013
TeO_2	Thermal evaporation	Nanowires array	Wu et al. 2014
YIG (yttrium iron garnet)	PLD	Single-phase polycrystalline film	Zheng et al. 2014
WO_3	Electrochemical deposition, thermal W oxidation	Crystalline film, pore decorating nanowires array	Ma et al. 2013; Mendoza-Agüero and Agarwal 2013
ZnO	RF-sputtering, electrochemical deposition, sol-gel spin coating	Continuous polycrystalline films	Alaya et al. 2009; Liu et al. 2003b; Cai et al. 2009; Balucani et al. 2011
ZnSe	Electrochemical deposition	Polycrystalline film	Chubenko et al. 2009

7.3.5 OTHER MATERIALS

Various materials have also been tried to be grown on the PSi substrates by non-MBE or CVD based methods excepting the above-mentioned semiconductors. Fabricated films have usually had polycrystalline nature. Under this approach, PSi influence on the structure of the deposited films has been decreased but not excluded. Many works in this direction have mostly been related to the fabrication of nanocomposites *"something*/PSi" and their results may be found in Chapter 9 of this book. Nevertheless, for an understanding of the number of papers published on this research, some of the results are presented in Table 7.3. The table does not contain polymers and organic compounds introduced in PSi because this information is described in detail in Chapter 9.

7.4 SUMMARY

PSi can be used as a buffer layer both for homo- and heteroepitaxial growth of semiconductor films on silicon substrates. Processes of homoepitaxial growth of silicon films on the PSi are applied as part of layer transfer technology used in industrial production of SOI structures (ELTRAN) and thin-film silicon solar cells (see corresponding chapters of this book). These processes were successfully optimized for fabricating the high-quality homoepitaxial Si films for production of commercial products.

Beneficial effects of PSi buffer layer on crystalline quality have been demonstrated for heteroepitaxial films of PbS, PbSe, ZnSe, ZnO, GaAs, and GaN. Although this approach did not find an industrial application, we can expect further research in this direction, which is arguably very important to develop the technology, providing formation of the single-crystal films of compound semiconductors on silicon substrates. PSi layers can be also used for depositing the polycrystalline and nanostructured films of various semiconductors and the metal oxides materials on silicon substrates for fabrication of prospective optoelectronic devices and sensors.

ACKNOWLEDGMENTS

The authors thank Dr. V. Yakovtseva and Dr. V. Levchenko for helpful discussions.

REFERENCES

Alaya, A., Nouiri, M., Ben Ayadi, Z., Djessas, K., Khirouni, K., and El Mir, L. (2009). Elaboration and characterization of Si(n)/PS/ZnO(n) structure obtained by rf-magnetron sputtering from aerogel nanopowder target material. *Mater. Sci. Eng. B* **159–160**, 2–5.

Aouassa, M., Escoubas, S., Ronda, A. et al. (2012). Ultra-thin planar fully relaxed Ge pseudo-substrate on compliant porous silicon template layer. *Appl. Phys. Lett.* **101**, 233105.

Arora, S., Chhoker, S., Sharma, N., Singh, V., and Vankar, V. (2008). Growth and field emission characteristics of diamond films on macroporous silicon substrate. *J. Appl. Phys.* **104**, 103524.

Balucani, M., Nenzi, P., Chubenko, E., Klyshko, A., and Bondarenko, V. (2011). Electrochemical and hydrothermal deposition of ZnO on silicon: From continuous films to nanocrystals. *J. Nanopart. Res.* **13**, 5985–5997.

Baranauskas, V., Li, B.B., Peterlevitz, A.C., Tosin, M.C., and Durrant, S.F. (1999). Structure and properties of diamond films deposited on porous silicon. *Thin Solid Films* **355**, 233–238.

Baranauskas, V., Tosin, M., Peterlevitz, A., Ceragioli, H., and Durrant, S. (2000). Microcrystalline diamond deposition on a porous silicon host matrix. *Mater. Sci. Eng. B* **69**, 171–176.

Baranauskas, V., Peterlevitz, A.C., Chang, D.C., and Durrant, S.F. (2001). Method of porous diamond deposition on porous silicon. *Appl. Surf. Sci.* **185**, 108–113.

Beale, M.I.J., Chew, N.G., Cullis, A.G. et al. (1985). A study of silicon MBE on porous silicon substrates. *J. Vac. Sci. Technol. B* **3**, 732–735.

Beaucarne, G., Duerinckx, F., Kuzma, I., Van Nieuwenhuysen, K., Kim, H.J., and Poortmans, J. (2006). Epitaxial thin-film Si solar cells. *Thin Solid Films* **511–512**, 533–542.

Beaumont, B., Venne, P., and Gibart, P. (2001). Epitaxial lateral overgrowth of GaN. *Phys. Stat. Sol. (b)* **43**, 1–43.

Belyakov, L.V., Zakharova, I.B., Zubkova, T.I., Musikhin, S.F., and Rykov, S.A. (1997). Study of PbTe photodiodes on a buffer sublayer of porous silicon. *Semiconductors* **31**, 76–77.

Bergmann, R.B., Berge, C., Rinke, T.J., Schmidt, J., and Werner, J.H. (2002). Advances in monocrystalline Si thin film solar cells by layer transfer. *Sol. Energy Mater. Sol. Cells* **74**, 213–218.

Bilyalov, R., Solanki, C.S., Poortmans, J. et al. (2002). Crystalline silicon thin films with porous Si backside reflector. *Thin Solid Films* **403–404**, 170–174.

Blanchard, N.P., Boucherif, A., Regreny, P.H. et al. (2011). Engineering pseudosubstrates with porous silicon technology. In: Nazarov A., Colinge J.-P., Balestra F., Raskin J.-P., Gamiz F., and Lysenko V.S. (Eds.). *Semiconductor on Insulator Materials for Nanoelectronics*. Springer-Verlag, Berlin, Heidelberg, pp. 47–67.

Bomchil, G., Halimaoui, A., and Herino, R. (1989). Porous silicon: The material and its applications in silicon-on-insulator technologies. *Appl. Surf. Sci.* **41–42**, 604–613.

Bondarenko, V.P., Vorozov, N.N., Dikareva, V.V. et al. (1994). Heteroepitaxial growth of lead sulfide on silicon. *Tech. Phys. Lett.* **20**, 410–411.

Bondarenko, V.P., Bondarenko, A.V., Dolgyi, L.N. et al. (1996). Porous silicon based heterostructures: Formation, properties, and application. In: *Proc. of International Semiconductor Conference CAS'96*, Oct. 9–12, 1996, Sinaia, Romania, pp. 229–232.

Bondarenko, V., Troyanova, G., Balucani, M., and Ferrari, A. (2005). Porous silicon based SOI: History and prospects. In: Flandre D., Nazarov A., and Hemment P. (Eds.) *Science and Technology of Semiconductor-On-Insulator Structures and Devices Operating in a Harsh Environment*. Kluwer Academic Publishers, the Netherlands, pp. 53–64.

Boufaden, T., Chaaben, N., Christophersen, M., and El Jani, B. (2003). GaN growth on porous silicon by MOVPE. *Microelectr. J.* **34**, 843–848.

Brendel, R. (1997). A novel process for ultrthin monocrystalline silicon solar cells on glass. In: *Proc. of 14th European Photovoltaic Solar Energy Conference*, June 30–July 04, 1997, Barcelona, Spain, 1354.

Brendel, R., Auer, R., and Artmann, H. (2001). Textured monocrystallinr thin-film Si cells from the porous silicon (PSI) process. *Prog. Photovolt: Res. Appl.* **9**, 217–221.

Cai, H., Shen, H., Yin, Y., Lu, L., Shen, J., and Tang, Z. (2009). The effects of porous silicon on the crystalline properties of ZnO thin films. *J. Phys. Chem. Solids* **70**, 967–971.

Cai, H., Shen, H., Zhang, L. et al. (2010). Silicon epitaxy on textured double layer porous silicon by LPCVD. *Phys. B* **405**, 3852–3856.

Chaaben, N., Boufaden, T., Christophersen, M., and El Jani, B. (2004). Structural and optical characterization of GaN grown on porous silicon substrate by MOVPE. *Microelectr. J.* **35**, 891–895.

Chaaben, N., Yahyaoui, J., Christophersen, M., Boufaden, T., and El Jani, B. (2006). Morphological properties of AlN and GaN grown by MOVPE on porous Si(111) and Si(111) substrates. *Superlattices Microstruct.* **40**, 483–489

Chang, C.C. and Lee, C.H. (2000a). Characterization and fabrication of ZnSe epilayer on porous silicon substrate. *Thin Solid Films* **379**, 287–291.

Chang, C.C. and Lee, C.H. (2000b). Study and fabrication of PIN photodiode by using ZnSe/PS/Si Structure. *IEEE Trans. Electron Dev.* **47**, 50–54.

Chang, C.C. and Lee, C.H. (2001). The study of highly crystalline ZnSe growth on porous silicon. *J. Mater. Sci.* **36**, 3801–3803.

Chen, G., Cai, R., Song, X., and Deng, J. (2004). Preparation and field electron emission of microcrystalline diamond deposited on a porous silicon substrate. *Mater. Sci. Eng. B* **107**, 233–236.

Chen, Y.-J., Young, T.-F., Lee, S.-L. et al. (2006). Electrochemical studies of nano-structured diamond thin-film electrodes grown by microwave plasma CVD. *Vacuum* **80**, 818–822.

Christiansen, S., Albrecht, M., Michler, J., and Strunk, H.P. (2006). Elastic and plastic relaxation in slightly undulated misfitting epitaxial layers—A quantitative approach by three-dimensional finite element calculations. *Phys. Stat. Sol. (a)* **156**, 129–150.

Chubenko, E.B., Klyshko, A.A., Petrovich, V.A., and Bondarenko, V.P. (2009). Electrochemical deposition of zinc selenide and cadmium selenide onto porous silicon from aqueous acidic solutions. *Thin Solid Films* **517**, 5981–5987.

Dai, R., Chen, N., Zhang, X.W., and Penga, C. (2007). Net-like ferromagnetic MnSb film deposited on porous silicon substrates. *J. Cryst. Growth* **299**, 142–145.

Daoudi, K., Sandu, C.S., Moadhen, A. et al. (2003). ITO spin-coated porous silicon structures. *Mater. Sci. Eng. B* **101**, 262–265.

Dariani, R.S. (2014). Heteroepitaxy on porous silicon. In: Canham L. (Ed.) *Handbook of Porous Silicon*. Springer International Publishing, Switzerland, pp. 581–588.

Dong, Z., Xue, C., Zhang, H., Gao, H., Tian, D., and Wu, Y. (2006). Synthesis of GaN films on porous silicon substrates. *Rare Met.* **25**, 96–98.

Duerinckx, F., Kuzma-Filipek, I., Van Nieuwenhuysen, K., Beaucarne, G., and Poortmans, J. (2006). Reorganized porous silicon bragg reflectors for thin-film silicon solar cells. *IEEE Electron Dev. Lett.* **27**, 837–839.

Duttagupta, S.P. and Fauchet, P.M. (1997). Microhardness of porous silicon. In: Canham L. (Ed.) *Properties of Porous Silicon*. INSPEC, London, pp. 132–137.

Farag, A.A.M., Ashery, A., and Terra, F.S. (2008). Fabrication and electrical characterization of n-InSb on porous Si heterojunctions prepared by liquid phase epitaxy. *Microelectr. J.* **39**, 253–260.

Fave, A., Quoizola, S., Kraiem, J., Kaminski, A., Lemiti, M., and Laugier, A. (2004). Comparative study of LPE and VPE silicon thin film on porous sacrificial layer. *Thin Solid Films* **451–452**, 308–311.

Fernandes, A., Ventura, P., Silva, R., and Carmo, M. (1999). Porous silicon capping by CVD diamond. *Vacuum* **52**, 215–218.

Ferreira, N., Azevedo, A., Beloto, A. et al. (2005). Nanodiamond films growth on porous silicon substrates for electrochemical applications. *Diamond Relat. Mater.* **14**, 441–445.

Ghosh, S., Kim, H., Hong, K., and Lee, C. (2002). Microstructure of indium tin oxide films deposited on porous silicon by rf-sputtering. *Mater. Sci. Eng. B* **95**, 171–179.

Gouder, S., Mahamdi, R., Aouassa, M. et al. (2014a). Investigation of microstructure and morphology for the Ge on porous silicon/Si substrate hetero-structure obtained by molecular beam epitaxy. *Thin Solid Films* **550**, 233–238.

Gouder, S., Mahamdi, R., Aouassa, M. et al. (2014b). FTIR and AFM studies of the Ge on Porous Silicon/Si substrate hetero-structure obtained by molecular beam epitaxy. *J. New Technol. Mater.* **4**, 112–115.

Hasegawa, S., Maehashi, K., and Nakashima, H. (2000). Growth and characterization of GaAs films on porous Si. *J. Cryst. Growth* **95**, 113–116.

Herino, R., Perio, A., Barla, K., and Bomchil, G. (1984). Microstructure of porous silicon and its evolution with temperature. *Mater. Lett.* **2**, 519–523.

Huangfu, Y., Zhan, W., Hong, X., Fang, X., Ding, G., and Ye, H. (2013). Heteroepitaxy of Ge on Si(001) with pits and windows transferred from free-standing porous alumina mask. *Nanotechnology* **24**, 185302.

Imai, K. (1981). A new dielectric isolation method using porous silicon. *Solid-State Electron.* **24**, 159–164.

Ishikawa, H., Shimanaka, K., Tokura, F., Hayashi, Y., Hara, Y., and Nakanishi, M. (2008). MOCVD growth of GaN on porous silicon substrates. *J. Cryst. Growth* **310**, 4900–4903.

Ito, T., Yasumatsu, T., and Hiraki, A. (1990). Homoepitaxial growth of silicon on anodized porous silicon. *Appl. Surf. Sci.* **44**, 97–102.

Iyer, S.B. and Srinivas, S. (1997). Diamond deposition on as-anodized porous silicon: Some nucleation aspects. *Thin Solid Films* **305**, 259–265.

Jin, S., Bender, H., Stalmans, L. et al. (2000). Transmission electron microscopy investigation of the crystallographic quality of silicon films grown epitaxially on porous silicon. *J. Cryst. Growth* **212**, 119–127.

John, J., Fach, A., Masek, J., Müller, P., Paglino, C., and Zogg, H. (1996). IR-sensor array fabrication in Pb$_{1-x}$Sn$_x$Se-on-Si heterostructures. *Appl. Surf. Sci.* **102,** 346–349.

Kang, T.W., Oh, Y.T., Leem, J.Y., and Kim, T.W. (1992). Growth and optical studies of a GaAs epitaxial layer on porous Si (100) grown by molecular beam epitaxy. *J. Mater. Sci. Lett.* **11**, 392–395.

Kang, T.W., Leem, J.Y., and Kim, T.W. (1996). Growth of GaAs epitaxial layers on porous silicon. *Microelectr. J.* **27**, 423–436.

Karim, M., Martini, R., Radhakrishnan, H.S. et al. (2014). Tuning of strain and surface roughness of porous silicon layers for higher-quality seeds for epitaxial growth. *Nanoscale Res. Lett.* **9**, 348.

Khan, M., Haque, M., Naseem, H., Brown, W., and Malshe, A. (1998). Microwave plasma chemical vapor deposition of diamond films with low residual stress on large area porous silicon substrates. *Thin Solid Films* **332**, 93–97.

Kim, J., Li B., and Xie, Y.H. (2007). A method for fabricating dislocation-free tensile-strained SiGe films via the oxidation of porous Si substrates. *Appl. Phys. Lett.* **91**, 252108.

Kim, M.S., Kim, D.Y., Cho, M.Y. et al. (2012a). Effects of buffer layer thickness on properties of ZnO thin films grown on porous silicon by plasma-assisted molecular beam epitaxy. *Vacuum* **86**, 1373–1379.

Kim, M.S., Kim, S., Nam, G., Lee, D.-Y., and Leem, J.-Y. (2012b). Effects of growth temperature for buffer layers on properties of ZnO thin films grown on porous silicon by plasma-assisted molecular beam epitaxy. *Opt. Mater.* **34**, 1543–1548.

Kim, M.S., Nam, G., Son, J.-S., and Leem, J.-Y. (2012c). Photoluminescence studies of ZnO thin films on porous silicon grown by plasma-assisted molecular beam epitaxy. *Curr. Appl. Phys.* **12**, S94–S98.

Konaka, S., Tabe, M., and Sakai, T. (1982). A new silicononinsulator structure using a silicon molecular beam epitaxial growth on porous silicon. *Appl. Phys. Lett.* **41**, 86–88.

Kovyazina, T., Kutas, A., Khitko, V. et al. (1994). Heteroepitaxy of GaAs on porous silicon: The structure of the interface. *Mater. Sci. Forum* 147-147, 583–586.

Kovyazina, T.V., Prokhorenko, T.A., Sobolev, N.A. et al. (1995). Properties of MOCVD GaAs films grown on porous Si. In: Borisenko, V.E., Filonov, A.B., Gaponenko, S.V. et al. (Eds.). *Physics, Chemistry and Application of Nanostrucrures*, World Scientific, Singapore, pp. 97–100.

Krinke, J., Kuchler, G., Brendel, R. et al. (2001). Microstructure of epitaxial layers deposited on silicon by ion assisted deposition. *Sol. Energy Mater. Sol. Cells* **65**, 503–508.

Kuzma-Filipek, I., Duerinckx, F., Van Nieuwenhuysen, K., Beaucarne, G., Poortmans, J., and Mertens, R. (2007). Porous silicon as an internal reflector in thin epitaxial solar cells. *Phys. Stat. Sol. A* **204**, 1340–1345.

Kuzma-Filipek, I.J., Duerinckx, F., Van Kerschaver, E., Van Nieuwenhuysen, K., Beaucarne, G., and Poortmans, J. (2008). Chirped porous silicon reflectors for thin-film epitaxial silicon solar cells. *J. Appl. Phys.* **104**, 073529.

Kuzma-Filipek, I., Duerinckx, F., Van Nieuwenhuysen, K., Beaucarne, G., Poortmans, J., and Mertens, R.. (2009). A porous silicon intermediate reflector in thin film epitaxial silicon solar cells as a gettering site of impurities. *Phys. Stat. Sol. (c)* **6**, 1745–1749.

Kuzma-Filipek, I., Dross, F., Baert, K. et al. (2012). >16% thin-film epitaxial silicon solar cells on 70-cm^2 area with 30-μm active layer, porous silicon back reflector, and Cu-based top-contact metallization. *Prog. Photovolt: Res. Appl.* **20**, 350–355.

Labunov, V.A., Bondarenko, V.P., Glinenko, L.K., and Basmanov, I.N. (1983). Process of formation of porous silicon and autoepitaxy on its surface. *Sov. J. Microelectron.* **12**, 11–16.

Labunov, V.A., Lobanovich, E.F., Bondarenko, V.P., and Glinenko, L.K. (1984). Effective lattice misfit in a substrate-porous material-epitaxial layer system. *Sov. Phys. Crystallogr.* **29**, 216–218.

Labunov, V., Bondarenko, V., Glinenko, L., Dorofeev, A., and Tabulina, L. (1986). Heat treatment effect on porous silicon. *Thin Solid Films* **137**, 123–134.

Labunov, V.A., Bondarenko, V.P., Borisenko, V.E., and Dorofeev, A.M. (1987). High-temperature treatment of porous silicon. *Phys. Stat. Sol. (a)* **102**, 193–198.

Lajnef, M., Chtourou, R., and Ezzaouia, H. (2010). Electric characterization of GaAs deposited on porous silicon by electrodeposition technique. *Appl. Surf. Sci.* **256**, 3058–3062.

Lamedica, G., Balucani, M., Ferrari, A., Bondarenko, V., Yakovtseva, V., and Dolgyi, L. (2002). X-ray diffractometry of Si epilayers grown on porous silicon. *Mater. Sci. Eng. B* **91–92**, 445–448.

Levchenko, V.I., Postnova, L.I., Bondarenko, V.P., Vorozov, N.N., Yakovtseva, V.A., and Dolgyi, L.N. (1999). Heteroepitaxy of PbS on porous silicon. *Thin Solid Films* **348**, 141–144.

Levchenko, V., Postnova, L., Truhanova, E., and Bondarenko, V. (2015). Epitaxial ZnSe films grown on porous silicon. *Reports of BSUIR* (in press).

Liang, J., Hong, S.-K., Kouklin, N., Beresford, R., and Xu, J.M. (2003). Nanoheteroepitaxy of GaN on a nano-pore array Si surface. *Appl. Phys. Lett.* **83**, 1752–1754.

Liao, Y., Ye, F., Shao, Q., Chang, C., Wang, G., and Fang, R. (2000). Study of diamond film growth mechanism on porous silicon during hot-filament chemical vapor deposition. *Thin Solid Films* **368**, 211–215.

Lin, T.L., Chen, S.C., Kao, Y.C., Wang, K.L., and Iyer, S. (1986). 100-μm-wide silicon-on-insulator structures by Si molecular beam epitaxy growth on porous silicon. *Appl. Phys. Lett.* **48**, 1793–1795.

Lin, T.L., Sadwick, L., Wang, K.L. et al. (1987). Growth and characterization of molecular beam epitaxial GaAs layers on porous silicon. *Appl. Phys. Lett.* **51**, 814–816.

Liu, Z., Zong, B., and Lin, Z. (1995). Diamond growth on porous silicon by hot-filament chemical vapor deposition. *Thin Solid Films* **254**, 3–6.

Liu, W., Xie, X., Zhang, M. et al. (2003a). Microstructure and crystallinity of porous silicon and epitaxial silicon layers fabricated on p^+ porous silicon. *J. Vac. Sci. Technol., B* **21**, 168–173.

Liu, Y.L., Liu, Y.C., Yang, H. et al. (2003b). The optical properties of ZnO films grown on porous Si templates. *J. Phys. D: Appl. Phys.* **36**, 2705–2708.

Liu, W., Xing, S., Lian, J. et al. (2004). Microstructure investigation of $Ba_xSr_{1-x}TiO_3$ thin film grown on porous silicon substrate. *Mater. Sci. Semicond. Process.* **7**, 253–258.

Luryi, S. and Suhir, E. (1986). New approach to the high quality epitaxial growth of lattice-mismatched materials. *Appl. Phys. Lett.* **49**, 140–142.

Ma, S., Hu, M., Zeng, P., Li M., Yan, W., and Li, C. (2013). Synthesis of tungsten oxide nanowires/porous silicon composites and their application in NO_2 sensors. *Mater. Lett.* **112**, 12–15.

Matoussi, A., Boufaden, T., Missaoui, A. et al. (2001). Porous silicon as an intermediate buffer layer for GaN growth on (100) Si. *Microelectr. J.* **32**, 995–998.

Matoussi, A., Ben Nasr, F., Salh, R. et al. (2008). Morphological, structural and optical properties of GaN grown on porous silicon/Si(100) substrate. *Mater. Lett.* **62**, 515–519.

Matoussi, A., Ben Nasr, F., Boufaden, T. et al. (2010). Luminescent properties of GaN films grown on porous silicon substrate. *J. Lumin.* 130, 399–403.

Matte, E., van Nieuwenhuysen, K., Depauw, V., Govaerts, J., and Gordon, I. (2013). Layer transfer process assisted by laser scribing for thin film solar cells based on epitaxial foils. *Phys. Status Solidi A* **210**, 682–686.

Mendoza-Agüero, N. and Agarwal, V. (2013). Optical and structural characterization of tungsten oxide electrodeposited on nanostructured porous silicon: Effect of annealing atmosphere and temperature. *J. Alloys Compd.* **581**, 596–601.

Mii, Y.J., Lin, T.L., Kao, Y.C. et al. (1988). Studies of molecular-beam epitaxy growth of GaAs on porous Si substrates. *J. Vac. Sci. Technol.* **6**, 696–698.

Miranda, C., Braga, N., Baldan, M., Beloto, A., and Ferreira, N. (2010). Improvements in CVD/CVI processes for optimizing nanocrystalline diamond growth into porous silicon. *Diamond Relat. Mater.* **19**, 760–763.

Müller, G. and Brendel, R. (2000). Simulated annealing of porous silicon. *Phys. Stat. Sol. (a)* **182**, 313–318.

Naderi, N. and Hashim, M.R. (2013). Nanocrystalline SiC sputtered on porous silicon substrate after annealing. *Mater. Lett.* **97**, 90–92.

Nishida, S., Nakagawa, K., Iwane, M. et al. (2001). Si-film growth using liquid phase epitaxy method and its application to thin-film crystalline Si solar cell. *Sol. Energy Mater. Sol. Cells* **65**, 525–532.

Novikov, P.L. (1999). Simulation of porous silicon formation and silicon epitaxy on its surface. *Russ. Phys. J.* **42**, 282–287.

Novikov, P.L., Aleksandrov, L.N., Dvurechenski, A.V., and Zinov'ev, V.A. (1998). Mechanism of silicon epitaxy on porous silicon layers. *JETP Letters* **67**, 539–544.

Novikov, P.L., Bolkhovityanov, Yu.B., Pchelyakov, O.P., Romanov, S.I., and Sokolov, L.V. (2003). Specific behaviour of stress relaxation in Ge$_x$Si$_{1-x}$ films grown on porous silicon based mesa substrates: Computer calculations. *Semicond. Sci. Technol.* **18**, 39–44.

Oules, C., Halimaoui, A., Regolini, J.L. et al. (1989). Epitaxial silicon growth on porous silicon by reduced pressure, low temperature chemical vapour deposition. *Mater. Sci. Eng. B* **4**, 435–439.

Oules, C., Halimaoui, A., Regolini, J.L., Perio, A., and Bomchil, G. (1992). Silicon on insulator structures obtained by epitaxial growth of silicon over porous silicon. *J. Electrochem. Soc.* **139**, 3595–3599.

Radhakrishnan, H.S., Ahny, C., Van Hoeymissen, J. et al. (2012). Gettering of transition metals by porous silicon in epitaxial silicon solar cells. *Phys. Stat. Sol. (a)* **209**, 1866–1871.

Radhakrishnan, H.S., Martini, R., Depauw, V. et al. (2014). Improving the quality of epitaxial foils produced using a porous silicon-based layer transfer process for high-efficiency thin-film crystalline silicon solar cells. *IEEE J. Photovolt.* **4**, 70–77.

Raiko, V., Spitzl, R., Engemann ,J., Borisenko, V., and Bondarenko, V. (1996). MPCVD diamond deposition on porous silicon pretreated with the bias method. *Diamond Relat. Mater.* **5**, 1063–1069.

Samsudin, M.E.A., Zainal, N., and Hassan, Z. (2014). Controlled porosity of GaN using different pore size of Si (100) substrates. *Superlattices Microstruct.* **73**, 54–59.

Saravanan, S., Hayashi, Y., Soga, T., Jimbo, T., Umeno, M., and Sato, N. (2002). Growth of GaAs epitaxial layers on Si substrate with porous Si intermediate layer by chemical beam epitaxy. *J. Cryst. Growth* **237–239**, 1450–1454.

Sato, N., Sakaguchi, K., Yamagata, K., Fujiyama, Y., and Yonehara, T. (1995). Epitaxial growth on porous Si for a new bond and etchback silicon-on-insulator. *J. Electrochem. Soc.* **142**, 3116–3122.

Sato, N., Sakaguchi, K., Yamagata, K., Fujiyama, Y., Nakayama, J., and Yonehara, T. (1996). Advanced quality in epitaxial layer transfer by bond and etch-back of porous Si. *Jpn. J. Appl. Phys.* **35**, 973–977.

Severino, A., D'Arrigo, G., Leone, S. et al. (2006). Heteroepitaxial growth of 3C-SiC on Silicon-Porous Silicon-Silicon (SPS) substrates. *ECS Trans.* **3**, 287–298.

Solanki, C.S., Bilyalov, R.R., Poortmans, J. et al. (2004). Characterization of free-standing thin crystalline films on porous silicon for solar cells. *Thin Solid Films* **451–452**, 649–654.

Spitzl, R., Raiko, V., and Engemann, J. (1994). Diamond deposition on porous silicon by plasma-assisted CVD. *Diamond Relat. Mater.* **3**, 1256–1261.

Spitzl, R., Raiko, V., Heiderhoff, R., Gnaser, H., and Engemann, J. (1995). MPCVD diamond deposition on bias pretreated porous silicon. *Diamond Relat. Mater.* **4**, 563–568.

Takai, H. and Itoh, T. (1983). Isolation of silicon film grown on porous silicon layer. *J. Electron. Mater.* **12**, 973–982.

Tayanaka, H. and Matsushita, T. (1996). Separation of thin epitaxial films on porous Si for solar cells. In: *Proc. of the 6th Sony Research Forum*, Nov. 27, 1996, Tokyo, Japan, p. 556.

Tayanaka, H., Yamauchi, K., and Matsushita, T. (1998). Thin-film crystalline silicon solar cells obtained by separation of a porous silicon sacrificial layer. In: *Proc. of 2nd World Conference on photovoltaic solar energy conversion*, July 6-10, 1998, Vienna, Austria, pp. 1272–1277.

Terheiden, B., Hensen, J., Wolf, A., Horbelt, R., Plagwitz, H., and Brendel, R. (2011). Layer transfer from chemically etched 150 mm porous Si substrates. *Materials* **4**, 941–952.

Tsao, S.S., Fleming, J.G., Guilinger, T.R., Han, J., and Kelly, M.J. (1999). Porous silicon as a stress relief layer for growing gallium nitride-based compound semiconductor devices. *Sandia National Laboratories Disclosure of Technical Advance*, July 14.

Unagami, T. and Seki, M. (1978). Structure of porous silicon layer and heat-treatment effect. *J. Electrochem. Soc.* **125**, 1339–1344.

Van Hoeymissen, J., Depauw, V., Kuzma-Filipek, I. et al. (2011). The use of porous silicon layers in thin-film silicon solar cells. *Phys. Stat. Sol. (a)* **208**, 1433–1439.

Vescan, L., Bomchil, G., Halimaoui, A., Perio, A., and Herino, R. (1988). Low-pressure vapor-phase epitaxy of silicon on porous silicon. *Mater. Lett.* **7**, 94–98.

Wang, L., Xia, Y., Ju, J., Fan, Y., Mo, Y., and Shi, W. (2000). Efficient luminescence from CVD diamond film-coated porous silicon. *J. Phys. Condens. Matter* **12**, L257–L260.

Watanabe, Y. and Sakai, T. (1972). Semiconductor device and method of producing the same. US Patent 3640806.

Wu, Y., Hu, M., Qin, Y., Wei, X., Ma, S., and Yan, D. (2014). Enhanced response characteristics of p-porous silicon (substrate)/p-TeO$_2$(nanowires) sensor for NO$_2$ detection. *Sens. Actuators, B.* **195**, 181–188.

Xie, Y.H. and Bean, J.C. (1990a). Heteroepitaxy of Si_xGe_{1-x} on porous Si substrates. *J. Appl. Phys.* **67**, 792–795.

Xie, Y.H. and Bean, J.C. (1990b). From porous Si to patterned Si substrate: Can misfit strain energy in a continuous heteroepitaxial film be reduced? *J. Vac. Sci. Technol. B* **8**, 227–231.

Yakovtseva, V., Vorozov, N., Dolgyi, L. et al. (2000). Porous silicon: A buffer layer for PbS heteroepitaxy. *Phys. Stat. Sol. (a)* **182**, 195–199.

Yasumatsu, T., Ito, T., Nishizawa, H., and Hiraki, A. (1991). Ultrathin Si films grown epitaxially on porous silicon. *Appl. Surf. Sci.* **48/49**, 414–418.

Ye, H. and Yu, J. (2014). Germanium epitaxy on silicon. *Sci. Technol. Adv. Mater.* **15**, 024601.

Yonehara, T. (2002). ELTRAN (SOI-Epi Wafer) Technology. In: *Progress in SOI Structures and Devices Operating at Extreme Conditions*, Balestra F., Nazarov A., and Lysenko V.S. (Eds.). NATO Science Series Vol. **58**, Dordrecht, the Netherlands, 39–86.

Yonehara, T. and Sakaguchi, K. (2001). ELTRAN: Novel SOI wafer technology. *JSAP International* **4**, 10–16.

Yonehara, T., Sakaguchi, K., and Sato, N. (1994) Epitaxial layer transfer by bond and etch back of porous Si. *Appl. Phys. Lett.* **64**, 2108–2110.

Zang, K.Y., Wang, Y.D., Chua, S.J., Wang, L.S., Tripathy, S., and Thompson, C.V. (2006). Nanoheteroepitaxial lateral overgrowth of GaN on nanoporous Si (111). *Appl. Phys. Lett.* **88**, 141925.

Zettner, J., Thoenissen, M., Hier, T., Brendel, R., and Schulz, M. (1998). Novel porous silicon backside light reflector for thin silicon solar cells. *Prog. Photovolt. Res. Appl.* **6**, 423–432.

Zhang, P., Kim, P.S., and Sham, T.K. (2001). XANES studies of CdS nano-structures on porous silicon. *J. Electron Spectrosc. Relat. Phenom.* **119**, 229–233.

Zheng, H., Zhou, J.J., Deng, J.X. et al. (2014). Preparation of two-dimensional yttrium iron garnet magnonic crystal on porous silicon substrate. *Mater. Lett.* **123**, 181–183.

Zimin, S.P., Preobrazhensky, M.N., Zimin, D.S., Zaykina, R.F., Borzova, G.A., and Naumov, V.V. (1999). Growth and properties of PbTe films on porous silicon. *Infrared Phys. Technol.* **40**, 337–342.

Zubia, D. and Hersee, S.D. (1999). Nanoheteroepitaxy: The application of nanostructuring and substrate compliance to the heteroepitaxy of mismatched semiconductor materials. *J. Appl. Phys.* **85**, 6492–6496.

Porous Silicon and Cold Cathodes

Ghenadii Korotcenkov and Beongki Cho

8

CONTENTS

8.1   Cold Cathodes and Their Application   166

8.2   PSi-Based Cold Cathodes   169

8.3   Cold Cathodes with Silicon Tips Modified by Porous Layer   173

8.4   Porous Silicon as Substrate for New Emitter Materials   178

Acknowledgments   179

References   179

8.1 COLD CATHODES AND THEIR APPLICATION

Emission of electrons into vacuum by the surface of metals and semiconductors is considered to be a promising way of building cold cathodes of electronic components for special applications such as fine-resolution electron microscopy, atomic-resolution electron holography, ultrafast processing of signals, sensors, microwave power amplifiers, high-temperature applications or work in hazardous and radioactive environments, and so on (Nation et al. 1999; Temple 1999; Fursey 2003). In the case of gas sensor applications, field emission devices are used as ionization sources. However, the greatest interest to the cold cathodes appeared in connection with the development of flat panel displays (FPDs) (Talin et al. 2001; Uchikoga 2006). It was found that based on microbaricated cold cathodes, called field emitter arrays (FEAs), field emission flat panel displays (FEDs) with high consumer parameters could be manufactured (Uchikoga 2006). The FED (see Figure 8.1) is a vacuum electron device, sharing many common features with the vacuum fluorescent display (VFD) and the cathode ray tubes (CRTs). Just like in a VFD or a CRT, the image in an FED is created by impinging electrons from a cathode onto a phosphor-coated screen. FEAs, aimed for application in FEDs, are electron sources that have the form of arrays of microfabricated sharp tips, typically with an integrated extraction electrode (gate). When the gate is biased to a large enough positive potential with respect to the tips, electron emission takes place from the tip into vacuum—a phenomenon known as field emission. The Fowler–Nordheim Law (Fowler and Nordheim 1928) explaining field emission as a quantum effect became the basis for research of cold cathodes. A potential barrier at the surface of a metallic conductor called the "work function" binds electrons to the material. For an electron to leave the material, the electron must gain an energy that exceeds the work function. This can be accomplished in a variety of ways, including thermal excitation (thermionic emission), electron and ionic bombardment (secondary emission), and the absorption of photons (photoelectric effect). Fowler–Nordheim emission or field emission differs from these other forms of emission in that the emitted electrons do not gain an energy that exceeds the material work function. Field emission occurs when an externally applied electric field at the material surface thins the potential barrier to the point where electron tunneling occurs, and thus differs greatly from thermionic emission. Since there is no heat involved, field emitters are a "cold cathode" electron source. More detailed description of the principle of cold cathode operation can be found in Gomer (1993), Temple (1999), and Fursey (2003).

Because of such possibility to design FEDs, during the last decades field-emission cathodes of various configurations have been fabricated using a variety of materials and by a variety of methods (Huq et al. 1998a; Nation et al. 1999; Temple 1999; Marrese 2000; Talin et al. 2001; Xu et al. 2003; Xu and Huq 2005; Nagao 2006). However, taking into account the technological capabilities of silicon and the availability of silicon fabrication lines for integrated circuits,

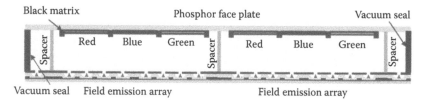

FIGURE 8.1 Schematic diagram of FED. CRTs and FEDs share many common features, including a glass vacuum envelope, and phosphor coated anode, and a cathode electron source. If in a CRT electrons from a triad of thermionic emitters are scanned across the phosphor screen with electromagnetic deflection coils, in a FED electrons from an addressable array of cold cathode impinge onto a precisely aligned phosphor anode. Field emission array (FEA) is placed in close proximity (0.2–2.0 mm) to a phosphor faceplate and is aligned such that each phosphor pixel has a dedicated set of field emitters. At a 1-mm gap and anode voltage of 5 kV, proximity focusing is sufficient to produce color pixels with dimensions below 100 μm. In addition to the anode and cathode, a FED contains ceramic spacers to prevent the structure from collapsing under atmospheric pressure, a frame coated on both sides with low-melting glass frit, a getter used to remove residual gases inside the package, row and column drivers, and an anode power supply. (Adapted from Talin A.A. et al., *Sol.-St. Electron.* 45, 963, 2001. Copyright 2001: Elsevier. With permission.)

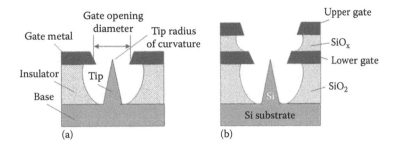

FIGURE 8.2 Typical view of a cell in a gated FEA (a) without and (b) with an integrated beam-focusing electrode. (Adapted from Temple D., *Mater. Sci. Eng. R* 24, 185, 1999. Copyright 1999: Elsevier. With permission.)

the most active interest was attracted to the possibility of manufacturing the Si microcathodes using photolitography and chemical etching (Thomas and Nathanson 1972; Marcus et al. 1990; Huq et al. 1998a,b; Lee et al. 1999). Various strategies have been suggested for the fabrication of silicon microtips and associated electrodes using a number of etch and deposition steps. As a rule, cathodes have structure as shown in Figure 8.2. One should note that the Spindt-type cathode shown in Figure 8.2 is the most mature FE cathode technology. Microtips can be either wet or dry etched. Examples of silicon-based cold cathodes realization are shown in Figure 8.3. During development, it was taken into account that in addition to having a sharp

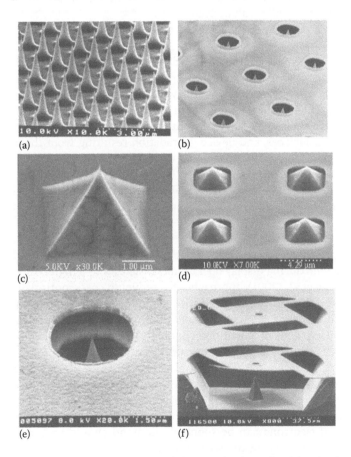

FIGURE 8.3 SEM images of various variants of designed Si-based cold cathodes. ([a,b] From Huq S.E. et al., *J. Vac. Sci. Technol. B* 16, 796, 1998. Copyright 1998: American Vacuum Society. With permission. [c,d] From Wisitsora A. et al., *Appl. Phys. Lett.* 71, 3394, 1997. Copyright 1997: AIP Publishing. With permission. [e] From Xu N.S. and Huq S.E., *Mater. Sci. Eng. R* 48, 47, 2005. Copyright 2005: Elsevier. With permission. [f] From Debski T. et al., *J. Vac. Sci. Technol. B* 18, 896, 2000. Copyright 2000: American Vacuum Society. With permission.)

microtip, the electron extractor gate with an aperture hole must be placed very close to the emitter. Some redundancies are also built into the array to take account of the fact that not all the microtips will be producing an equal amount of current. In order to enhance the lifetime of microfabricated emitters, FEA configuration should provide efficient operation at low current levels (fraction of a μA/microtip).

Experiment has shown that the field emitter cathode can be compact, simple to operate, and emission current density from single tips can be much larger than 10^8A/cm^2. It was also established that Spindt-type cathodes can operate in higher-pressure environments if operated below the voltage threshold for sputtering and negative electron affinity (NEA) films have proven to be incredibly robust in simulated electric propulsion (EP) environments (Marrese 2000). Postemission acceleration schemes have been proposed and developed to decouple electron energies from gate electrode voltages. With these schemes electron energies can be increased without sacrificing cathode lifetime to increase space-charge current limits. Electrode configurations have also been proposed to shield the electron-emitting surface from bombardment by ions originating near the thruster. This configuration can significantly increase tolerable operating voltages and cathode lifetime. Current limiting architectures have also been developed to increase cathode lifetime by suppressing arcs between the emitting surface and the gate electrode.

As for the disadvantages of cold cathodes, identified during their operation, they are as follows (Xu and Huq 2005; Komoda and Koshida 2009):

- Complicated technology. As a result, FED displays can potentially suffer from manufacturing problems that will result in dead pixels.
- Cold cathodes normally need focus electrodes in order to gather emitted electrons to the target.
- The efficiency of the field emitters is based on the extremely small radii of the tips, but this small size renders the cathodes susceptible to damage by ion impact. The ions are produced by the high voltages interacting with residual gas molecules inside the device. Unfortunately, FEDs require high vacuum levels, which are difficult to attain. In addition, emitter damage can be due to uncontrolled emission current, electron stimulated desorption from surfaces, and device degradation resulting from the gas–emitter surface interactions; e-beam induced desorption of gas from phosphor and subsequent adsorption on tips; and so on.
- Degradation of cold cathodes is also possible due to the dielectric breakdown between the cathode and anode, excessive leakage between cathode to anode electrical paths, and so on.

As a result, the precision required to fabricate a large array of microtips with aligned gates and maintenance of vacuum during display operation translates into a high cost.

According to Talin et al. (2001), approaches for reducing costs, optimizing parameters, and improving scalability of FEAs during fabrication generally fall into two categories: (1) replacing the field emitter with an alternate material, and (2) producing field emitter designs that do not require fine photolithography or thin film technology. Research has shown that the most promising direction is the development of cold cathodes based on carbon-based materials, such as carbon nanotubes and diamond (Talin et al. 2001; Xu et al. 2003; Xu and Huq 2005). For example, it was found that special surface treatments of diamond (111) crystal plane produce a property called negative electron affinity (Diederich et al. 1998; Takeuchi et al. 2006). Electrons may be emitted from this surface into vacuum with no barrier, but the conduction band must first be populated. If electrons can be supplied, diamond should emit electrons at very low applied fields, greatly simplifying the structure of an FED. However, in research carried out first by Koshida's team (Koshida et al. 1995, 1999; Koshida and Matsumoto 2003) it was found that the use of PSi is also promising for cold cathode design. It was established that PSi (1) may be the basis of original instruments with improved parameters, (2) can act as a substrate in the deposition of advanced carbon-based materials acting as the cathode, and (3) can be used in the fabrication technology of conventional silicon field-emitters. Consider in more detail these approaches to the development of cold cathodes.

8.2 PSi-BASED COLD CATHODES

During experiments with PSi diode (Al/n^+-Si/PSi(40 μm)/Au) placed into vacuum and biased in excess of 20 V, it was established that when a positive voltage was applied to the Au electrode, electrons as well as photons were uniformly emitted through the Au contacts (Koshida et al. 1995). PSi layers were formed in an ethanoic solution of 50% HF:ethanol = 1:1. Thus, it was found that the PSi diode could act as a cold cathode (Figure 8.4). As a result, in 1998, a novel cold cathode technology was reported (Komoda et al. 1998; Koshida et al. 1999). One and a half micrometers of nondoped polysilicon layer on n-type (100) silicon wafer was anodized in a solution of HF(50%):ethanol = 1:1 at a current density of 10 mA/cm^2 for 30 sec under illumination by a 500-W tungsten lamp from a distance of 20 cm. Emission current was about 10^{-4} A/cm^2 at 20 V bias, and no fluctuation of emission current was observed as a function of time.

The emission mechanism has been explained as follows (Koshida et al. 1995, 1999). Figure 8.5 illustrates the electron emission process for cathodes fabricated using porosification of (a) single

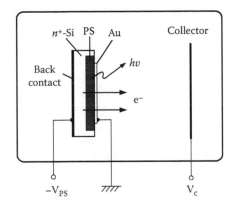

FIGURE 8.4 Device structure and experimental configuration for emission characteristics measurements. (From Sheng X. et al., *Thin Solid Films* 297, 314, 1997. Copyright 1997: Elsevier. With permission.)

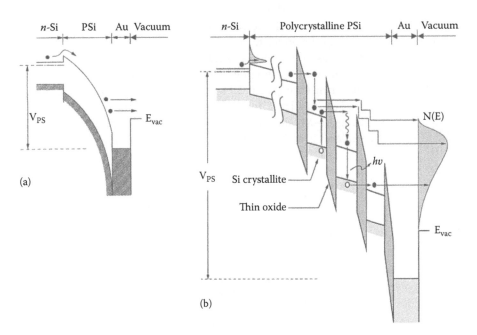

FIGURE 8.5 Band diagram showing generation and emission processes of quasiballistic electrons in PSi diodes under biased condition: (a) single crystalline PSi; (b) polycrystalline and oxidized PSi. (From Koshida N. et al., *Jpn. J. Appl. Phys.* 34, L705, 1995. Copyright 1995: The Japan Society of Applied Physics. With permission. From Koshida N. et al., *Appl. Surf. Sci.* 146, 371, 1999. Copyright 1999: Elsevier. With permission.)

crystalline Si and (b) polycrystalline Si with following oxidation. Under positive bias, a large potential drop is produced in the PSi layer, especially near the PSi surface. Electrons are thermally injected from the heavy doped n-type substrate and drifted through PSi toward the top contact. In PSi, electrons are accelerated and in the region near the outer surface, electrons become hot enough, and are ejected quasi-ballistically into vacuum through a thin Au film. The PSi layer plays a role as a semi-insulating semiconductor in which the high-electric-field electronic transport takes place. In oxidized porous polycrystalline layer, this process is more effective. Polycrystalline PSi is composed of electrically isolated or interconnected Si nanocrystals surrounded by a thin, wide bandgap-layer. This latter layer consists of Si oxide. Therefore, under positive bias, a large potential drop is produced near intercrystallite interfaces (see Figure 8.5b). As a result, electrons in polycrystalline PSi receive additional acceleration, due to a multitunneling process across regions between Si nanocrystals, and eventually become ballistic electrons (Koshida et al. 1995, 1999; Uno et al. 2003). This process is supported by Fowler–Nordheim analysis, numerical analysis of electron drift length from time-of-flight measurements, and voltage dependence of the energy of emitted electrons (Koshida et al. 1999; Kojima and Koshida 2001; Uno et al. 2003; Sakai et al. 2008). The peak energy shifts toward the higher energy side in accordance with increasing bias voltages (see Figure 8.6). Impact ionization processes occur as well. At sufficiently high bias, the situation near the Au-PSi interface can be considered a state of voltage-controlled negative electron affinity. Electrons in PSi near the outer surface by the field-induced carrier generation cascade can be emitted into vacuum through a thin oxide layer and this Au film. The lack of the effect of the emitter cessation on the energy distribution in the emission current, observed by Mimura et al. (2003), confirms that the emission mechanism of the PSi emitter is conventional field emission from the nanocrystals in the PSi.

A somewhat different interpretation of the mechanism of ballistic electrons emission in PSi was proposed by Jessing et al. (1996). They believed that in oxidized macroporous silicon covered by metal (see Figure 8.7) the nanoscale tips at the base of the pores serve as the cathodes, while the evaporated metal film acts as the anode. In this diode structure, the low voltage emission is possible because the ultra sharp silicon fibrils formed in the base of the pores geometrically enhance the electric field. In addition, the operating voltage of the structure can be low due to small cathode-to-anode distance (the thickness of the porous layer), which can be changed over a large range during the fabrication of the structure. Furthermore, stable emission occurs due to the high density of emission sites beneath the metal anode.

It was established that the dynamics of electron emission are limited by the capacitance of PSi, which depends on the surface of the emitter. Experiment has shown that the response time of a device with a 0.2-cm² surface was in the microsecond range.

Subsequent research has allowed optimizing both the structure PSi-based diodes and the technology of their manufacturing (Sheng et al. 1997). It was established that the best thicknesses for Pt and Au electrodes are 5–8 nm. The films are probably discontinuous at this thickness. The emission was enhanced using Al–Mg alloy owing to its lower work function. It was also shown that the thickness of the PSi layer is an important determining factor of the emission properties.

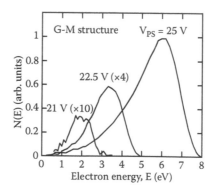

FIGURE 8.6   Measured energy distribution curves of a PSi diode with a graded-multilayer structure. (From Koshida N. et al., *Appl. Surf. Sci.* 146, 371, 1999. Copyright 1999: Elsevier. With permission.)

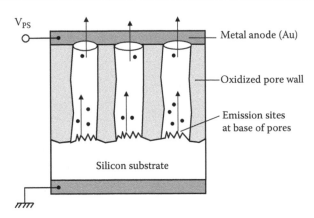

FIGURE 8.7 Microscopic representation of a cross-section of an oxidized PSi structure. (Idea from Jessing J.R. et al., *J. Vac. Sci. Technol. B* 14(3), 1899, 1996.)

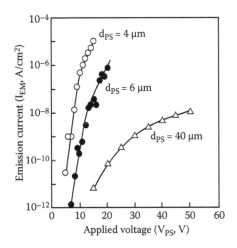

FIGURE 8.8 Diode current and emission current versus applied bias voltage for three PSi diodes with different PSi layer thicknesses. The top electrode is Pt with a thickness of 5 nm. (From Sheng X. et al., *Thin Solid Films* 297, 314, 1997. Copyright 1997: Elsevier. With permission.)

When the thickness of the PSi layer was reduced less than 10 μm, the emission was remarkably increased (see Figure 8.8). The efficiency of the electron emission has also been significantly improved by oxidizing PSi using a rapid thermal oxidation (RTO) procedure (Sheng et al. 1998). After the anodization, the samples were treated in a dry O_2 gas for 70 min at 900°C. Further improvements of the surface-emitting cold cathodes based on PSi diodes have been achieved using a nanocrystallized porous polycrystalline silicon instead of single crystals (Komoda et al. 1999, 2000; Sheng et al. 2001; Kim et al. 2002), a multilayered PSi structure (Sheng et al. 1998), and a multilayered graded PSi structure (Koshida et al. 1999), in addition to the RTO process. In this case, the structures of PSi layers were controlled by modulating the anodization current such that the multilayered and graded-band structures with high- and low-porosity layers were formed periodically. Kim et al. (2002) found that in the case of polycrystalline silicon using, the PPSi diode emitter with 2.0-μm polysilicon anodized at the current density 10 mA/cm² for 15 s, has higher efficiency, lower diode current, and higher emission current compared to that of PPS diode emitter prepared in other conditions. As the anodizing current density increased, the turn-on voltage and diode current were increased. A schematic illustration of multilayered and graded-multilayered PSi diodes is shown in Figure 8.9. Using these structures, an extremely high efficiency of 12% was obtained, which is much higher than that of the conventional PSi diodes and other surface-emitting cold cathodes.

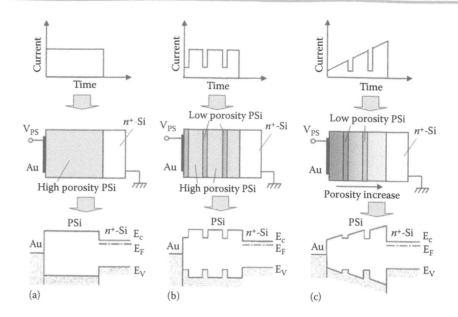

FIGURE 8.9 Anodization current modulation and corresponding PSi diode structure and band diagram: (a) normal structure, (b) multilayer structure, and (c) graded-multilayer structure. (Adapted from Sheng X. et al., *J. Vac. Sci. Technol. B* 16(2), 793, 1998. Copyright 1998: American Vacuum Society, and Koshida N. et al., *Appl. Surf. Sci.* 146, 371, 1999. Copyright 1999: Elsevier. With permission.)

Further research showed that the PSi diode as surface-emitted cold cathode in compared to conventional cold emitters has several advantageous features (Ohta et al. 2007; Gelloz and Koshida 2009): (1) electrons are emitted perpendicularly from the diode surface, (2) the cathode operates at relatively low voltage, (3) energetic emission has a small angular dispersion and quick dynamic response, (4) availability for large-area emitter array configuration on glass substrates, (5) fabrication is compatible with silicon planar processing at low temperatures, and (6) the emission current is not strongly sensitive to ambient pressure. In particular, Ohta et al. (2005, 2007) have shown that cold cathodes based on nanocrystalline PSi diode can operate in atmospheric pressure of various gases. Moreover, the electron emission characteristics of the PSi emitter at an atmospheric pressure of 10 Pa have been shown to be similar to the electron emission characteristics of the PSi emitter operating in a higher vacuum (Komoda et al. 1998). This is an important advantage of PSi-based emitters. However, the same authors stated that both the complete passivation of PSi surfaces and the suppression of interfacial traps between neighboring nanocrystalline Si dots are key issues for enhancement in the efficiency and stability. It was also established by Nakajima et al. (2002) that by placing a fluorescent film directly on top of surface-emitted cold cathode, it becomes possible to create an element of FPD that is referred to as a "vacuum-less cathode-ray tube." According to Nakajima et al. (2002), these devices could be easily integrated into a flat-screen device array. In addition, because of their simple design, uniform light emission, low power consumption, high resolution, and most importantly, being based on silicon, their process compatibility with existing large-scale microelectronics production, they offer many potential advantages over other flat-screen display technologies currently under investigation. Thus, these features of PSi-based cold cathodes testify that application of these devices is beneficial to increase the stability and performance of the micro-cathodes and FPDs (Komoda and Koshida 2009).

For the implementation of these advantages of PSi-based cold cathode, the group of Koshida, in collaboration with Japanese companies, has developed a ballistic electron surface-emitting display (BSD) (Koshida et al. 2003; Komoda and Koshida 2009). A schematic illustration of the BSD configuration is shown in Figure 8.10. The flat panel display was 168 (RGB) × 126 pixels, 2.6 in. diagonal full-color. Subpixel size was 320 × 107 μm. Since RTO requires temperature, which is very high for the glass substrate, a low-temperature process, the ECO technique, was used (Komoda et al. 2000). This technique was based on electrochemical oxidation. It was shown that

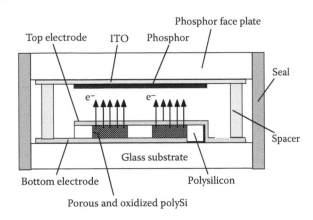

FIGURE 8.10 A schematic of the BSD cross-section formed on a glass substrate. (Adapted from Koshida N. et al., *Electrochem. Soc. Interface* 12(2), 52, 2003. Copyright 2003: The Electrochemical Society. With permission.)

the BSD could operate at relatively low vacuum level (10 Pa) and without any focusing electrodes even when the distance between pixels was only 40 μm. It was also shown that the BSD had excellent thermal stability and a frit-sealed model was fabricated. Later, the same group reported a prototype of a large display (7.6 in. in diagonal) (Komoda et al. 2004). A prototype ballistic electron surface-emitting display was fabricated on a TFT (thin film transistor) or PDP (plasma display panel) glass substrates. A 84 × 63 pixel, full-color BSD showed excellent performance, comparable to the previously reported 2.6-in. model (Komoda et al. 2000). Komoda and Koshida (2009) believe that larger panel BSD also can be fabricated using the proposed technology. However, one should note that although BSD exhibits various excellent characteristics as a novel cathode, there are still a lot of points to clarify such as mechanism and implication of the electron emission from the structure.

8.3 COLD CATHODES WITH SILICON TIPS MODIFIED BY POROUS LAYER

Optimization of the parameters of conventional silicon cold cathodes using PSi usually occurs either by modifying the silicon tip's surface, or through the use of Si porosification technology for reducing the radius of silicon tips. It should be noted that this direction of Si cold cathode design is not as well developed as the previous one. There are only a few papers done in this area. For example, Wilshaw and Boswell (1994) and Evtukh et al. (1995, 2001) have shown that the emissive ability of cathodes formed at the Si surface by chemical etching may be essentially improved by covering it with a layer of PSi (see Figure 8.11).

In their study, they proceeded from the following. In cold cathodes, one of the most critical parts is the apex of the cathode from which electrons are emitted. For low voltage applications, its radius must be very small. In addition, if large arrays of emitting cathodes are to be produced with approximately uniform characteristics, then the emitting tip of each cathode must be manufactured to have the same consistently small radius. However, such processing required to produce very sharp uniform emitters is relatively difficult. As a rule, these processes have high complexity and low compatibility with CMOS technology. The conventional approach for resolving this problem is covering the cathodes with a thin layer of material of low work function so that field emission is obtained at low applied voltages while using the tips of a larger radius than would otherwise be required. An approach proposed by Wilshaw and Boswell (1994) is an alternative one in which, rather than fashion a single very sharp emitting point at the apex of each cathode, the cathode surface itself is treated in such a way as to cover it with a high density of naturally very sharp asperities. In this way, emission is now largely controlled by the asperities and not by the single "macroscopic" apex of the emitter. For these purposes, Wilshaw and Boswell (1994) and Jessing et al. (1998) have taken pyramidal cathodes manufactured by wet etching of *p*-type single crystal silicon and produced on them a thin surface layer of PSi. In this case, each fibril of PSi layer

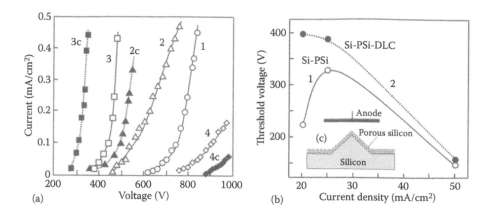

FIGURE 8.11   (a) Current-voltage characteristics of Si tip array with porous silicon layer (2–4) unmodified and (2c–4c) modified by diamond-like carbon (DLC) film: (1) Si tip array; (2, 2c-4, 4c) PSi layer on tip: (2) 2c-J/t = 25 mA/cm^2/5 s, (3, 3c) J/t = 25 mA/cm^2/30 s, (4, 4c) J/t = l0 mA/cm^2/60 s. (2–4) J/t is ratio of current density to time during the forming of porous silicon layer. (b) Dependence of the emission parameters of Si tips with porous silicon layer without (1) and with (2) DLC coating on the etching current density. (c) Schematic diagram of Si tip with PSi layer. (Data extracted from Evtukh A.A. et al., In: *Abstract book of MRS Spring Meeting,* San-Francisco, CA, April 17–21, p. 396, 1995; and Evtukh A.A. et al., U.S. Department of Energy, Preprint UCRL-JC-143779, 2001.)

produces a local increase of the electric field, acting like an emitter of nanometric dimensions—a "nano-emitter" (Boswell et al. 1996; Nicolaescu et al. 1996). For PSi forming a solution of 48% hydrofluoric acid and ethanol (1:1) and current density of 25–30 mA/cm^2 was used for 10–30 s to give the correct PSi morphology (Wilshaw and Boswell 1994; Jessing et al. 1998). Such treatment allowed increasing the maximum emission current from 1 to 9 mA at a voltage about 1 kV between tip and anode to some 25–90 mA with voltage reduced to some 200 V. Subsequently, the same process has been applied in the manufacture of polysilicon-based emitters (Pullen et al. 1996). Polysilicon layers of 2 μm and 0.7 μm were deposited undoped and subsequently doped by ion implantation. Field emission characterization after the anodization of polysilicon tips showed a significant starting voltage reduction of 40–50% in each of the samples tested.

Kleps et al. (1997) have shown that porosification of silicon blunt emitters also helps to improve their emission properties. Studies of Sotgiu and Schirone (2005) were also conducted in this direction. Using surface microstructuring by electrochemical oxidation in organic solutions containing HF, Sotgiu and Schirone (2005) achieved a significant increase in field emission of electrons from silicon surfaces. For example, in the most favorable conditions, the threshold field for the emission of an electron current I_{th} = 10^{-10} A was only 11.1 V/μm. These results appear quite interesting, if compared to the best threshold field obtained for diamond-like carbon and polycrystalline HFCVD diamond samples. In Table 8.1, the best threshold field values obtained for different materials by means of the same experimental setup are reported. At that, they found that the emission threshold was strongly correlated with the overall charge exchanged during

TABLE 8.1   Threshold Field for Some Materials Considered for the Fabrication of Si-Based Cold Cathodes

Material	E_{th} (V/μm)
Diamond like carbon (DLC)	20
CVD diamond	17
Microstructured silicon	11

Source: Sotgiu G. and Schirone L., *Appl. Surf. Sci.* 240, 424, 2005. Copyright 2005: Elsevier. With permission.

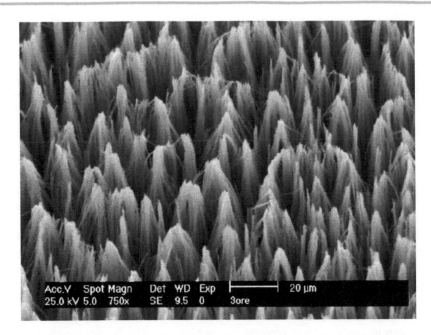

FIGURE 8.12   Anodization at high etching rates gives rise to "spaghetti-like" silicon structures. (From Sotgiu G. and Schirone L., *Appl. Surf. Sci.* 240, 424, 2005. Copyright 2005: Elsevier. With permission.)

electrochemical oxidation, that is, the morphology of macroporous PSi layer. Better results were obtained at high current densities, when almost constant silicon etching takes place along the pore wall. At the high current, high exchanged-charge end of the investigated preparation conditions, the spaghetti-like structures shown in Figure 8.12 are obtained. The maximum efficiency of indicated structures is quite expected because just such structures provide maximum curvature (minimum radius) of formed PSi tips.

According to Evtukh et al. (2001), the porosification of Si tips should also improve the stability of the emission properties of cathodes. Due to the distribution of radii of the tips in the cold cathode array, most of the emission current flows through the tips having small radii. If the tip is very small, its thermal resistance is very high and the equilibrium temperature necessary to dissipate the power may be greater than the melting point. As a result, some tips can be overheated and blunted. In the case of a porous layer on the silicon tip surface, the blunting and destruction of some asperities does not influence the emission significantly because other asperities presented in the PSi layer can begin to contribute to the emission process. According to estimations, the density of asperities in PSi, which act as separate emission centers, can be in the range of 10^8–10^{11} asperities per mm^2, while the density of the emission tips in conventional Spindt Si cathodes is about 10^5 tips per mm^2.

During the experiments, it was also established that the PSi cover improved the uniformity of emission from different cathodes (Wilshaw and Boswell 1994). Wilshaw and Boswell (1994) assumed that the mechanism of improvement of emissive capabilities is due to the concentration of the electric field near surface microfibers, which work as microcathodes at one Si tip. However, Boswell et al. (1995) believe that this effect takes place due to the very high resistivity of the *p*-type PSi layer. Thus, each cathode can be considered to be composed of many nanoscale emitters, each of which is due to a single asperity, connected to an underlying conducting substrate by a thin highly resistive layer, namely the *p*-type PSi. In this way, the current emitted from each fibril is to some extent limited by the voltage drop produced by the current flowing through the underlying resistive layer. This acts as a negative feedback mechanism so that even fibrils with otherwise different field enhancement factors will tend to emit similar currents, and so this will lead to an increase in the uniformity of emission between different fibrils. In this way, it is suggested that simultaneous emission from many asperities on a single cathode of *p*-type PSi material will allow much higher currents to be obtained than for the equivalent plain cathode. This is consistent with the results presented elsewhere (Wilshaw and Boswell 1994) that show the

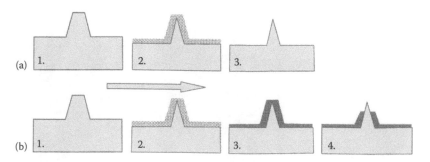

FIGURE 8.13 (a) "PSi sharpening" steps: (1) microtip as obtained, (2) PSi formation over microtip, and (3) PSi etch to obtain apex reduction. (b) "Ox-PSi sharpening" steps: (1) microtip as obtained, (2) PSi obtaining over microtip, (3) thermal oxidation of PSi layer, and (4) oxidized-PSi etch to obtain apex reduction. The parameters of short anodization were: $J = 20$ mA/cm^2, $t = 2$ min, HF chemical solution (48%wt). On "PSi sharpening," the thin PSi layer over the microtip was etched by very low concentration KOH chemical solution (1% wt), with $t = 2$ min at room temperature. To promote PSi oxidation on "Ox-PSi sharpening" process, samples were oxidized with $T = 1200°C$ (dry O_2 environment) for $t = 10$ h. In fact, due to its high reactivity, it is possible to oxidize thick PS layers (with tens of micrometers in depth) in a very short time (5–20 min). The relative long-time oxidation process applied in this work (10 h), besides obtaining tip apex reduction, aims to promote an electrical insulation between cathode and anode in FE devices with these two structures integrated on the same chip. Finally, the Ox-PSi layer was etched from the apex of the microtips by HF chemical solution. (Adapted from Dantas M.O.S. et al., *ECS Trans.* 9(1), 481. 2007. Copyright 2007: The Electrochemical Society. With permission.)

maximum emission current from p-type PSi emitters decreased as the thickness of the PSi layer was reduced

As for the approach based on improving the form of single crystalline silicon tip using the technology of Si porosification, this was proposed by Dantas (Dantas et al. 2007, 2008; Dantas 2008). The sharpening of microtips (emitters) was achieved using additional optimization processes (named "PSi-Sharpening" and "Ox-PSi-Sharpening," respectively). This process is illustrated in Figure 8.13. Because of such postprocessing, a reduction in the emitter apex diameter by around 40% was achieved in comparison with as-obtained ones, and the decrease in the energy to promote electron emission was observed. Table 8.2 summarizes the obtained E values for various tested devices. Thus, the proposed technique showed to be a good alternative for microtips sharpening by simple and low complexity processes, which can be applied in postprocessing of field emitters obtained by various techniques.

One should note that silicon tips optimized in the present research were also fabricated using technology of silicon porosification named as hydrogen ion implantation-porous silicon (HI-PSi) technique (Galeazzo et al. 2001; Dantas et al. 2003, 2004, 2008). In this technology, the patterning was obtained by blocking PSi formation on selective areas with a high resistivity buried layer as a mask. This high resistivity layer was formed by hydrogen ion implantation and adequate annealing. The technological process of tips fabrication is illustrated in Figure 8.14. Dantas et

TABLE 8.2 Comparison between Si-Based Field Emission Devices

Device	E (V/μm)
Planar Si	55 ± 10
Si microtips as obtained	28 ± 4
Microtips after "PSi Sharpening"	23 ± 2
Microtips after "Ox-PSi Sharpening"	11.0 ± 0.6

Source: Data from Dantas M.O.S., PhD Thesis, Universidade de São Paulo, Brazil, 2008.

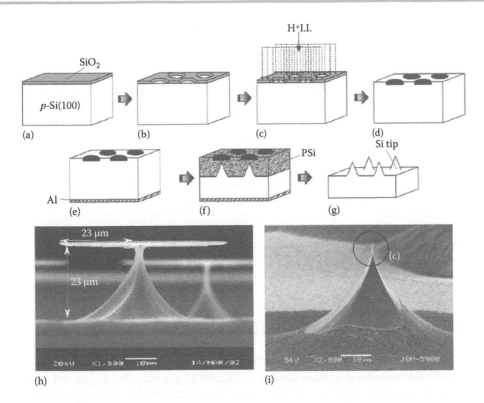

FIGURE 8.14 (a–g) Process sequence for Si tips fabrication, and (h,i) the SEM images of fabricated tips: (a) thermal growth of 1-μm thickness silicon oxide (SiO$_2$) as a barrier against the HII; (b) definition of regions (by lithography) in which H$^+$ is implanted; (c) HII with energy of 50 keV and dose of 1 × 10^{16} H$^+$/cm^2; (d) SiO$_2$ etching. The black regions are the regions with H$^+$I.I., where PSi is not desired. The white ones are the PSi formation regions. The masks designed explore the isotropic PSi formation; (e) Al evaporation to form ohmic contact; (f) PSi formation on selective areas by anodization process (HF chemical solution [48% wt.], current density [J] of 20 mA/cm^2). The PSi layer will be formed under the HII region and after the removal of PS, only the HII regions remains as thin membranes; (g) PSi etching by diluted aqueous KOH (1% concentration) at room temperature for Si tip obtaining. SEM images showing: (h) "pedestals" and (i) tip obtained after PSi etching by KOH (1%). (From Dantas M.O.S. et al., *Sens. Actuators A* 115, 608, 2004. Copyright 2004: Elsevier. With permission.)

al. (2004) stated that applying HII–PSi process, an array of Si microtips could be produced with different tip shapes. Mask geometries and anodization time define the height and shape of the Si tips. Then, it is possible to fabricate large quantities of Si tips in the same process with good repeatability.

It is important that modified HII-PSi technology for Si tip fabrication was also used by Yu et al. (2004), who reported about efficient field emission with extremely low turn-on field of about 3.2 V/μm at current density of 0.1 μA/cm^2. Patterned uniformly orientated cone-like silicon nanocrystallite (Si-nC) films were formed on a single crystalline p-Si:B(100) wafer ($\rho = 0.01$ Ω·cm) by a typical electrochemical anodic etching method, used for macroporous silicon forming, and an ion implantation technique. The samples were laterally anodized in an electrochemical cell in darkness for 10 min at a constant current density of 1 mA/cm^2 in the anodizing solution, the mixtures of hydrofluoric acid (40 wt%), ethanol (99.7 wt%), and deionized water with volume ratio of 1:2:1. Formed homogeneous Si-nC films had the thickness of about 1.5 μm. Comparison with the field emission properties of the tips fabricated using different approaches (Table 8.3) testifies that the technology discussed above is really promising for manufacturing cold cathodes. The porous structure of Si type was also formed during Si hydrothermal etching (Fu and Li 2006). One should note that this approach gives possibility to designing planar field emission devices without using any pyramidal Si tips.

TABLE 8.3   Comparison of Turn-On Fields for Different Silicon-Based Emitters

Tip's Structure	Method	Turn-On Field	Emission Current	Ref.
Nanocrystalline porous Si films	HII and AE	0.32 V/μm	0.1 μA/cm²	Yu et al. 2004
Macroporous Si	QE, ECO	11 V/μm	0.1 nA	Sotgiu and Schirone 2005
Si nano-wires arrays	CVD	14 V/μm	10 μA/cm²	Lu et al. 2004
CNTs-sheathed SiNWs	CVD	7 V/μm	10 μA/cm²	Lu et al. 2003
CNTs/Si tips	CVD	3.2 V/μm	10 pA	Wong et al. 2003
Au-Si alloy nano-cones	VEBE	10 V/μm	1 μA	Wan et al. 2004
Single crystalline Si tips	PE	35 V/μm	10 μA	Gunther et al. 2003
M/Si tips (M=W, Pt, Cr)	MS	~70 V/μm		
α-CN$_x$/Si	P	1.7 V/μm	–	Lin et al. 2008
Si-nanoporous arrays (Si-NPA)	HTE	1.5 V/μm	0.1 μA	Fu and Li 2006

Note: AE: anodic etching, ECO: electrochemical oxidation, HII: hydrogen ion implantation, HTE: hydrothermal etching, MS: magnetron sputtering, P: pyrolysis, PE: plasma etching, VEBE: vacuum electron-beam evaporation.

8.4 POROUS SILICON AS SUBSTRATE FOR NEW EMITTER MATERIALS

As mentioned previously, the search for new materials is the main direction of current research in the field of cold cathodes design. In these studies, a lot of material was tested and it was found that the carbon-based materials are the most promising materials for such applications. In particular, it was shown that CNTs provide extremely high curvature like ultra sharp emitters and can ensure excellent emission conditions. Beside this, CNT possesses many properties that are favorable for field emitters, such as high chemical stability, high aspect ratio, and high mechanical strength. In this regard, currently a large number of studies are conducted with the purpose to use carbon-based materials for optimization of already developed emitters. Based on this approach and by the usage of different fabrication methods, various cold cathodes were fabricated. PSi-based cathodes modified by carbon-based materials were also designed. In particular, Evtukh et al. (1996) studied silicon field emitting tip arrays with PSi layers on the surface of the tip, formed by electrochemical etching, and found that such structures covered by diamond-like carbon (DLC) film had improved emission efficiency. However, the observed effect of surface modification by DLC film was highly dependent on the parameters of silicon porosification (see Figure 8.11). In the following studies, Evtukh et al. (2004) found that the coating of PSi with thin (60 nm) DLC film allowed significant decrease of the long-term instability, and also decrease of the fluctuations of the emission current. The arrays of initial silicon emitter tips were fabricated using microelectronics technology for the formation of silicon points by wet chemical etching. The sharpening technique allowed the production of tips with a curvature radius of 10–20 nm. Evtukh et al. (2004) believe that thin DLC film protects the pores of the PSi layer from penetration of atoms and molecules of gas and their interaction with the PSi layer thus stabilizes the emission properties of the cathode.

Li et al. (2004) for the surface modification of PSi have used multiwalled carbon nanotubes (MWCNTs) grown by pyrolysis of acetylene on iron catalytic particles within a macroporous PSi template via CVD at 700°C. CNTs had diameter in the range from 75 to 100 nm. Li et al. (2004) established that cathodes fabricated on the PSi with pore size about 2–4 μm had a turn-on field of 4 V/μm and the average fluctuation of the emission current density was less than 5%. The same approach was used by Cichy and Gorecka-Drzazga (2008). They also found that the procedure of carbon MWNT deposition brought down the threshold field to 1.4 V/μm with comparable maximum current of 250 μA at about 850 V. A meso-micro PSi layer has been obtained by electrochemical dissolution of n-Si(111) in ethanoic HF solution (HF:C_6H_6O = 1:1) at the current density of 70 mA/cm² for 6 min.

Cahay et al. (2008) have shown that PSi substrates can also be used for deposition of other materials. In particular, they established that the Au nanostructures formed on the nanoporous silicon templates by evaporation are stable and efficient field emitters with large field enhancement factors. However, the improvement of stability, reliability, and efficiency is required. For these purposes, Cahay et al. (2008) proposed to cover these structures with refractory materials of low work function (e.g., LaS). They believe that such covering allows ameliorating the effects of surface adsorbed molecules that cause thermal runaway during the process of field emission.

ACKNOWLEDGMENTS

This work was supported by the Ministry of Science, ICT and Future Planning (MSIP) of the Republic of Korea, and partly by the National Research Foundation (NRF) grants funded by the Korean government (No. 2011-0028736 and No. 2013-K000315).

REFERENCES

Boswell, E., Seong, T.Y., and Wilshaw, P.R. (1995). Studies of porous silicon field emitters. *J. Vac. Sci. Technol. B* 13, 437–440.

Boswell, E.C., Huang, M., Smith, G.D.W., and Wilshaw, P.R. (1996). Characterization of porous silicon field emitter properties. *J. Vac. Sci. Technol. B* 14(3), 1895–1898.

Cahay, M., Garre, K., Fraser, J.W. et al. (2008). Field emission properties of metallic nanostructures self-assembled on nanoporous alumina and silicon templates. *J. Vac. Sci. Technol. B* 26(2), 885–890.

Cichy, B. and Gorecka-Drzazga, A. (2008). Electron field emission from microtip arrays. *Vacuum* 82, 1062–1068.

Dantas, M.O.S. (2008). Desenvolvimento de dispositivos de emissão por efeito de campo elétrico fabricados pela técnica HI-PS. PhD Thesis, Universidade de São Paulo, Brazil, pp. 75–79.

Dantas, M.O.S., Galeazzo, E., Peres, H.E.M., and Ramirez-Fernandez, F.J. (2003). Silicon micromechanical structures fabricated by electrochemical process. *IEEE Sensors J.* 3(6), 722–727.

Dantas, M.O.S., Galeazzo, E., Peres, H.E.M., Ramirez-Fernandez, F.J., and Errachid, A. (2004). HI–PS technique for MEMS fabrication. *Sens. Actuators A* 115, 608–616.

Dantas, M.O.S., Galeazzo, E., Peres, H.E.M., and Ramirez-Fernandez, F.J. (2007). Silicon microtips sharpening method for field emission applications. *ECS Transactions* 9(1), 481–488.

Dantas, M.O.S., Galeazzo, E., Peres, H.E.M., Kopelvski, M.M., and Ramirez-Fernandez, F.J. (2008). Silicon field-emission devices fabricated using the hydrogen implantation–porous silicon (HI–PS) micromachining technique. *J. Microelectromech. Syst.* 17(5), 1263–1269.

Debski, T., Volland, B., Barth, W. et al. (2000). Field emission arrays by silicon micromachining. *J. Vac. Sci. Technol. B* 18, 896–899.

Diederich, L., Kuttel, O.M., Aebi, P., and Schlapbach, L. (1998). Electron affinity and work function of differently oriented and doped diamond surfaces determined by photoelectron spectroscopy. *Surf. Sci.* 418(1), 219–239.

Evtukh, A.A., Litovchenko, V.G., Marchenko, R.I., and Romanuyk, B.N. (1995). Field emitter tip arrays with high efficiency for flat panel display application. In: *Abstract Book of MRS Spring Meeting*, San Francisco, April 17–21, p. 396.

Evtukh, A.A., Litovchenko, V.G., Marchenko, R.I., Klyui, N.I., Popov, V.G., and Semenovich, V.A. (1996). Peculiarities of the field emission with porous Si surfaces, covered by ultrathin DLC films. *J. Physique IV* 6, C5–119–124.

Evtukh, A.A., Litovchenko, V.G., Litvin, Yu.M. et al. (2001). Porous silicon coated with ultrathin diamond-like carbon film cathodes. U.S. Department of Energy, Preprint UCRL-JC-143779.

Evtukh, A.A., Hartnagel, H., Litovchenko, V.G., Semenenko, M.O., and Yilmazoglu, O. (2004). Enhancement of electron field emission stability by nitrogen-doped diamond-like carbon film coating. *Semicond. Sci. Technol.* 19, 923–929.

Fowler, R.H. and Nordheim, L.W. (1928). Electron emission in intense electrical fields. *Proc. Royal Soc. Lond.* 119, 173–181.

Fu, X.-N. and Li, X.-J. (2006). Enhanced field emission from well-patterned silicon nanoporous pillar arrays. *Chin. Phys. Lett.* 23(8), 2172–2174.

Fursey, G.N. (2003). Field emission in vacuum micro-electronics. *Appl. Surf. Sci.* 215, 113–134.

Galeazzo, E., Salcedo, W.J., Peres, H.E.M., and Ramirez-Fernandez, F.J. (2001). Porous silicon patterned by hydrogen ion implantation. *Sens. Actuators B* 76, 343–346.

Gelloz, B. and Koshida, N. (2009). Nanocrystalline Si EL devices. In: Koshida N. (Ed.) *Device Applications of Silicon Nanocrystals and Nanostructures.* Springer, New York, pp. 25–70.

Gomer, R. (1993). *Field Emission and Field Ionization.* American Institute of Physics, New York.

Gunther, B., Kaldasch, F., Muller, G. et al. (2003). Uniformity and stability of field emission from bare and metal coated Si tip arrays. *J. Vac. Sci. Technol. B* 21, 427–432.

Huq, S.E., Grayer, G.H., Moon, S.W., and Prewett, P.D. (1998a). Fabrication and characterisation of ultra sharp silicon field emitters. *Mater. Sci. Eng. B* 51, 150–153.

Huq, S.E., Huang, M., Wilshaw, P.R., and Prewett, P.D. (1998b). Microfabrication and characterization of gridded polycrystalline silicon field emitter devices. *J. Vac. Sci. Technol. B* 16, 796–798.

Jessing, J.R., Parker, D.L., and Weichold, M.H. (1996). Porous silicon field emission cathode development. *J. Vac. Sci. Technol. B* 14(3), 1899–1901.

Jessing, J.R., Kim, H.R., Parker, D.L., and Weichold, M.H. (1998). Fabrication and characterization of gated porous silicon cathode field emission arrays. *J. Vac. Sci. Technol. B* 16(2), 777–779.

Kim, H., Park, J.-W., Lee, J.-W. et al. (2002). Electron emission characteristics of the porous polycrystalline silicon diode. *Curr. Appl. Phys.* 2, 233–235.

Kleps, I., Nicolaescu, D., Lungu, C., Musa, G., Bostan, C., and Caccavale, F. (1997). Porous silicon field emitters for display applications. *Appl. Surf. Sci.* 111, 228–232.

Kojima, A. and Koshida, N. (2001). Evidence of enlarged drift length in nanocrystalline porous silicon layers by time-of-flight measurements. *Jpn. J. Appl. Phys.* 40, 366–368.

Komoda, T. and Koshida, N. (2009). Nanocrystalline silicon ballistic electron emitter, In: Koshida N. (Ed.) *Device Applications of Silicon Nanocrystals and Nanostructures*. Springer, New York, pp. 251–292.

Komoda, T., Sheng, X., and Koshida, N. (1998). Characteristics of surface-emitting cold cathode based on porous polysilicon. *MRS Proc.* 509, 187 (doi: http://dx.doi.org/10.1557/PROC-509-187).

Komoda, T., Sheng, X., and Koshida, N. (1999). Mechanism of efficient and stable surface-emitting cold cathode based on porous polycrystalline silicon films. *J. Vac. Sci. Technol. B* 17, 1076–1079.

Komoda, T., Ichihara, T., Honda, Y., Aizawa, K., and Koshida, N. (2000). Ballistic electron surface-emitting cold cathode by porous polycrystalline silicon film formed on glass substrate. *MRS Proc.* 638 (http://dx.doi.org/10.1557/PROC-638-F4.1.1).

Komoda, T., Ichihara, T., Honda, Y. et al. (2004). Fabrication of a 7.6-in.-diagonal prototype ballistic electron surface-emitting display on a glass substrate. *J. Soc. Inform. Display* 12(1), 29–35.

Koshida, N. and Matsumoto, N. (2003). Fabrication and quantum properties of nanostructured silicon. *Mater. Sci. Eng. R* 40, 169–205.

Koshida, N., Ozaki, T., Sheng, X., and Koyama, H. (1995). Cold electron emission from electroluminescent porous silicon diodes. *Jpn. J. Appl. Phys.* 34(1995), L705–L707.

Koshida, N., Sheng, X., and Komoda, T. (1999). Quasiballistic electron emission from porous silicon diodes. *Appl. Surf. Sci.* 146, 371–376.

Koshida, N., Kojima, A., Nakajima, Y., Ichihara, T., Watabe, Y., and Komoda, T. (2003). Application of nanocrystalline silicon and ballistic electron emitter to flat panel display devices. *The Electrochemical Society, Interface*, 12(2), 52–56.

Lee, J.D., Shim, B.C., Uh, H.S., and Park, B.G. (1999). Surface morphology and I-V characteristics of single-crystal, polycrystalline, and amorphous silicon FEA's. *IEEE Electron Dev. Lett.* 20, 215–218.

Li, J., Lei, W., Zhang, X., Wang, B., and Ba, L. (2004). Field emission of vertically-aligned carbon nanotube arrays grown on porous silicon substrate. *Solid-State Electon.* 48, 2147–2151.

Lin, L., Niu, H.J., Zhang, M.L., Song, W., Wang, Z., and Bai, X.D. (2008). Electron field emission from amorphous carbon with N-doped nanostructures pyrolyzed from polyaniline. *Appl. Surf. Sci.* 254, 7250–7254.

Lu, M., Li, M.K., Zhang, Z.J., and Li, H.L. (2003). Synthesis of carbon nanotubes/Si nanowires core-sheath structure arrays and their field emission properties. *App. Surf. Sci.* 218, 196–202.

Lu, M., Li, M.K., Kong, L.B., Guo, X.Y., and Li, H.L. (2004). Synthesis and characterization of well-aligned quantum silicon nanowires arrays. *Composites B* 35, 179–184.

Marcus, R.B., Ravi, T.S., Gmitter, T. et al. (1990). Formation of silicon tips with <1 nm radius. *Appl. Phys. Lett.* 56, 236–238.

Marrese, C.M. (2000). A review of field emission cathode technologies for electric propulsion systems and instruments. *Aerospace Conf. Proc., IEEE* 4, 85–98.

Mimura, H., Miyajima, K., and Yokoo, K. (2003). Electron emission from porous silicon planar emitters. *J. Vac. Sci. Technol. B* 21(4), 1612–1615.

Nagao, M. (2006). Cathode technologies for field emission displays. *IEEJ Trans.* 1, 171–178.

Nakajima, Y., Kojima, A., and Koshida, N. (2002). Generation of ballistic electrons in nanocrystalline porous silicon layers and its application to a solid-state planar luminescent device. *Appl. Phy. Lett.* 81, 2472–2474.

Nation, J.A., Schachter, L., Mako, F.M. et al. (1999). Advances in cold cathode physics and technology. *Proc. IEEE* 87(5), 865–889.

Nicolaescu, D., Filip, V., and Wilshaw, P.R. (1996). Modelling of the field emission microtriode with emitter covered with porous silicon. *Appl. Surf. Sci.* 94/95, 79–86.

Ohta, T., Kojima, A., Hirakawa, H., Iwamatsu, T., and Koshida, N. (2005). Operation of nanocrystalline silicon ballistic emitter in low vacuum and atmospheric pressures. *J. Vac. Sci. Technol. B* 23, 2336–2339.

Ohta, T., Kojima, A., and Koshida, N. (2007). Emission characteristics of nanocrystalline porous silicon ballistic cold cathode in atmospheric ambience. *J. Vac. Sci. Technol. B.* 25(2), 524–527.

Pullen, S.E., Huang, M., Huq, S.E. et al. (1996). Enhanced field emission from poly silicon emitters using porous silicon. In: *Proceedings of 9th International Vacuum Microelectronics Conference,* July 7–12, St. Petersburg, Russia, pp. 211–214.

Sakai, D., Oshima, C., Ohta, T., and Koshida, N. (2008). Specific spectral features in electron emission from nanocrystalline silicon quasi-ballistic cold cathode detected by an angle-resolved high resolution analyzer. *J. Vac. Sci. Technol. B* 26, 1782–1786.

Sheng, X., Ozaki, T., Koyama, H. et al. (1997). Operation of electroluminescent porous silicon diodes as surface-emitting cold cathodes. *Thin Solid Films* 297, 314–316.

Sheng, X., Koyama, H., and Koshida, N. (1998). Efficient surface-emitting cold cathodes based on electroluminescent porous silicon diodes. *J. Vac. Sci. Technol. B* 16(2), 793–795.

Sheng, X., Kojima, A., Komoda, T., and Koshida, N. (2001). Efficient and ballistic cold electron emission from porous polycrystalline silicon diodes with a porosity multilayer structure. *J. Vac. Sci. Technol. B* 19, 64–67.

Sotgiu, G. and Schirone, L. (2005). Microstructured silicon surfaces for field emission devices. *Appl. Surf. Sci.* 240, 424–431.

Takeuchi, D., Nebel, C.E., and Yamasaki, S. (2006). Photoelectron emission properties of hydrogen terminated intrinsic diamond. *J. Appl. Phys.* 99(8), 086102.

Talin, A.A., Dean, K.A., and Jaskie, J.E. (2001). Field emission displays: A critical review. *Sol.-St. Electron.* 45, 963–976.

Temple, D. (1999). Recent progress in field emitter array development for high performance applications. *Mater. Sci. Eng. R* 24, 185–239.

Thomas, R.N. and Nathanson, H.C. (1972). Photosensitive field emission from silicon point arrays. *Appl. Phys. Lett.* 21, 384–386.

Uchikoga, S. (2006). Future trend of flat panel displays and comparison of its driving methods. In: *Proceedings of the 18th International Symposium on Power Semiconductor Devices & IC's,* June 4–8, Naples, Italy, IEEE Xplore (doi: 10.1109/ISPSD.2006.1666053).

Uno, S., Nakazato, K., Yamaguchi, S., Kojima, A., Koshida, N., and Mizuta, H. (2003). New insights in high-energy electron emission and underlying transport physics of nanocrystalline Si. *IEEE Trans. Nanotechnol.* 2(4), 301–307.

Wan, Q., Wang, T.H., and Lin, C.L. (2004). Self-assembled Au–Si alloy nanocones: Synthesis and electron field emission characteristics. *Appl. Surf. Sci.* 221, 38–42.

Wilshaw, P.R. and Boswell, E.C. (1994). Field emission from pyramidal cathodes covered in porous silicon. *J. Vac. Sci. Technol. B* 12, 662–665.

Wisitsora, A., Kang, W.P., Davidson, J.L., and Kerns, D.V. (1997). A study of diamond field emission using micro-patterned monolithic diamond tips with different sp^2 contents. *Appl. Phys. Lett.* 71, 3394–3396.

Wong, Y.M., Kang, W.P., Davidson, J.L. et al. (2003). Field emitter using multiwalled carbon nanotubes grown on the silicon tip region by microwave plasma-enhanced chemical vapor deposition. *J. Vac. Sci. Technol. B* 21, 391–394.

Xu, N.S. and Huq, S.E. (2005). Novel cold cathode materials and applications. *Mater. Sci. Eng. R* 48, 47–189.

Xu, N.-S., Deng, S.-Z., and Chen, J. (2003). Nanomaterials for field electron emission: Preparation, characterization and application. *Ultramicroscopy* 95, 19–28.

Yu, K., Zhang, Y., Luo, L. et al. (2004). Efficient field emission from electrochemically fabricated silicon nanocrystallite films. *Physica B* 348, 391–396.

Porous Silicon as Host and Template Material for Fabricating Composites and Hybrid Materials

Eugene Chubenko, Sergey Redko, Alexey Dolgiy,
Hanna Bandarenka, Sergey Prischepa, and Vitaly Bondarenko

CONTENTS

9.1 Introduction 184
9.2 Formation, Properties, and Applications of Metal/PSi Composites 184
    9.2.1 Ferromagnetic Metals 185
    9.2.2 Noble Metals 189
    9.2.3 Conclusions 190
9.3 Formation, Properties, and Applications of Semiconductor/PSi Composites 191
    9.3.1 II-VI and IV-VI Compound Semiconductors 191
    9.3.2 Ge and Si 194
    9.3.3 Conclusion 195
9.4 Hybrid Materials Formed by Infiltration of PSi with Organic Substances 195
    9.4.1 Organic Substances for Optoelectronic Devices Based on PSi 197
    9.4.2 Organic Substances for Passivation and Functionalization of the PSi Surface 198
    9.4.3 Organic Substances for Electrical Contacts to PSi 199
    9.4.4 Conclusions 200
9.5 Deposition of Carbon Materials on/in PSi 200
    9.5.1 Fullerenes C60 200
    9.5.2 Carbon Nanotubes 201
    9.5.3 Graphite/Graphene 201
    9.5.4 Conclusions 202
9.6 Summary 202
References 202

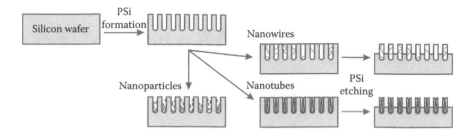

FIGURE 9.1   Formation of nanostructures with PSi template.

9.1 INTRODUCTION

The open porous structure and very large specific surface area of porous silicon (PSi) have driven scientists to introduce different foreign substances into its pores to fabricate PSi-based composite and hybrid materials (Hérino et al. 1997; Hérino 2000). Under this approach, PSi plays a role of host material, which receives foreign substances as "guests": metals, semiconductors, and carbon-containing materials (polymers, fullerenes, carbon nanotubes, and graphene). Nowadays in most cases, PSi is considered a template, that is, a type of host material with the structure that defines a shape of a "guest" substance. Elongated pores of PSi provide a formation of nanowires (NWs) and nanotubes (NTs) with high aspect ratio (more than 1:100) or formation of nanoparticles (NPs) at initial stages of the filling process. Process flow for fabrication of NWs, NTs, and NPs using PSi as template is shown schematically in Figure 9.1. The aims of filling of PSi pores are: (1) to obtain 1D nanostructures with high aspect ratio and anisotropy of properties; (2) to fabricate 1D and 0D structures showing unusual properties caused by quantum size effects; (3) to provide an electrical contact to the whole PSi surface; and (4) to integrate composite and hybrid materials with CMOS, MEMS, and NEMS devices. Composite material is a mixture of similar kinds of materials while hybrid consists of different kinds of materials (organic and inorganic) with chemical bonds between them. The main distinction between composite and hybrid is that the second one possesses properties, which do not exist in either of its parent components (Drisko and Sanchez 2012).

Deposition of guest substances in PSi can be carried out by vacuum-based methods such as thermal evaporation, magnetron sputtering, and chemical vapor deposition (CVD) or from the liquid phase, for example, the electrochemical and chemical deposition, sol-gel method. Most vacuum-based techniques provide the deposition of guest substance predominantly on the top of the PSi layer. Nevertheless, the optimization of CVD technologies provides readily filling pores of 20–120 nm in diameter excepting a blockade of pore entrances that earlier occurred because pore constrictions suppressed the depositing process. Atomic layer deposition (ALD) provides great scope even for filling the pores of thick meso-PSi (Grigoras et al. 2014). The advantage of the deposition from the liquid phase is low temperature as it prevents PSi sintering, which takes place at heating up to Tammant's temperature (572°C for Si) (Labunov et al. 1986), and easy infiltration of the pores with deposition medium defined by wettability and capillary effects contributing to filling pores with guest substance.

Chapter 9 is a review of recent research in the field of fabrication, properties, and applications of PSi-based composites and hybrid materials. For detailed information on the early research, it is advised to refer to the primary sources or reviews of Hérino et al. (1997) and Hérino (2000).

9.2 FORMATION, PROPERTIES, AND APPLICATIONS OF METAL/PSi COMPOSITES

For the formation of metal/PSi composites, the deposition from the liquid phase is more often used. Chemical reactions between PSi and deposited metals hardly ever occur providing distinct silicon/metal interface. Nevertheless, in some cases the formation of silicides is observed (Dolgiy et al. 2012a). The metal deposition from the liquid phase can be performed electrochemically or without applying electric current (Zhang 2001; Ogata et al. 2006). Both processes are electrochemical by nature in that the deposition of metal atoms is a reduction reaction involving a charge transfer. In the electrochemical deposition method, electrons are delivered by electric current flowing through the electrolyte/substrate interface. The electrochemical method can provide uniform pore filling, especially when PSi skeleton is fully depleted, because in this case the current localization at the pore bottoms takes place. In the electroless deposition, electrons are provided either by a reducing agent or by the silicon substrate due to the corrosion processes. In the last case, the displacement of Si atoms with metal ions occurs. Only several metals (Cu, Hg, Ag, Pd, Pt, Au) that have the electrochemical potential higher than +0.187 V (the reduction potential of Si in HF-based solutions) can be deposited on bulk Si by the displacement process. Compared to bulk Si, the reduction of metals on PSi occurs more easily due to the presence of Si–H species on the pore walls and elemental silicon in the skeleton. The reduction potential of either of these species can reduce many metal ions not limited by those listed above down to their elemental state (Coulthard et al. 1993; Tsuboi et al. 1998; Parbukov et al. 2001; Harraz et al. 2002; Xu et al. 2007a,b). As for the electrochemical method, metals placed to the right of Al in the metal activity series can be deposited in PSi from the aqueous solutions. Other metals, in particular rare earth elements (REEs), tend to deposit in the form of hydroxides, forcing the use of nonaqueous solutions (Petrovich et al. 2000a,b). PSi doping with REEs by the electrodeposition method is described in Chapter 9, Section 9.7.2, in *Porous Silicon: Formation and Properties*.

9.2.1 FERROMAGNETIC METALS

Ni, Co, and Fe are the most interesting ferromagnetic metals to fabricate PSi-based composites. They have high Curie temperature (1043 K, 1403 K, and 631 K for Fe, Co, and Ni correspondingly), conditioning inherent magnetization in the wide temperature range. Other ferromagnetic metals can have high spontaneous magnetization (for example, 0.2713 T for Tb against 0.17352 T for Fe) but quite low Curie temperatures (223 K for Tb), so additional cooling is required to explore their magnetic properties. Magnetic NWs and NPs, grown in PSi, demonstrate magnetic anisotropy, giant magnetoresistance, and coherent spin waves. These properties are important for various magnetic applications such as magnetic recording, magneto-optics, sensor technology, and spintronics (Fert and Piraux 1999; O'Brien et al. 2009; Fernández-Pacheco et al. 2013)

The pioneering study into the Ni electrochemical deposition in PSi was directed to the formation of thick Ni silicides (Hérino et al. 1985). Uniform Ni deposition was achieved only for thin 0.5-μm PSi layer. The Ni silicides formation occurred after annealing in dry nitrogen. The idea of the thick silicide formation gained further progress in the works by Borisenko et al. (1989, 1990) where silicides based on electrodeposited Co were fabricated and studied. Uniaxial anisotropy of the magnetic properties of Ni NWs formed in PSi was first shown by Gusev et al. (1994). In this work, Ni was deposited electrochemically in the 30-μm thick meso-PSi layer in the potentiodynamic regime with the frequency of 50 Hz. It provided uniform filling of PSi with Ni along pores.

Recent Ni/PSi composites formation study is aimed to develop a reliable technology for the deposition of Ni in PSi with a high filling factor F_f, uniformity, and reproducibility. The research group of the University of Graz (Granitzer et al. 2005) electrochemically deposited Ni in PSi in the way described by Gusev et al. (1994). As a result, an irregular array of Ni NWs arranged with ~2.5 × 10⁹ mm⁻² surface density was formed. Lagging of magnetization M behind variations of the magnetic field, H (a hysteresis loop) was found by SQUID magnitometric study as a dependence $M(H)$ in the range of weak magnetic fields to be dependent on the field direction (in or out of plane). The Ni electrochemical deposition in a pulse mode with an alternation of direct and returning pulses improved pore filling with Ni that was deposited as nanostructures of varying size and shape (Rumpf et al. 2006). As shown in Figure 9.2, shape of the deposited Ni-nanostructures varied from spheres (which diameters were defined by the pore diameters) through ellipsoids to high aspect ratio needle-like structures by decreasing the pulse duration from 40 to 5 s (Granitzer and Rumpf 2010).

Hysteresis in the $M(H)$ dependence in the range of strong magnetic fields (~5 T) was found to depend on the size of NWs. Surface magnetism of giant orbital interaction was also observed. It led to paramagnetic effect in strong magnetic fields. The $M(H)$ curve behavior in the range of

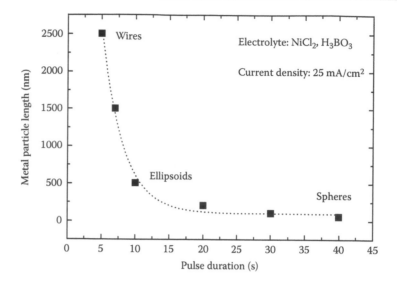

FIGURE 9.2   Relation between pulse duration of current density and elongation of the Ni-structures deposited in the PSi in the same conditions. The length of the Ni structures increases from 60 nm (for sphere-like NPs) to 2.5 μm (for NWs). The diameter of the Ni structures corresponds to pore-diameter, which in average is 60 nm. (From Granitzer P. and Rumpf K., *Materials* 3, 943, 2010. Published by MDPI as open access. With permission.)

weak magnetic fields has been explained by the presence of fine Ni NPs in pores (Granitzer et al. 2006, 2007a,b; Rumpf et al. 2008). Recently, this research group has reported new results on Ni/PSi composites aimed at formation of uniform metal NWs by minimization of roughness of the PSi pore walls. Granitzer et al. (2012) fabricated PSi by anodization with applied external 8 T magnetic field, which was transversely directed to the sample surface and prevented etching of the pore walls. Then Ni was deposited into PSi by the electrochemical method. An enhanced magnetic anisotropy of the final Ni NWs in PSi was demonstrated as shown in Figure 9.3. This practical result is very important for the magnetic memory devices based on Ni NWs formed

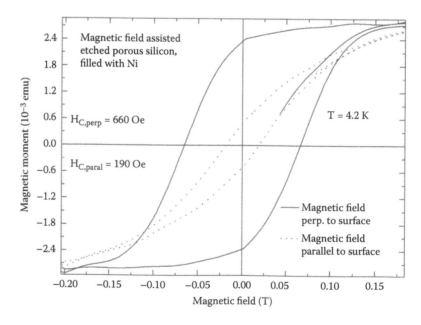

FIGURE 9.3   Magnetic anisotropy observed in a Ni-PS composite for the two magnetization directions parallel (full line) and perpendicular (dotted line) to the Ni nanowires. The PS template was prepared by magnetic anodization. (From from Granitzer P. et al., *Appl. Phys. Lett.* 101, 033110, 2012. Copyright 2012: AIP Publishing. With permission.)

in PSi. The coercivity and remanent magnetization of such structures were studied by Rumpf et al. (2014), which showed that F_f of Ni is 40–50%, and the shape of Ni NWs is defined by the pore channels.

The beneficial influence of smooth pore walls on the coercivity of Ni-filled PSi is demonstrated in Figure 9.4. The results obtained by the research team from the University of Graz are summarized in Granitzer and Rumpf (2010) and Granitzer et al. (2014).

The research group from BSUIR (Dolgiy et al. 2012a,b, 2013) used the galvanostatic electrochemical deposition to deposit Ni in meso-PSi with the fully depleted skeleton with the rough surface. The Ni deposition was shown to start from the NPs formation on endings of the PSi skeleton branches at a depth of 8–10 µm. Then the number of NPs increased in the middle of the PSi layer, and after the prolonged deposition the pores were fully filled with Ni (Figure 9.5).

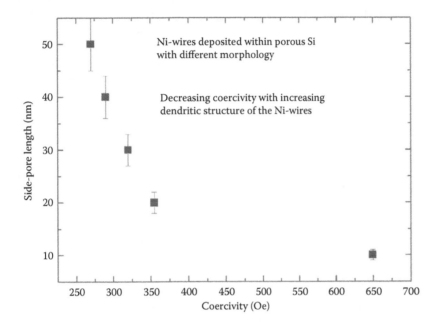

FIGURE 9.4   Coercivity of Ni-filled PSi versus side-pore length of the templates. Decreasing side-pore length is concomitant with an increase of the pore diameter conventionally etched samples. The sample offering a side-pore length of 10 nm has been prepared by magnetic field-assisted etching. (From Rumpf K. et al., *Nanoscale Res. Lett.* 9, 412, 2014. Published by Springer as open access. With permission.)

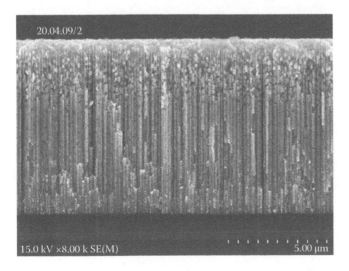

FIGURE 9.5   SEM cross-section image of Ni/PSi composite fabricated by the electrodeposition of Ni into the 10 µm thick and 72% porosity meso-PSi layer.

It was revealed that contact phenomena at the electrolyte/PSi interface play an important role at the electrochemical deposition of Ni in PSi. It happens at the initial stage of the electrochemical process and can lead to the formation of NPs of Ni silicides on the needle parts of the pore walls (Dolgiy et al. 2012b). NPs of Ni silicides are not ferromagnetic; thus, their presence deteriorates magnetic properties of Ni/PSi composite. In such structures, the lattice parameter of Ni was shown to be expanded up to 0.4–0.5% of that of bulk Ni (Prischepa et al. 2014). Magnetic anisotropy of Ni NWs correlated with packing density and pore sizes was found. The dipole interaction was shown to decrease anisotropy and change "easy" axis of magnetization from "along NW" to "perpendicular to NW." The magnetization versus magnetic field measurements for Ni/PSi samples with different deposition time of Ni revealed the dependence of the saturation magnetization value (M_{sat}) on the deposition time (Figure 9.6). It was established that for Ni/PSi samples with longer deposition time starting from 15 min, the specific magnetization is equal to that of bulk Ni (Dolgiy et al. 2013). That moment corresponds to the start of Ni NWs formation in the PSi matrix. Note that the coercivity does not vary significantly with the deposition time.

Michelakaki et al. (2013) from NCSR Demokritos/IMEL compared pulse and galvanostatic regimes at the Ni electrochemical deposition in meso-PSi. The effect of the pulse duration, number of pulses, and total process time on pore filling was investigated for meso-PSi with different porosities and thickness varying in the range of 0.5–4 μm. The authors noticed that the full pore filling and continuous Ni NWs formation are achieved under the galvanostatic electrodeposition, and the results are quite similar to those obtained with the pulsed electrodeposition when the same total deposition time is used in both cases. However, an explanation of the obtained results was not proposed. Probably insensitivity of the Ni deposition to the current density mode is caused by the complete depletion of PSi with charge carriers. Using other electrolytes and PSi of alternative morphology can lead to great variations of the results of the Ni deposition in pulse and constant current density modes. In particular, significant differences may take place at the deposition in thick PSi layers where limited factor is supplying areas of the reduction reaction with the fresh electrolyte. In this case, the impulse mode presents a more effective way of pore filling with Ni.

The electrochemical deposition of Co and Fe in PSi in contrast to the Ni deposition is more compliated. Ronkel et al. (1996) showed that after the potentiostatic Fe deposition into macro-PSi the porous layer is filled with Fe but also contained large amounts of oxygen. The intense SiO_2 formation was supposed to take place while Fe was not oxidized. Renaux et al. (2000) studied the Fe deposition in p^--type PSi. Only the partial pore filling with single Fe clusters was observed. It can be supposed that the Fe deposition in PSi formed on p^--Si should begin on the pore bottoms. Nevertheless, the authors noticed that the Fe deposition occurred simultaneously over all surfaces of the pores.

The electrochemical deposition of alloys of ferromagnetic metals is also possible (Hamadache et al. 2002, 2003). The composition of the deposited Fe-Co varied with the change of ion

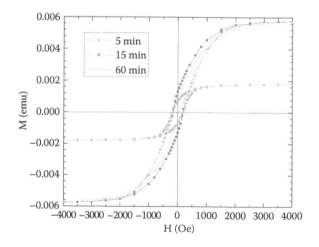

FIGURE 9.6   Magnetization M versus magnetic field H hysteresis loops for PS-Ni samples with different time of Ni deposition. Data are for $T = 5$ K. Magnetic field is oriented parallel to the substrate.

concentrations of these metals in the electrolyte. Pure Co starts depositing from the pore bottoms while the Fe deposition takes place on the pore wall surfaces to give the increased Fe concentration in the upper part of the PSi layer. Increasing Fe ions concentration in the electrolyte caused rising Fe deficiency in the alloy. This was explained by the fact that Co acts as a catalyst for the Fe deposition in the depth of the PSi layer. Even small amounts of Co in the solution (5%) was established to cause the Fe deposition in the whole pore depth rather than at the layer surface. The authors realized good PSi filling with Fe-Co alloys of various percentage compositions. The deposition of the Ni-Fe alloy in PSi by the potentiostatic regime was studied by Ouir et al. (2008). The deposition regimes were shown to influence not only percentage of the alloy composition but also predominant crystallographic orientations of the Ni-Fe compounds. For example, the alloy deposited at 1.25 V consisted of 87% of Ni and 13% of Fe and (111) orientation predominated while the deposition at 1.5 V formed the alloy consisting of 81% of Ni and 19% of Fe and the (111) phase amount was approximately equal to the (220) phase amount.

Co-Pt alloys deposited in PSi (Harraz et al. 2013) also demonstrated magnetic hysteresis when the magnetic field is applied transversely to the sample surface; the coercivity value was equal to 73.13 Oe at that. The sample annealing has resulted in the nearly doubled coercivity and saturation magnetization and the ratio of residual magnetization increased almost an order of magnitude.

Roughness of the PSi pore walls influences magnetic properties of NWs of both pure ferromagnetic metals and alloys. Aravamudhan et al. (2007) fabricated long (up to 100 µm) vertical and branched Ni–Fe nanostructures by the galvanostatic electrochemical deposition in the PSi template. Ni–Fe nanostructures were analyzed for temperature-dependent magnetization and magnetization versus magnetic field measurements. They revealed no magnetic anisotropy of the nanostructures probably due to a balance between "reduced" shape anisotropy from branched and rough pore surfaces and magnetocrystalline anisotropy.

Xu et al. (2007a,b) reported using the displacement deposition method allowing the transformation of the 200-µm thick PSi layer into porous Ni layer. Recently, Zhang et al. (2011) used the displacement method to transform 200-µm thick macro-PSi layer into Ni. The fabricated Ni/PSi composite is shown to be of interest to engineer the Si substrate impedance for microwave crosstalk isolation in mixed-signal integrated circuits (Zhang et al. 2011).

9.2.2 NOBLE METALS

The deposition of noble metals in PSi can be readily made by the displacement method from solutions of metal salts because the standard electrode reduction potential for noble metals in aqueous solutions is positive (Bard et al. 1985). This method has found increasing favor because noble metals are often used as catalysts, so there is no need to form continuous films of these metals.

The Pd deposition by the displacement method is usually carried out from the aqueous or aqueous-alcoholic $PdCl_2$ solutions (Coulthard et al. 1993; Lin et al. 2004; Pedrero et al. 2004). The Pd clusters deposition takes place in the regions of pore entrances and tops of the Si crystallites of PSi. Nevertheless, Razi et al. (2010) showed that very small amounts of metal penetrate into the pores. Large particles of the deposit had the sandwich structure consisting of the first SiO_2 layer, the second Pd_xSi layer, the third pure Pd layer, and the last PdO_2 layer at the particle surface. Prolong stay in air led to increase of the PdO_2 amount while the amount of the pure Pd decreased. The PSi/Pd structures obtained were used as hydrogen sensors allowing detecting hydrogen concentrations up to 0.05–0.17% in dry air.

As Pd, gold is used as a catalyst so Au/PSi composites are important as well. Brito-Neto et al. (2007) have used the displacement method to deposit Au to form a catalytic layer for microfuel cells. The presence of HF in the solution leads to the partial or full dissolution of PSi skeleton during the deposition depending on reactivity of Au containing complexes. The presence of high reactive $[AuCl_4]^-$ complex leads to the complete dissolution of the PSi layer and the chaotic deposition of porous Au in its place. While the displacement method provides the deposition of only a small amount of Au in the pores (Chourou et al. 2011), the electrochemical method allows covering pore walls of macro-PSi with the 100-nm thick Au layer and forming Au microtubes. The decrease of the deposition current density results in the longer deposition time but improves Au film quality.

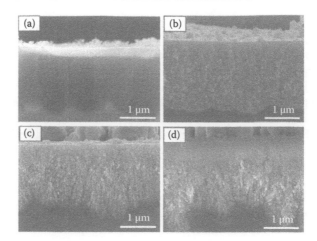

FIGURE 9.7   Cross-sectional SEM of micro-PSi after the electrodeposition of Pt (a, b) and Ag (c, d). The images in (a, c) and (b, d) show the samples of the hydrophilic and hydrophobic PSi, respectively. (From Koda R. et al., *Nanoscale Res. Lett.* 7, 330, 2012. Published by Springer as open access. With permission.)

Other noble metals can be deposited in PSi by the electrochemical method as well. Koda et al. (2012) have deposited Pt and Ag in PSi layers. For the Pt deposition, the deposition behavior was shown to depend on the PSi surface conditions. The deposition in hydrophilic PSi begins from the pore bottoms whereas the Pt deposition in hydrophobic PSi takes place predominantly at the surface. For the Ag deposition, in any case, the deposition begins from the porous layer surface and extends into pore depth independently on the surface type (Figure 9.7). The reason is that at the noble metal deposition in any case the displacement mechanism takes place. For Pt, the displacement is less active in comparison with the electrochemical deposition, and so the process is sensitive to the surface type. For Ag, the deposition by the displacement method predominates and so the surface type influence is insignificant. Moreover, the over voltage at the Pt deposition on the Si electrode is much higher than at the Ag deposition on the Si electrode. Ag does not occlude pore mouths but Pt can occlude them in the case of the hydrophobic surface.

A comparison of Pt, Pd, and Au deposition processes into the ordered macro-PSi of various thicknesses from the solution containing additional amounts of NaCl or Na_2SO_4 was carried out by Fukami et al. (2008). For the Pt and Pd depositions from the solutions containing Na_2SO_4, the deposition has been shown to take place mainly in the pore tops whereas with the NaCl containing solutions the deposition begins from the pore bottoms. For the Au deposition, the attempts to find the regime providing full pore filling with Au have not met with success. With the Na_2SO_4 containing solution, the deposition begins in the pore mouths, and with the NaCl containing solution, the deposition begins from the upper half of the pores and continues to the Au outcropping at the surface. It has been also shown that when deposited in 25-μm thick PSi, the occlusion of pore mouths with Pd deposited occurred in a short time while Pt deposited occluded pore mouths later but pores had not had time to be filled. The Pd deposition in 50-μm thick PSi demonstrated similar behavior while the 50-μm thick PSi layer can be filled with Pt fully. A tenfold decrease in the metal ion concentrations in the solutions has resulted in the Pt spongy deposit and Pd tree-like deposit formation.

9.2.3 CONCLUSIONS

The electrochemical deposition has been used with advantage for the deposition of ferromagnetic metals into PSi. Several teams have already demonstrated Ni NWs in the PSi template with the aspect ratio >1:100 and the filling factor F_f higher than 50%. However, the morphological changes of the pore walls like dendritic branches as well as hydrogen evolution lead to the inhomogeneity of the metal deposit preventing homogeneous filling of the pores. Another problem is related to the exchange of reagents within the pore, which influences the filling factor F_f. If this exchange is limited, the channels are blocked by the accumulation of the deposited metal. This prevents the uniform filling through the whole length of the pores. From the technological point of view, the pulse mode of the metal electrodeposition seems a more reliable regime to provide uniform pore filling with metals while the galvanostatic regime can be applied for high porosity PSi with the fully depleted skeleton.

For the successful filling of pores with metals by the electrochemical technique, monitoring the potential of the PSi during the electrochemical deposition is of importance to provide *in situ* control of the deposition process stages and to stop the process in the time moments defined by the characteristic points in the potential curve to secure the desired extent of the pore filling with metal. To fabricate cylindrical NWs in the PSi template, the roughness of the PSi pore walls should be minimized. The anodization under the magnetic field action offers a certain scope for that. The pore wall roughness also may be decreased by the use of the pulse modes in the Si anodization as shown by Cheng et al. (2003) and Chubenko et al. (2014). To provide good pore filling with metals, it is very important to provide the full carrier depletion of the PSi skeleton to prevent the current passing through the skeleton and thereby to have the current passing mainly through the pore bottoms, promoting the progressive pore filling with metal from bottoms.

Table 9.1 gives a comprehensive listing of the various metals deposited into PSi and the methods utilized.

9.3 FORMATION, PROPERTIES, AND APPLICATIONS OF SEMICONDUCTOR/PSi COMPOSITES

The deposition of both elementary and compound semiconductors has been made into PSi. The aims of the deposition of elementary Ge and Si were either to form an electric contact to PSi, to create electroluminescent devices, or to obtain SiGe (Halimaoui et al. 1995). The deposition of compound semiconductors mostly presented by II-VI and IV-VI has two goals, which are schematically shown in Figure 9.8. The first one is associated with the formation of the electric contact to PSi layer by the deposition of optically transparent films for the PSi-based electroluminescent optoelectronic devices with the luminescence intensity maximum at 600–900 nm (Hérino et al. 1997). Most compound II-VI and IV-VI semiconductors meet this requirement. The second goal is a creation of composite and a formation of these compounds using PSi as a passive template. Those structures utilize inherent optical properties of II-VI and IV-VI semiconductors since most of them have direct bandgap and high luminescence efficiency (Singh et al. 2007; Balucani et al. 2011).

In most cases, the deposition of elementary and compound semiconductors was carried out in meso-PSi of 70–80% porosity with the thickness not exceeding 1–2 μm. The presence of semiconductors in the PSi pores is usually confirmed by the EDS analysis or Auger electron spectroscopy. SEM images showing actual distribution of deposited semiconductors in the porous layer were not often presented.

9.3.1 II-VI AND IV-VI COMPOUND SEMICONDUCTORS

Dücsö et al. (1996) and Utriainen et al. (1997) made first attempts of the PSi filling with II-VI and IV-VI semiconductors. In their works, the atomic layer deposition (ALD) of the SnO_2 into the PSi layer has been studied. The controlled conformal pore filling of PSi with SnO_2 or SnO_x with the tin concentration gradient across the thickness has been successfully realized by the control of temperature and impulses of the reagent supplying. SnO_2 nanostructures with the aspect ratio 1:140 have been formed. A beneficial effect of deposited SnO_2 coatings on the stability of PSi luminescence and electrical properties was determined.

The research group from the Grenoble University (France) introduced the electrochemical deposition of compound semiconductors in the PSi template from the aqueous solutions. The prime object of their work was the development of transparent conductive electrodes to obtain reliable electric contact to the whole PSi surface including walls and bottom parts of the pores. The electrochemical deposition allows depositing binary compound semiconductors in accordance with the induced codeposition mechanism (Kröger 1978). Montès et al. (1997) made the

TABLE 9.1   Deposition of Metals into PSi Template

Metal	Silicon (Type, Orientation, ρ, Ω·cm)	Deposition Method	Porous Silicon			Result	Application	Ref.
			UIPAC Class	d, μm	P, %			
Ni	p^+, (100), 0.01	Electrochemical	Meso	10	45	Partial filling	Silicides	Hérino et al. 1985
Ni	p^+, (100), 0.001	Electrochemical	Meso	30	–	Good filling	–	Gusev et al. 1994
Ni	p, (100), 1–10	Displacement	Macro	200	–	Transformation of PSi layer into Ni	–	Xu et al. 2007a,b
Ni	n^+, (100), 0.01	Electrochemical, pulse	Meso	10–40	45–80	Ni nanowires and separate particles	Magnetic sensors, magneto-optical and spintronic devices	Rumpf et al. 2006, 2014; Granitzer et al. 2012, 2014
Ni	n^+, (100), 0.01	Electrochemical	Meso	10	72	Ni nanowires; good filling	Magnetic sensors, magneto-optical devices	Dolgiy et al. 2012a,b, 2013
Ni	p^+, (100), 0.001–0.005	Electrochemical, pulse	Meso	1–3	70–88	Ni nanowires; good filling	–	Michelakaki et al. 2013
Co	n^+, (100), 0.01	Electrochemical	Meso	1–5	30–60	Partial filling	Silicides	Borisenko et al. 1989, 1990
Fe	n, (111), 4.8–7.1	Electrochemical	Macro	1.5	20–30		Electrical contacts	Ronkel et al. 1996
Fe	p^+, (100)	Electrochemical	Micro	0.3	60–80	Partial filling	–	Renaux et al. 2000
Fe-Co	p^+, (100)	Electrochemical	Micro	0.3	60–80	Good filling	–	Hamadache et al. 2002, 2003
Fe-Ni	n, (100), 1–10	Electrochemical	Meso	–	–	Partial filling	–	Ouir et al. 2008
Co-Pt	n^+, (100), 0.01–0.02	Electrochemical	Meso	1	–	Partial filling	–	Harraz et al. 2013
Pd	p, (100), 3	Immersion	–	10	–	Clusters	Catalyst	Coulthard et al. 1993
Pd	p, (100), 0.6–1	Immersion	–	–	–	Clusters near surface	Gas sensors	Lin et al. 2004
Pd	p, (100), 1–5	Immersion	Meso	–	–	Clusters, film on surface	Gas sensors	Pedrero et al. 2004
Pd	p, (100)	Immersion	Macro	5	–	Big clusters	Gas sensors	Razi et al. 2010
Au	n^+, (100), 0.007–0.02	Displacement	Meso	10.5	–	Good filling; porous metal	Catalyst layers for micro fuel cells	Brito-Neto et al. 2007
Au	p/p^+, (100), 10–20/0.01–0.02	Displacement, Electrochemical	Macro, Meso	5–50	–	Gold microtubes; different	–	Chourou et al. 2011
Ag, Pt	p, (100), 10–20	Electrochemical	–	2, 7	–	Good filling; deposition on surface	–	Koda et al. 2012
Au, Pd, Pt	p, (100), 10–20	Electrochemical	Macro	25–100	–	Good filling; partial filling; deposition on surface	–	Fukami et al. 2008

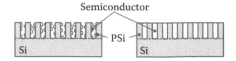

FIGURE 9.8   Major semiconductor/PSi composite strategies of design.

first attempts of the PSi pore filling with CdTe by the electrochemical deposition. CdTe/PSi struc-tures with the CdTe continuous film on the PSi surface were obtained. This approach was also used for the ZnSe deposition (Hérino et al. 1997). However, the conformal pore filling with these materials was not achieved because the deposition of the compounds takes place more intensely at the PSi surface. That problem could be solved by two ways (Montès et al. 2000). The first one is the increase of the electrolyte concentration to increase the ion diffusion rate along the pore channels during the deposition. However, it is complicated for II-VI semiconductors since sto-chiometric compounds are formed in a limited range of deposition parameters values. The sec-ond way is the stimulation of the deposition process in the depth of the pores. It could be done, for example, by substrate illumination with light for which the PSi is transparent and which will be absorbed by the bulk silicon below PSi layer avoiding generation of the excess charge carriers in the top part of the PSi layers. The successful ZnSe deposition into PSi has been made at the continuous illumination of the substrate with 830 nm laser (Montès et al. 2000). As a result, ZnSe was observed only at the bottom part of PSi layer. The tenfold increase of the photoluminescence intensity associated with the defect passivation of the silicon crystallites surface was observed. The thermal annealing of the ZnSe/PSi structures caused the PSi photoluminescence intensity to decrease, red shift of photoluminescence maximum, and improvement of PSi/ZnSe contact electric parameters (Montès et al. 2000; Montès and Hérino 2000).

ZnO can also be deposited in the PSi template by the cathodic electrochemical deposition (Balucani et al. 2011; Chubenko et al. 2011) from the non-aqueous DMSO-based solution. The PSi layer thickness was shown to have significant impact on the ZnO deposition process. At the about 1-μm PSi thickness, the ZnO deposition preferably takes place on the surface of the porous layer; and at the 10-μm PSi layer thickness, the ZnO crystals are formed in the pores to form the composite structure (Figure 9.9). The change in the crystal formation process is associated with an increase in the length and so in the resistance of nanosized crystallites of the PSi skeleton. The structures formed have shown the PL in the visible spectrum range associated with the ZnO crystal lattice defects.

The research of the chemical bath deposition (CBD) of CdS in the PSi layer was made as well (Hérino et al. 1997; Gros-Jean et al. 1998). The deposition was performed from the alkaline cadmium acetate and thioacetamide solutions. With good CdS pore filling, the instability of optical properties of the CdS/PSi composite associated with the Si-H bond breaking at the PSi surface and PSi partial oxidation with the formation of the nonradiation recombination centers has been shown (Gros-Jean et al. 2000). The methyl grafting preceding the CdS deposition allowed the PSi surface to be pro-tected at the CBD of CdS, preventing the PSi luminescence degradation (Gros-Jean et al. 2000).

The ZnSe and CdSe deposition into the PSi template pores can be also fulfilled by the evapora-tion of metal salts incorporated in PSi followed by selenisation (Belogorokhov et al. 1998, 1999). Obtained composites demonstrated complex PL spectra with bands associated with the emission both from the PSi and from the compound semiconductor clusters in the pores. After the ZnSe and CdSe incorporation into the pores, the PSi emission intensity decreases by 20–50%. Formed

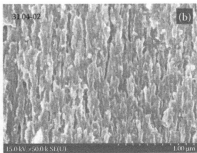

FIGURE 9.9 SEM images of the middle section (depth approximately 3 μm from surface) of meso-PSi layer of 70% porosity with ZnO particles electrochemically deposited from the nonaqueous DMSO based solution: (a) corresponds to shorter deposition time and hence distinct separated ZnO nanocrystals are visible; (b) corresponds to longer deposition time, individual ZnO crystals merged into a continuous film covering the walls of the pores.

film consisted of 3–5 nm sized ZnSe or CdSe quantum dots. A small shift of the photoluminescence maximum positions into the short-wave range relative to these for the bulk crystals of these semiconductors has been observed at that. Non-rectifying CdSe/PSi junction is suitable as the electric contact to the porous layer that is of importance for the PSi-based devices.

Recently the isothermal close space sublimation (ICSS) technique was used for PSi filling with ZnTe and CdSe (Torres-Costa et al. 2012; Melo et al. 2014). Obtained compound semiconductor clusters with polycrystalline structure consisted of crystallites of the stoichiometric or close to stoichiometric composition with the oxygen atom inclusions. The PSi filling has occurred at the whole depth of the PSi layer. The increase of binary semiconductors bandgap caused by quantum-confinement effects has been observed. The self-regulation of the ICSS deposition process due to the equality of substrate and component sources temperature allows a formation of very thin polycrystalline films of compounds in the PSi (Melo et al. 2014).

II-VI and IV-VI oxide semiconductors such as SnO_2 and ZnO can be readily synthesized by the sol-gel deposition. With this technique, sol in the liquid state could penetrate deeply into the pores. However, two conditions should be met at that. The first condition is the assurance of the pore walls wetting with the solution, and the second one is the limitation of the maximum particle size that should not exceed the diameter of PSi pore mouths. For the first time this approach was demonstrated by the example of SnO_2 (Cobianu et al. 1997). Sols used have been prepared on the base of tin ethoxide ethanol solutions. Tin ethoxide based sol has allowed forming SnO_2 in the depth to 1 μm from the PSi layer surface. The sol-gel technique also enables doped SnO_2 films to be formed by the addition of dopant containing additives into the sol (Chatelon et al. 1994). These films doped with Sb or F may be used either as transparent conducting electrodes (Cobianu et al. 1997; Moadhen et al. 2003; Elhouichet et al. 2005; Garcés et al. 2012) or as array that includes optically active particles emitting light in the visible range, for example, rare-earth ions (Elhouichet et al. 2003; Moadhen et al. 2003; Dabboussi et al. 2006). Noteworthy is the SnO_2 films doped with rare-earth ions Eu^{3+} and Tb^{3+} demonstrated optical effects associated with the nonradiative transfer of absorbed optical energy from rare-earth ions to Si crystallites (Elhouichet et al. 2003) or vice versa (Dabboussi et al. 2006). This results in the widening of the absorption spectra of composite material and increasing of the luminescence efficiency.

To form ZnO on the PSi substrates by the sol-gel method, zinc acetate ethanol solutions with the addition of monoethanolamine for the sol stabilization were used (Singh et al. 2007). The composite structures formed have demonstrated a wide emission spectrum in the range of 450–850 nm composed of the PSi (650–850 nm) and ZnO (450–650 nm) emission bands. The ZnO emission is associated with oxygen vacancies and interstitial atoms in the crystal lattice (Singh et al. 2007; Singh et al. 2009a). The healing of defects in the ZnO crystal lattice and at the ZnO/PSi interface allows increasing the luminescence efficiency of composite material (Singh et al. 2009b). Martínez et al. (2014) used the ZnO sol-gel deposition from the same solutions in two-layer PSi for the fabrication of memristive devices. The PSi application has allowed defining formation of ZnO clusters of the required size and increasing the number of oxygen defects in the ZnO crystal lattice responsible for memristive properties of ZnO/PSi structures.

9.3.2 Ge AND Si

The Ge and Si deposition into PSi in the first time was carried out by the chemical vapor deposition in ultra-high vacuum (UHV-CVD) (Halimaoui et al. 1995). For Ge, the full filling of PSi pores has been made at low pressure of $GeH_4 + H_2$ operating gas of 10^{-3} Torr, as revealed by the Ge distribution in depth of the PSi layer. The achieved deposition rate has been 0.01–0.07 nm/min to be two orders lower as compared with the deposition rate on the smooth substrates. At high pressure, the mouth bridging effect has been observed resulting in partial pore filling. For the Si deposition, the effect of pore downsizing in PSi due to the silicon layer formed on the pore walls has been achieved. This result showed promise for obtaining ordered pores of smaller diameter unachievable by conventional anodization methods. However, Halimaoui et al. (1995) have not presented the Ge and Si distribution in pores. Apart from the creation of the Ge electric contact to PSi, the use of PSi as a template for the formation of luminescent Ge clusters was of interest (Liu et al. 1998). The vapor transport deposition method has been used for the Ge deposition in the PSi layer with

13-nm diameter pores. Ge cluster size has been about 5 nm. The Ge clusters showed intense photoluminescence with maximum in the 700-nm range associated with the quantum-sized effects in the Ge clusters and transition of the nonequilibrium carriers generated by radiation from the PSi skeleton into Ge clusters. Unfortunately, the low time stability of the characteristics of the structures obtained that quickly degraded because of gas adsorption from atmosphere should be noted.

The deposition of elementary semiconductors also was made in the composition of more complex compounds intended for "white" light sources. In the work (Abd Rahim et al. 2012), the ability of the ZnO/Ge/PSi structure formation was demonstrated. A sequential deposition of Ge and ZnO has been made by the method of vapor transport deposition in vacuum. As a result, at the least the full filling of pore mouths with materials deposited has been achieved. The structures obtained have shown complex photoluminescence spectra with bands at about 380, 520, and 770 nm associated with the processes of ZnO band-to-band transitions, recombination in nanosize Ge clusters, and defect-related luminescence in ZnO, respectively. Unfortunately, electroluminescent characteristics of Ge/PSi and ZnO/Ge/PSi structures that are critically important for the LED formation on their basis were not shown in the work (Liu et al. 1998; Abd Rahim et al. 2012).

9.3.3 CONCLUSION

The peculiarities of the electrochemical deposition of metals in PSi (Section 9.2.3) are also valid for the electrodeposition of semiconductors in PSi. Nevertheless, it is necessary to note that this method shows worse structural perfection of the deposited semiconductor in contrast to CVD or ALD.

The PSi filling with semiconductor materials to fabricate the electrical contacts for LED did not gain broad development. Composite materials based on PSi, filled with particles of optically active semiconductors, showed the PL spectra covering the wide range of electromagnetic radiation. Interesting effects of nonradiative transfer of energy from Si crystallites to particles of semiconductors and vice versa were observed in a number of experiments. However, LEDs based on such materials have not been presented yet. Use of semiconductor/PSi composite materials to fabricate the heterostructural photodetectors and sensors seems to be more in perspective.

Table 9.2 gives a comprehensive listing of the various compound and elementary semiconductors deposited into PSi and the methods utilized.

9.4 HYBRID MATERIALS FORMED BY INFILTRATION OF PSi WITH ORGANIC SUBSTANCES

There are various options for the formation of hybrid organic-PSi materials. Bonanno and Segal (2011) presented the most obvious possible results (Figure 9.10). Unfortunately, there are no regularities and common approaches to determining the criteria by which one or another option should be used. Each researcher is forced to determine his or her own.

To infiltrate the PSi matrix with polymers, two infiltration procedures can be used. The first infiltration procedure consists of diffusion of the polymers into the pores. Immediately following anodization, a sample is immersed in a polymer solution containing various common organic solvents such as ethanol, acetone, and chloroform. The sample is left in the solution for a specific period of time (from minutes to days) in order to allow the polymers to diffuse into the pores. The diffusion process takes advantage of the highly porous structure of PSi and the created capillary forces. The polymers have a tendency to go into the pores due to the attraction between the porous surface and the polymer solutions. However, this is a slow process because the polymers have large and bulky molecules. In addition, due to the poor solubility of the polymers in the solvents used, the concentration of the solution is dilute.

The second procedure consists of infiltrating the corresponding monomer, as well as radical initiator, into the pores and polymerizing the polymer within the PSi matrix. Similar to the first procedure, capillary forces attract the small organic molecules enabling diffusion into the pores. Since monomers and radical initiators are extremely small when compared to the polymers, infiltration should be more effective.

TABLE 9.2   Deposition of Compound and Elementary Semiconductors into PSi Template

Semiconductor	Silicon (Type, Orientation, ρ, Ω·cm)	Deposition Method	Porous Silicon			Result	Application	Ref.
			UIPAC Class	d, μm	P, %			
SnO_2	p^+, (100), 0.001	Atomic layer deposition	Meso	2	70	Conformal coverage of pore walls	Gas sensors, optoelectronic devices	Dücsö et al. 1996; Utriainen et al. 1997
SnO_2:Sb	p, (100), 8–10	Sol-gel	Meso	2	70–80	SnO_2:Sb penetrate whole porous structure	Transparent electrodes, rectifying contacts to PSi	Elhouichet et al. 2005
SnO_2:F	p, (100), 25.5–42.5	Sol-gel	Macro	30	–	Partial penetration of the pores	Transparent electrodes	Garcés et al. 2012
SnO_2:Eu^{3+} SnO_2:Tb^{3+} SnO_2:(Eu^{3+}+ Tb^{3+})	p, (100), 8–10	Sol-gel	Meso	2	70	Pore filling	Rare-earth and PSi-based light emitting devices	Elhouichet et al. 2003; Dabboussi et al. 2006
ZnO	p, (100), 0.8–1.2	Sol-gel	Meso	–	–	Pore filling with ZnO layer on top	White light luminescence devices	Singh et al. 2007, 2009a,b
ZnO	p^+, (100), 0.002–0.005	Sol-gel	Meso	0,3	–	Pore filling with ZnO layer on top	Memristive devices	Martínez et al. 2014
ZnO	n^+, (100), 0.01	Electro-chemical	Meso	1–10	70	Partial of or full pore filling depending on PS layer thickness	White light luminescence devices	Balucani et al. 2011; Chubenko et al. 2011
ZnSe	p, (100), 10	Chemical	Meso	10–15	70	Pores filled with nanometer size clusters	Optoelectronic devices	Belogorokhov et al. 1999
ZnSe	p, (100), 8	Electro-chemical	Meso	0.2–2	70	Pores filled with ZnSe from the bottom	Transparent electrodes	Montès et al. 2000
ZnTe	p^+, (100)	Isothermal close space sublimation	Meso	1–1.5	–	Nanocomposite with good pore filling	PSi-based optoelectronic devices	Torres-Costa et al. 2012; Melo et al. 2014
CdSe	p, (100), 10	Chemical	Meso	10–15	70	Pores filled with nanometer size clusters	Optoelectronic devices, ohmic contacts	Belogorokhov et al. 1998, 1999
CdSe	p^+, (100)	Isothermal close space sublimation	Meso	1–1.5	–	Nanocomposite with good pore filling	PSi-based optoelectronic devices	Torres-Costa et al. 2012
CdTe	n^+, (100), ≈1	Electro-chemical	Meso	0.1–0.3	70	Partial penetration of the pores	Transparent electrodes	Montès et al. 1997
CdS	p, (100), ≈1	Chemical bath deposition	Meso	1	75	Good penetration of the pores	Transparent electrodes	Hérino et al. 1997; Gros-Jean et al. 1998
Si	p^+, (100), 0.01	UHV-CVD	Meso	0.52	80	Full pore filling	Electrical contact to PSi for PSi-based luminescent devices	Halimaoui et al. 1995
Ge	p^+, (100), 0.01	UHV-CVD	Meso	0.52	80	Partial pore filling	Pore dimensions decreasing	Halimaoui et al. 1995
Ge	p, (100), 1–3	Thermal evaporation in Ar atmosphere	Meso	–	–	Deposition of 5 nm Ge clusters possibly on pore walls	Ge based light emitting devices	Liu et al. 1998
Ge/ZnO	n, (100), 1–10	Thermal evaporation in vacuum	Meso	–	–	Partial pore filling with crystalline Ge and ZnO	PSi/Ge/ZnO based broadband light emitting devices	Abd Rahim et al. 2012

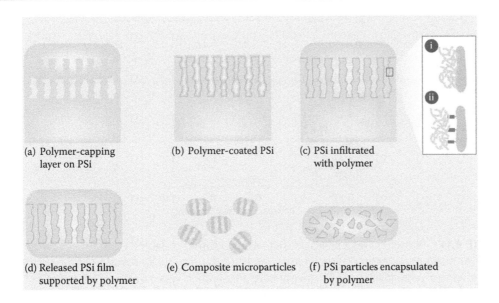

(a) Polymer-capping
     layer on PSi

(b) Polymer-coated PSi

(c) PSi infiltrated
     with polymer

(d) Released PSi film
     supported by polymer

(e) Composite microparticles

(f) PSi particles encapsulated
     by polymer

FIGURE 9.10   Strategies for design of polymer/PSi hybrids. (a–f) Depict different formats of hybrid device design as described above each illustration. Inset illustrates interfacial chemistry where polymer is (i) not attached to PSi and (ii) is attached to a chemical modification layer on the surface area of the PSi. Schematics are not drawn to scale. (From Bonanno L.M. and Segal E., *Nanomedicine* 6, 1755, 2011. Copyright 2011: Future Science Group. With permission.)

9.4.1 ORGANIC SUBSTANCES FOR OPTOELECTRONIC DEVICES BASED ON PSi

The scientific community in the 1980s was interested in the possibility of using organic dyes to create tunable lasers in the form of solid-state electronic devices. Canham (1993) used organic dyes: coumarin 47, coumarin 535, fluorescin, rhodamine 6G, and oksazin 750 to improve PSi luminescence. Before an impregnation of PSi with the organic substanses, it was thermally oxidated. Presence of all organic dyes in PSi except fluorescin allowed reaching the effectiveness of the luminescence up to 1%. An increase of the oxidation level of PSi caused rising luminescence effectiveness. Li et al. (1996) studied luminescence kinetics for different dyes introduced into PSi and proposed an idea of energy transfer from PSi exposed with light to dyes' molecules. Letant and Vial (1997) presented a model of the energy exchange between dye molecules and PSi. Several groups (Yin et al. 1998; Gole et al. 1999; Setzu et al. 1999; Li et al. 2000a; Salcedo et al. 2004; Acikgoz et al. 2011) impregnated PSi with different dyes and observed an improvement of the PSi photoluminescence. Pranculis et al. (2013) studied PSi-based nanocomposites with rhodamine and oksazin and proved an ability of the energy exchange between PSi and molecules of dyes. Elhouichet and Oueslati (2001) initiated the thorough works on PSi/dyes nanocomposites. It was shown that the energy transfer from PSi to molecules of dyes actually takes place (Moadhen et al. 2002; Chouket et al. 2007a,b, 2009a,b, 2013; Gelloz et al. 2010). Piryatinski et al. (2007) chose an alternative way. They proposed an introduction of liquid crystals into PSi. It was shown that liquid crystals like other organic substances, interact with PSi promoting the increase of its luminescence. In those days, mechanisms of the interaction of PSi with organic substances were well studied. That allowed the authors to propose a model that explained the results of their experimental work. Later Piryatinski et al. (2010) experimentally demonstrated the improvement of the photoluminescence of PSi caused by the introduction of liquid crystals. Ma et al. (2011) observed that the variation of a number of the liquid crystals introduced into the PSi matrix allows managing a color of PL. Furthermore, they demonstrated a possibility of white light PL.

Along with the research of the fundamental properties of the organic substances/PSi nanocomposites for optical applications works on the development of devices, which can be successfully used in practice, have been conducted. For instance, Lopez et al. (1999) studied the changing PL of PSi. They showed that it is possible not only increase the effectiveness of PL of PSi but also to control PL peak position (Figure 9.11). Guendouz et al. (2003) filled the PSi pores with polymer,

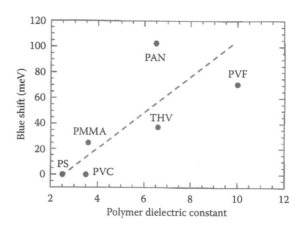

FIGURE 9.11 Photoluminescence blue shift versus polymer dielectric constant at a frequency of 60 Hz. PSi-polymer nanocomposites were produced (74% porosity) by diffusion of the polymers for a period of 2–3 days. Initial peak PL energy before infiltration was 1.63 eV. (From Lopez H.A. et al., *J. Lumin.* 80, 115, 1999. Copyright 1999: Elsevier. With permission.)

which has nonlinear optical properties. Later, they studied the ways of modification of dielectric permittivity of composite while preserving nonlinear optical properties of polymer.

9.4.2 ORGANIC SUBSTANCES FOR PASSIVATION AND FUNCTIONALIZATION OF THE PSi SURFACE

The freshly prepared PSi surface is terminated with hydrogen. Si-H bonds are broken in conditions of atmospheric air (not to mention the more aggressive media) and the surface of the PSi begins adsorbing different atoms and molecules. For this reason, any practical applications deal with passivated PSi. Moreover, for sensors or biomedical devices applications the PSi surface needs to be functionalized. The easiest and most obvious way to passivate the surface—the oxidation of PSi—obviously negates all the qualities inherent to PSi, turning the material into the porous silica. Therefore, the scientific community has taken steps to search for other processing methods of the hydrogenated PSi surface. Lauerhaas et al. (1992) found a high sensitivity of the PSi luminescence to organic substances using solvents. Other parameters of PSi are very sensitive to the organic substances; however, this was revealed later.

Canham (1995) was the first who proposed to use PSi as a bioactive and biocompatible material. Nowadays these peculiar features of PSi have been actively studied. Li et al. (2000b) studied a possibility of using PSi as bioactive material, which delivers Pt-based complex compound into a cancer cell for its destruction. Li et al. (2003) used the PSi as a matrix for the fabrication of the nanostructured polymer biofuntionalized sensors. They underlined the simplicity and accessibility of the process in comparison with nanolithography. Li et al. (2005) proposed a technology to fabricate the photonic crystals based on the nanocomposites polymer/PSi. Stability and high quality of the obtained photonic crystals was denoted. However, no comparison with analogous structures fabricated by other methods has been made.

Chirvony et al. (2006), basing on the formed understanding of the mechanism of the energy interaction of PSi with organic substances, proposed to use porphyrine inbuilt to the PSi matrix as a sensibilizator for the photodynamic therapy of cancer. The function of the sensibilizator is localization in the cancer cells and generation of the singlet oxygen under near IR excitation in the range of transparency of the living cells of the human body.

Sailor research group from University of California actively studies the biomedical application of PSi and reviewed applications of PSi-based composites for the drug delivery into the treating area of the human body at the required time (Anglin et al. 2008). The advantages of PSi biocompatibility and its wide functional possibilities have been demonstrated. Wu et al. (2008) reported about Si microparticles functionalized by hydrosililation and impregnated with drugs toxic for cancer cells. These particles have a controlled delivery rate to the human body. The control is provided by managing the rate of the composite material oxidation. Further research works on biomedical application of PSi were mostly connected with the use of particles of this porous material (McInnes and Voelcker 2012).

Jane et al. (2009) reviewed the most interesting sensors based on PSi for biomedicine. Various techniques of fabrication of functionalized PSi-based nanocomposites were mentioned as well as registration methods of useful sensors' signal. Bonanno and Segal (2011) presented a review devoted to the biomedical application of hybrid materials polymer/PSi. The review is characterized by a logical systematization of works on the topic and convincible conclusion remarks.

9.4.3 ORGANIC SUBSTANCES FOR ELECTRICAL CONTACTS TO PSi

Since the discovery of the effective visible PL of PSi, the attention of the scientific community has been prevalently directed to the development of the light-emitting devices based on PSi. The greatest expectations have been placed on the use of electroluminescence while it requires electrical injecting contacts. Thermal and magnetron sputtering as well as CVD have not resulted in the fabrication of the electrical contacts to PSi with low resistance and reproducible characteristics. That is why research has turned to the electrochemical methods and organic substances such as polymers.

Matveeva et al. (1993) proposed using conductive polymers (polyaniline, piolifuran, polipirol, polythiophene) as electrical contacts to PSi. Nevertheless, the luminescence has not been studied in the work. One year later, Parkhutik et al. (1994) performed a detailed study of the luminescence of PSi/polyaniline. They investigated the photoluminescence variation in a wide temperature range (from 4 to 300 K). It was found that polyaniline improves the PL of PSi especially in the range of 100–150 K. Koshida et al. (1993) and Li et al. (1994) studied using polyaniline and polipirol for contacting to PSi and showed that the electroluminescence is several times better than in the case of a metallic contact. Halliday et al. (1996) proposed to form the contact and a light-emitting junction simultaneously. In earlier works, polyaniline has been used just as the electrical contact to PSi with the previously formed p-n junction.

Wakefield et al. (1997) made attempts of the fabrication of the effective light emitter based on PSi by using the multilayer heterostructure metal oxide/polymer/PSi. A principle of the method is based on the injection of holes in *n*-type PSi through the conductive polymer. As a result, the quantum effectiveness increased. Furthermore, an operating voltage rose up to 50–60 V.

Kim and Laibinis (1999) proposed a method of the fabrication of a contact polypirole/PSi with more high quality. They used organolithium compound as initial material for forming low-loose polypirole. The resistance of the polypirole contact to PSi was studied in dependence on the pore filling level. It was found that the filling level is influenced by the current density at the polypirole deposition (Moreno et al. 1999).

Antipán Lara and Kathirgamanathan (2000) used polythiophene as a polymer contact to PSi. The final diodes with rectifying contacts emitted white light. Hérino (2000) presented a paper on nanocomposites based on PSi where he reviewed applications of the organic compounds as electrical contacts to PSi and considered the fabrication of the heterostructures organic substances/ PSi for the study of some physical phenomenon. There were described known features of the polymer contacts (improved luminescence and pore filling) in comparison to metallic contacts. Disadvantages of the polymer contacts such as high operating voltage, instability, and nonreproducibility were not mentioned in Hérino's review (Hérino et al 1997; Hérino 2000).

Nguyen et al. (2003a,b) developed a modified technology of the PSi filling with polyphenylenevinylene, which provided good filling pores of 10-μm length. The expected improvement of the specific characteristics of the electrical contact was not experimentally proved in the work. The significance of the work is in the finding method of the reversible electrochemical switching studied heterostructure from the conductive to nonconductive state accompanied by changing polymer color. Other polymers such as polyaniline and polythiophene were deposited as electrical contacts to PSi (Harraz 2011). Schultze and Jung (1995) published a paper devoted to the fabrication of heterostructures based on PSi and polymers. It was the first work where PSi was considered as a material-base for the creation of the nanostructured objects (Figure 9.12). There was given a description of the technology and proposed models of the regularities revealed at

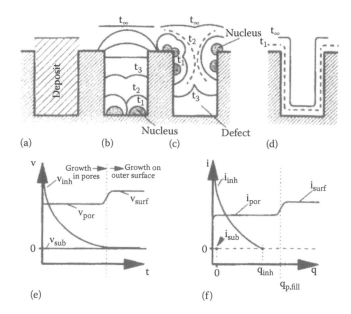

FIGURE 9.12 Mechanistic scheme of pore deposition (a) ideal case; (b) nucleation at the bottoms of the pores; (c) nucleation at the walls of the pores; (d) self inhibiting layer growth; (e) reaction rates v_j versus time t; (f) current density i_j versus charge q. (From Schultze J. and Jung K., *Electrochim. Acta* 40, 1369, 1995. Copyright 1995: Elsevier. With permission.)

the fabrication of the nanosized heterostructures polymer/SiO$_2$/PSi. The papers on realization of different devices using structure-forming properties of PSi had appeared much later. In 2002, Steinhart et al. (2002) published a short paper on the use of PSi as a matrix-base at fabrication of micro- and nanostructured polymer materials. Their work with more details appeared later (Steinhart et al. 2003, 2004). Although there are many reports on polymer/PSi hybrid fabrication, their application is still unclear (Palacios et al. 2009).

D'arrigo et al. (2003) reported the use of PSi filled with nafiol as a catalytic membrane in miniature fuel cells. It was supposed that the developed specific surface of PSi positively influences specific energy parameters of the final device. Nevertheless, the results of the study of the experimental device were not presented. Mattei and Valentini (2003) proposed their own original method of fabrication of the composite structures based on PSi. The advantage of this method is in simultaneously forming PSi with the required structure and its functionalization. It was realized by the addition of the functionalizing reagents into the solution for the anodization.

9.4.4 CONCLUSIONS

Despite the huge efforts of the scientific community, until now PSi did not become the basis for creating commercial light-emitting devices. Polymer electrical contacts were unreliable and organic dyes did not help. Only a theme of biomedical applications of hybrid organic-PSi materials continues to develop actively. Today appear more and more designs of sensors, therapies, and other practical implementations of useful features of hybrid organic-PSi materials.

9.5 DEPOSITION OF CARBON MATERIALS ON/IN PSi

9.5.1 FULLERENES C60

A growing interest in nanotechnology in general and fullerenes in particular has pushed some research groups to form and study fullerene films on PSi. These works have had just an exploratory character which has not implied solving defined problems. Feng and Miller (1998) used a method of fixing organic molecules with the fullerene C60 on the surface of PSi by hydrosilila-tion. Photoelectrical and chemical properties of the formed composite material were interesting but not unique. Dattilo et al. (2006) also applied the hydrosililation to fix the molecules of C60 to PSi and studied a wettability of the obtained samples. It was found that the final material demonstrates an increased hydrophoby in comparison to bulk Si. However, the authors noticed that there are alternative chemical compounds, which have much higher hydrophoby on bulk Si and PSi.

9.5.2 CARBON NANOTUBES

Carbon nanotubes (CNTs) are an essential object in nanotechnology. That is why a number of attempts to combine properties of CNTs and PSi have been made to date. Among different features of CNTs the most attention has been paid to the ability of cold emission because this peculiarity is the closest to the luminescent properties of PSi. Fan et al. (1999) reported a fabrication of organized arrays of CNTs on the PSi substrates. According to the proposed model, the main role of PSi is improving ethylene supplying into the growth zone of CNTs. At the same time, CNTs grow on the PSi surface. Later Xu et al. (1999) realized the growth of CNTs directly from the pores of PSi. They showed that dimensions of CNTs depend on a diameter of the pores of the meso- and micro-PSi. The successful results on the cold emission of the formed arrays of CNTs were also reported.

9.5.3 GRAPHITE/GRAPHENE

Carbon in the form of graphite has been applied as an anode of lithium-ion batteries for a long time. It is also known that a theoretical specific energy density (with respect to weight) of the graphite anodes is about one order of magnitude lower than that of Si analogues (370 mA·h/g versus 4200 mA·h/g). However, great variations of the Si anode volume (up to four times) at charge-discharge cycles leading to its fast destruction limit their capabilities. Nevertheless, the specific energy density of Si is so high that research has not left the attempts to fabricate the lithium-ion batteries with the Si-based anode by now. There are also other electrical energy storage graphite-based devices–ionistors or supercapacitors. These devices have electrodes made of graphite with very developed surfaces. Detailed information on PSi-based electrodes of the lithium-ion batteries and supercapacitors is reported in corresponding chapters of this book. Subsections presented next are devoted only to the ways of the PSi/graphite composite fabrication.

An understanding of a shady future of the Si anodes of the lithium-ion batteries led Zheng et al. (2007) to attempt to form carbon-Si composites. Their efforts were directed to the fabrication of composite micro-PSi/graphite/carbon by the mechanochemical reduction for the development of anodes with increased capacity. Specific capacity of the experimental cell was two times more in contrast to commercial lithium-ion batteries. Kim et al. (2008) applied an alternative method of the thermal reduction and formed PSi NPs, which than were passivated by CVD of graphite. As a result, the authors achieved nine times increasing specific capacity of the experimental cells in comparison to commercial analogues. Later, Lv et al. (2009) accurately investigated a synthesis of nanocomposites by the mechanochemical reduction. The authors suggested that precisely developed structure of PSi is responsible for the improved properties of the formed composite. Very recently, Feng et al. (2014) proposed new composite material for the anodes of lithium-ion batteries consisting of micro-PSi, CNTs, and graphite. This composite was fabricated by the magnesium thermal reduction of Si from SiO_2 and following CVD of carbon. The understanding that record levels of Si anodes' capacity has a small presence without evaluating the possibility of their practical application has come soon. The Mg thermal reduction is cheaper than special mills for the fabrication of NPs but still quite expensive for the manufacturing. Jia et al. (2011) used commercially available material SBA-15 as a source for the fabrication of composite anode graphite/PSi. An advantage of the proposed method is an initial structuring SBA-15. Therefore, the production of the anode requires just chemical treating SBA-15. Wang et al. (2012) applied natural diatomite as initial material for the formation of the composite carbon/PSi. Specific capacity of the experimental cells was four times greater than that of commercial lithium-ion batteries. Moreover, their reliability was higher in comparison to anodes based on SBA-15.

Szczech and Jin (2011) presented a review devoted to research into the application of the nano-structured Si as anodes of lithium-ion batteries. The authors revealed some disadvantages of the performed research such as an absence of a comparison of bulk specific energy densities of the novel and existing anodes. It is connected with the fact that the relation of the stored energy to volume in powerful batteries is more important than the relation of the stored energy to weight. The review ended with the recommendation to use more complex materials for the fabrication of anodes with improved characteristics. Nevertheless, analysis of the composites developed to 2011 has revealed that a main problem preventing their successful practical application is high cost and complexity of the manufacturing. Yin et al. (2012) turned their attention to the problem of Si degradation in composite materials that have been developed for new anode types. An essence of the problem is the presence of fluorine ions in the electrolyte of lithium-ion batteries. An electrolyte aging causes a continuous Si etching.

It has been already mentioned that supercapacitors are devices that are also applied to store electrical energy as lithium-ion batteries. Oakes et al. (2013) reported a supercapacitor based on graphite and PSi. They formed composite material by CVD of graphene-like graphite thin films on PSi. The obtained supercapacitor was compared with the PSi electrode without graphite. An advantage of the composite electrode was obvious while the comparison to other types of supercapacitors was not presented. One year later, Chatterjee et al. (2014) demonstrated the effect of the PSi passivation with graphen to apply it as electrodes of supercapacitors.

9.5.4 CONCLUSIONS

Today, there are two promising applications of hybrid carbon-PSi materials: improved anodes for lithium batteries and electrodes for electrochemical studies. Despite progress in both directions, such materials are still very far from widespread use due to the high cost and complexity of their fabrication.

9.6 SUMMARY

We can now finally summarize that there has been performed a great number of research works on the fabrication and investigation of composite and hybrid materials by the PSi filling with foreign substances. The most remarkable results have been achieved in the deposition of metals from liquids by electrochemical and displacement methods as well as infiltration with polymers. Although, we can expect further research activity on using low cost electrochemical, chemical displacement and infiltration methods, it is clear that an advanced low-temperature CVD and ALD method can provide the highly controlled rate of PSi filling with different foreign substances. These methods also offer a favorable solution for progress on the way from fundamental research to application of composite and hybrid materials based on PSi.

REFERENCES

Abd Rahim, A., Hashim, M., Rusop, M., Ali, N., and Yusuf, R. (2012). Room temperature Ge and ZnO embedded inside porous silicon using conventional methods for photonic application. *Superlattices Microstruct.* **52**, 941–948.

Acikgoz, S., Sarpkaya, I., Milas, P., Inci, M.N., Demirci, G., and Sanyal, R. (2011). Investigation of fluorescence dynamics of BODIPY embedded in porous silicon and monitoring formation of a SiO_2 layer via a confocal FLIM-based NSET method. *J. Phys. Chem. C* **115**, 22186–22190.

Anglin, E.J., Cheng, L., Freeman, W.R., and Sailor, M.J. (2008). Porous silicon in drug delivery devices and materials. *Adv. Drug Delivery Rev.* **60**, 1266–1277.

Antipán Lara, J. and Kathirgamanathan, P. (2000). White light electroluminescence from PSi devices capped with poly (thiophene) as top contact. *Synth. Met.* **110**, 233–240.

Aravamudhan, S., Luongo, K., Poddar, P., Srikanth, H., and Bhansali, S. (2007). Porous silicon templates for electrodeposition of nanostructures. *Appl. Phys. A* **87**, 773–780.

Balucani, M., Nenzi, P., Chubenko, E., Klyshko, A., and Bondarenko, V. (2011). Electrochemical and hydro-thermal deposition of ZnO on silicon: From continuous films to nanocrystals. *J. Nanopart. Res.* **13**, 5985–5997.

Bard, A.J., Parsons, R., and Jordan, J. (Eds.) (1985). *Standard Potentials in Aqueous Solution.* Marcel Dekker, New York.

Belogorokhov, A., Belogorokhova, L., Pérez-Rodríguez, A., Morante, J., and Gavrilov, S. (1998). Optical characterization of porous silicon embedded with CdSe nanoparticles. *Appl. Phys. Lett.* **73**, 2766–2768.

Belogorokhov, A., Belogorokhova, L., and Gavrilov, S. (1999). Investigation of properties of porous silicon embedded with ZnSe and CdSe. *J. Cryst. Growth* **197**, 702–706.

Bonanno, L.M. and Segal, E. (2011). Nanostructured porous silicon-polymer-based hybrids: From biosensing to drug delivery. *Nanomedicine* **6**, 1755–1770.

Borisenko, V., Bondarenko, V., and Raiko, V. (1989). Electrochemical deposition of cobalt on porous silicon. *Reports of Academy of Sciences of Belarus Republic* **1**, 528–530.

Borisenko, V., Bondarenko, V., and Raiko, V. (1990). Structure of cobalt disilicides films formed on porous silicon by solid phase reaction. *Rus. J. Surf.: Phys. Chem. Mech.* **4**, 84–90.

Brito-Neto, J.G.A., Kondo, K., and Hayase, M. (2007). Porous gold structures built on silicon substrates. *ECS Trans.* **6**, 223–227.

Canham, L. (1993). Laser dye impregnation of oxidized porous silicon on silicon wafers. *Appl. Phys. Lett.* **63**, 337–339.

Canham, L.T. (1995). Bioactive silicon structure fabrication through nanoetching techniques. *Adv. Mater.* **7**, 1033–1037.

Chatelon, J.P., Terrier, C., Bernstein, E., Berjoan, R., and Roger, J.A. (1994). Morphology of SnO_2 thin films obtained by the sol-gel technique. *Thin Solid Films* **247**, 162–168.

Chatterjee, S., Carter, R., Oakes, L., Erwin, W.R., Bardhan, R., and Pint, C.L. (2014). Electrochemical and corrosion stability of nanostructured silicon by graphene coatings; toward high power porous silicon supercapacitors. *J. Phys. Chem. C* **118**, 10893–10902.

Cheng, X., Feng, Z., and Luo, G. (2003). Effect of potential steps on porous silicon formation. *Electrochim. Acta* **48**, 497–501.

Chirvony, V., Bolotin, V., Matveeva, E., and Parkhutik, V. (2006). Fluorescence and $^1O_2$ generation properties of porphyrin molecules immobilized in oxidized nano-porous silicon matrix. *J. Photochem. Photobiol. A* **181**, 106–113.

Chouket, A., Elhouichet, H., Boukherroub, R., and Oueslati, M. (2007a). Porous silica-laser dye composite: Physical and optical properties. *Phys. Stat. Sol. (a)* **204**, 1518–1522.

Chouket, A., Elhouichet, H., Oueslati, M., Koyama, H., Gelloz, B., and Koshida, N. (2007b). Energy transfer in porous-silicon/laser-dye composite evidenced by polarization memory of photoluminescence. *Appl. Phys. Lett.* **91**, 211902.

Chouket, A., Charrier, J., Elhouichet, H., and Oueslati, M. (2009a). Optical study of planar waveguides based on oxidized porous silicon impregnated with laser dyes. *J. Lumin.* **129**, 461–464.

Chouket, A., Gelloz, B., Koyama, H., Elhouichet, H., Oueslati, M., and Koshida, N. (2009b). Effect of high-pressure water-vapor annealing on energy transfer in dye-impregnated porous silicon. *J. Lumin.* **129**, 1332–1335.

Chouket, A., Cherif, B., Salah, N.B., and Khirouni, K. (2013). Optical and electrical properties of porous silicon impregnated with Congo Red dye. *J. Appl. Phys.* **114**, 243105.

Chourou, M.L., Fukami, K., Sakka, T., and Ogata, Y.H. (2011). Gold electrodeposition into porous silicon: Comparison between meso- and macroporous silicon. *Phys. Status Solidi C* **8**, 1783–1786.

Chubenko, E., Klyshko, A., Bondarenko, V., Balucani, M., Belous, A., and Malyshev, V. (2011). ZnO films and crystals on bulk silicon and SOI wafers: Formation, properties and applications. *Adv. Mater. Res.* **276**, 3–19.

Chubenko, E., Redko, S., Dolgiy, A., and Bondarenko, V. (2014). Fabrication of highly ordered uniform mesoporous silicon for nanocomposite templates. In: *Proc. of E-MRS Fall Meeting*, Sept. 15–19, 2014, Warsaw, Poland, B11–5.

Cobianu, C., Savaniu, C., Buiu, O. et al. (1997). Tin dioxide sol–gel derived thin films deposited on porous silicon. *Sens. Actuators B* **43**, 114–120.

Coulthard, I., Jiang, D.-T., Lorimer, J.W., Sham, T.K., and Feng, X.-H. (1993). Reductive deposition of Pd on porous silicon from aqueous solutions of $PdCl_2$: An X-ray absorption fine structure study. *Langmuir* **9**, 3441–3445.

D'arrigo, G., Spinella, C., Arena, G., and Lorenti, S. (2003). Fabrication of miniaturised Si-based electrocatalytic membranes. *Mater. Sci. Eng. C* **23**, 13–18.

Dabboussi, S., Elhouichet, H., Ajlani, H., Moadhen, A., Oueslati, M., and Roger, J.A. (2006). Excitation process and photoluminescence properties of Tb^{3+} and Eu^{3+} ions in SnO_2 and in SnO_2: Porous silicon hosts. *J. Lumin.* **121**, 507–516.

Dattilo, D., Armelao, L., Maggini, M., Fois, G., and Mistura, G. (2006). Wetting behavior of porous silicon surfaces functionalized with a fulleropyrrolidine. *Langmuir* **22**, 8764–8769.

Dolgiy, A., Redko, S.V., Bandarenka, H. et al. (2012a). Electrochemical deposition and characterization of Ni in mesoporous silicon. *J. Electrochem. Soc.* **159**, D623–D627.

Dolgiy, A., Bandarenka, H., Prischepa, S. et al. (2012b). Electrochemical deposition of Ni into mesoporous silicon. *ECS Trans.* **41**, 111–118.

Dolgiy, A.L., Redko, S.V., Komissarov, I., Bondarenko, V.P., Yanushkevich, K.I., and Prischepa, S.L. (2013). Structural and magnetic properties of Ni nanowires grown in mesoporous silicon templates. *Thin Solid Films* **543**, 133–137.

Drisko, G.L. and Sanchez, C. (2012). Hybridization in materials science—Evolution, current state, and future aspirations. *Eur. J. Inorg. Chem.* **2012**, 5097–5105.

Dücső, C., Khanh, N.Q., Horváth, Z. et al. (1996). Deposition of tin oxide into porous silicon by atomic layer epitaxy. *J. Electrochem. Soc.* **143**, 683–687.

Elhouichet, H. and Oueslati, M. (2001). Photoluminescence properties of porous silicon nanocomposites. *Mater. Sci. Eng. B* **79**, 27–30.

Elhouichet, H., Moadhen, A., Férid, M., Oueslati, M., Canut, B., and Roger, J.A. (2003). High luminescent Eu^{3+} and Tb^{3+} doped SnO_2 sol-gel derived films deposited on porous silicon. *Phys. Stat. Sol. (a)* **197**, 350–354.

Elhouichet, H., Moadhen, A., Oueslati, M., Romdhane, S., Roger, J.A., and Bouchriha, H. (2005). Structural, optical and electrical properties of porous silicon impregnated with SnO_2:Sb. *Phys. Stat. Sol. (c)* **2**, 3349–3353.

Fan, S., Chapline, M.G., Franklin, N.R., Tombler, T.W., Cassell, A.M., and Dai, H. (1999). Self-oriented regular arrays of carbon nanotubes and their field emission properties. *Science* **283**, 512–514.

Feng, W. and Miller, B. (1998). Fullerene monolayer-modified porous Si. Synthesis and photoelectrochemistry. *Electrochem. Solid-State Lett.* **1**, 172–174.

Feng, X., Yang, J., Bie, Y., Wang, J., Nuli, Y., and Lu, W. (2014). Nano/micro-structured Si/CNT/C composite from nano-SiO_2 for high power lithium ion batteries. *Nanoscale* **6**, 12532–12539.

Fernández-Pacheco, A., Serrano-Ramón, L., Michalik, J.M. et al. (2013). Three dimensional magnetic nanowires grown by focused electron-beam induced deposition. *Sci. Rep.* **3**, 1492.

Fert, A. and Piraux, L. (1999). Magnetic nanowires. *J. Magn. Magn. Mater.* **200**, 338–358.

Fukami, K., Kobayashi, K., Matsumoto, T., Kawamura, Y.L., Sakka, T., and Ogata, Y.H. (2008). Electrodeposition of noble metals into ordered macropores in p-type silicon. *J. Electrochem. Soc.* **155**, D443–D448.

Garcés, F., Acquaroli, L., Urteaga, R., Dussan, A., Koropecki, R., and Arce, R. (2012). Structural properties of porous silicon/SnO_2:F heterostructures. *Thin Solid Films* **520**, 4254–4258.

Gelloz, B., Harima, N., Koyama, H., Elhouichet, H., and Koshida, N. (2010). Energy transfer from phosphorescent blue-emitting oxidized porous silicon to rhodamine 110. *Appl. Phys. Lett.* **97**, 171107.

Gole, J., DeVincentis, J., and Seals, L. (1999). Optical pumping of dye-complexed and -sensitized porous silicon increasing photoluminescence emission rates. *J. Phys. Chem. B* **103**, 979–987.

Granitzer, P. and Rumpf, K. (2010). Porous silicon—A versatile host material. *Materials* **3**, 943–998.

Granitzer, P., Rumpf, K., Surnev, S., and Krenn, H. (2005). Squid-magnetometry on ferromagnetic Ni-nanowires embedded in oriented porous silicon channels. *J. Magn. Magn. Mater.* **290–291**, 735–737.

Granitzer, P., Rumpf, K., and Krenn, H. (2006). Micromagnetics of Ni-nanowires filled in nanochannels of porous silicon. *Thin Solid Films* **515**, 735–738.

Granitzer, P., Rumpf, K., Pölt, P., Reichmann, A., Hofmayer, M., and Krenn, H. (2007a). Magnetization of self-organized Ni-nanowires with peculiar magnetic anisotropy. *J. Magn. Magn. Mater.* **316**, 302–305.

Granitzer, P., Rumpf, K., Pölt, P., Reichmann, A., and Krenn, H. (2007b). Quasi-regular self-organized porous silicon channels metallized with Ni-structures of strong anisotropy. *J. Magn. Magn. Mater.* **310**, e838–e840.

Granitzer, P., Rumpf, K., Ohta, T., Koshida, N., Reissner, M., and Poelt, P. (2012). Enhanced magnetic anisotropy of Ni nanowire arrays fabricated on nano-structured silicon templates. *Appl. Phys. Lett.* **101**, 033110.

Granitzer, P., Rumpf, K., Strzhemechny, Y., and Chapagain, P. (2014). Transient surface photovoltage studies of bare and Ni-filled porous silicon performed in different ambients. *Nanoscale Res. Lett.* **9**, 423.

Grigoras, K., Keskinen, J., Uli-Ranta, E. et al. (2014). ALD modified porous silicon electrodes for supercapacitors. In: *Extended Abstracts of the 9th International Conference on Porous Semiconductors—Science and Technology*, Apr. 9–13, 2014, Alicante-Benidorm, Spain, pp. 314–315.

Gros-Jean, M., Hérino, R., and Lincot, D. (1998). Incorporation of cadmium sulfide into nanoporous silicon by sequential chemical deposition from solution. *J. Electrochem. Soc.* **145**, 2448–2452.

Gros-Jean, M., Hérino, R., Chazalviel, J.-N., Ozanam, F., and Lincot, D. (2000). Formation and characterization of CdS: Methyl-grafted porous silicon junctions. *Mater. Sci. Eng. B* **69–70**, 77–80.

Guendouz, M., Pedrono, N., Etesse, R. et al. (2003). Oxidised and non oxidised porous silicon/disperse red 1composite: Physical and optical properties. *Phys. Stat. Sol. (a)* **197**, 414–418.

Gusev, S.A., Korotkova, N.A., Rozenstein, D.B., and Fraerman, A.A. (1994). Ferromagnetic filaments fabrication in porous Si matrix. *J. Appl. Phys.* **76**, 6671–6672.

Halimaoui, A., Campidelli, Y., Badoz, P., and Bensahel, D. (1995). Covering and filling of porous silicon pores with Ge and Si using chemical vapor deposition. *J. Appl. Phys.* **78**, 3428–3430.

Halliday, D., Holland, E., Eggleston, J., Adams, P., Cox, S., and Monkman, A. (1996). Electroluminescence from porous silicon using a conducting polyaniline contact. *Thin Solid Films* **276**, 299–302.

Hamadache, F., Duvial, J.-L., Scheuren, V. et al. (2002). Electrodeposition of Fe–Co alloys into nanoporous p-type silicon: Influence of the electrolyte composition. *J. Mater. Res.* **17**, 1074–1084.

Hamadache, F., Renaux, C., Duvail, J.-L., and Bertrand, P. (2003). Interface investigations of iron and cobalt metallized porous silicon: AES and FTIR analyses. *Phys. Stat. Sol. (a)* **197**, 168–174.

Harraz, F.A. (2011). Impregnation of porous silicon with conducting polymers. *Phys. Stat. Sol. (c)* **8**, 1883–1887.

Harraz, F.A., Tsuboi, T., Sasano, J., Sakka, T., and Ogata, Y.H. (2002). Metal deposition onto a porous silicon layer by immersion plating from aqueous and nonaqueous solutions. *J. Electrochem. Soc.* **149**, C456–C463.

Harraz, F.A., Salem, A.M., Mohamed, B.A., Kandil, A., and Ibrahim, I.A. (2013). Electrochemically deposited cobalt/platinum (Co/Pt) film into porous silicon: Structural investigation and magnetic properties. *Appl. Surf. Sci.* **264**, 391–398.

Hérino, R. (2000). Nanocomposite materials from porous silicon. *Mater. Sci. Eng. B* **69**, 70–76.

Hérino, R., Jan, P., and Bomchil, G. (1985). Nickel plating on porous silicon. *J. Electrochem. Soc.* **132**, 2513–2514.

Hérino, R., Gros-Jean, M., Montès, L., and Lincot, D. (1997). Electrochemical and chemical deposition of II-VI semiconductors in porous silicon. *Mat. Res. Soc. Symp. Proc.* **452**, 467–472.

Jane, A., Dronov, R., Hodges, A., and Voelcker, N.H. (2009). Porous silicon biosensors on the advance. *Trends Biotechnol.* **27**, 230–239.

Jia, H., Gao, P., Yang, J., Wang, J., Nuli, Y., and Yang, Z. (2011). Novel three-dimensional mesoporous silicon for high power lithium-ion battery anode material. *Adv. Energy Mater.* **1**, 1036–1039.

Kim, N.Y. and Laibinis, P.E. (1999). Improved polypyrrole/silicon junctions by surfacial modification of hydrogen-terminated silicon using organolithium reagents. *J. Am. Chem. Soc.* **121**, 7162–7163.

Kim, H., Han, B., Choo, J., and Cho, J. (2008). Three-dimensional porous silicon particles for use in high-performance lithium secondary batteries. *Angew. Chem.* **120**, 10305–10308.

Koda, R., Fukami, K., Sakka, T., and Ogata, Y.H. (2012). Electrodeposition of platinum and silver into chemically-modified microporous silicon electrodes. *Nanoscale Res. Lett.* **7**, 330.

Koshida, N., Koyama, H., Yamamoto, Y., and Collins, G.J. (1993). Visible electroluminescence from porous silicon diodes with an electropolymerized contact. *Appl. Phys. Lett.* **63**, 2655–2657.

Kröger, F. (1978). Cathodic deposition and characterization of metallic or semiconductor binary allows or compounds. *J. Electrochem. Soc.* **125**, 2028–2034.

Labunov, V., Bondarenko, V., Glinenko, L., Dorofeev, A., and Tabulina, L. (1986). Heat treatment effect on porous silicon. *Thin Solid Films* **137**, 123–134.

Lauerhaas, J.M., Credo, G.M., Heinrich, J.L., and Sailor, M.J. (1992). Reversible luminescence quenching of porous silicon by solvents. *J. Am. Chem. Soc.* **114**, 1911–1912.

Letant, S. and Vial, J. (1997). Energy transfer in dye impregnated porous silicon. *J. Appl. Phys.* **82**, 397–401.

Li, K., Diaz, D.C., He, Y., Campbell, J.C., and Tsai, C. (1994). Electroluminescence from porous silicon with conducting polymer film contacts. *Appl. Phys. Lett.* **64**, 2394–2396.

Li, P., Li, Q., Ma, Y., and Fang, R. (1996). Photoluminescence and its decay of the dye/porous-silicon composite system. *J. Appl. Phys.* **80**, 490–493.

Li, H., Xu, D., Guo, G. et al. (2000a). Intense and stable blue-violet emission from porous silicon modified with alkyls. *J. Appl. Phys.* **88**, 4446–4448.

Li, X., John, J.S., Coffer, J.L. et al. (2000b). Porosified silicon wafer structures impregnated with platinum anti-tumor compounds: Fabrication, characterization, and diffusion studies. *Biomed. Microdev.* **2**, 265–272.

Li, Y.Y., Cunin, F., Link, J.R. et al. (2003). Polymer replicas of photonic porous silicon for sensing and drug delivery applications. *Science* **299**, 2045–2047.

Li, Y.Y., Kollengode, V.S., and Sailor, M.J. (2005). Porous silicon/polymer nanocomposite photonic crystals formed by microdroplet patterning. *Adv. Mater. (Weinheim, Ger.)* **17**, 1249–1251.

Lin, H., Gao, T., Fantini, J., and Sailor, M.J. (2004). A porous silicon-palladium composite film for optical interferometric sensing of hydrogen. *Langmuir* **20**, 5104–5108.

Liu, F.Q., Wang, Z.G., Li, G.H., and Wang, G.H. (1998). Photoluminescence from Ge clusters embedded in porous silicon. *J. Appl. Phys.* **83**, 3435–3437.

Lopez, H.A., Linda Chen, X., Jenekhe, S.A., and Fauchet, P.M. (1999). Tunability of the photoluminescence in porous silicon due to different polymer dielectric environments. *J. Lumin.* **80**, 115–118.

Lv, R., Yang, J., Gao, P., NuLi, Y., and Wang, J. (2009). Electrochemical behavior of nanoporous/nanofibrous Si anode materials prepared by mechanochemical reduction. *J. Alloys Compd.* **490**, 84–87.

Ma, Q., Xiong, R., and Huang, Y.M. (2011). Tunable photoluminescence of porous silicon by liquid crystal infiltration. *J. Lumin.* **131**, 2053–2057.

Martínez, L., Ocampo, O., Kumar, Y., and Agarwal, V. (2014). ZnO-porous silicon nanocomposite for possible memristive device fabsrication. *Nanoscale Res. Lett.* **9**, 437.

Mattei, G. and Valentini, V. (2003). In situ functionalization of porous silicon during the electrochemical formation process in ethanoic hydrofluoric acid solution. *J. Am. Chem. Soc.* **125**, 9608–9609.

Matveeva, E., Parkhutik, V., Diaz Calleja, R., and Martinez-Duart, J. (1993). Growth of polyaniline films on porous silicon layers. *J. Lumin.* **57**, 175–180.

McInnes, S.J., and Voelcker, N.H. (2012). Porous silicon-based nanostructured microparticles as degradable supports for solid-phase synthesis and release of oligonucleotides. *Nanoscale Res. Lett.* **7**(1), 1–10.

Melo, C., Larramendi, S., Torres-Costa, V. et al. (2014). Enhanced ZnTe infiltration in porous silicon by isothermal close space sublimation. *Microporous Mesoporous Mater.* **188**, 93–98.

Michelakaki, E., Valalaki, K., and Nassiopoulou, A.G. (2013). Mesoscopic Ni particles and nanowires by pulsed electrodeposition into porous Si. *J. Nanoparticle Res.* **15**, 1499.

Moadhen, A., Elhouichet, H., and Oueslati, M. (2002). Stokes and anti-stokes photoluminescence of Rhodamine B in porous silicon. *Mater. Sci. Eng. C* **21**, 297–301.

Moadhen, A., Elhouichet, H., Romdhane, S., Oueslati, M., Roger, J.A., and Bouchriha, H. (2003). Structural, optical and electrical properties of SnO_2:Sb:Tb^{3+}/porous silicon devices. *Semicond. Sci. Technol.* **18**, 703–707.

Montès, L. and Hérino, R. (2000). Luminescence and structural properties of porous silicon with ZnSe intimate contact. *Mater. Sci. Eng. B* **69–70**, 136–141.

Montès, L., Muller, F., Gaspard, F., and Hérino, R. (1997). Investigation on the electrochemical deposition of cadmium telluride in porous silicon. *Thin Solid Films* **297**, 35–38.

Montès, L., Muller, F., and Hérino, R. (2000). Localized photo-assisted electro-deposition of zinc selenide into *p*-type porous silicon. *J. Porous Mater.* **7**, 77–80.

Moreno, J., Marcos, M., Agulló-Rueda, F. et al. (1999). A galvanostatic study of the electrodeposition of polypyrrole into porous silicon. *Thin Solid Films* **348**, 152–156.

Nguyen, T., Le Rendu, P., Lakehal, M. et al. (2003a). Filling porous silicon pores with poly (p phenylene vinylene). *Phys. Stat. Sol. (a)* **197**, 232–235.

Nguyen, T.-P., Rendu, P.L., and Cheah, K.W. (2003b). Optical properties of porous silicon/poly (p phenylene vinylene) devices. *Phys. E* **17**, 664–665.

O'Brien, L., Read, D.E., Zeng, H.T., Lewis, E.R., Petit, D., and Cowburn, R.P. (2009). Bidirectional magnetic nanowire shift register. *Appl. Phys. Lett.* **95**, 232502.

Oakes, L., Westover, A., Mares, J.W. et al. (2013). Surface engineered porous silicon for stable, high performance electrochemical supercapacitors. *Sci. Rep.* **3**, 3020.

Ogata, Y., Kobayashi, K., and Motoyama, M. (2006). Electrochemical metal deposition on silicon. *Curr. Opt. Sol. St. Mater. Sci.* **10**, 163–172.

Ouir, S., Sam, S., Fortas, G., Gabouze, N., Beldjilali, K., and Tighilt, F. (2008). FeNi alloys electroplated into porous (*n*-type) silicon. *Phys. Stat. Sol. (c)* **5**, 3694–3697.

Palacios, R., Formentín, P., Santos, A. et al. (2009). Synthesis of ordered polymer micro and nanostructures via porous templates. In: *Proc. of the 2009 Spanish Conference on Electron Devices, CDE 09*, Feb. 11–13, 2009, Santiago de Compostela, Spain, pp. 424–427.

Parbukov, A.N., Beklemyshev, V.I., Gontar, V.M., Makhonin, I.I., Gavrilov, S.A., and Bayliss, S.C. (2001). The production of a novel stain-etched porous silicon, metallization of the porous surface and application in hydrocarbon sensors. *Mater. Sci. Eng. C* **15**, 121–123.

Parkhutik, V., Diaz Calleja, R., Matveeva, E., and Martinez-Duart, J. (1994). Luminescent structures of porous silicon capped by conductive polymers. *Synth. Met.* **67**, 111–114.

Pedrero, L.O., Rena-Sierra, R., and Romero Paredes, G.R. (2004). Gas sensors based on porous silicon and palladium oxide clusters. In: *1st International Conference on Electrical and Electronics Engineering (ICEEE)*, June 24–27, 2004, Acapulco, Mexico, pp. 276–281.

Petrovich, V., Volchek, S., Dolgyi, L. et al. (2000a). Deposition of erbium containing film in porous silicon from ethanol solution of erbium salt. *J. Porous Mater.* **7**, 37–40.

Petrovich, V., Volchek, S., Dolgyi, L. et al. (2000b). Formation features of deposits during a cathode treatment of porous silicon in aqueous solutions of erbium salts. *J. Electrochem. Soc.* **147**, 655–658.

Piryatinski, Y.P., Dolgov, L., Yaroshchuk, O., and Lazarouk, S. (2007). Fluorescence of porous silicon filled with liquid crystal 5CB. *Mol. Cryst. Liq. Cryst.* **467**, 195–202.

Piryatinski, Y.P., Dolgov, L., Yaroshchuk, O., Gavrilko, T., and Lazarouk, S. (2010). Enhancement of fluorescence of porous silicon upon saturation by liquid crystal. *Opt. Spectrosc.* **108**, 70–79.

Pranculis, V., Šimkien, I., Treideris, M., and Gulbinas, V. (2013). Excitation energy transfer in porous silicon/laser dye composites. *Phys. Stat. Sol. (a)* **210**, 2617–2621.

Prischepa, S.L., Dolgiy, A.L., Bandarenka, A.V. et al. (2014). Synthesis and properties of Ni nanowires in porous silicon tempaltes. In: Wilson L.J. (Ed.) *Nanowires: Synthesis, Electrical Properties and Uses in Biological Systems*. Nova Science Publishers, New York, pp. 89–129.

Razi, F., Irajizad, A., and Rahimi, F. (2010). Investigation of hydrogen sensing properties and aging effects of Schottky like Pd/porous Si. *Sens. Actuators B* **146**, 53–60.

Renaux, C., Scheuren, V., and Flandre, D. (2000). New experiments on the electrodeposition of iron in porous silicon. *Microelectron. Reliab.* **40**, 877–879.

Ronkel, F., Schultze, J., and Arens-Fischer, R. (1996). Electrical contact to porous silicon by electrodeposition of iron. *Thin Solid Films* **276**, 40–43.

Rumpf, K., Granitzer, P., Pölt, P., Reichmann, A., and Krenn, H. (2006). Structural and magnetic characterization of Ni-filled porous silicon. *Thin Solid Films* **515**, 716–720.

Rumpf, K., Granitzer, P., and Krenn, H. (2008). Beyond spin-magnetism of magnetic nanowires in porous silicon. *J. Phys. Condens. Matter.* **20**, 454221.

Rumpf, K., Granitzer, P., Koshida, N., Poelt, P., and Reissner, M. (2014). Magnetic interactions between metal nanostructures within porous silicon. *Nanoscale Res. Lett.* **9**, 412.

Salcedo, W.J., Fernandez, F.J., and Rubim, J.C. (2004). Nano-composite of porous silicon and organic dye molecules for optical gas sensor and lasing medium. *Phys. Stat. Sol. (c)* **1**, S26–S30.

Schultze, J. and Jung, K. (1995). Regular nanostructured systems formed electrochemically: Deposition of electroactive polybithiophene into porous silicon. *Electrochim. Acta* **40**, 1369–1383.

Setzu, S., Létant, S., Solsona, P., Romestain, R., and Vial, J. (1999). Improvement of the luminescence in *p*-type as-prepared or dye impregnated porous silicon microcavities. *J. Lumin.* **80**, 129–132.

Singh, R., Singh, F., Agarwal, V., and Mehra, R. (2007). Photoluminescence studies of ZnO/porous silicon nanocomposites. *J. Phys. D: Appl. Phys.* **40**, 3090–3093.

Singh, R., Singh, F., Kanjilal, D., Agarwal, V., and Mehra, R. (2009a). White light emission from chemically synthesized ZnO–porous silicon nanocomposite. *J. Phys. D: Appl. Phys.* **42**, 062002.

Singh, R.G., Singh, F., Sulania, I. et al. (2009b). Electronic excitations induced modifications of structural and optical properties of ZnO-porous silicon nanocomposites. *Nucl. Instrum. Methods Phys. Res., Sect. B* **267**, 2399–2402.

Steinhart, M., Wendorff, J., Greiner, A. et al. (2002). Polymer nanotubes by wetting of ordered porous templates. *Science* **296**, 1997.

Steinhart, M., Wendorff, J., and Wehrspohn, R. (2003). Nanotubes à la carte: Wetting of porous templates. *Chem. Phys. Chem.* **4.11,** 1171–1176.

Steinhart, M., Wehrspohn, R., Gösele, U., and Wendorff, J. (2004). Nanotubes by template wetting: A modular assembly system. *Angew. Chem. Int. Edit.* **43.11**, 1334–1344.

Szczech, J.R. and Jin, S. (2011). Nanostructured silicon for high capacity lithium battery anodes. *Energy Environ. Sci.* **4**, 56–72.

Torres-Costa, V., Melo, C., Climent-Font, A., Argulló-Rueda, F., and Melo, O. (2012). Isothermal close space sublimation for II-VI semiconductor filling of porous matrices. *Nanoscale Res. Lett.* **7**, 409.

Tsuboi, T., Sakka, T., and Ogata, Y.H. (1998). Metal deposition into a porous silicon layer by immersion plating: Influence of halide ions. *J. Appl. Phys.* **83**, 4501–4506.

Utriainen, M., Lehto, S., Niinistö, L. et al. (1997). Porous silicon host matrix for deposition by atomic layer epitaxy. *Thin Solid Films* **297**, 39–42.

Wakefield, G., Dobson, P., Foo, Y., Loni, A., Simons, A., and Hutchison, J. (1997). The fabrication and characterization of nickel oxide films and their application as contacts to polymer/porous silicon electroluminescent devices. *Semicond. Sci. Technol.* **12**, 1304.

Wang, M.-S., Fan, L.-Z., Huang, M., Li, J., and Qu, X. (2012). Conversion of diatomite to porous Si/C composites as promising anode materials for lithium-ion batteries. *J. Power Sources* **219**, 29–35.

Wu, E.C., Park, J.-H., Park, J., Segal, E., Cunin, F., and Sailor, M.J. (2008). Oxidation-triggered release of fluorescent molecules or drugs from mesoporous Si microparticles. *ACS Nano* **2**, 2401–2409.

Xu, D., Guo, G., Gui, L. et al. (1999). Controlling growth and field emission property of aligned carbon nanotubes on porous silicon substrates. *Appl. Phys. Lett.* **75**, 481–483.

Xu, C., Zhang, X., Tu, K.-N., and Xie, Y. (2007a). Nickel displacement deposition of porous silicon with ultrahigh aspect ratio. *J. Electrochem. Soc.* **154**, D170–D174.

Xu, C., Li, M., Zhang, X., Tu, K.-N., and Xie, Y. (2007b). Theoretical studies of displacement deposition of nickel into porous silicon with ultrahigh aspect ratio. *Electrochim. Acta* **52**, 3901–3909.

Yin, F., Xiao, X.R., Li, X.P. et al. (1998). Photoluminescence enhancement of porous silicon by organic cyano compounds. *J. Phys. Chem. B* **102**, 7978–7982.

Yin, Y., Wan, L., and Guo, Y. (2012). Silicon-based nanomaterials for lithium-ion batteries. *Chin. Sci. Bull.* **57**, 4104–4110.

Zhang, X.G. (2001). *Electrochemistry of Silicon and Its Oxide.* Kluwer Academic/Plenum Publishers, New York.

Zhang, X., Xu, C., Chong, K., Tu, K.-N., and Xie, Y.-H. (2011). Study of Ni metallization in macroporous Si using wet chemistry for radio frequency cross-talk isolation in mixed signal integrated circuits. *Materials* **4**, 952–962.

Zheng, Y., Yang, J., Wang, J., and NuLi, Y. (2007). Nano-porous Si/C composites for anode material of lithium-ion batteries. *Electrochim. Acta* **52**, 5863–5867.

Energy Technologies

Porous Si and Si Nanostructures in Photovoltaics

10

Valeriy A. Skryshevsky and Tetyana Nychyporuk

CONTENTS

10.1   Introduction   212

10.2   Gettering of Si Wafers and Purification of Metallurgical Grade Si   213

10.3   Macroporous Texturization of Si Wafers   214

     10.3.1   Comparison of Alkaline and Macroporous Texturization   214

     10.3.2   Influence of Texturization Techniques on Solar Cell Performance   217

10.4   Nano PSi as Antireflection Coating   217

     10.4.1   Requirements for Antireflection Coatings on Si Solar Cells   217

     10.4.2   Technology of Nanoporous Si ARC   218

     10.4.3   Influence of Porous Si ARC on Solar Cell Performance   219

     10.4.4   Influence of PSi Emission on Quantum Efficiency of Solar Cells   221

10.5   PSi Selective Emitter   221

10.6   Surface Passivation by Porous Layers   222

     10.6.1   Estimation of Surface Recombination Velocity in PSi/Si Interface   222

     10.6.2   Light-Induced Passivation on PSi/Si Interface   224

10.7   PSi as Optical Reflector   225

     10.7.1   Application of Rear Bragg Reflector for Si Solar Cells   225

     10.7.2   PSi Rear Reflectors for Thin Film Solar Cell   226

10.8   Application of Porous Silicon for Layer Transfer   228

10.9   Modern Tendencies in Si PV   229

    10.9.1   Solar Cells with Si Nanowires   229

    10.9.2   Tandem Si Solar Cells with Si Quantum Dots   231

References   233

10.1 INTRODUCTION

Nowadays, photovoltaics (PV) is growing rapidly, a total global installed capacity achieved of 139 GW at the end of 2013 (and 38 GW was installed for the last year). PV now covers 3% of the electricity demand and 6% of the peak electricity demand in Europe (EPIA 2014). Monocrystalline, multicrystalline, and thin film Si cells and modules actually draw up more than 88% from the total PV production (Aulich et al. 2010). However, the PV electricity still remains more expensive compared with traditional nuclear or thermal power engineering. That is why now the main efforts of solar power engineering are directed to improve the cell efficiency and to reduce the cell cost, particularly by developing new structures and materials.

The simple sketch of an Si solar cell design with diffusion p-n junction can be presented as follows (Figure 10.1): heavy doped n^+-emitter ($N_d \sim 10^{20}$ cm^{-3}) of 0.3–0.5 μm thick is formed on standard textured p-type Si wafer (base) of 150–250 μm thick with resistivity of 0.5–5 Ω·cm. On the surface of n^+-emitter the dielectrical passivating layer and antireflection coating (ARC) are formed as well as metal terminals. Heavy doped p^+-layer (back surface field, BSF), passivating layer, and metal terminals are created on the rear side of a cell. By using low-cost Si grade materials (upgraded or purified metallurgical Si), the conversion efficiencies of 16% were obtained at cell dimensions of 150 × 150 or 200 × 200 mm^2 (Halm et al. 2010). If as a substrate the electronic quality Si is used, then the conversion efficiency can be increased up to 17–19% (Burgers et al. 2008). At that time, the laboratory solar cells with small area show 25% for monocrystalline Si and 20.4% for multicrystalline Si (Green et al. 2011).

Porous Si (PSi) formed on crystalline Si wafer by electrochemical or chemical etching exhibits several optical, electrical, and morphological properties that can be useful in the Si solar cells processing (Dzhafarov 2014). The research reports concerning the application of PSi layers for PV began to appear in 1982 (Prasad et al. 1982). Some potential advantages of using PSi in Si solar cell technology were first stated by Tsuo et al. (1993) and now include the following ideas:

1. Due to the highly textured morphology of PSi, it can be used as efficient antireflection coating and texturing technique to minimize the optical losses in single- and multicrystalline Si solar cells.
2. The bandgap of nano PSi can be easily adjusted in the range 1.1–2.0 eV by varying the porosity. The optimum bandgap for efficient photoconversion* thus can be achieved.
3. Wide bandgap PSi can be used like a material for the top layer of a heterojunction cell.
4. Strong luminescence of PSi can be used to convert ultraviolet and blue light into longer wavelength emission that has better quantum efficiency in an Si solar cell.
5. PSi layers on Si wafer can act as effective gettering centers for impurity atoms that improve the lifetime of minority charge carriers in Si bulk.
6. PSi can play the role of a passivating layer for the front and the back sides of an Si cell.
7. Thin film solar cells can be created using thin film layer transfer with sacrificial porous layers.
8. To enhance long wavelength light absorption in silicon solar cells, the multiporous layers like Bragg reflector can be formed on the back side of cells.
9. "All-Si" tandem solar cells can be formed using Si quantum wires and quantum dots.
10. Very simple technology of PSi fabrication has high potential for large area applications.

---

* For one junction cell, the theoretical limit of efficiency is 31% (or 40.8%) in the case of one solar (or maximal concentrated) flux at optimal semiconductor band gap $E_g = 1.3$ (or 1.1) eV (Conibeer 2007).

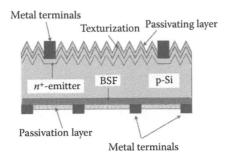

FIGURE 10.1   Design of Si solar cell.

However, there are also many disadvantages associated with using porous Si in PV. Among them the most important for PV are: (1) high resistivity (in the range of 10^7–10^{12} Ω·cm) of nanoporous layers (Ben-Chorin 1997), which eliminates its application to form the junction and (2) chemical composition modification during aging in ambient (Canham 1997), which needs the use of supplementary protective layers. Next we consider how PSi, nano Si dots, and wires can be used in Si PV.

10.2 GETTERING OF Si WAFERS AND PURIFICATION OF METALLURGICAL GRADE Si

It is well known that metallic impurities in solar grade Si create deep electronic levels inside the bandgap, which act as recombination centers. The presence of large defect densities in Si wafer reduces the lifetimes of minority charge carriers and their diffusion lengths and, as a result, strongly limits the efficiency of solar cells. In Si device processing, a *gettering* has been widely used to minimize the impact of different contamination (in the first place, the transition metal impurities dissolved in the Si matrix) on device performance. Gettering is a method of reducing impurity concentration in wafer by removing them or localizing them in regions away from the active device regions. The effectiveness of the gettering process depends on the establishment of sinks for absorbing impurities, on the diffusion coefficients of the impurities in bulk semiconductor and in the getter phase, and on the segregation coefficient describing the ability of gettering sites to absorb impurities (Wolf and Tauber 1986).

Nanometer-scaled cavities in PSi provide effective gettering sites (Shieh and Evans 1993; Wong-Leung et al. 1995; Efremov et al. 2000) due to the reaction of the metallic contaminations with Si dangling bonds on the huge developed internal surface of PSi. The probability of tying up contaminants enhances during the postfabrication annealing step. The gettering of Si wafers consists in the creation of a thin porous layer followed by a thermal annealing in nitrogen or oxygen atmosphere (Tsuo et al. 1996; Khedher et al. 2002; Derbali et al. 2011). The porous layers are removed, along with the gettered impurities, after gettering. Each PSi gettering process removes up to about 10 μm of wafer thickness. The process can be repeated so that the desired purity level is obtained (Tsuo et al. 1996).

The sufficient diffusion length enhancement was obtained at the combination of PSi etching and phosphorus diffusion into PSi layer (Zhang et al. 2010; Derbali and Ezzaouia 2012) or Si tetrachloride treatment (Hassen et al. 2004). The last technique consists of heating a p-Si substrate with a thin porous layer (formed on both sides of the wafer by stain-etching in a HF/HNO$_3$/H$_2$O solution) under SiCl$_4$/N$_2$ atmosphere. For example, if for untreated reference sample the diffusion length L_d = 86 μm, the sample treated at 950°C in N$_2$ atmosphere for 60 min shows L_d = 150 μm, the sample treated at 950°C in SiCl$_4$/N$_2$ atmosphere for 90 min displays L_d = 210 μm. The L_d increasing of p-Si wafer (from 63 to 134 μm) was obtained when PSi etching combines with the Al layer deposition followed by annealing at 750°C (Evtukh et al. 2001).

Gettering properties of PSi open the possibility to apply it for the *purification* of metallurgical grade Si. Their importance for the PV industry today is obvious. Now, most commercial Si solar cells are made on wafers sliced from ingots grown from molten Si by Czochralski, casting

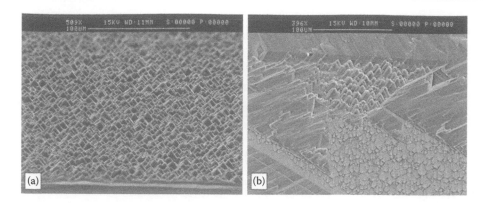

FIGURE 10.2   Alkaline texturization of (a) mono- and (b) multicrystalline Si wafers. The SEM images were kindly provided by Photowatt Technologies.

methods, or Si ribbons grown directly from the Si melt. High-purity poly Si rods are normally used as the feed materials for the Si melt. Poly Si is usually obtained by reducing high-purity trichlorosilane, which is converted from metallurgical grade Si. The development of a low-cost supply of Si feedstock still remains an important task because the cost of purifying the trichlorosilane and converting it by CVD to poly Si rods is high (Menna et al. 1998).

Solar grade Si can be obtained by purifying metallurgical grade Si using pyro- and hydrometallurgical processes, where typical treatments of the melted Si such as casting, grinding, leaching, remelting, and controlled solidification are performed (Lian et al. 1992). An alternative technologically simple method of purification of metallurgical grade Si by PSi etching was proposed (Menna et al. 1998). It includes the sequence of PSi etching in $HF/HNO_3/H_2O$ solution and conventional (or light) annealing at temperature of 1000°C for 30 min in argon of 600 torr. The accumulation of the impurities (B, Al, Cr, Fe, Cu) near the surface in first 1–2 μm is observed, when the metallurgical grade Si wafer undergoes the process sequences of PSi etching/annealing. Furthermore, the combined effect of PSi etching and light-annealing has a marked influence over the diffusion of heavy metal impurities and boron from the bulk to the surface.

10.3 MACROPOROUS TEXTURIZATION OF Si WAFERS

10.3.1 COMPARISON OF ALKALINE AND MACROPOROUS TEXTURIZATION

One of the factors that lead to an increase in solar cell efficiency is the reduction of light losses in cell by lowering the surface reflectivity and improving the scheme of light trapping. It can be achieved by surface *texturization*, which plays an important role in production both of mono- and multicrystalline Si solar cells (Springer et al. 2000).

Traditionally the standard alkaline texturization (NaOH) has been used in the solar cell industry. Although the usual alkaline etching procedures work well on monocrystalline Si (Figure 10.2a), on multicrystalline wafers (Figure 10.2b) they are less efficient. Indeed, the etch rate of Si in alkaline solutions is highly anisotropic, that is, it is strongly dependent on the crystallographic grain orientation. Etch rate of Si (111) is considerably lower (100–200 times) than that of Si (100) and Si (110). Alkaline etching of multicrystalline Si wafers having the random grain orientation results in coarse and nonuniform surface morphology (Figure 10.2b). The alkaline texturization leads to the effective reflectivity (R_E)* of about 20% at best for multicrystalline substrates and 12% for (100) monocrystalline ones. It is expected that uniform texturization of multicrystalline Si wafer can be obtained by technology of macropore formation.

Electrochemical texturization of *p*-type Si in a mixture of hydrofluoric acid with some organic solvents leads to the isotropic formation of macroporous layers (Lévy-Clément et al. 2003;

---

* The effective reflectivity is the reflectance $R(\lambda)$ normalized to the solar flux $N(\lambda)$: $R_{eff} = \int R(\lambda)N(\lambda)d\lambda \Big/ \int N(\lambda)d\lambda$.

Bastide et al. 2004), which can be very efficient in lowering the reflectivity of Si wafers. This type of texturization can be performed from a chemically polished surface as well as a saw damaged surface. In the latter case, the saw damage removal and the texturization are done simultaneously, which is very important for industrial applications. Figure 10.3 presents a SEM image of macroporous textured multicrystalline Si for (100) and (111) oriented grains. This type of isotropic texturization assures an effective reflectivity of 9–11% on multicrystalline Si wafers. It was found that the main parameter affecting the texturization efficiency is the substrate resistivity. The best results of R_E = 9% on p-type multicrystalline Si were obtained on low resistivity substrates (0.4 Ω·cm) (Bastide et al. 2004). The macroporous texturization is more efficient for monocrystalline Si. R_E < 5% on p-type monocrystalline Si was obtained on (100) substrate with resisitivity of 0.2 Ω·cm (Lévy-Clément et al. 2003).

n-Type doped multicrystalline Si presents potential advantages over p-type Si, in particular higher diffusion lengths at a given impurity concentration and improved long-term stability (Martinuzzi et al. 2005). Micrometer-sized macropores have also been isotropically formed on n-type multicrystalline Si wafers under white light illumination, leading to an overall effective reflectivity as low as 8.5% (Grigoras et al. 1998; Bastide et al. 2006; Tena-Zaera et al. 2007).

Despite rather good results in decreasing of reflectivity loses, the electrochemical method possesses several considerable disadvantages, which limit its industrial implementation. The necessity to create a good conductive contact on the backside of each sample as well as a development of corresponding equipment to form a complete electrode system are some of them.

In turn, the macroporous formation by *chemical etching* does not require the electrical bias applying and corresponding good rear contact formation. The technology of this type of texturization is rather simple and requires only two steps. First, the samples are dipped into electrolyte on the base of HF:HNO$_3$ to carry out the etching process and then wafer rising to remove the etching residuals. Contrary to the electrochemical method, the macroporous formation occurs on both sides of the wafer (Yerokhov et al. 2002; Marrero et al. 2008). Figure 10.4a and b present the honeycomb surface texturing obtained by this type of isotropical texturization on monocrystalline and multicrystalline Si (Yerokhov et al. 2002). This type of texturing becomes more and more expanded in the PV industry. One of the derivatives of the present method is the chemical vapor etching technique (Ben Rabha et al. 2005).

The comparison of experimental spectral dependence of the global reflectance of polished silicon, random pyramids, and chemically and electrochemically prepared macroporous textures, respectively, are presented in Figure 10.5.

Metal assisted etching is another promising approach for surface texturization of Si wafers. Usually metal assisted etching occurred in HF/H$_2$O$_2$/H$_2$O solutions with Ag nanoparticles as catalyst agent (Yae et al. 2005; Chartier et al. 2007; Chaoui et al. 2008; Bastide et al. 2009). This

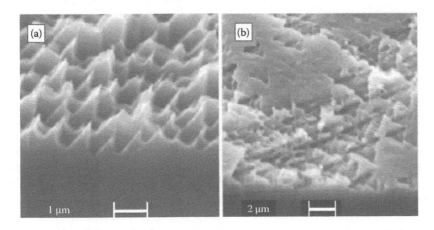

FIGURE 10.3 SEM images of anodized 0.4 Ω·cm multicrystalline Si (POLIX) with an optimized morphology for photovoltaic application. 4 M HF/ dimethylsulfoxide, Q = 4.5 C/cm². (a) (100) oriented grain and (b) (111) oriented grain. (From Lévy-Clément C. et al., *Phys. Stat. Sol. (a)* **197**, 27, 2003. Copyright 2003: John Wiley & Sons. With permission.)

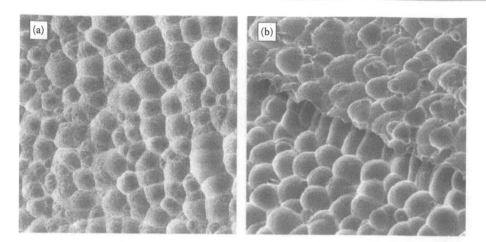

FIGURE 10.4   SEM images of chemical texturing obtained on p-type monocrystalline (a) and multi-crystalline (b) silicon. (From Yerokhov V.Y. et al., *Solar Energy Mater. Solar Cells* **72**, 291, 2004. Copyright 2004: Elsevier. With permission.)

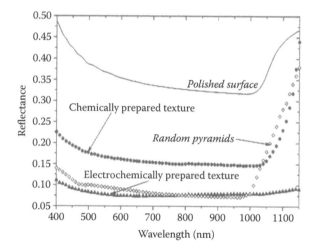

FIGURE 10.5   Measured spectral dependence of the global reflectance of polished silicon, random pyramids, and chemically and electrochemically prepared macroporous textures, respectively. (From Yerokhov V.Y. et al., *Solar Energy Mater. Solar Cells* **72**, 291, 2004. Copyright 2004: Elsevier. With permission.)

method implied several steps. The first one is electroless deposition of Ag nanoparticles on the surface of Si wafer. Next is the immersion of Ag-loaded Si samples into an $HF/H_2O_2/H_2O$ solution. Ag nanoparticles sink into Si and pore openings with diameters similar to those of the nanoparticles are obtained.

Metal assisted etching forms a meso PSi layer that is subsequently dissolved by NaOH to reveal a macrotextured surface. It should be noted that this texturization technique is anisotropic. Rounded macropores with inversed pyramids at the bottom are observed for (100) oriented Si. The (111) Si surface exhibits craters with a triangular form. The best effective reflectivity result of 5.8% (Chartier et al. 2007) was obtained for (100) oriented monocrystalline samples, compared to a standard NaOH etching of 12%. For multicrystalline substrates, the lowest obtained effective reflectivity was 12% (Figure 10.6a) (Bastide et al. 2009). One of the disadvantages of this texturization technique for PV applications is the necessity to dissolve residual Ag nanoparticles.

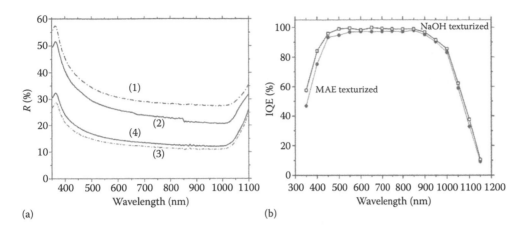

FIGURE 10.6 (a) Reflectivity spectra of mc-Si: (1) as cut (R_E = 30 %); (2) after NaOH texturization (25 %); (3) after metal assisted etching (MAE) + PSi dissolution for a time t (12 %); and (4) for a time t + 2 min (15 %). (b) Internal quantum efficiency (IQE) as a function of the wavelength in a similar grain of MAE (solid circles) and NaOH texturized solar cells (open square). (From Bastide S. et al., *Phys. Stat. Sol. (c)* **6**, 1536, 2009. Copyright 2009: John Wiley & Sons. With permission.)

10.3.2 INFLUENCE OF TEXTURIZATION TECHNIQUES ON SOLAR CELL PERFORMANCE

Reducing of light losses by lowering the surface reflectivity results in an increase of short circuit current (I_{sc}), which is one of the basic solar cell parameters. It was found that the I_{sc} of electrochemically textured samples can reach 37 mA/cm² against 24 mA/cm² for the control (no texturing) samples (Gamboa et al. 1998). However, despite the benefit of reduced reflectivity, the internal quantum efficiency (IQE) of the metal assisted etched and electrochemically textured cells are lower over the entire spectral region with a significant difference in a short wavelength region (Bastide et al. 2009) (Figure 10.6b). The higher recombination losses in the emitter are probably due to the larger surface area of macroporous textured surfaces that results in a larger volume of emitter. This statement is in agreement with the higher fill factor (FF) obtained in the case of macroporous textured cells. Higher FF can be explained by the increase of the contact surface. The lower spectral response of the macroporous textured cell eliminates partially the benefit of their reduced reflectivity. I_{sc} is only 2% higher (vs. NaOH cell) while +6.5% could be expected from the difference in effective reflectance R_E.

With respect to solar cell performance, the electrochemical texturization leads to an improvement in solar cell efficiency by 0.8% (Bastide et al. 2004) when compared to commercial alkaline etched cells. However, this improvement is not due to an increase of photocurrent response as expected but to a higher FF. Additional optimization of the surface passivation of macroporous textured cells should be performed.

10.4 NANO PSi AS ANTIREFLECTION COATING

10.4.1 REQUIREMENTS FOR ANTIREFLECTION COATINGS ON Si SOLAR CELLS

The reduction of reflection losses can be realized also by forming *antireflection coating* (ARC) in the form of a quarter wavelength dielectric layer on the front surface of a cell. The principle of a quarter wavelength ARC is well known. Light reflected from the dielectric layer–semiconductor interface arrives back at the dielectric–air interface 180° out of phase with that reflected from the dielectric–air interface, cancelling it out to some extent. Refractive index of dielectric layer should obey to

(10.1)
$$n = \sqrt{n_{air} n_{Si}}$$

where $n_{air} = 1$, $n_{Si} = 3.8$. Then a single layer ARC on Si has a refractive index of about $n = 1.9$ (Green 1992).

To prevent the penetration of humidity into PV modules and to ensure their long-term stability in the standard PV modules, the cells are normally embedded in a glass or encapsulated by different materials like ethylene vinyl acetate or resin (Nagel et al. 1999) with a similar refractive index to glass ($n = 1.5$). This increases the optimum value of the index of the ARC to about 2.3. Usually, one-layer coatings are chosen to produce minimum reflection at about 600 nm. The use of multiple layers of ARC materials can improve performance. The design of such coatings is more complex, but it possible to reduce reflection over a broad band.

10.4.2 TECHNOLOGY OF NANOPOROUS Si ARC

Since 1997 and up to 2003 an active research work has been paid to find an alternative to TiO_2, ZnS, and MgF_2 historically used as ARC for Si cells (Nagel et al. 1999). Indeed, three requirements should be fulfilled when applied in multicrystalline solar cells manufacturing: (1) ARC should be low-absorbing in the visible/near IR spectral region, (2) provide surface passivation, and (3) induce bulk passivation of the Si wafers.

The use of nanoporous Si as ARC was explored since at least 1997. A thin porous layer is grown on emitter (n^+ or p^+) of preliminary textured mono- or multicrystalline (p/n^+ or n/p^+) Si junctions. The thickness and porosity of the porous layer (and, respectively, the refractive indices) are chosen to produce a minimum reflection at 600 nm. Usually, the thicknesses of thin porous layers do not exceed 100 nm for three reasons:

1. In order to decrease the influence of the parasitic absorption in porous layer itself
2. Taking into account that the thickness of the standard shallow emitter is about 300 nm, the electrochemical or stain etched PSi layer should not damage the emitter
3. In order to reduce the influence of high resistivity of PSi on series resistance

There are three main benefits in using nano PSi as ARC. First, the refractive index of this material covers continuously the range between that of crystalline Si and air. Second, it is expected that chemically treated porous layer can assure a good passivation of emitter. And third, during the PSi formation, the superficial part of the emitter is etched back thus removing the dead layer from the highly doped surface.[*]

Two main techniques of thin porous layer formation have been employed: electrochemical (Krotkus et al. 1997; Strehlke et al. 1997; Lipinski et al. 2003) and chemical ("stain etching") (Bilyalov et al. 1997; Schirone et al. 1998; Schnell et al. 2000; Lipinski et al. 2003). They both have relative merits: in particular, the main advantage of stain etching is related to the lack of any electrical contact during the process. This makes it particularly suitable for large area processing, which is a prerequisite for perspective scale-up in industrial production. On the other hand, PSi layers obtained by chemical etching exhibit a porosity gradient (Strehlke et al. 1997; Lipinski et al. 2003), which causes the shift of the minimum of reflectivity curve. The electrochemical approach is so preferred, as it gives better reproducibility over the stain etching process.

The images of PSi layers formed by electrochemical and chemical methods on the surface of n^+ emitter are presented in Figure 10.7a and b, respectively. The effective reflectance coefficient can be reduced to about 1–3% in the whole visible range (Schirone et al. 1998; Lipinski et al. 2003).

A lot of efforts have been made in determining the formation conditions that led to the PSi layers with optimal optical properties resulting in efficiencies comparable to those obtained for commercial multicrystalline solar cells with classical ARC (Strehlke et al. 1997). Single or multi-layer structures on the base of PSi (Striemer and Fauchet 2003; Aroutiounian et al. 2004) have been calculated and realized. For example, in diamond-like carbon (DLC)/PSi double ARC, the DLC layer supplementary acts as a protective coating for PSi, increases solar cell stability relating

---

[*] In n^+/p Si solar cells, regions of excess phosphorus would lie near the surface of the cell. This can produce a "dead layer" near the surface where light-generated carriers have very little chance of being collected because of the very low minority-carrier lifetimes. This "dead layer" normally should be etched back.

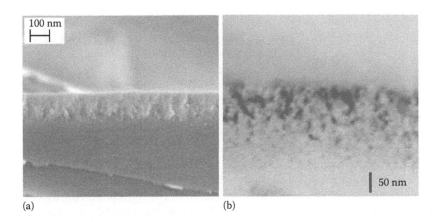

(a)　　　　　　　　　　　(b)

FIGURE 10.7   Cross-section images of a PSi layer formed by (a) electrochemical etching (SEM micrograph) and by (b) chemical etching (TEM micrograph). (From Lipinski M. et al., *Phys. Stat. Sol. (a)* **197**, 512, 2003 Copyright 2003: John Wiley & Sons. With permission.)

to the effect of proton and UV irradiation, making such solar cells especially promising for space application (Litovchenko and Kluyi 2001; Aroutiounian et al. 2004).

10.4.3 INFLUENCE OF POROUS Si ARC ON SOLAR CELL PERFORMANCE

PSi film that is etched in the shallower portion of the n^+/p junction gives a positive contribution to photocarrier collection efficiency. It reduces optical losses due to the ARC effect and decreases the absorption in the dead layer, allowing the photocurrent increase (Bilyalov et al. 1997; Lipinski et al. 2000).

Besides Fresnel interference, thin PSi layers also display components of diffuse scattering. Indeed, by analyzing the influence of the porous surface on the IQE of Si solar cells, it was shown that the porous layer of 160-nm thick acts as a perfect light diffuser with an effective entrance angle of 60° for the light entering the cell structure after passing through the porous surface layer (Figure 10.8, insert) (Stalmans et al. 1998). This increases the effective length path of incoming photons and IQE in the long wavelength region (Figure 10.8). Diffuser effect can be explained by Rayleigh's scattering, which occurs in medium if diameters of particles are no larger than about 1/10 of the wavelength of incident light and spatial distribution of these particles is fully randomized. Indeed, the dimension of nano PSi obeys the above-mentioned criteria. Theoretical analysis of light scattering in PSi for PV applications was performed (Abouelsaood et al. 2000).

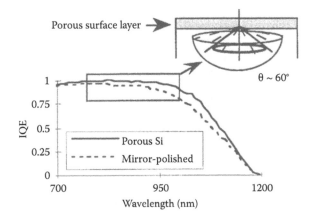

FIGURE 10.8   Measured IQE data for conventional cell structure with a mirror-polished and porous top surface. (From Stalmans L. et al., *Prog. Photovolt. Res. Appl.* **6**, 233, 1998 Copyright 1998: John Wiley & Sons. With permission.)

The possible improvement of multicrystalline Si solar cell parameters at the creation in the n^+-emitter the selective regions in form of PSi pipes (Skryshevsky and Laugier 1999; Skryshevsky et al. 2000) was also examined. It is expected that the light scattering in PSi leads to: (1) growth of incoming photon density in parts of the p-n junction between pipes (concentrator effect), (2) increasing of the effective length path of low adsorbed photons of long wavelength part of the sun illumination, and (3) possibility of partial absorption of high energy photons in low doped p-base instead top n^+ emitter that increases the photogeneration carrier collection in the short wavelength region. The difference in crystal orientation of Si grains results in the dispersion of the measured curves of external quantum efficiency (EQE) after PSi forming. However, the general tendency of EQE increasing in the long wavelength region of the sun illumination is observed for the cells with different PSi pipes density. In the short wavelength region, the behavior of EQE is more complicated due to the partial light absorption in PSi, light scattering by PSi, and change of surface recombination rate in PSi/Si interface. The growth of the PSi pipe area increases series and shunt resistances (due to the partial replacing of the top part of the emitter by the high resistive porous layer).

However, technology of PSi ARC displays few disadvantages. During the PSi fabrication, the most heavily doped portion of the n^+ region is also etched away, reducing the internal electrical field and eventually the open circuit voltage V_{oc} of the cell. A degradation of the spectral response in the short wavelength (Figure 10.9) (Bilyalov et al. 2000) was rather important for the cells with a PSi ARC compared to cells with classical dielectric ARC. This degradation was caused by the recombination of photogenerated charge carriers at the surface or at PSi/bulk Si interface. A lot of efforts have been undertaken to passivate PSi ARC (see Section 10.6).

FF degradation of about 10% is usually observed after formation of PSi. Indeed, if the PSi is made after the contacts, the metallic grids are etched in HF. So, the contacts should be protected by a polymeric film and short etching times (~10 s) should be applied. Another solution is to fabricate the PSi before the contacts. It was shown (Matic et al. 2000) that the high temperature annealing required for contact firing slightly increased the porosity and the thickness of porous layers. As a result, no significant changes in reflectance characteristics are observed. Therefore, firing of the contacts through PSi does not destroy its antireflection properties and can be successfully used as an alternative way for preparation of PSi ARC for Si solar cells.

Since 2004, hydrogenated Si nitride (SiN_x:H) has become a dominant material for the ARC in the modern PV technology. Indeed, SiN_x:H layers perfectly fulfill three requirements when

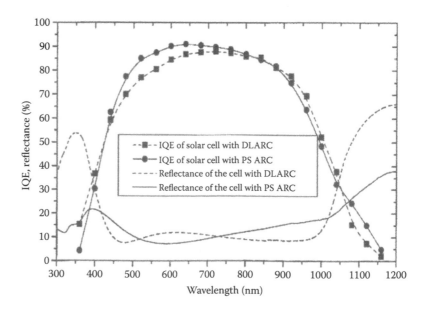

FIGURE 10.9   IQE and reflectance characteristics of the cells with PSi and double layer (TiO_2/MgF_2) ARC. (From Bilyalov R.R. et al., *Solar Energy Mater. Solar Cells* **60**, 391, 2000. Copyright 2000: Elsevier. With permission.)

applied in solar cells manufacturing: act as nonabsorbing ARC, provide surface passivation, and induce the bulk passivation of the Si wafer. The letter effect can be accomplished by a thermal activation of the layer (e.g., during a firing of a screen-printed metallization) by which hydrogen diffuses from the SiN_x:H into the Si wafer where it passivates crystal defects and impurities (Duerinckx and Szlufcik 2002). At the same time, the surface passivation is improved by reducing the number of dangling bonds and by creating a positive field effect due to positive charges naturally present in SiN_x:H layers (Aberle 2000). Now there is not a strong activity in PSi implementation as ARC of industrial solar cells; however, the articles devoted to the optimization of PSi mono and multilayers ARC we can find in the literature (Ben Rabha et al. 2011; Hsueh et al. 2011; Osorio et al. 2011; Ramizy et al. 2011; Dzhafarov et al. 2012; Salman et al. 2012; Thogersen et al. 2012; Dubey 2013).

10.4.4 INFLUENCE OF PSi EMISSION ON QUANTUM EFFICIENCY OF SOLAR CELLS

Since nanoporous Si displays efficient red-orange photoluminescence at room temperature, it will be very attractive to use the transformation of UV-blue part of sun flux absorbed by the PSi layer into long wavelength light having higher conversion efficiency for Si solar cells. Indeed, the efficient excitation of photoluminescence in nano PSi occurrs at the absorption of photon with energy $h\nu > 2.8$ eV ($\lambda < 450$ nm), which lies in the region where Si solar cell photoresponse strongly reduces (Tsuo et al. 1993). Numerically simulation and experimental verification of influence of this reemission effect on solar cell parameters was fulfilled (Skryshevsky et al. 1996a,b). As it was shown, the effect of reemission is relatively small and well observed only at high photoluminescence yield and for thick (0.25–2 µm) PSi layers that are not compatible with the routine PSi ARC. Moreover, the application of this effect for Si solar cells is restricted also by difficulty to obtain the efficient PSi photoluminescence on n^+ emitter and strong photoluminescence quenching at various thermal annealing at cell processing (e.g., at firing of contacts).

10.5 PSi SELECTIVE EMITTER

The optimal emitter parameters (thickness, doping) of Si solar cells with *homogeneous emitter* is a result of a compromise of the requirements of low dark current and low contact resistance. For passivated surface emitter, the surface dopant concentration must be moderated and the emitter must be shallow to keep the dark emitter saturation current density low. However, for nice electrical contact the surface dopant concentration must be higher than 10^{19}–10^{20}cm^{-3} for n-Si and emitter must be relatively thick (>0.3 µm). Therefore, the homogeneous emitter needs such high surface dopant concentration and emitter thickness that makes the surface passivation almost impossible because recombination increases with doping. For a nonscreen-printed contact, the optimal homogeneous emitter has a sheet resistivity of 60 Ω/sq and a doping level of 2.10^{19}cm^{-3} (Kuthi 2004). However, such a highly doped homogeneous emitter decreases the response to short wavelength illumination due to high recombination velocity.

In solar cell devices with *selective emitter*, the high doping regions are just under the metallization and sheet resistivity can be increased up to 70–200 Ω/sq between the contacts. Therefore, the recombination losses decrease and a good Ohmic contact and high FF can be obtained along with a low dark current and a good carrier collection yield in short wavelength region due to well passivation between the metal contacts.

The selective emitters are formed in many ways (Bilyalov et al. 2000; Moon et al. 2009). There are two types: technologies that need scheme aligning (more expensive and more difficult) and self-aligning technologies. The selective emitter based on self-aligning PSi technique uses that after metallization of heavy doped n^{++}-Si emitter, the emitter is etched back forming a porous layer (Figure 10.10). Since, it was assumed that PSi will act as ARC and passivating layer, no further layers are needed (Stalmans et al. 1997; Kuthi 2004). This is a strongly simplified processing scheme, aimed to replace the conventional texturing, thermal oxidation, and the ARC. In first works, the optimistic expectations of PSi selective emitter were announced for multi- and polycrystalline Si materials because, as was discussed in Section 10.3.1, conventional alkaline

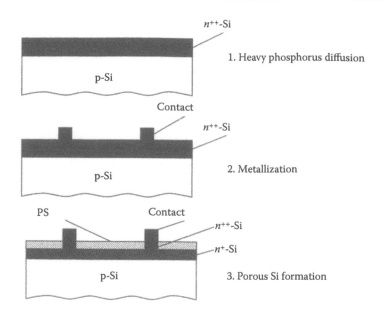

FIGURE 10.10   Process of etchback selective emitter formed by PSi.

texturing only functions well for monocrystalline Si. Efficiencies between 13 and 14% have been reached on multicrystalline substrates (Strehlke et al. 1997; Bilyalov et al. 2000) in earlier works.

The disadvantages of PSi selective emitter are the same as for the application of PSi as ARC. For the case of screen-printed contacts, the underetching of the metallization and of the underlying Si surface takes place due to the presence of HF in the electrochemical etching solution, which leads to the increase of the contact resistivity and to the decrease of FF. For the case of sputtered metal contacts, the etching does not strongly degrade the quality of the contacts; however, this technique is more expensive (Kuthi 2004).

Now, the effective selective emitter is proposed to form when PSi creation is used as an interim technological step at the standard screen-printing process. Process sequence uses a heavy diffusion, which is masked in the area that will be contacted. The emitter is then etched back in an acidic solution to the desired sheet resistance. Then the PSi and masking layer are subsequently removed in an alkaline solution. The following process of cell fabrication is the standard: PECVD deposition of SiN_x, screen metallization, cofiring, and edge isolation. Cz-Si solar cells of 5″ prepared by this technology shows V_{oc} = 640 mV, J_{sc} = 37.6 mA/cm², fill factor FF = 78.4%, and photoconversion efficiency of 18.9%. It was shown that for an etch back emitter as created in this process, the saturation dark current could be reduced to 28 fA/cm² for a 68 Ω/sq emitter by etching back from a very low sheet resistance of 17 Ω/sq (Book et al. 2009).

10.6 SURFACE PASSIVATION BY POROUS LAYERS

10.6.1 ESTIMATION OF SURFACE RECOMBINATION VELOCITY IN PSi/Si INTERFACE

Large surface recombination velocity strongly reduces the IQE in the short wavelength region of the cell. For conventional Si solar cell of n^+/p type the front surface recombination is important for the spectral region of 400–600 nm while the back surface recombination is revealed in IR part of the spectrum in the case if the diffusion length of minority charge carriers in the base is bigger than its thickness. The search of *passivating* materials and processes that are able to suppress the negative impact of the surface recombination has been one of the important tasks of PV. Application of PSi for the front surface passivation seemed to be very attractive in the technology in which these layers are used as ARC. The first published results on PSi passivation properties were rather contradictory. In some papers, PSi has revealed excellent passivation properties (Zheng et al. 1992) while in others the surface recombination velocity on PSi/Si interface was higher than on SiO_2/Si interface or on the free Si surface (Smestad et al. 1992).

The surface passivating capabilities of a porous layer on a top of n^+-emitter can be extracted from the IQE measurement in the short wavelength region. However, the PSi on the top of the cell absorbs a part of the incident light without photogeneration in the cell (the charge carriers are trapped on dangling bonds on the developed surface). Hence, the IQE should be corrected according to

(10.2) $$IQE^* = IQE \exp(\alpha d),$$

where α and d are the absorption coefficient and the thickness of PSi, respectively. For example, the absorption from 160-nm porous layer results in decreasing of the short circuit current I_{sc} by more than 2%. The estimated value of surface recombination velocity of Si wafer with porous top surface was found to be $S_p = 3.10^5$ cm/s in comparison with that one for unpassivated mirror-polished surface $S_p = 3.10^6$ cm/s (Figure 10.11a) (Stalmans et al. 1997, 1998).

The surface recombination velocity can also be extracted from the analysis of dark I–V characteristics. If the emitter thickness is lower than the diffusion length of minority charge carriers, then the emitter saturation current depends on the surface recombination velocity S_p at the front side of the cell. In this case, according to Sze (1981), the saturation current density for the diffusion current could be written as a function of S_p (when recombination at rear cell side is neglected). The good fitting of experimental I–V characteristics of Si cell with porous coating was obtained at the $S_p = 0.8–3.10^5$ cm/s (Skryshevsky et al. 2000). However, these values are higher in comparison with PECVD Si nitride and Si dioxide passivation (Nagel et al. 1999).

The other drawback of PSi is its fast degradation, which leads to the increase of the surface recombination velocity. The photoconductivity decay measurements showed the degradation of the effective lifetime τ_{eff} during aging of p-type Si wafer with porous top surface in ambient (Figure 10.11b) (Stalmans et al. 1998). In the case when the rear side recombination can be neglected (rear side of p-type Si wafer is well passivated), the effective lifetime τ_{eff} is determined by the bulk lifetime τ_b and front recombination velocity:

(10.3) $$\frac{1}{\tau_{eff}} = \frac{1}{\tau_b} + \frac{S_p}{W},$$

where W is the substrate thickness. Since no bulk changes occurred during aging, the observed decrease of τ_{eff} (from 40 μsec to 10 μsec for 40 h) was attributed to changes at the porous surface (Stalmans et al. 1998). To stabilized surface passivation capabilities of PSi several oxidation and hydrogenation treatments have been proposed: rapid thermal oxidation (RTO), electrochemical oxidation, plasma-nitride treatment, and electrochemical hydrogenation (Stalmans et al. 1997; Yerokhov and Melnyk 1999). Note the improvement of blue response after the rapid thermal oxidation of the porous surface is observed. Furthermore, after the thermal treatment

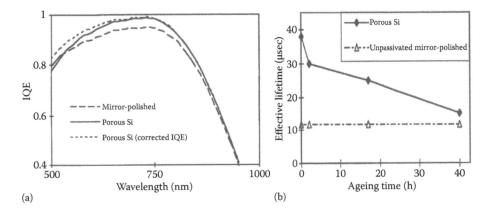

(a)   (b)

FIGURE 10.11   (a) Effect of a porous surface on the IQE in c-Si thin-film cells, (b) degradation in ambient with time of τ_{eff} of p-type Si wafer (unpassivated and passivated with PSi). (From Stalmans L. et al., *Prog. Photovolt. Res. Appl.* **6**, 233, 1998. Copyright 1998: John Wiley & Sons. With permission.)

of the cell with the PSi top layer, its IQE in the near IR region still remains higher than that of the reference cell without PSi coating (Stalmans et al. 1997). Haddadi et al. (2011) showed the reduction of the surface recombination velocity and an improvement of the diffusion length after the immersion PSi in a lithium bromide (LiBr) aqueous solution. Ben Rabha et al. (2013) studied the effect of Al_2O_3/PSi combined treatment on the surface passivation of monocrystalline silicon c-Si and the effective minority carrier lifetime increasing from 2 μs to 7 μs is observed.

10.6.2 LIGHT-INDUCED PASSIVATION ON PSi/Si INTERFACE

The light-induced passivation of the p-type Si surface by a thin ($d \approx 0.1$ μm) porous layer, formed by electrochemical etching was observed for thin film solar cells with interdigitated back contacts (IBC) (Nichiporuk et al. 2005, 2006). The illuminated I–V characteristics of IBC cells with different front surface coatings are presented in Figure 10.12a. The best passivation was obtained with UV–CVD Si nitride passivation, and the worst with chemically formed SiO_2 because of the poor quality of this oxide. The short circuit current I_{sc} of the solar cell with PSi passivation is two times better than that of unpassivated Si surface. Indeed, the maximum I_{sc} gain with optical confinement (PSi serves as ARC and light diffuser) for the selected cell configuration is about 40% (calculated with PC1D numerical simulation). Therefore, the significant gain in I_{sc} after PSi elaboration can be explained by the decreasing of the front surface recombination.

However, the laser beam induced current (LBIC) measurement of spatial I_{sc} distribution at scanning the front surface of the IBC cell with a modulated red laser beam shows that photocurrent is much lower in the PSi area (right part of the LBIC image in Figure 10.12b) in comparison with an unpassivated Si surface (left part of the same image). When the cell is scanned by a red laser beam in the presence of a permanent (nonmodulated) pumping light with λ < 470 nm ($P \sim$ 0.5 mW/cm²), a strong increase of the photocurrent in the PSi area is observed (Figure 10.12c). This I_{sc} growth can be provoked by a decrease of the front surface recombination at the PSi covered area in the presence of pumping light. The light-dependent PSi passivation phenomenon can be explained by a significant negative charge accumulation at the PSi/p-type Si interface traps under illumination and formation of an induced hi-low (p^+/p) junction at the front surface of the cell (Mizsei et al. 2004).

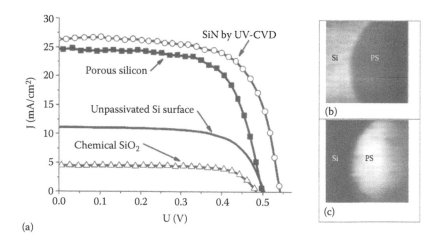

FIGURE 10.12   (a) I–V characteristics under AM1.5G illumination of the same IBC cell with different front surface coatings; (b) LBIC images of the same IBC solar cell with PSi layer at the front surface for two different measurement conditions without additional illumination; (c) in the presence of permanent blue light. Lighter areas correspond to large values of the induced photocurrent. (From Nichiporuk O. et al., *Thin Solid Films* **511–512**, 248, 2006. Copyright 2006: Elsevier. With permission.)

10.7 PSi AS OPTICAL REFLECTOR

10.7.1 APPLICATION OF REAR BRAGG REFLECTOR FOR Si SOLAR CELLS

Although Si-based solar cells dominate in the PV industry, some of their drawbacks still are not eliminated. Thus, Si displays a relatively low absorption coefficient in near-IR spectral region. Si-based solar cells must be thick (up to millimeters) to absorb 90% of the incident light at wavelengths from 700 to 1100 nm, respectively, which comprise about one-half of the solar energy available above the bandgap of Si (Green and Keevers 1995). However, the cell thickness increasing is not desirable because it leads to growth of bulk recombination losses and Si consumable. Thereby, it is desirable to construct cells with improved schemes of light trapping (e.g., with the elongation of the light path inside the cell) without increasing the wafer thickness. As was shown in Section 10.4, the forming of thin PSi diffuser on top of the cell does not solve this problem.

One of the methods to enhance long wavelength light absorption without thickness increasing is the use of a reflector at the backside of the solar cell. Sunlight is partially reflected backward into the cell that results in its additional absorption leading to the supplementary electron/hole generation. The rear surface reflector in n^+/p solar cell should provide both the multipassing of low adsorbed photons inside a cell and a good passivation. For the reason of strong recombination at metal/Si interface, the application of continuous metallic layers evaporated directly on p/p^+ base of Si solar cell is an obstacle. Rear side reflector can be formed as multilayers dielectrical films on Si substrate (SiN, SiO_2) with windows for back Ohmic contacts (Figure 10.13a).

A simple backside reflector based on PSi is a Bragg mirror, which can be easily realized by electrochemical etching. A Bragg mirror consists of alternating sequences of quarter wavelength layers of two porosities, corresponding with two refractive indexes n_H (high) and n_L (low) (Figure 10.13b). If individual porous layers have the correct refractive index and thickness, constructive interference can lead to a high internal reflection for perpendicular striking photons. The wavelength λ_0 in the maximum reflection obeys the following equation:

$$(10.4) \qquad m\lambda_0 = 2(d_1 n_H + d_2 n_L)$$

The important characteristics of the Bragg mirror are the value of reflection coefficient R_{max} at λ_0 and full width at half maximum (FWHM). Figure 10.14 presents the calculated reflection coefficient in maximum and FWHM versus n_H/n_L ratio for different number of bilayers, respectively. The increasing numbers of layer pairs results in higher reflection coefficient R_{max} for lower ratio value n_H/n_L, and the value of reflection coefficient $R_{max}(\lambda_0) = 99\%$ can be obtained for bilayer number $N_{bi} = 4$ when $n_H/n_L > 1.8$. The increasing number bilayers leads to an important FWHM enlargement. For example, at $n_H/n_L = 2$, the FWHM increases by 31% when the bilayer number

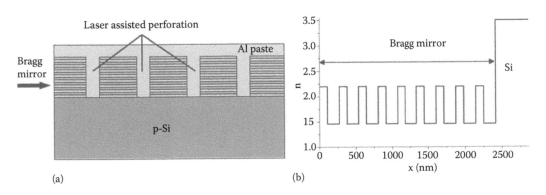

(a)   (b)

FIGURE 10.13   (a) Scheme of rear side of SC with Bragg mirror and laser assisted perforation in mirror, (b) profile of refractive index at rear side of SC with Bragg mirror.

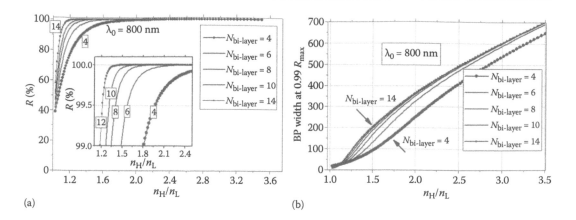

(a)

(b)

FIGURE 10.14   Dependence of (a) R_{max} reflection coefficient (at $\lambda_0 = 800$ nm) and (b) FWHM on n_H/n_L ratio for different bi-layer number (digits show quantity of bi-layers). (From Ivanov I.I. et al., *Renewable Energy* **55**, 79, 2013. Copyright 2013: Elsevier. With permission.)

changes from $N_{bi} = 4$ to $N_{bi} = 6$. However, further increasing the bilayer number up to $N_{bi} = 14$ does not lead to considerable FWHM enlargement (Ivanov et al. 2009, 2013).

Simulation results show that optical light path increasing in long-wavelength spectral region caused by internal multiple bouncing inside of standard cell of 200-μm thick leads to increasing of I_{sc} on 1–2 mA/cm². Solar cell conversion efficiency increases 1% in absolute when internal reflection coefficients grow from 20% up to 95% (Ivanov et al. 2009).

10.7.2 PSi REAR REFLECTORS FOR THIN FILM SOLAR CELL

The application of a backside reflector has more profit for thin solar cells. Indeed, compared to bulk Si solar cells, in thin Si devices similar levels of V_{oc} and FF can be obtained. The I_{sc}, however, is held back by the optically thin active layer. Light that traverses the epitaxial layer is lost for collection in the highly doped low quality substrate. Consequently, a typical value of I_{sc} in epitaxial thin film solar cells is 26 mA/cm². In order to improve the optical pathlength for long wavelength light, the intermediate reflector between the epilayer and the substrate can be applied. The concept of an intermediate reflector in thin film epitaxial cell technology is illustrated in Figure 10.15 (Duerinckx et al. 2006; Kuzma-Filipek et al. 2008). The photons reaching epi/substrate interface can now be reflected and pass a second time through the active layer.

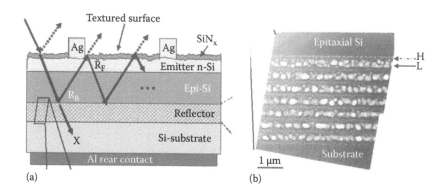

(a)

(b)

FIGURE 10.15   (a) Schematic cross-section of a thin-film epitaxial cell including a Bragg porous mirror; (b) TEM image of a reorganized porous Bragg reflector. (From Kuzma-Filipek I.J. et al., *J. Appl. Phys.* **104**, art. 073529, 2008. Copyright 2008: AIP Publishing LLC. With permission.)

As well as for thick Si solar cells the Bragg reflector can be made by electrochemical growth of a porous stack of alternating high and low porosity layers. Note that the rest of the solar cell fabrication process remains very similar to that of standard multicrystalline Si process. The main difference is the implementation of the electrochemical etching process for PSi formation after saw damage removal and the deposition of the active layer by CVD. Numerical modeling has demonstrated that PSi multilayer reflector is particularly useful for extremely thin Si cells on textured substrate (Zettner et al. 1998).

The use of PSi single layer and multilayer structures as an optical reflector for thin film epitaxial growth on Si substrate has been reported in numerous research reports (Zettner et al. 1998; Kuzma-Filipek et al. 2007, 2008, 2009; Van Hoeymissen et al. 2011). A new method for formation of PSi with a multilayer structure in one step has also been developed, which makes the up scaling of PSi into an industrial level (Matic et al. 2000) much easier.

There are three ways to use a PSi back reflector in thin film devices (Bilyalov et al. 2001). The first one is the high temperature road (>800°C) resulting in quasi-monocrystalline Si and good quality epitaxial layers. During the epitaxial growth on top of the porous stack, the individual layers reorganize into quasi-monocrystalline Si. Using this method, efficiency of more than 10% has been reached in a 10-µm thick epitaxial cell with PSi reflector on a highly doped multicrystalline Si substrate (Bilyalov et al. 2001). A medium temperature road (700–800°) allows preserving the PSi structure (Jin et al. 2000), but results in pore filling effect during the epitaxy. A low-temperature road (<700°C) would be an attractive solution if deposition on a PSi layer could be done with an epitaxial quality and sufficient growth rate.

Reflectance values up to 85% have been obtained using PSi rear surface reflector. Together with the total internal reflection on the front surface, this leads to an additional optical path length enhancement of 7 (Van Hoeymissen et al. 2008). On a cell level, the PSi reflector has given considerable increase in I_{sc} (Duerinckx et al. 2006) resulting in efficiencies of 13.5% on low-cost Si substrate.

Recently, a multilayer PSi Bragg reflector was improved by chirping the structure, that is, varying the periodicity in the depth of the alternating PSi sublayers (Kuzma-Filipek et al. 2008; Van Hocymissen et al. 2008) (Figure 10.16a). The experimentally measured reflectance curves of samples with Bragg reflectors, with and without chirping the structure, is presented in Figure 10.16b. The wavelength band with high reflectance was broadened by 50–80%. The cells with chirped reflectors contain 60 or 80 sublayers. Efficiencies of large area epitaxial cells on low-cost Si substrates including conventional porous reflectors have reached values of around 13%. With the implementation of alternative designs of PSi reflectors such as chirped structures, those efficiencies have increased to almost 14% with the standard screen-printing technology.

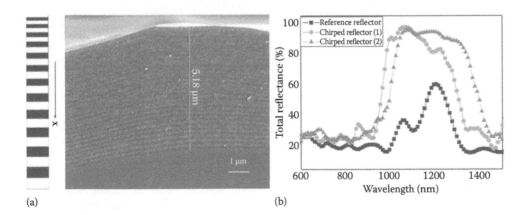

(a)   (b)

FIGURE 10.16   (a) Linear chirped structure of PSi stack as grown. The thickness of PSi layers increases with the etching depth of the silicon substrate. (b) Experimental reflectance spectrum of various fabricated reflectors as grown porous Si, single mirror Bragg stack of 15 layers, and unconventional chirped reflector of 80 layers (substrate is monocrystalline Si). (From Kuzma-Filipek I.J. et al., *J. Appl. Phys.* **104**, art. 073529, 2008. Copyright 2008: AIP Publishing LLC. With permission.)

10.8 APPLICATION OF POROUS SILICON FOR LAYER TRANSFER

PSi can be applied for layer transfer processes to get a thin monocrystalline silicon film on a foreign substrate for solar cell application (see Figure 10.17). The top surface of the monocrystalline silicon wafer is made porous by electrochemical etching. The PSi layer has a double porosity structure: low-porosity layer at the top and high-porosity one at the bottom. The remaining material in the low-porosity layer is of monocrystalline quality and allows the growth of a high-quality epitaxial silicon layer on it after annealing the porous structure at high temperature. The epitaxial layer deposited on the PSi layer is detachable from the original silicon substrate through a high-porosity layer. The deposited epitaxial layer is transferred onto a foreign substrate. Device fabrication can be obtained before or after the layer transfer onto a foreign substrate and needs a different processing sequence and special requirements for the substrate. Therefore, the porous structure serves two purposes. First, it allows the growth of a high quality epitaxial layer and, second, it allows separation of the epitaxial layer from the starting substrate and transfer onto a foreign substrate (Brendel 2004; Solanki et al. 2004).

The active layer deposition is generally preceded by the high-temperature treatment of the double porosity structure. The high temperature treatment (>1050°C) of the PSi layer in an H_2 atmosphere causes the reorganization of the porous structure—transforms the 10–50 nm pores to spherical voids ranging from 50 nm to 1 μm depending on the morphology and porosity of the initial PSi layer and top surface of the low-porosity layer closes and becomes smooth. This smooth surface provides a perfect seeding layer for the epitaxial layer deposition. Besides, the high-temperature annealing of a high-porosity layer results in the formation of big voids with remaining thin silicon pillars connecting the initial silicon substrate and the low-porosity layer. This layer with silicon pillars serves as the separation layer (Rinke et al. 1999; Solanki et al. 2004).

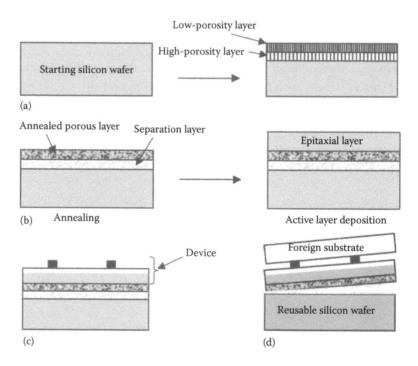

FIGURE 10.17   Steps of porous layer transfer processes using PSi as a sacrificial layer for obtaining thin monocrystalline silicon films on cost-effective substrates: (a) a double porosity structure (high-porosity layer beneath low-porosity layer) formation on starting silicon substrate by anodization, (b) thermal annealing of PSi and active layer deposition: annealed low-porosity layer acts as a good seeding layer for epitaxial layer deposition and voids with weak silicon pillars forms in high-porosity layer acts as a separation layer. (c) Device fabrication. (d) Separation of epitaxial layer from the starting silicon substrate and transfer onto foreign substrate by gluing it with an adhesive layer and applying mechanical force. (From Solanki C.S. et al., *Solar Energy Mater. Solar Cells* **83**, 101, 2004. Copyright 2004: Elsevier. With permission.)

There are many processes developed or being developed which use PSi as a sacrificial layer for the layer transfer process. Among these processes are ELTRAN (Yonehara et al. 1994), ψ-process (Brendel 1997), quasi-monocrystalline silicon (QMS) process (Rinke et al. 1999), LAST process (Solanki et al. 2002), freestanding monocrystalline thin film silicon (FMS) process (Solanki et al. 2004), and others. The mentioned layer transfer processes mostly differ in the sequence of the steps that are used to complete the cell fabrication and transfer onto a foreign substrate.

The QMS process (Rinke et al. 1999; Werner et al. 2003) consists of a double porous layer that restructures after annealing at high temperature under hydrogen. The upper low porosity layer becomes quasi monocrystalline and can be used as the active layer of the solar cell and the high porosity layer permits the detachment. The ψ-process uses the transfer of a texturized monocrystalline layer using a texturized silicon wafer for porous layers formatting. Efficiencies of 13.3% (Brendel 2004) have been obtained using this process and 15.4% (Feldrapp et al. 2003) have been reached by texturizing the wafer after the epitaxy. The main drawbacks of this process are the high consumption of the silicon wafer (12 µm per cycle) (Horbelt et al. 2005).

In general, layer transfer processes for silicon solar cells are promising candidates to reduce cell costs because kerf losses are greatly reduced since one substrate wafer can be reused many times. Recently, Petermann et al. (2012a,b) demonstrated the high efficiency potential of this material by reporting a new independently confirmed record efficiency value of 19.1% for this type of solar cell.

10.9 MODERN TENDENCIES IN Si PV

10.9.1 SOLAR CELLS WITH Si NANOWIRES

Advances in semiconductor materials and technology, naturally, have appreciable influence on the modern R&D in the PV field. In the last years, a new solar cell structure with Si nanowires, quantum dots, and quantum wells attracted special attention as a promising way to improve the solar cells' efficiency and reduce the Si consumable.

The collection efficiency of charge carriers generated in a cell outside of a p-n junction depends on the diffusion length of minority charge carriers in the quasi-neutral regions, which is limited by recombination losses. To minimize charge carrier recombination, cells must be created on material with a large diffusion length of minority charge carriers, which is very expensive for routine Si cell fabrication. A proposed solution to this problem relies on decoupling the long extinction distance of Si and the proximity of generated charge carriers to the p-n junction in cells based on Si nanowire (NW) carpets (Sivakov et al. 2009; Hochbaum and Yang 2010). By comparing classical thin film cells based on multicrystalline Si with NW solar cells, the last ones are expected to have higher efficiency due to: (1) the possibility to produce the perfect single crystalline structure of (2) wires, the short distances needed for charge separation, and (3) the perfect light trapping in NW arrays.

There exist two methods for production Si NW, a bottom up and a top down approach. The bottom up way relies on the vapor liquid solid (VLS) method, which uses metal nanotemplates on Si wafer or on Si thin film (Tian et al. 2007; Tsakalakos et al. 2007; Kelzenberg et al. 2008; Stelzner et al. 2008). Most frequently gold is taken as a template which forms a eutectic with Si at 370°C. The Au nanotemplates are prepared from gold colloids or by deposition of a thin Au film followed by a heating step above the eutectic temperature during which the nanodroplets are formed. The Au nanotemplates act as catalysts to decompose silane at $T > 500°C$. Si NWs grow with a diameter similar to that of the template droplet. As a result, a carpet of perfect single crystalline NWs of 10–200 nm in diameter and several micrometers in length (with a rate of several 100 nm/min) can be grown on the crystalline substrate (Andra et al. 2008).

In a top down approach, well-aligned Si NW arrays are obtained by electroless metal-assisted chemical etching in HF/AgNO$_3$ solution (Srivastava et al. 2010) (see also Section 10.3.1). Basically, a noble metal is deposited on the surface in the form of nanoparticles which act as catalyst for Si etching in HF solution containing an oxidizing agent. As a consequence, the etching only occurs in the vicinity of the metal nanoparticles and results in the formation of well-defined mesopores (20–100 nm in diameter) (Peng et al. 2005; Fang et al. 2008). Tapering the NWs by post-KOH dipping achieved separation of each NW from the bunched NW (Figure 10.18), resulting in a strong enhancement of broadband optical absorption. As electroless etching time increases,

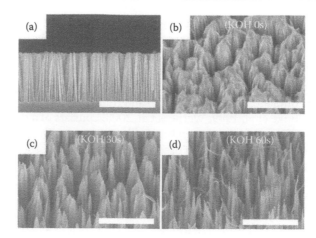

FIGURE 10.18   Cross-sectional SEM images showing the SiNW array after metal-assisted electroless etching (a). Scale bar is 10 µm. 30°-tilted SEM images also show the morphological change of the SiNW tips according to the postetching time of KOH: (b) 0 s, (c) 30 s, (d) 60 s. Scale bars in (b)–(d) are 5 µm. (From Jung J.-Y. et al., *Opt. Express* **18**, A286, 2010. Copyright 2010: The Optical Society. With permission.)

the optical crossover feature was observed in the tradeoff between enhanced light trapping (by graded-refractive index during initial tapering) and deteriorated reflectance (Jung et al. 2010).

Independently of the nanowire preparation method, two designs of NW solar cells are now under consideration with *p-n* junction either radial or axial. In the radial case, the *p-n* junction covers the whole outer cylindrical surface of the NWs (Figure 10.19a) (Kayes et al. 2005; Hochbaum and Yang 2010). This was achieved either by gas doping or by CVD deposition of a shell oppositely doped to the wire (Peng et al. 2005; Tian et al. 2007; Fang et al. 2008).

In the axial variant, the *p-n* junction cuts the NW in two cylindrical parts and requires minimal processing steps (Andra et al. 2008). However, cells that absorb photons and collect charges along orthogonal directions meet the optimal relation between the absorption values and minority charge carrier diffusion lengths. Figure 10.19b shows the variant of axial *p-n* junction fabrication process on *p*-substrate when *n*-region is formed on a nanowire array. Since charge extraction occurs through the nanowires, decreasing I_{SC} is observed due to the larger series resistance of the nanowires and their contacts (Hochbaum and Yang 2010). In order to assure high conversion

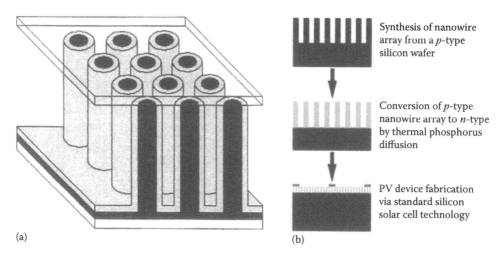

Synthesis of nanowire array from a *p*-type silicon wafer

Conversion of *p*-type nanowire array to *n*-type by thermal phosphorus diffusion

PV device fabrication via standard silicon solar cell technology

(a)   (b)

FIGURE 10.19   (a) Schematic cross-section of the radial p-n junction nanorod cell. Light is incident on the top surface. The light gray area is *n* type, the dark gray area is *p* type. (From Kayes B.M. et al., *J. Appl. Phys.* **97**, art. 114302, 2005. Copyright 2005: AIP Publishing LLC. With permission.) (b) Schematic of a subsurface p-n junction device fabrication process. (From Hochbaum A.I. and Yang P., *Chem. Rev.*, **110**, 527, 2010. Copyright 2010: American Chemical Society. With permission.)

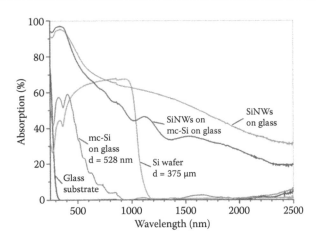

FIGURE 10.20   Optical absorption of SiNWs on a glass substrate and on mc-Si on glass compared to the absorption of a 375-μm thick Si wafer. (From Stelzner T. et al., *Nanotechnol.* **19**, 295203, 2008. Copyright 2008: IOP Publishing. With permission.)

efficiencies, the NW solar cell parameters should be carefully optimized. The vertical array geometry scatters light efficiently, especially at short wavelengths, and can absorb more light than a comparably thick solid crystalline film. Optical absorption in cells is influenced by NW diameter and their density (NW diameter governs the energy band structure due to the quantum confinement effect). Series resistance depends on NW diameter, length, and density. The interface between the nanowires and the encapsulating material impacts photogenerated carriers' recombination at the nanowire's surface.

In Figure 10.20 the optical absorption (A) of Si NWs on glass substrate and on mc-Si on glass, derived from transmission and reflection data (A = 1–T–R), is compared to the absorption of a 375-μm thick Si wafer. A strong broadband optical absorption is observed in the relatively thin Si NW (3–6 μm) due to strong light trapping of the NW. Pay special attention to the strong absorption of light with an energy below the bandgap energy due to light trapping together with absorption by defect states and plasmon coupling of light with the NW and underlying nanocrystalline Au-Si (Stelzner et al. 2008).

It should be also noted that from economical point of view in VLS technology, it is attractive to use low cost or flexible substrates like glass or metal foil instead of Si for NW growth (Tsakalakos et al. 2007; Andra et al. 2008). The technique of electroless metal-assisted chemical etching is preferable to create vertically aligned Si NWs of high electronic quality with desirable crystallographic orientation and doping characteristics. These Si NW arrays can be used as an efficient ARC for Si solar cells incorporating this array (Peng et al. 2005). The tapered NW solar cells (Figure 10.18) demonstrated superior photovoltaic characteristics, such as a short circuit current 17.7 mA/cm² and a cell conversion efficiency of ~6.6% under 1.5 AM illumination (Jung et al. 2010).

10.9.2 TANDEM Si SOLAR CELLS WITH Si QUANTUM DOTS

In standard solar cells, when photons with energy $2E_g > h\nu > E_g$ are absorbed it generates one electron-hole pair. After the photogeneration, the conduction band electron and valence band hole quickly lose any energy in excess of the semiconductor band gap (thermal relaxation loses). This loss mechanism alone limits conversion efficiency to 44% (Green 2000) when the one bandgap semiconductor is used.[*]

Among the proposed concepts, the tandem cells approach is the only one that has already permitted realizing photovoltaic structures with efficiency exceeding the limit for a single bandgap

---

[*] The thermodynamic limit on solar energy conversion to electricity is 93% (Luque and Marti 1997).

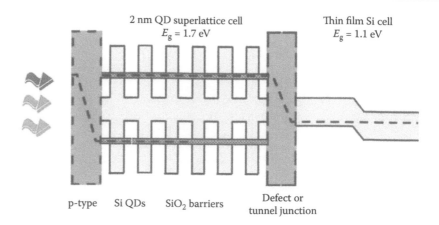

FIGURE 10.21   "All Si" tandem solar cell. Nanostructured cell consists of Si quantum wells or quantum dots in an amorphous dielectric matrix connected by a defect tunnel junction to Si cell. (From Conibeer G., *Materials Today* **10**, 42, 2007. Copyright 2007: Elsevier. With permission.)

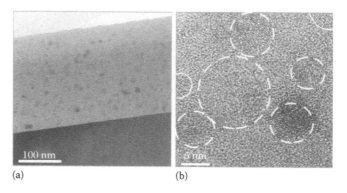

FIGURE 10.22   TEM images of Si QDs in a SiO_2 matrix: (a) low-magnification image and (b) high-resolution image. (From Cho E.C. et al., *Nanotechnol.* **19**, 245201 2008. Copyright 2008: IOP Publishing. With permission.)

device (Green 2003). In the tandem cell approach, the cells are stacked on top of one another. By placing the largest band gap cell uppermost, this cell will absorb the highest energy photon, allowing photons of lower energy to pass through to underlying cells, arranged in order of decreasing band gap. The cells can be connected together in series by tunnel junctions and each of them should generate the same current.[*]

Up to now, the tandem cell approach is based on monolithic integration of III-V materials by means of rather expensive technologies of fabrication such as molecular beam epitaxy. Thus, the concept of "all-Si" tandem solar cells appears as an attractive alternative permitting to replace III-V materials by Si and its dielectric compounds.

The concept of "all-Si" tandem solar cell with Si quantum wells or quantum dots sandwiched between layers of a dielectric based on Si compounds such as SiO_2, Si_3N_4, and SiC is shown in Figure 10.21 (Conibeer 2007; Conibeer et al. 2012). If quantum dots or wells are close enough to each other so that neighbors interact, quantum levels broaden out into bands. For quantum dots of 2 nm (or quantum wells of 1 nm), the miniband of larger bandgap of 1.7 eV is formed, which is optimal for tandem cells on top of Si.

Optical and structural properties of Si quantum dots in dielectric matrix were studied (Nychyporuk et al. 2008; Delachat et al. 2009). A simple approach to prepare superlattices of Si quantum dots is reported (Zacharias et al. 2002). The Si QDs were formed by alternate deposition of SiO_2 and silicon-rich SiO_x with magnetron co-sputtering, followed by high-temperature

---

[*] The theoretical limiting efficiency of tandem cell depends on the number of subcells in the device. For 1, 2, 3, 4, and ∞ subcells, the theoretical limiting efficiency η are 31.0%, 42.5%, 48.6%, 52.5%, and 68.2%, respectively, for unconcentrated sun light (Conibeer 2007).

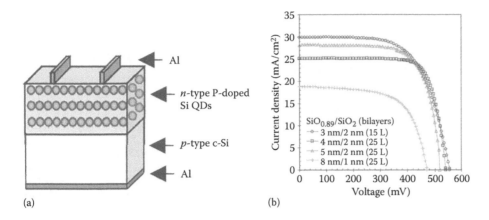

FIGURE 10.23 (a) Schematic diagram of an (*n*-type) Si QD/(*p*-type) c-Si photovoltaic device. (b) One-sun illuminated I–V curves of four different (*n*-type) Si QD/(*p*-type) c-Si solar cells measured at 298 K. (From Cho E.C. et al., *Nanotechnol.* **19**, 245201 2008. Copyright 2008: IOP Publishing. With permission.)

annealing (Figure 10.22). On heating, surface energy minimization favors the precipitation of Si into approximately spherical quantum dots (Cho et al. 2008). The matrix remains amorphous, thus avoiding some of the problems of lattice mismatch. Other dielectrics (Si nitride, Si carbide) are of interest because their lower band gap compared to SiO_2 should greatly increase current transport in these materials.

Schematic diagram of an (*n*-type) Si QD/(*p*-type) c-Si photovoltaic device is presented in Figure 10.23a. Solar cells consist of phosphorus-doped Si QDs in an SiO_2 matrix deposited on *p*-type crystalline Si substrates (c-Si). Current tunnelling through the QD layer was observed from the solar cells with a dot spacing of 2 nm or less. To get the required current densities through the devices, the dot spacing in the SiO_2 matrix had to be 2 nm or less. The open-circuit voltage was found to increase proportionally with reductions in QD size, which may relate to a bandgap widening effect in Si QDs. One-sun illuminated I–V curves of four different (*n*-type) Si QD/(*p*-type) c-Si solar cells measured at 298 K are presented in Figure 10.23b. The best cell parameters obtained were an open-circuit voltage V_{oc} of 556 mV, short-circuit current J_{sc} of 29.8 mA/cm², fill factor FF of 64%, and conversion efficiency of 10.6% from 3 nm Si QDs with a 2-nm SiO_2 layer (Cho et al. 2008).

REFERENCES

Aberle, A.G. (2000). Surface passivation of crystalline silicon solar cells: A review. *Prog. Photovolt.: Res. Appl.* **8**, 473–487.

Abouelsaood, A.A., Ghannam, M.Y., Stalmans, L., Poortmans, J., and Nijs, J.F. (2000). Theoretical analysis of light scattering in porous Si for PV applications. In: *Proceedings of 16th European Photovoltaic Solar Energy Conference*, May 1–5, Glasgow, UK, pp. 199–202.

Andra, G., Pietsch, M., Sivakov, V. et al. (2008). Thin film solar cells based on single crystalline silicon nanowires. In: *Proceedings of 23rd European Photovoltaic Solar Energy Conference*, September 1–5, Valencia, Spain, pp. 163–167.

Aroutiounian, V., Martirosyan, Kh.S., and Soukiassian, P. (2004). Low reflectance of diamond-like carbon/porous silicon double layer antireflection coating for silicon solar cells, *J. Phys. D: Appl. Phys.* **37**, L25–L28.

Aulich, H.A., Schulze, F.-W., and Anspach, O. (2010). Large scale crystallisation and wafer production: The way to 2020. In: *Proceedings of 25th European Photovoltaic Solar Energy Conf.*, September 6–10, Valencia, Spain, pp. 1066–1072.

Bastide, S., Le Quang, N., Lust, S., Korte, L., Vard, C., and Lévy-Clément, C. (2004). Electrochemical macroporous texturization of Polix mc-Si solar cells. In: *Proceedings of 19th European Photovoltaic Solar Energy Conference*, June 7–11, Paris, France, pp. 943–946.

Bastide, S., Tena-Zaera, R., Lévy-Clément, C., Barakel, D., Palais, O., and Martinuzzi, S. (2006). Photoelectrochemical texturization of *n*-type crystalline Si. In: *Proceedings of 21st European Photovoltaic Solar Energy Conference*, September 4–8, Dresden, Germany, pp. 911–914.

Bastide, S., Le Quang, N., Monna, R., and Lévy-Clément, C. (2009). Chemical etching of Si by Ag nano-catalysts in HF-H$_2$O$_2$: Application to multicrystalline Si solar cell texturization. *Phys. Stat. Sol. C.* **6**, 1536–1540.

Ben-Chorin, M. (1997). Resistivity of porous silicon, In: Canham, L. (Ed.) *Properties of Porous Silicon*, EMIS data reviews series, Vol. 18, INPEC, Dera, Malvern, pp. 165–175.

Ben Rabha, M., Boujmil, M.F., Saadoun, M., and Bessaïs, B. (2005). Chemical vapour etching-based porous silicon and grooving: Application in silicon solar cell processing. *Phys. Stat. Sol. C.* **2**, 3505–3509.

Ben Rabha, M., Mohamed, S.B., Dimassi, W., Gaidi, M., Ezzaouia, H., and Bessais, B. (2011). Reduction of absorption loss in multicrystalline silicon via combination of mechanical grooving and porous silicon. *Phys. Stat. Sol. C.* **8**(3), 883–886.

Ben Rabha, M., Salem, M., El Khakani, M.A., Bessais, B., and Gaadi, M. (2013). Monocrystalline silicon surface passivation by Al$_2$O$_3$/porous silicon combined treatment. *Mat. Sci. Eng. B.* **178**, 695–697.

Bilyalov, R.R., Lautenschlager, H., Schetter, C., Schomann, F., Schubert, U., and Schindler, R. (1997). Porous silicon as an antireflection coating for multicrystalline silicon solar cells. In: *Proceeding of 14th European Photovoltaic Solar Energy Conference*, June 30–July 4, Barcelona, Spain, pp. 788–791.

Bilyalov, R.R., Ludemann, R., Wettling, W. et al. (2000). Multicrystalline silicon solar cells with porous silicon emitter. *Solar Energy Mater. Solar Cells* **60**, 391–420.

Bilyalov, R., Stalmans, L., Baucarne, G., Loo, R., Caymax, M., Poortmans, J., and Nijs, J. (2001). Porous silicon as an intermediate layer for thin-film solar cell. *Solar Energy Mater. Solar Cells* **65**, 477–485.

Book, F., Dastgheib-Shirazi, A., Raabe, B., Haverkamp, H., Hahn, G., and Grabitz, P. (2009). Detailed analysis of high sheet resistance emitters for selectively doped silicon solar cells. In: *Proceeding of 24th European Photovoltaic Solar Energy Conference*, September 21–25, Hamburg, Germany, pp. 1719–1722.

Brendel, R. (1997). A novel process for ultrathin monocrystalline silicon solar cells on glass. In: *Proceeding of Proc. 14th European Photovoltaic Solar Energy Conf.*, June 30–July 4, Barcelona, Spain, pp. 1354–1357.

Brendel, R. (2004). Thin-film crystalline silicon mini-modules using porous Si for layer transfer. *Solar Energy* **77**, 969–982.

Burgers, A.R., Naber, R.C.G., Carr, A.J. et al. (2008). 19% efficient n-type Si solar cells made in pivot production, In: *Proceeding of 23rd European Photovoltaic Solar Energy Conference*, September 1–5, Valencia, Spain, pp. 1106–1109.

Canham, L.T. (1997). Storage of porous silicon. In: Canham L. (Ed.) *Properties of Porous Silicon*. EMIS data reviews series, Vol. 18. INPEC publ., Dera, Malvern, pp. 44–50.

Chaoui, R., Mahmoudi, B., and Si, A.Y. (2008). Porous Si antireflection layer for solar cells using metal-assisted chemical etching. *Phys. Stat. Sol. A.* **205**, 1724–1728.

Chartier, C., Bastide, S., and Lévy-Clément, C. (2007). Metal-assisted wet chemical etching of crystalline Si. In: *Proceeding of 22nd European Photovoltaic Solar Energy Conference*, September 3–7, Milan, Italy, pp. 1231–1234.

Cho, E.C., Park, S., Hao, X., Song, D., Conibeer, G., Park, S.C., and Green, M.A. (2008). Silicon quantum dot/crystalline silicon solar cells. *Nanotechnol.* **19**, 245201.

Conibeer, G. (2007). Third-generation photovoltaics. *Mater. Today* **10**, 42–50.

Conibeer, G., Perez-Wurfl, I., Hao, X., Di, D., and Lin, D. (2012). Si solid-state quantum dot-based materials for tandem solar cells. *Nanoscale Res. Letter.* **7**, art.193.

Delachat, F., Carrada, M., Ferblantier, G. et al. (2009). Spectrocscopic ellipsometry analysis of silicon-rich silicon nitride layers for photovoltaic applications. In: *Proceeding of 24th European Photovoltaic Solar Energy Conference*, September 21–25, Hamburg, Germany, pp. 520–523.

Derbali, L. and Ezzaouia, H. (2012). Phosphorus diffusion gettering process of multicrystalline silicon using a sacrificial porous silicon layer. *Nanoscale Res. Lett.* **7**, 338.

Derbali, L., Dimassi, W., and Ezzaouia, H. (2011). Improvement of the minority carrier mobility in low-quality multicrystalline silicon using a porous silicon-based gettering under an O$_2$ atmosphere. *Energy Procedia* **10**, 243–248.

Dubey, R.S. (2013). Electrochemical fabrication of porous silicon structures for solar cells. *Nanosci. Nanoeng.* **1**(1), 36–40.

Duerinckx, F. and Szlufcik, J. (2002). Defect passivation of industrial multicrystalline solar cells based on PECVD silicon nitride. *Solar Energy Mater. Solar Cells* **72**, 231–246.

Duerinckx, F., Kuzma, F.I., Van Nieuwenhuysen, K. et al. (2006). Optical path length enhancement for >13% screenprinted thin film silicon solar cells. In: *Proceeding of 21st European Photovoltaic Solar Energy Conference*, September 4–8, Dresden, Germany, pp. 726–729.

Dzhafarov, T. (2014). Porous silicon and solar sells. In: Canham L. (Ed.) *Handbook of Porous Silicon*. Springer, Switzerland. doi: 10.1007/978-3-319-04508-5_95-1.

Dzhafarov, T.D., Aslanov, S.S., Ragimov, S.H., Sadigov, M.S., and Aydin Yuksel, S. (2012). Effect of nano-porous silicon coating on silicon solar cell performance. *Vacuum* **86**, 1875–1879.

Efremov, A.A., Klyui, N.I., Litovchenko, V.G., Popov, V.G., Romanyuk, A.B., and Romanyuk, B.N. (2000). Development of gettering process for the preparation of the solar silicon material. *Opto-Electron. Rev.* **8**, 410–413.

European Photovoltaic Industry Association (EPIA). (2014). Global market outlook for photovoltaics 2014–2018, Brussels, Belgium, http://www.epia.org.

Evtukh, A.A., Litovchenko, V.G., Oberemok, A.S. et al. (2001). Investigations of impurity gettering in multi-crystalline silicon. *Semicond. Phys. Quantum Electron. Optoelectron.* **4**, 278–282.

Fang, H., Li, X., Song, S., Xu, Y., and Zhu, J. (2008). Fabrication of slantingly-aligned silicon nanowire arrays for solar cell applications. *Nanotechnol.* **19**, 255703.

Feldrapp, K., Horbelt, R., Auer, R., and Brendel, R. (2003). Thin-film (25.5 μm) solar cells from layer transfer using porous silicon with 32.7 mA/cm^2 short-circuit current, *Progress in Photovoltaics* **11**, 105–112.

Gamboa, R., Martins, M., Serra, J.M. et al. (1998). First solar cells on electrochemically textured macroporous silicon, In: *Proceeding of 2nd World Conf. and Exhibition on Photovoltaic Solar Energy Conversion*, July 6–10, Vienna, Austria, pp. 1669–1672.

Green, M.A. (1992). *Solar Cells: Operating Principles, Technology and System Applications.* Prentice Hall, Englewood Cliffs, Kensington.

Green, M.A. (2000). Potential for low dimensional structures in photovoltaics. *Mat. Sci. Eng. B.* **74**, 18–124.

Green, M. (2003). *Third Generation Photovoltaics (Advanced Solar Energy Conversion).* Springer, Berlin.

Green, M.A. and Keevers, M.J. (1995). Optical properties of intrinsic silicon at 300K. *Prog. Photovoltaics* **3**, 189–192.

Green, M., Emery, K., Hishikawa, Y., and Warta, W. (2011). Solar cell efficiency tables (version 37). *Prog. Photovolt: Res. Appl.* **19**, 84–92.

Grigoras, K., Härkönen, J., Jasutis, V. et al. (1998). Porous silicon emitter formation from spin-on glasses, In: *Proceeding of 2nd World Conf. and Exhibition on Photovoltaic Solar Energy Conversion*, July 6–10, Vienna, Austria, pp. 1717–1720.

Haddadi, I., Dimassi, W., Bousbih, R., Hajji, M., Ali Kanzari, M., and Ezzaouia, H. (2011). Improvement of solar cells performances by surface passivation using porous silicon chemically treated with LiBr solution. *Phys. Stat. Sol. C* **8**(3), 755–758.

Halm, A., Jourdan, J., Nichol, S., Ryningen, B., Tathgar, H., and Kopecek, R. (2010). Detailed study on large area 100% SOG Silicon MC solar cells with efficiencies exceeding 16%. In: *Proceeding of 25th European Photovoltaic Solar Energy Conf.*, September 6–10, Valencia, Spain, pp. 1210–1215.

Hassen, M., Hajji, M., Ben Jaballah, A. et al. (2004). Improvement of minority carrier diffusion length in solar grade silicon: A new method. In: *Proceeding of 19th European Photovoltaic Solar Energy Conf.*, June 7–11, Paris, France, pp. 766–768.

Hochbaum, A.I. and Yang, P. (2010). Semiconductor nanowires for energy conversion. *Chem. Rev.* **110**, 527–546.

Horbelt, R., Terheiden, B., Auer, R., and Brendel, R. (2005). Manifold use of growth substrate in the porous silicon—Layer transfer process. In: *Record 31st IEEE Photovoltaic Specialist Conf.*, January 3–7, Orlando, FL, pp. 1193–1196.

Hsueh, T.-J., Chen, H.-Y., Tsai, T.-Y. et al. (2011). A microridge-like structured c-Si solar cell prepared by reactive ion etching. *J. Electrochem. Soc.* **158**(1), H35–H37.

Ivanov, I.I., Nychyporuk, T., Skryshevsky, V.A., and Lemiti, M. (2009). Thin silicon solar cells with SiO$_x$/SiN$_x$ Bragg mirror rear surface reflector. *Semicond. Phys. Quantum Electron. Optoelectron.* **12**, 406–411.

Ivanov, I.I., Skryshevsky, V.A., Nychyporuk, T. et al. (2013). Porous silicon Bragg mirrors on single- and multi-crystalline silicon for solar cells. *Renewable Energy* **55**, 79–84.

Jin, S., Bender, H., Stalmans, L. et al. (2000). Transmission electron microscopy investigation of the crystallographic quality of silicon films grown epitaxially on porous silicon. *J. Crystal Growth.* **212**, 119–127.

Jung, J.-Y., Guo, Z., Jee, S.-W., Um, H.-D., Park, K.-T., and Lee, J.-H. (2010). A strong antireflective solar cell prepared by tapering silicon nanowires. *Opt. Express.* **18**, A286–A292.

Kayes, B.M., Atwater, H.A., and Lewis, N.S. (2005). Comparison of the device physics principles of planar and radial *p-n* junction nanorod solar cells. *J. Appl. Phys.* **97**, 114302.

Kelzenberg, M.D., Turner-Evans, D.B., Kayes, B.M. et al. (2008). Photovoltaic measurements in single-nanowire silicon solar cells. *Nano Lett.* **8**, 710–714.

Khedher, N., Hajji, M., Bouaïcha, M. et al. (2002). Improvement of transport parameters in solar grade monocrystalline silicon by application of a sacrificial porous silicon layer. *Solid State Comm.* **123**, 7–10.

Krotkus, A., Grigoras, K., Pacebutas, V. et al. (1997). Efficiency improvement by porous silicon coating of multicrystalline solar cells. *Solar Energy Mater. Solar Cells* **45**, 267–273.

Kuthi, E.B. (2004). Crystalline silicon solar cells with selective emitter and the self-doping contact. *Híradástechnika* **59**, 21–31.

Kuzma-Filipek, I., Duerinckx, F., Van Nieuwenhuysen, K., Baucarne, G., Poortmans, J., and Mertens, R. (2007). Porous silicon as an internal reflector in thin epitaxial solar cells. *Phys. Stat. Sol. (a)* **204**, 1340–1345.

Kuzma-Filipek, I.J., Duerinckx, F., Van Kerschaver, E., Van Nieuwenhuysen, K., Beaucarne, G., and Poortmans, J. (2008). Chirped porous silicon reflectors for thin-film epitaxial silicon solar cells. *J. Appl. Phys.* **104**, 073529.

Kuzma-Filipek, I., Recaman-Payo, M., Van Nieuwenhuysen, K. et al. (2009). Rear junction epitaxial thin film solar cells with diffused front surface field and porous Si back reflectors. In: *Proceeding of 22nd European Photovoltaic Solar Energy Conf.*, September 21–25, Hamburg, Germany, pp. 2584–2588.

Lévy-Clément, C., Lust, S., Bastide, S., Le Quang, N., and Sarti, D. (2003). Macropore formation on p-type multicrystalline Si and solar cells. *Phys. Stat. Sol. A.* **197**(1), 27–33.

Lian, S.S., Kammel, R., and Kheiri, M.J. (1992). Preliminary study of hydrometallurgical refining of MG-silicon with attrition grinding. *Solar Energy Mater. Solar Cells* **26**, 269–276.

Lipinski, M., Panek, P., Bielanska, E., Weglowska, J., and Czternastek, H. (2000). Influence of porous silicon on parameters of silicon solar cells. *Optoelectron. Rev.* **8**(4), 418–420.

Lipinski, M., Bastide, S., Panek, P., and Lévy-Clément, C. (2003). Porous Si antireflection coating by electrochemical and chemical etching for Si solar cells manufacturing. *Phys. Stat. Sol. A.* **197**, 512–517.

Litovchenko, V.G. and Kluyi, N.I. (2001). Solar cells based on DLC film-Si structures for space application. *Solar Energy Mater. Solar Cells* **68**, 55–70.

Luque, A. and Marti, A. (1997). Entropy production in photovoltaic conversion. *Phys. Rev. B* **55**, 6994–6999.

Marrero, N., Guerrero-Lemus, R., Gonzalez-Diaz, B., and Borchert, D. (2008). Effect of porous silicon stain etched on large area alkaline textured crystalline silicon solar cells. *Thin Solid Films* **517**, 2648–2650.

Martinuzzi, S., Palais, O., Pasqunelli, M., and Farrazza, F. (2005). N-type multicrystalline silicon wafers and rear junction solar cells. *Eur. Phys. J. Appl. Phys.* **32**, 187–192.

Matic, Z., Bilyalov, R.R., and Poortmans, J. (2000). Firing through porous silicon antireflection coating for silicon solar cells. *Phys. Stat. Sol. A.* **182**, 457–460.

Menna, P., Tsuo, Y.S., Al-Jassim, M.M., Asher, S.E., Matson, R., and Ciszek, T.F. (1998). Purification of metallurgical grade silicon by porous silicon etching. In: *Proceeding of 2nd World Conf. and Exhibition Photovoltaic energy Conversion*, July 6–10, Vienna, Austria, pp. 1232–1235.

Mizsei, J., Shrair, J.A., and Zolomy, I. (2004). Investigation of Fermi-level pinning at silicon/porous-silicon interface by vibrating capacitor and surface photovoltage measurements. *Appl. Surf. Sci.* **235**, 376–388.

Moon, I., Kim, K., Thamilselvan, M. et al. (2009). Selective emitter using porous silicon for crystalline silicon solar cells. *Solar Energy Mater. Solar Cells* **93**, 846–850.

Nagel, H., Aberle, A.G., and Hezel, R. (1999). Optimised antireflection coatings for planar silicon solar cells using remote PECVD silicon nitride and porous silicon dioxide. *Prog. Photovolt: Res. Appl.* **7**, 245–260.

Nichiporuk, O., Kaminski, A., Lemiti, M., Fave, A., and Skryshevsky, V. (2005). Optimisation of interdigitated back contacts solar cells by two-dimensional numerical simulation. *Solar Energy Mater. Solar Cells* **86**, 517–526.

Nichiporuk, O., Kaminski, A., Lemiti, M., Fave, A., Litvinenko, S., and Skryshevsky, V. (2006). Passivation of the surface of rear contact solar cells by porous silicon. *Thin Solid Films* **511–512**, 248–251.

Nychyporuk, T., Marty, O., Rezgui, B., Sibai, A., Lemiti, M., and Bremond, G. (2008). Towards the 3rd generation photovoltaic: Absorption properties of silicon nanocrystals embedded in silicon nitride matrix. In: *Proceeding of 23rd European Photovoltaic Solar Energy Conf.*, September 1–5, Valencia, Spain, pp. 491–494.

Osorio, E., Urteaga, R., Acquaroli, L.N., Garcia-Salgado, G., Juarez, H., and Koropecki, R.R. (2011). Optimization of porous silicon multilayer as antireflection coatings for solar cells. *Solar Energy Mater. Solar Cells* **95**, 3069–3073.

Peng, K.Q., Xu, Y., Wu, Y., Yan, Y., Lee, S.-T., and Zhu, J. (2005). Aligned single-crystalline Si nanowire arrays for photovoltaic applications. *Small* **1**, 1062–1067.

Petermann, J.H., Ohrdes, T., Altermatt, P.P., Eidelloth, S., and Brendel, R. (2012a). 19% Efficient thin-film crystalline silicon solar cells from layer transfer using porous silicon: A loss analysis by means of three-dimensional simulations. *IEEE Trans. Electron Dev.* **59**(4), 909–917.

Petermann, J.H., Zielke, D., Schmidt, J., Haase, F., Garralaga Rojas, E., and Brendel, R. (2012b). 19%-efficient and 43 µm-thick crystalline Si solar cell from layer transfer using porous silicon. *Progr. Photovol. Res. App.* **20**(1), 1–5.

Prasad, A., Balakrishnan, S., Jain, S.K., and Jain, G.C. (1982). Porous silicon oxide anti-reflection coating for solar cells. *J. Electrochem. Soc.* **129**, 596–599.

Ramizy, A., Hassan, Z., Omar, K., Al-Douri, Y., and Mahdi, M.A. (2011). New optical features to enhance solar cell performance based on porous silicon surfaces. *Appl. Surf. Sci.* **257**, 6112–6117.

Rinke, T.J., Bergmann, R.B., Brugemann, R., and Werner, J.H. (1999). Ultrathin quasi-monocrystalline silicon films for electronic devices. *Solid State Phenomena* **67–68**, 229–236.

Salman, K.A., Omar, K., and Hassan, Z. (2012). Effective conversion efficiency enhancement of solar cell using ZnO/PS antireflection coating layers. *Solar Energy.* **86**, 541–547.

Schirone, L., Sotgiu, G., Montecchi, M., Righini, G., and Zanoni, R. (1998). Stain etched porous silicon technology for large area solar cells. In: *Proceedings of 2nd World Conf. and Exhibition on Photovoltaic Solar Energy Conversion*, July 6–10, Vienna, Austria, pp. 276–279.

Schnell, M., Lüdemann, R., and Schaefer, S. (2000). Stain etched porous silicon—A simple method for the simultaneous formation of selective emitter and ARC. In: *Proceedings of 16th European Photovoltaic Solar Energy Conf.*, May 1–5, Glasgow, UK, pp. 1482–1485.

Shieh, S.Y. and Evans, J.W. (1993). Some observations of the effect of porous silicon on oxidation induced stacking faults. *J. Electrochem. Soc.* **140**, 1094–1096.

Sivakov, V., Andrä, G., Gawlik, A. et al. (2009). Silicon nanowire-based solar cells on glass: Synthesis, optical properties, and cell parameters. *Nano Lett.* **9**(4), 1549–1554.

Skryshevsky, V.A. and Laugier, A. (1999). Improved thin film solar cell with Rayleigh's scattering in porous silicon pipes. *Thin Solid Films.* **346**, 254–258.

Skryshevsky, V.A., Laugier, A., Vikulov, V.A., Strikha, V.I., and Kaminski, A. (1996a). Effect of porous silicon layer re-emission on silicon solar cell photocurrent. In: *Proceedings of 25th IEEE Photovoltaic Spec. Conf.*, May, Washington, pp. 589–592.

Skryshevsky, V.A., Laugier, A., Strikha, V.I., and Vikulov, V.A. (1996b). Evaluation of quantum efficiency of porous silicon photoluminescence. *Mat. Sci. Eng. B* **40**, 54–57.

Skryshevsky, V.A., Kilchitskaya, S.S., Kilchitskaya, T.S. et al. (2000). Impact of recombination and optical parameters on silicon solar cell with the selective porous silicon antireflection coating. In: *Proceedings of 16th European Photovoltaic Solar Energy Conf.*, May 1–5, Glasgow, UK, pp. 1634–1637.

Smestad, G., Kunst, M., and Vial, C. (1992). Photovoltaic response in electrochemically prepared photoluminescent porous silicon, *Solar Energy Mater. Solar Cells* **26**, 277–283.

Solanki, C.S., Bilyalov, R.R., Poortmans, J., and Nijs, J. (2002). Transfer of a thin silicon film on to a ceramic substrate. *Thin Solid Films* **403–404**, 34–38.

Solanki, C.S., Bilyalov, R.R., Poortmans, J., Nijs, J., and Mertens, R. (2004). Porous silicon layer transfer processes for solar cells. *Solar Energy Mater. Solar Cells* **83**, 101–113.

Springer, J., Poruba, A., Fejfar, A., and Vanecek, M. (2000). Nanotextured thin film silicon solar cells: Optical model. In: *Proceedings of 16th European Photovoltaic Solar Energy Conf.*, May 1–5, Glasgow, UK, pp. 434–437.

Srivastava, S.K., Kumar, D., Singh, P.K., Kar, M., Kumar, V., and Husain, M. (2010). Excellent antireflection properties of vertical silicon nanowire arrays. *Solar Energy Mater. Solar Cells* **94**, 1506–1511.

Stalmans, L., Laureys, W., Said, K. et al. (1997). Effect of a porous silicon surface layer on the on the internal quantum efficiency of crystalline silicon solar cells. In: *Proceedings of 14th European Photovoltaic Solar Energy Conf.*, June 30–July 4, Barcelona, Spain, pp. 2484–2487.

Stalmans, L., Poortmans, J., Bender, H. et al. (1998). Porous silicon in crystalline silicon solar cells: A review and the effect on the internal quantum efficiency. *Prog. Photovolt. Res. Appl.* **6**, 233–246.

Stelzner, T., Pietsch, M., Andra, G., Falk, F., Ose, E., and Christiansen, S. (2008). Silicon nanowire-based solar cells. *Nanotechnol.* **19**, art. 295203.

Strehlke, S., Sarti, D., Krotkus, A. et al. (1997). Porous Si emitter and high efficiency multicrystalline silicon solar cells. In: *Proceedings of 14th European Photovoltaic Solar Energy Conf.*, June 30–July 4, Barcelona, Spain, pp. 2480–2483.

Striemer, C. and Fauchet, P.M. (2003). Dynamic etching of silicon for solar cell applications. *Phys. Stat. Sol. A.* **197**, 502–506.

Sze, S.M. (1981). *Semiconductor Devices.* John Wiley, New York.

Tena-Zaera, R., Bastide, S., and Lévy-Clément, C. (2007). Photoelectrochemical texturization of n-type multicrystalline Si. *Phys. Stat. Sol. A.* **204**, 1260–1265.

Thogersen, A., Selj, J.H., and Marstein, E.S. (2012). Oxidation effects on graded porous silicon anti-reflection coatings. *J. Electrochem. Soc.* **159**(5) D276–D281.

Tian, B., Zheng, X., Kempa, T.J. et al. (2007). Coaxial silicon nanowires as solar cells and nanoelectronic power sources. *Nature* **449**, 885–889.

Tsakalakos, L., Balch, J., Fronheiser, J., Korevaar, B.A., Sulima, O., and Rand, J. (2007). Silicon nanowire solar cells. *Appl. Phys. Lett.* **91**, 233117.

Tsuo, Y.S., Xiao, Y., Heben, M.J., Wu, X., Pern, F.J., and Deb, S.K. (1993). Potential applications of porous silicon in photovoltaics. In: *Record 23th IEEE Photovoltaic Specialists Conf.*, May, Louisville, KY pp. 287–393.

Tsuo, Y.S., Menna, P., Pitts, J.R. et al. (1996). Porous silicon gettering. In: *Record 25th IEEE Photovoltaic Specialist Conf.*, May, Washington, pp. 461–464.

Van Hoeymissen, J., Kuzma-Filipek, I., Van Nieuwenhuysen, K., Duerinckx, F., Baucarne, G., and Poortmans, J. (2008). Thin film epitaxial solar cells on low-cost Si substrates: Closing the efficiency gap with Si cells using advanced photonic structures and emitters. In: *Proc.23th European Photovoltaic Solar Energy Conf.*, September 1–5, Valencia, Spain, pp. 2037–2041.

Van Hoeymissen, J., Depauw, V., Kuzma-Filipek, I. et al. (2011). The use of porous silicon layers in thin-film silicon solar cells. *Phys. Stat. Sol. A* **208**(6), 1433–1439.

Werner, J.H., Wagner, T.A., Gemmer, C.E.M., Berge, C., Brendle, W., and Schubert, M.B. (2003). Recent progress on transfer-Si solar cells at IPE-Stuttgart. In: *Proceedings of 3rd World Conf. Photovoltaic Energy Conversion*, May 11–18, Osaka, Japan, pp. 1272–1275.

Wolf, S. and Tauber, R.N. (1986). *Silicon Processing for the VLSI Era*, Vol. 1. Lattice Press, Sunset Beach, CA.

Wong-Leung, J., Ascheron, C.E., Petravic, M., Elliman, R.G., and Williams, J.S. (1995). Gettering of copper to hydrogen-induced cavities in silicon. *Appl. Phys. Lett.* **66**, 1231–1233.

Yae, S., Tanaka, H., Kobayashi, T., Fukumuro, N., and Matsuda, H. (2005). Porous Si formation by HF chemical etching for antireflection of solar cells. *Phys. Stat. Sol. C.* **2**, 3476–3480.

Yerokhov, V.Y. and Melnyk, I.I. (1999). Porous silicon in solar cell structures: A review of achievements and modern directions of further use. *Renewable Sustainable Energy Rev.* **3**, 291–322.

Yerokhov, V.Y., Hezel, R., Lipinski, M. et al. (2002). Cost-effective methods of texturing for silicon solar cells. *Solar Energy Mater. Solar Cells* **72**(1–4), 291–298.

Yonehara, T., Sakaguchi, K., and Sato, N. (1994). Epitaxial layer transfer by bond and etch back of porous Si. *Appl. Phys. Lett.* **64**, 2108–2110.

Zacharias, M., Heitmann, J., Scholz, R., Kahler, U., Schmidt, M., and Bläsing, J. (2002). Size-controlled highly luminescent silicon nanocrystals: A SiO/SiO$_2$ superlattice approach. *Appl. Phys. Lett.* **80**, 661–663.

Zettner, J., Thoenissen, M., Hierl, Th., Brendel, R., and Schulz, M. (1998). Novel porous silicon backside light reflector for thin silicon solar cells. *Prog. Photovoltaics: Res. Appl.* **6**, 423–432.

Zhang, C., Liu, S., Wang, Y., and Chen, Y. (2010). Performance improvements of solar-grade crystalline silicon by continuously variable temperature phosphorus gettering process using a porous silicon layer. *Mater. Sci. Semicond. Process.* **13**, 209–213.

Zheng, J.P., Jiao, K.L., Shen, W.P., Anderson, W.A., and Kwok, H.S. (1992). Highly sensitive photodetector using porous silicon. *Appl. Phys. Lett.* **61**, 459–462.

PSi-Based Betavoltaics

Ghenadii Korotcenkov and Vladimir Brinzari

CONTENTS

11.1   Betavoltaics: A General View   240
11.2   Optimization via Porous Silicon   242
11.3   Radioisotopes for PSi-Based Microbatteries   245
Acknowledgments   247
References   247

11.1 BETAVOLTAICS: A GENERAL VIEW

Betavoltaic devices are one type of so-called radioisotope generator, which transform radioactive decay energy into electricity (Corliss and Harvey 1964; Olsen 1973; Linder and Reddy 2002; Duggirala et al. 2010). This type of device refers to a group of nonthermal devices, whose output power is not a function of a temperature difference between the source and the outside world. The betavoltaics effect is the creation of excess electron-hole pairs by impinging beta particles. Another group is formed by thermal devices in which the output power depends on the thermal power of the sources of ionizing radiation.

The main feature of radioisotope generators, which stimulated development for approximately one century, is their ability to produce electricity during years or even dozens of years depending on the half-life of the radioisotope. The second advantage of radioisotope generators is high energy density, which can be around ten times higher than hydrogen fuel cells, and a thousand times more than a chemical battery. In addition, radioisotope generators do not depend on environmental conditions. They function over a large range of temperature and pressure, and can work in space or under water. Radioisotope generators are autonomous, and so do not need remounting, refilling, or recharging.

Betavoltaics were invented over 50 years ago. They have been developed since the 1950s (Rappaport 1954; Garrett 1956). Betavoltaic devices are not nuclear reactors in the traditional sense. Unlike typical nuclear power generating devices, betavoltaic power cells do not rely on a nuclear reaction (fission/fusion) or chemical processes (as in most batteries) and do not produce radioactive waste products. The atomic nuclei (protons and neutrons) is not split apart or fused with other nuclei. Rather, this process takes advantage of beta (electron) emissions that occur when a neutron decays into a proton. The functioning of a betavoltaic device is somewhat similar to a solar panel, which converts photons into electric current. The surface in both devices has an identical configuration, but the voltage in betavoltaic devices comes from captured beta particles from radioactive isotopes. That is why betavoltaic devices are sometimes called radioactive batteries (see Figure 11.1). Internally, the impact of the beta electron on the *p-n* junction material causes a forward bias in the semiconductor.

The I-V curve for the circuit model of a betavoltaic microbattery with a load R can be derived as

$$(11.1) \qquad I = I_\mathrm{p} - I_0 \left[\exp\left(\frac{eV}{kT} \right) - 1 \right]$$

and open circuit voltage V_oc as

$$(11.2) \qquad V_\mathrm{oc} = \frac{kT}{e} \ln\left(1 + \frac{I_\mathrm{p}}{I_0} \right)$$

where I_p is the current generated by radioisotope and I_0 is the leakage current of the *p-n* junction device.

One should note that, fundamentally, the betavoltaic effect is similar to the photovoltaic effect, but the development of nuclear microbatteries is more difficult than that of solar cells

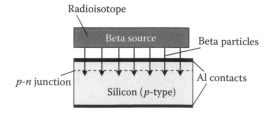

FIGURE 11.1   Schematic diagram of conventional betavoltaic microbattery.

(Guo et al. 2008). The main reason is that electron flux density in nuclear batteries is 10^2–10^4 lower than photon flux density in solar cells. For microbatteries, it becomes even worse because the utilization of very low activity results in further decreasing electron flux density for energy conversion. The actual incident current is also affected by loss due to the reflected electrons and secondary electrons when the emitted electrons from radioisotopes incident on the surface of the *p-n* junction device. Therefore, to get more power output, optimum design of the *p-n* junction device and microfabrication are needed to make electron-hole pairs (EHPs) be able to be collected into the depletion region as much as possible. In addition, from the above equations, one can see that a very small leakage current of *p-n* junction is required for large V_{oc} in the microbattery. To achieve a maximum power output, I_p/I_0 should be greater than 1000 (Guo et al. 2008). This requires reducing the leakage current as much as possible. For photovoltaic cells and high-power betavoltaic battery, the value of I_p is usually on the order of several mA to 100 mA. Therefore, the leakage current I_0 is not important and could be on the order of nA or even μA. However, for the microbattery using very low radioactivity (1–100 mCi), I_p is on the order of pA or nA. If the leakage current I_0 is on the order of nA or μA, there will be minimal power output. One should note that the decrease of the leakage current I_0 in *p-n* junctions formed in porous material is a great problem.

Although betavoltaics use a radioactive material as a power source, it is important to note that beta particles are low energy and easily stopped by shielding, as compared to the gamma rays generated by more dangerous radioactive materials. With proper device construction (i.e., shielding), a betavoltaic device would not emit any dangerous radiation. Leakage of the enclosed material would of course engender health risks, just as leakage of the materials in other types of batteries leads to significant health and environmental concerns.

These devices were initially designed to meet the high-voltage, high-current draw requirements of electrically powered space probes and satellites. Such power generators have been used in a few satellites, but never commercialized. More recently, however, as early as 1973, betavoltaics were suggested for use in low-voltage, long-term medical devices such as pacemakers. The appearance of a large number of micro-devices, requiring long-term power supply, such as micro-sensors or sensor networks for environmental monitoring, also stirs interest in this class of devices (Liu et al. 2008; Tin et al. 2008; Duggirala et al. 2010; Olsen et al. 2012). The batteries could also potentially power electrical circuits that protect military planes and missiles from tampering by destroying information stored in the systems, or by sending out a warning signal to a military center. The radioisotopes have the energy density 10^2–10^4 times greater than fossil or chemical fuels, and long lifetime nuclear microbatteries can be realized if a proper radioisotope is selected. In addition, in a betavoltaic cell, which is completely solid-state, we have direct nuclear-to-electric conversion without using any moving parts. Moreover, its operation does not require any additional energy sources. As it is known, conventional generators, including classical chemical batteries, do not satisfy these requirements. For example, micro-combustion based energy generation and micro fuel cells still require external microfluidics and external energy to drive the engine and to supply fuel into their working chamber, or to ignite the chemical reaction for energy conversion. Micro lithium batteries have also been investigated but can suffer from low energy density and lifetime. Micro solar cell arrays have also been explored but need light.

Regarding disadvantages of betavoltaic devices, which may limit the use of these devices, they include the following:

1. The highly energetic electrons tend to wear down or break apart the internal components (semiconductors) of the power cell. Therefore, betavoltaic devices suffer internal damage to their components because of the energetic electrons (Lei et al. 2014).
2. As the radioactive material emits, it slowly decreases in activity (refer to half-life). Thus, over time a betavoltaic device will output less and less power. However, this decrease occurs over a period of many years.

This means that in device design, one must account for what battery characteristics are required at end-of-life, and ensure that the beginning-of-life properties take into account the desired useable lifetime.

11.2 OPTIMIZATION VIA POROUS SILICON

A conventional betavoltaic cell is a planar silicon *p-n* junction shown in Figure 11.1. Silicon is used for this purpose because silicon has a relatively high-energy bandgap, high lattice damage threshold (~200–250 keV), and relatively low cost. In addition, Si is a well-known material, suitable for various microfabrication techniques designed during the last decade (Luo et al. 2010). SiC, GaN, and diamond also have attractive properties (see Figure 11.2). However, these materials have manufacturing problems that are more difficult.

However, experiment has shown that planar configuration of betavoltaic cells does not allow achieving high efficiency. The energy conversion efficiency of such batteries was low. First, one limitation of betavoltaic power cells is the reabsorption of electrons in the radioactive source itself. In order to reduce the self-absorption of beta energy, the radioactive isotope must be incorporated into the lattice of a semiconductor. Second, in a planar cell, the source is unidirectional, and efficiency losses are caused by emission of beta particles in the "wrong" direction. Moreover, with a planar battery, half the particles are going the "wrong way." In the third, this structure has a small surface area. Due to high energy of the electron, for complete separation and collection of the charge carriers generated by beta particles, the thickness of the space charge region in Si should be more than 100–200 μm. In contrast, the thickness of an average *p-n* junction is roughly 1 μm. Therefore, most of the beta radiation escapes with a planar junction.

Research carried out by Sun et al. (2005) has shown that the use of porous silicon (PSi) allowed resolving many of the above-mentioned problems of conventional betavoltaic batteries. 3D porous structure vastly increases the exposed surface area. This configuration is excellent for absorbing essentially all the kinetic energy of the source electrons. Instead of generating current by absorbing electrons at the outermost layer of a thin sheet, surfaces deep within these PSi wafers accommodate a much larger amount of incoming radiation. In tests carried out by Sun et al. (2005), nearly all electrons emitted during the tritium's beta decay were absorbed. A schematic diagram of the porous structure used in this experiment is shown in Figure 11.3. In this device, after silicon pores formed by electrochemical anodization, diffusion of dopant into the PSi was done to create the *p-n* junction over the entire surface of the device. The pore channels with diameter of ~1 μm were reasonably cylindrical with distinct small-scale variability along the pore walls. The testing of the *p-n* junction characteristics demonstrated that the porous diode behavior was comparable to that of the planar diode. Duggirala et al. (2007) have shown that dry etching also can be used for fabrication of high efficiency 3D silicon betavoltaics.

Experiments have shown that 3D design was allowed to increase efficiency of 8–10 times over planar structures by increasing the radioisotope-betavoltaic interface surface area, and a 1.5–2 times increase in the conversion efficiency realized by efficiently utilizing the β-electrons radiation emitted from both sides of the radioisotope source. According to Sun et al. (2005), the efficiency increased from 0.023% (planar) to 0.22% (porous) in terms of power per activity (W/Ci).

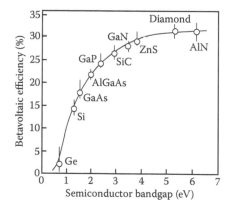

FIGURE 11.2 Dependence of limiting betavoltaic efficiency versus semiconductor bandgap. (Data extracted from Olsen L.C., In: *Processing of the 12th Space Photovoltaic Research and Technology Conference*, p. 256, 1993.)

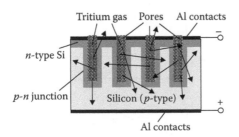

FIGURE 11.3 Schematic diagram of the betavoltaic microbattery where the silicon pores are used for storing gaseous tritium. The beta particles emitted by decaying tritium atoms are absorbed when they strike the *n*-type silicon, generating a voltage difference between the two layers. Arrows show the direction of the beta. (Idea from Sun W. et al., *Adv. Mater.* 17(10), 1230, 2005.)

This means that the resulting betavoltaics can enable smaller microbatteries with useful power output levels, using smaller quantities of radioisotope fuel for lower cost and radiation dose rate.

Guo et al. (2008) proposed and developed another type of betavoltaic microbattery using PSi. Schematically, the structure of this cell is shown in Figure 11.4. It is known that solid-source diffusion in PSi has not been thoroughly investigated. Therefore, instead of the formation of a PSi layer first, the planar *p-n* junction device was created by diffusing boron into the *n*-type silicon wafer, and then the electrochemistry method was used to form the PSi layer. Guo et al. (2008) believed that the pores in the scale of nanometer formed by using the electrochemistry method are more stable and easily controlled than the ones using the chemical method, and this microfabrication procedure can ensure the availability of the desired devices with improved parameters. However, in reality the improvement of a microbattery's parameters was very small. Typical I-V characteristics of planar and porous *p-n* junctions are shown in Figure 11.5.

Guo et al. (2008) also believe that the power output and conversion efficiency can be increased via optimization of pore size. It is known that the energy bandgap of the PSi can be greatly affected by the diameter of pores, and if the diameter of the pores is in the scale of several nanometers or even smaller, the quantum effect may dominate and energy bandgap may increase sharply; if the diameter of the pores is in the scale of several tens of nanometers, the effect of energy bandgap increased may not exist but the pores can be regarded as the macroscopic wells to trap the electrons emitted from the radioisotope, greatly reducing their bounce-back. Bandgap influence on the betavoltaic efficiency is shown in Figure 11.2.

With regard to our opinion, then, from our point of view the formation of macroporous silicon with the following formation of the *p-n* junctions is the most promising approach to the development of PSi-based betavoltaic microbattery. This configuration provides the best quality of *p-n* junctions that is extremely necessary for reducing reverse (leakage) current and increasing the output voltage, and enabling the formation on the pore walls of coatings containing radioisotopes (Figure 11.6). This configuration also allows fabricating isotope sources separately from the PSi-based matrix and conducting the aggregation of the battery on the final stage (see Figure 11.7). In

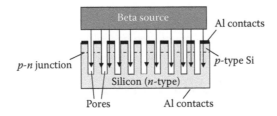

FIGURE 11.4 Schematic of PSi-based betavoltaic microbattery designed by Guo et al. (2008). The junction depth is 2 μm, and PSi layer is 3 μm deep, penetrating through the depletion region of the *p-n* junction. Aluminum is sputtered on both sides of the chip. (Idea from Guo H. et al., In: *Proceedings of 9th International Conference on Solid-State and Integrated-Circuit Technology, ICSICT 2008*, p. 2365, 2008.)

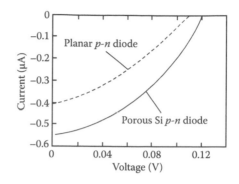

FIGURE 11.5   Tested result of betavoltaic microbattery using PSi and Pm-147. It is noted that the silicon pores were not uniform in size. Most of them were of 10~20 nm in diameter, but some of them were around 100 nm, and for a silicon pore, it had many tiny branches. (Data extracted from Guo H. et al., In: *Proceedings of IEEE 20th International Conference on Micro Electro Mechanical Systems, MEMS 2007, 867, 2007.*)

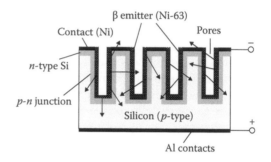

FIGURE 11.6   PSi-based betavoltaic battery with a *p-n* junction and solid metal radioisotope on pores. Arrows show the direction of the beta.

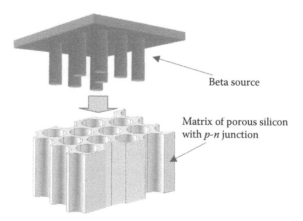

FIGURE 11.7   Variant of PSi-based betavoltaic battery with polymer beta source. (Idea from Miley G.H. and Lou N., In: *Proceedings of 9th Annual International Energy Conversion Engineering Conference,* AIAA 2011-5981, 2011.)

addition, a PSi betavoltaic that uses a solid metal radioisotope rather than gaseous tritium should have increased power and efficiency.

Development of stacked multi-*p-n* junction structures proposed by Luo et al. (2010) and Miley and Lou (2011) is also a promising approach to design of microbatteries especially when high-energy sources are used. Variants of such multilayered structures are shown in Figure 11.8. It is clear that it is difficult to materialize this configuration: many steps in silicon processing, relatively low yields, and so on. However, the increase in the number of layers improves strongly the efficiency of the use of such high-energy sources. The additional benefit of the stacked structure

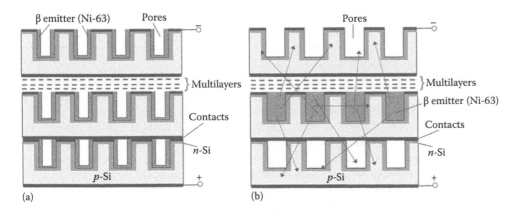

FIGURE 11.8   Variants of a stacked multi-*p-n* junction structure: (a) the assembly of identical structures; (b) the assembly of structures with one beta source. Arrows show the direction of the beta.

is higher voltages that match practical electronics. A single betavoltaic *p-n* junction typically gives 100 mV at normal load. A 30-layer stack should raise this up to 3 V, obviating the need for additional voltage step-up circuits.

11.3 RADIOISOTOPES FOR PSi-BASED MICROBATTERIES

Selection of a radiation source is a critical aspect for nuclear microbatteries (Olsen 1993; Guo et al. 2007, 2008; Yakubova 2010). It is based on the types of radiation, safety, energy, specific activity, cost, and half-life. Safety is always a great concern when using radioisotopes. Gamma ray has strong penetration ability and requires considerable external shielding to reduce the radiation dose rate. Alpha particles can be used to create electron-hole pairs in semiconductor, but they cause severe damage to the lattice. Pure beta emitters are best suited for nuclear microbatteries. Therefore, the choice of radiation source should take into account the required time of operation (long lifetime), the possibility of obtaining the required power (high specific power), and the absence of high-energy radiation, which can generate radiation defects in the semiconductor (below radiation damage threshold of silicon, 200–250 keV). In other words, the radiation source should be beta emitters (keV of energy) without alpha emission (MeV of energy) and minimum gamma emission. Table 11.1 shows the pure beta source candidates that could be used for this application.

TABLE 11.1   **Radioisotopes Which Can Be Used as Beta Source in PSi-Based Microbatteries**

Radioisotope	Specific Activity (g/mCi)	Half-Life Time	Average (Maximum) Energy (keV)
Tritium-3	1.03×10^{-7}	12.3 years	5.7 (18.6)
Nickel-63	1.8×10^{-5}	100.2 years	17.4 (67)
Promethium-147	1.06×10^{-6}	2.6 years	62 (250)
Stronium-90	7.25×10^{-6}	28.8 years	195.8 (900)
Krypton-85	2.56×10^{-6}	10.8 years	251.6 (670)
Ruthium-106	3.03×10^{-7}	1.06 years	93 (400)
Calcium-45	5.06×10^{-8}	162 days	77 (256)
Sulfur-35	2.4×10^{-8}	87.2 days	49 (167)

Source: Data extracted from Guo et al., In: *Proceedings of 9th International Conference on Solid-State and Integrated-Circuit Technology, ICSICT 2008*, p. 2365, 2008.

For safety reasons, it is attractive to use tritium (T) as a low energy beta particle emitter (with 5.7 keV average energy). Very thin materials, such as a sheet of paper, can shield these beta particles (electrons). However, low energy of beta electrons then limits considerably the power density of such a device. The majority of the kinetic energy is absorbed within ~200 nm. In addition, we have low activity and thus low output power due to low-density gas. However, such a low power is compensated by the relatively long T half-life time 12.3 years. Furthermore, T is industrially available, relatively cheap, and can be embedded, if needed, in a solid matrix (Liu et al. 2008; Lee et al. 2009). For example, tritium can be incorporated in hydrogenated amorphous materials, in particular in Si (Kosteski et al. 1998). Tritiated amorphous silicon films are mechanically stable, free for flaking or blistering, with good adherence to the substrate, and may be simultaneously deposited onto both conducting and insulating substrates using a discharge in tritium plasma. The silicon layer sputtered in a tritium/argon ambient at temperatures below 300°C results in a tritiated amorphous silicon film with the tritium concentration being variable from 5 to 30% depending on deposition conditions (Kosteski 2001). Lee et al. (2009) have found that titanium tritide also is stable, provides the increase tritium density, and can be used in 3D p-n junction microbattaries. Lee et al. (2009) have shown that titanium tritide can also be used in microbatteries in the form of films or powders. Miley and Lou (2011) proposed another approach. They proposed to use tritium-based source in the form of polymer matrix. Liu et al. (2009) have shown that the increase of gas tritium pressure also provides the increase of the output power. For example, the change in pressure from 0.05 to 33 atm tritium was accompanied by an increase in an isotope power density from 4.35 µW/cm^3 to 2.87 mW/cm^3.

The another radioisotope Ni-63 is very attractive as its lifetime is over 100 years, which means for most practical applications it could be used forever (Luo et al. 2010). In particular, the radioisotope Ni-63 was first selected as the source for energy conversion. In addition, Ni-63 can be provided in a liquid solution of Ni63Cl mixed with HCl, which allows it to be poured into various micromachined 3D p-n junction devices. For example, it was demonstrationed that a metallic isotope nickel-63 could be deposited in a porous-like structure (Luo et al. 2010; Miley and Lou 2011). However, the kinetic energy carried by the electrons emitted from Ni-63 is low, with the average energy of 16.7 keV and the highest energy of 67 keV, which limits the power of the batteries developed (Guo et al. 2008). On the other hand, electrons with the maximum kinetic energy of 66.7 keV cannot penetrate the outer layer of human skin, which guarantees the safety of people in operation. Therefore, in many cases, radioisotope Pm-147 is used as a radioactive source. The average kinetic energy of electrons emitted from Pm-147 is 62 keV, and the highest energy is 250 keV (Guo et al. 2008; Luo et al. 2010). This provides at least a threefold increase in short-circuit current I_s or I_p. The ratio of (I_p/I_0) is also greatly increased, which leads to the increase of open circuit voltage as well, according to Equation 11.2. Thus, the utilization of Pm-147 will greatly increase power output of the microbattery, compared with the one using Ni-63. At the same time, this energy is safe for both silicon devices and the people handling it. Thus, a p-n junction device to avoid the crystal damage caused by the energetically charged electrons does not need to be specially designed and fabricated. Although the half-life time of Pm-147 is 2.7 years, it is still very long compared with other technologies developed for micropower generation. For example, micro fuel cells can only work for several hours. According to Luo et al. (2010) and Miley and Lou (2011), strontium-90 also has high potential for application in betavoltaic microbatteries. The use of Sr-90 for radioactive power could serve the dual purpose of recycling a "waste" while producing an extremely valuable power supply. However, the maximum energy of beta electrons (546 keV) exceeds the radiation damage threshold of silicon, which creates problems with the use of this material in long lifetime batteries.

One should note that potassium-40, molybdenum-100, and zinc-70 are also promising fuel sources for stimulated beta-decay (http://www.peswiki.com/index.php/PowerPedia:BetaVoltaic). These isotopes have a decay rate of many thousands of years. For this reason, they are not regulated by the US government as are many more energetically radioactive materials (those with short half-lives i.e., rapid decay rates). These isotopes have been found to have significant beta decay energy. Many of these are light metals and can be inexpensively plated.

ACKNOWLEDGMENTS

This work was supported by the Ministry of Science, ICT and Future Planning (MSIP) of the Republic of Korea, and partly by the Moldova Government under grant 15.817.02.29F and ASM-STCU project #5937.

REFERENCES

Corliss, W.R. and Harvey, D.J. (1964). *Radioisotopic Power Generation*. Prentice-Hall, Englewood Cliffs, NJ.

Duggirala, R., Tin, S., and Lal, A. (2007). 3D silicon betavoltaics microfabricated using a self-aligned process for 5 milliwatt/CC average, 5 year lifetime microbatteries. In: *Proceedings of 14th International Conference on Solid-State Sensors, Actuators and Microsystems, Transducers and Eurosensors 07*, pp. 279–282.

Duggirala, R., Lal, A., and Radhakrishnan, S. (2010). *Radioisotope Thin-Film Powered Microsystems*. Springer, New York.

Garrett, A.B. (1956). Nuclear batteries. *J. Chem. Education* **33**(9), 446–449.

Guo, H., Yang, H., and Zhang, Y. (2007). Betavoltaic microbatteries using porous silicon. In: *Proceedings of IEEE 20th International Conference on Micro Electro Mechanical Systems, MEMS 2007*, January 21–25, Kobe, Japan, pp. 867–870.

Guo, H., Li, H., Lal, A., and Blanchard, J. (2008). Nuclear microbatteries for micro and nano devices. In: *Proceedings of 9th International Conference on Solid-State and Integrated-Circuit Technology, ICSICT 2008*, October 20–23, Beijing, China, pp. 2365–2370.

Kosteski, T. (2001). Tritiated Amorphous Silicon Films and Devices. PhD thesis, University of Toronto.

Kosteski, T., Kherani, N.P., Gaspari, F., Zukotynski, S., and Shmayda, W.T. (1998). Tritiated amorphous silicon films and devices. *J. Vac. Sci. Technol. (a)* **16**, 893–896.

Lee, S.-K., Son, S.-H., Kim, K.-S. et al. (2009). Development of nuclear micro-battery with solid tritium source. *Appl. Radiat. Isot.* **67**, 1234–1238.

Lei, Y., Yang, Y., Liu, Y. et al. (2014). The radiation damage of crystalline silicon PN diode in tritium betavoltaic battery. *Appl. Radiar. Isot.* **90**, 165–169.

Linder, D. and Reddy, T.B. (2002). *Handbook of Batteries*, 3rd ed. McGraw-Hill, New York.

Liu, B., Chen, K.P., Kherani, N.P., Zukotynski, S., and Antoniazzi, A.B. (2008). Betavoltaics using scandium tritide and contact potential difference. *Appl. Phys. Lett.* **92**, 083511.

Liu, B., Chen, K.P., Kherani, N.P., and Zukotynski, S. (2009). Power-scaling performance of a three-dimentional tritium detanoltaic diode. *Appl. Phys. Lett.* **95**, 233112.

Luo, N., Ulmen, B., and Miley, G.H. (2010). Nanopore/multilayer isotope batteries using radioisotopes from nuclear wastes. In: *Proceedings of 8th Annual International Energy Conversation Emgineering Conference*, July 25–28, Nashvile, TN, AIAA 2010-7003.

Miley, G.H. and Lou, N. (2011). A nanopore multilayer isotope battery using radioisotopes from nuclear wastes. In: *Proceedings of 9th Annual International Energy Conversion Engineering Conference*, July 1–August 3, 2011, San Diego, CA, AIAA 2011-5981.

Olsen, L.C. (1973). Betavoltaic energy conversion. *Energy Conversion* **13**(4), 117–124.

Olsen, L.C. (1993). Review of betavoltaic energy conversion. In: *Processing of the 12th Space Photovoltaic Research and Technology Conference, SPRAT 12*, October 20–22, Cleveland, OH, pp. 256–267.

Olsen, L.C., Cabauy, P., and Elkind, B.J. (2012). Betavoltaic power sources. *Phys. Today* **65**(12), 35–38.

Rappaport, P. (1954). The electron-voltaic effect in *p-n* junctions induced by beta-particle bombardment. *Phys. Rev.* **93**(1), 246–247.

Sun, W., Kherani, N.P., Hirschman, K.D., Gadeken, L.L., and Fauchet, P.M. (2005). A three-dimensional porous silicon *p-n* diode for betavoltaics and photovoltaics. *Adv. Mater.* **17**(10), 1230–1233.

Tin, S., Duggirala, R., Polcawich, R., Dubey, M., and Lal, A. (2008). Self-powered discharge-based wireless transmitter, In: *Proceedings of the 21st IEEE International Conference on Micro Electro Mechanical Systems (MEMS), MEMS 2008*, January 13–17, Tucson, AZ, pp. 988–991.

Yakubova, G.N. (2010). Nuclear Batteries with Tritium and Promethium-147 Radioactive Sources. PhD Thesis, University of Illinois at Urbana-Champaign, IL, 2010.

Porous Silicon in Micro-Fuel Cells

12

Gael Gautier and Ghenadii Korotcenkov

CONTENTS

12.1 Introduction 250
12.2 General Introduction of Fuel Cells and Micro-Fuel Cells 250
12.3 General View on the Integration of Porous Silicon in Fuel Cells 253
12.4 PSi-Based Fuel Cells 255
    12.4.1 Integration of PSi in the Electrodes 255
        12.4.1.1 Mesoporous Silicon 256
        12.4.1.2 Macroporous Silicon 260
    12.4.2 Porous Silicon as a Proton Exchange Membrane 264
    12.4.3 Micro-Machined Silicon-Based Supports for Small Fuel Cells 264
12.5 Conclusion 269
Acknowledgment 269
References 269

12.1 INTRODUCTION

Energy consumption has increased dramatically during the past decades in order to functionalize electrical appliances, provide heat sources, generate light sources, and communicate in daily routine. With breakthroughs in technological developments, these electrical needs are intensified. All this requires an increase in the power of stationary generating stations. At the same time, the last decade has seen an explosion in the number of portable devices such as cell phones and notebooks, medical devices, military applications, microdevices, and microsystems, which integrate more complex and energy consuming functions like wireless communications enabling large data exchange off grid. These functions involve more power consumption even if a tremendous effort is done to lower application power needs. This means that in addition to the powerful stationary energy sources there is a great need for powerful portable energy sources. Research has shown that one of the most interesting and prospective approaches to resolving this problem is the generation of electrical energy from a chemical energy in fuel cells (Dyer 1990, 2002; Kundu et al. 2007), which have the following advantages: (1) better efficiency than combustion engines, (2) no moving parts, (3) quiet operation, (4) highly reliable, (5) long-lasting systems, and (6) no particular NO_x and SO_x emission. In addition, constructively, fuel cells have great potential for miniaturization, which is especially important for portable devices (Morse 2007). In addition, it is necessary to consider that the fuel cells are capable of providing much greater power density than existing energy sources, including lithium ion batteries (see Table 12.1). This means that the use of miniaturized fuel cells (FC) can provide continuous operation of portable devices, even with a small weight of energy source. Furthermore, recharging is instantaneous by replacement of the fuel cartridge.

12.2 GENERAL INTRODUCTION OF FUEL CELLS AND MICRO-FUEL CELLS

A fuel cell is a device that generates electricity by a chemical (electrochemical) reaction, usually a reaction of oxidation and reduction. Thus, the system needs a fuel to operate and can work as long as fuel is provided. Every fuel cell has two electrodes—one positive and one negative—called, respectively, the anode and the cathode (see Figure 12.1). The reactions that produce electricity take place at the electrodes. Every fuel cell also has an electrolyte, which carries electrically charged particles from one electrode to the other. In a majority of operations, catalysts (platinum [Pt], ruthenium [Ru], or palladium [Pd]) are also implemented to both electrodes to enhance a cell reaction.

Hydrogen is the basic fuel, but fuel cells also require oxygen. An example of H_2-O_2 fuel cells consists of two platinum electrodes dipped into sulfuric acid (an aqueous acid electrolyte). When hydrogen is introduced across an electrode, it is split into protons (H^+) and electrons. This reaction is referred to as a hydrogen oxidation, which occurs at the anode side where electrons are removed from a species. Consequently, the protons can flow through the electrolyte, which allows only protons to pass, and the electrons flow into the external electrical circuit through a piece of wire that connects to the opposite electrode. When an electron is at the other electrode,

TABLE 12.1   Energy Density Comparison of Various Fuels to Practical Battery Power Sources

Technology	Energy Density (W·h/l)	Energy Density (W·h/kg)
Primary batteries (non-rechargeable) Alkaline; Zn–air; Li/SOCl₂	330–1050	125–340
Secondary batteries (rechargeable) Lead acid; Ni–cad; NiMH; Li-ion; Li-polymer	70–350	35–200
Fuel cells (hydrocarbons) Methanol; Ethanol; Butane; Iso-octane	4384–8680	5600–12,600

Source: Data extracted from Morse (2007).

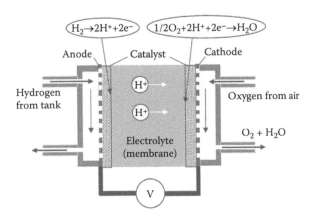

FIGURE 12.1 Schematics and working principle of air breathing hydrogen proton exchange membrane fuel cells (PEMFCs) with bubble inserts presenting the reactants, products, and the reactions at the anode and the cathode.

it is combined with protons and supplied oxygen to generate a product of water. This chemical reaction is called oxygen reduction at the cathode side where electrons are added to a species. The schematic diagram for a common fuel cell is shown in Figure 12.1.

Besides hydrogen, other gases such as natural gas, gasoline, methanol, and ethanol also can be used in fuel cells. Energy densities for indicated fuels are listed in Table 12.2. Several fuels can be used directly for fuel cell technologies, while other ones can be used indirectly only. The direct fuel cells are the cells in which a fuel is directly used to produce energy. The indirect fuel cells are the cells that implement a reformer to convert a chemical component into simple elements before feeding into the cell.

There are several kinds of fuel cells, and each operates a bit differently. The type of fuel cell depends mostly on the electrolyte used (Steele and Heinzel 2001). In general terms, all fuel cells operate on the same principles described above. Table 12.3 provides a useful summary of the different types of current fuel cell technologies and their most useful applications. Each type of fuel cell has advantages and drawbacks compared to the others (Morse 2007); however, none is yet cheap and efficient enough to widely replace traditional ways of generating power, such as coalfired, hydroelectric, or even nuclear power plants.

Regarding the possibility of miniaturization, the most suitable for this purpose are proton exchange membrane fuel cells (PEMFC) (Litster and McLean 2004; Barbir 2006; Morse 2007), capable of operating at room temperature. As a result, PEMFCs have been investigated in recent years most intensively, and, therefore, only this type of fuel cell will be analyzed in this chapter.

One should note that in addition to PEMFC, in the literature one can found mentions about direct hydrogen fuel cells (DHFC), reformed hydrogen fuel cells (RHFC), direct methanol fuel cell (DMFC), direct ethanol fuel cells (DEFC), and formic acid fuel cells (DFAFC), which also belong to the specified PEMFC-type of fuel cells. For a more extensive review focused on miniaturized fuel cells, one can see Zhao's book on fuel cells (Zhao 2009) and three recent reviews

TABLE 12.2   Fuel Energy Density

Storage Type	Volumetric Energy Density of Fuel	
	(W·h/l)	**(M·J/l)**
Natural gas	10	0.036
Hydrogen, liquid	2600	9.36
Hydrogen, gas at 700 bar	1555	5.6
Methanol	4390	15.80
Ethanol	6665	24.0
Gasoline	9700	34.92
Lithium ion	250	0.90

TABLE 12.3   Classification of Fuel Cells

Fuel Cell Type	The Fuel (Anode)	Mobile Ion	Operating T, °C	Applications and Notes
Direct methanol (DMFC)	Methanol, ethanol	H⁺	~80	Vehicles and mobile applications, and for lower power CHP systems; Suitable for portable electronic systems of low power, running for long times
Proton exchange membrane (PEMFC)	External reforming H_2	H⁺	20–100	
Phosphoric acid (PAFC)		H⁺	~220	Large numbers of 200-kW CHP systems in use
Alkaline (AFC)	H_2	OH⁻	50–200	Used in space vehicles, e.g., Apollo, Shuttle
Molten carbonate (MCFC)	Internal reforming H_2,	CO_3^{2-}	~650	Suitable for medium- to large scale CHP systems, up to MW capacity
Solid oxide (SOFC)	CO (natural gas, coal)	O^{2-}	500–1000	Suitable for all sizes of CHP systems, 2 kW to multi-MW

Source: Data extracted from Steele and Heinzel (2001) and Laramie and Dicks (2003).

Note: AFC: alkaline fuel cell, CHP: combined heat and power, MCFC: molten carbonate fuel cell, PAFC: phosphoric acid fuel cell, PEMFC: proton exchange membrane fuel cell, SOFC: solid oxide fuel cell.

on miniaturized fuel cells (Nguyen and Chan 2006; Kundu et al. 2007; Morse 2007). In addition, some focus reviews exist on DMFC (Kamarudin et al. 2007). Note that miniaturization work does exist on other types of fuel cells, in particular on SOFC, as mentioned by different authors (Jankowski et al. 2002; Baertsch et al. 2004; Chen et al. 2004; Srikar et al. 2004; Tang et al. 2005).

A schematic view of a fuel cell of PEMFC-type is given in Figure 12.2 illustrating the different necessary parts of the system (Morse 2007). The flow field is necessary for fuel delivery to the reaction sites and current collection. The gas diffusion layer allows a good distribution of the fuel on the catalyst layers. The membrane conducts the ions (protons) from the anode to the cathode when generated during the fuel cell operation. This membrane must also be impermeable to fuel and electrically isolating. A common electrolyte for PEMFCs is a thin polymer membrane that can transport proton as an ionic charge carrier. The thickness of the membranes is in the range of a few microns to a few hundred microns. The common materials for proton exchange membranes are sulfulnated polymers that will selectively allow only protons to pass through.

It is necessary also to keep in mind that the reaction at the anode involves three important steps: the access of gas to catalyst, the protonation and, simultaneously, the electron collection. Therefore, the fuel cell structure has to be optimized according to these steps, this involves at least that the catalyst, current collector, and proton exchange membrane are simultaneously in contact to perform the oxidation reaction. On the cathode side, electrons, oxygen, and protons

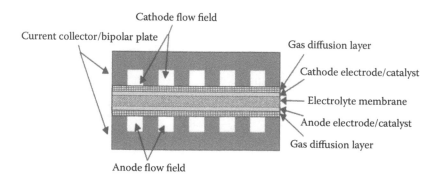

FIGURE 12.2   Fuel cell assembly with a solid electrolyte membrane with catalyst and gas diffusion layer on both sides and sandwiched by the two current collector bipolar plates acting also as the flow field. (From Morse J.D., *Int. J. Energy Res.* 31, 576, 2007. Copyright 2007: John Wiley & Sons. With permission.)

must be present simultaneously to allow reaction. For this reason, catalyst, proton exchange membrane, and current collector must also be simultaneously in contact. Moreover, at the cathode side, water is the product of the reaction and because this is a liquid, it has to be correctly removed to allow a continuous access of oxygen to the reaction sites.

12.3 GENERAL VIEW ON THE INTEGRATION OF POROUS SILICON IN FUEL CELLS

As it was shown in previous chapters, the use of the principles of MEMS (microelectromechanical systems) technology is the best way for miniaturization of any devices. Using this approach, many various silicon-based microsystems were fabricated. In this field, PSi has already been used to perform with success MEMS devices like airbag igniters, gas sensors, accelerometers, pressure sensors, microphones, photonics, and so on (Stewart and Buriak 2000; Armbruster et al. 2003; Lindroos et al. 2010). Information about these developments can be found in other chapters in this book. For the latter reason, and considering PSi properties, therefore, this material has been a natural candidate for miniaturization of fuel cells when starting from a silicon substrate (Morse 2007; Kundu et al. 2007). At first, infrastructure based on silicon wafers already exists due to the increasing demand for faster and cheaper microprocessors. Second, there are extensive researches in the area of silicon manufacturing and MEMS processing. Finally, conducted studies have generated hundreds of publications detailing PSi forming and its properties. As a result, there is flexibility in the parameters of PSi such as the porosity, thickness, and surface area of the membrane that affect performance of a fuel cell. For example, the possibility to tune the porosity can improve the fuel cell performance through better gas and liquid (water) management as reported by Tang et al. (2005) or demonstrated by Yao et al. (2006) with different TSV (through silicon via) diameters performed with deep reactive ion etching (DRIE). Moreover, with a wide variety of manufacturing options at hand, fabrication of PSi is simple and cost effective. This means that the low cost of PSi technology compared to DRIE can advantageously reduce the cost of the global system if industrial equipment (currently in development) can perform silicon etching with a good throughput. In addition, since silicon is resilient to most acids, the membrane is relatively stable in an environment used in fuel cells.

If one takes into account both the possibilities of PSi's technology, and the range of porosity, which can be achieved in PSi layers, then we come to the conclusion that PSi can be integrated in fuel cells in a different manner. In fact, it can be integrated as a hydrogen source in the gas feed system (Presting et al. 2004; Lysenko et al. 2005). For instance, Moghaddam et al. (2008) have carried out a self-regulating hydrogen generator for micro-fuel cells. Presting et al. (2004) performed a micro-sized reformer using PSi. However, this possibility is not developed in this chapter. This topic will be discussed in the following chapters of this book. In addition, PSi can be integrated in the core system (i.e., the MEA, membrane electrode assembly). PSi could be used like a drilled substrate to replace bipolar plates acting as the gas flow channel but can also act as the gas diffusion layer and catalyst support at the same time when the pore size is small enough. These ideas have been tested by different research groups in different ways, but also some concepts have been commercialized by startups like "NEAH power." Possible configurations of PSi-based fuel cells are shown in Figure 12.3.

Moreover, mesoporous silicon, after an efficient functionalization of its surface, can act as a proton exchange membrane (see Section 12.4.3). H_2 separating membranes also can be fabricated using technology of silicon porosification (Presting et al. 2004; Starkov et al. 2005). Figure 12.4 shows a SEM image of such a membrane. As it is known, after the dissociation of hydrocarbon fuels in a steam reactor, the product gases such as CO_2 and H_2 have to be separated. This is generally done with Pd or Pd-Ag foils, approximately 20 µm thick, because permeation through such membranes is only possible for the H-atoms, whereas any other atoms and molecules in the gas feed are held back. From this point, it is obvious that very thin membranes on a support layer can help to increase the permeation rate even at reduced permeation temperatures.

Of course, other materials such as metal, polymer, or ceramic can also be used for designing fuel cells. Metals have proven a good behavior concerning the conduction issue even though precision micromachining of these materials is sometimes complex and expensive (Hsieh et al.

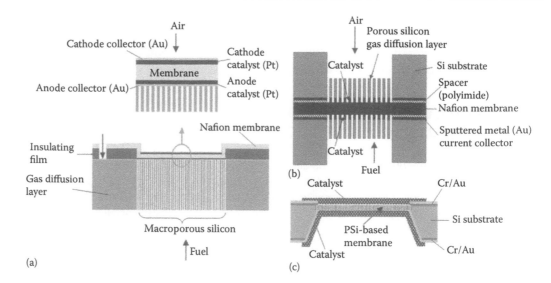

FIGURE 12.3   Schematic representation of PSi-based PEM micro-fuel cell. (a, b) PEMFC with Nafion proton exchange membrane. ((a) Idea from Desplobain (2009). (b) From Morse J.D., *Int. J. Energy Res.* 31, 576, 2007. Copyright 2007: John Wiley & Sons. With permission.) (c) PEMFC with PSi-based proton exchange membrane. (From Moghaddam S. et al. *Nature Nanotechnology* 5, 230, 2010. Copyright 2010: Macmillan Publishers. With permission.)

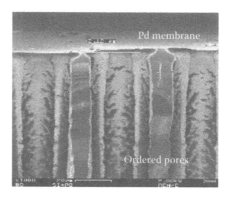

FIGURE 12.4   Cross-section SEM picture of randomly ordered pores capped with a thin Pd layer to act as H_2 separation membrane. The metal layer (Pd) deposited at the surface of the Si wafer acts as back contact for the subsequent anodic etching of the silicon substrate to produce a free standing Pd-membrane. (From Presting H. et al., *Mater. Sci. Eng. B* 108, 162, 2004. Copyright 2004: Elsevier. With permission.)

2006). Nevertheless, the most important drawback for metals is corrosion, which rapidly occurs at the operating temperature of PEMFCs (50 to 80°C). For the latter reason, the suitable metals must be noble or at least protected with a noble metal and are consequently expensive. When noble metal coating is envisaged to avoid corrosion, then metal supports are no longer required and other material resistant to corrosion and easier to shape with micromachining techniques like polymers can be considered. For the latter reasons, different polymers usually used in MEMS technology have been tested for small fuel cell prototyping in the literature like the thermoplastic polymer polymethyl methacrylate (PMMA) (Hsieh et al. 2004), the epoxy type negative photosensitive resin called SU8 (Hsieh et al. 2005), or polydimethylsiloxane (PDMS) (Shah et al. 2003). Even printed circuit boards (PCB), which are built up from a fiber glass epoxy composite (Schmitz et al. 2003), have been tested. These materials allow using many different techniques for the shaping process like photolithography, hot embossing, or laser machining. When polymers are used, different designs can be tested to perform good gas delivery and water management but for the conduction, a metal coating must be carried out. Therefore, this metal coating is the most important limiting factor for the cell performance because it must be conductive enough and continuous. Some prototypes involving ceramic and glass have also been carried out. Wainright et al. demonstrated the possibility to use ceramics with laser milled gas channels on an alumina wafer (Wainright et al. 2003). This substrate has been coated with inks for the different functional fuel cell layers (catalyst, PEM) to carry out a small fuel cell with jetting and spraying technologies. The peak power density was 2 mW/cm^2 with humidified hydrogen and air. Lee et al. (2002) also demonstrated the possibility to use glass as the substrate for fuel cells and, finally, Wang et al. (2006) presented a composite carbon black based fuel cell with a footprint of 5 cm^2. All these materials are interesting options for fuel cell miniaturization but when dealing with miniaturization of systems, the semiconductor industry experience on silicon has enabled large structuring possibility of this material down to the nanometer scale.

It should be particularly noted that it is hardly possible to find any other materials, such as carbon or ceramic, that can perform the functions such as a hard skeleton, gas diffusion layer (GDL), and the highly conductive electrode. PSi should also be stable at elevated temperatures, unlike many polymeric materials. For this reason, most of the prototypes have been performed on silicon.

Taking into account the functions that PSi can perform in a fuel cell, let us consider the properties required for these different parts of the fuel cell. The terms "supporting material" are used in this paper to underline the function of support ensured by the materials presented here. In fact, these terms will refer to the material covering one or more of the following functions: GDL (and sometimes catalyst support) or bipolar plate in the small fuel cell besides the function of the cell mechanical support. Requirements to ensure these functions in small size fuel cells are similar to large fuel cells. Barbir (2006) reported the required properties for gas diffusion layer and bipolar plate in the case of large fuel cells. These requirements will be the basis for material selection when working on miniaturization of fuel cells.

For the gas diffusion layer, the required properties are a high porosity to allow a flow of both reactant gases and produced water (note that these fluxes are in an opposite direction). It must be both electrically and thermally conductive. As we are dealing with thin layers, the interfacial or the contact resistance is typically more important than bulk conductivity (thermal and electrical). Since the catalyst layer is made of discrete small particles, the pores of the gas diffusion layer facing the catalyst layer must not be too large. Finally, it must be sufficiently rigid to support the "flimsy" membrane electrode assembly (MEA). In the state of the art of large fuel cell, these requirements are best met by carbon fiber based materials such as carbon-fiber papers and woven carbon fabrics or cloths. These diffusion media are generally made hydrophobic in order to avoid flooding in their bulk. Typically, both cathode and anode gas diffusion media are PTFE-treated to avoid flooding and to allow the gas diffusion (Mathias et al. 2003).

Concerning the bipolar plates, they ensure several functions in a large fuel cell stack. They must be electrically conductive, separate the gases in adjacent cells, provide structural support for the stack, and be thermally conductive. In addition, they must be corrosion resistant in the fuel cell environment. In fact, they are exposed to a very corrosive environment inside a PEM fuel cell (pH between 2 and 3, and temperature between 60 and 80°C). Nevertheless, they must not be made out of expensive materials. Finally, the manufacturing process must be suitable for mass production. Generally, two families of materials have been used for PEM fuel cell bipolar plates: graphite-composites and metals (Barbir et al. 1999).

12.4 PSi-BASED FUEL CELLS

12.4.1 INTEGRATION OF PSi IN THE ELECTRODES

Taking into account the structural features of mesoporous and macroporous silicon and specifics of their formation, a more detailed examination of the approaches used for the integration of PSi in fuel cells will be carried out for these materials separately. In particular, first we will present prototypes performed with mesoporous silicon and in second, prototypes performed with macroporous silicon. Structures mixing both mesoporous and macroporous silicon morphologies will be also discussed.

12.4.1.1 MESOPOROUS SILICON

Lucent technologies (Alcatel-Lucent, Inc. since 2006) (Meyers and Maynard 2002) proposed and patented in 2002 (Maynard and Meyers 2003) an innovative structure based on a standard bipolar plate with a mesoporous silicon layer performed on channels. This structure performed from p-type (Figure 12.5a) and p^+-type (Figure 12.5d) silicon can be used to manufacture a fuel cell in two ways. The first way is to use one single substrate in which side-by-side channels can act alternately as the anode on the cathode (Figure 12.5a). The second structure uses two stacked substrates, one for the anode placed on one side of the membrane (PEM) and the other for the cathode on the other side (Figure 12.5d). In the first case, the substrate is semi-insulating to avoid a short circuit between the anode and the cathode. In the second one, it is conducting to act as current collector. The channel performed with PSi on top is presented on the SEM image (Figure 12.5b). The channel and the porous layer on its top are obtained consecutively during etching in HF by changing the etching condition going from PSi formation (for the porous bridge) to electropolishing to form the channel. This is possible because the porous layer obtained is insulating and anchored on each side to the masking layer. Then, the electropolishing regime leads to the formation of the channel underneath without removal of the porous bridge due to its insulating nature. The authors have taken advantage of the work performed on MEMS structures at the University of Twente (Tjerkstra et al. 2009). The maximum peak performance obtained with H_2 and O_2 at 80 sccm is 300 mW for the global cell system corresponding to a power density of 62 mW/cm² (Figure 12.5c).

D'arrigo et al. (2003) presented an alternative way to perform the same structure on a silicon substrate with channels covered with a porous layer. The structure is obtained from a low-doped p-type silicon wafer with a particular <110> crystallographic orientation. At first, the substrate is etched using DRIE to perform trenches (see Figure 12.6a). Thereafter, an alkaline etching of the silicon permits producing channels with lozenge shape cross-sections presented in Figure 12.6b. This is due to the particular orientation of the silicon substrate. To seal the cavity, a chemical vapor deposition (CVD) treatment of $p+$ type silicon is carried out (Figure 12.6c). Finally, the silicon substrate is electrochemically etched producing a microporous silicon layer on the cavity due to $p+$ deposited silicon. The final result is presented in Figure 12.6d. Even if the process is complex, it allows very tiny structures to be carried out and therefore very small fuel cells are achieved

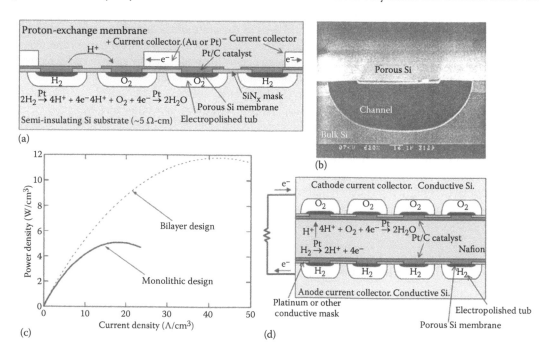

FIGURE 12.5 PSi on channels obtained by anodization: (a) and (d) present the two approaches evaluated by the authors, (b) presents a SEM image of the cross-section of the channel and (c) presents the power density performance of both prototypes versus the current output. (From Meyers J.P. and Maynard H.L., *J. Power Sources* 109, 76, 2002. Copyright 2002: Elsevier. With permission.)

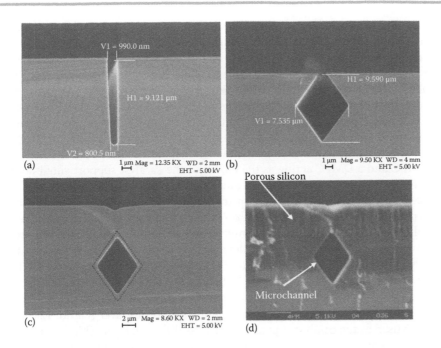

FIGURE 12.6   SEM images of PSi on a channel for fuel cells. First, DRIE is performed on a silicon <110> substrate to produce trenches (a), then alkaline etching permits producing a lozenge shape (b). Then, p^+-doped silicon is deposited using CVD (c). Finally, electrochemical etching is performed (d). (From D'arrigo G. et al., *Mater. Sci. Eng. C* 23, 3, 2003. Copyright 2003: Elsevier. With permission.)

with this process. While different variations have been patented (D'arrigo and Coffa 2006), to our knowledge, no micro-fuel cell prototype has been reported in the literature with this structure.

In both previous examples, PSi produced on channels is used as a gas exit toward the reaction site of the fuel cell. In the same way, considering the through silicon via approach, some authors have proposed using PSi as a gas diffusion layer and catalyst support on top of through silicon via or trenches. Therefore, this PSi layer leads to prototype improvement with the possibility to perform also the catalyst support with the silicon substrate for the fuel cell.

For example, Yamazaki (2004) proposed to perform a sealed channel using heterogeneous substrate with silicon and glass. The process is presented in Figure 12.7. First, silicon nitride deposited on the silicon substrate was patterned to open some regions (Figure 12.7a). These regions were anodized

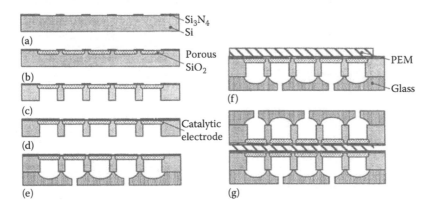

FIGURE 12.7   Process flow of fuel cell substrate with PSi performed on channels with heterogeneous substrates (silicon and glass) using anodic bonding. Fabrication process was proposed by Min et al. (2002). (a) Si_3N_4 forming and patterning, (b) anodization and thermal oxidation, (c) deep RIE, (d) catalytic electrode formation, (e) glass process and anodic bonding, (f) PEM attachment/coating, and (g) packaging and wiring. (From Yamazaki Y., *Electrochem. Acta* 50, 663, 2004. Copyright 2004: Elsevier. With permission.)

in an HF-ethanol mixture producing oriented mesoporous silicon (Figure 12.7b). These as-formed PSi were thermally oxidized to act as an etch stop during the DRIE step (Figure 12.7c). This process was performed on the wafer opposite side in areas situated in front of the porous regions. Then, a porous structure is performed on through silicon trenches on which the catalytic electrode was placed (Figure 12.7d). Finally, a glass substrate with adapted openings for gas inlets and outlets was anodically bonded on the silicon substrate silicon to seal the trenches (Figure 12.7e). This example shows that PSi performed on the sealed channels acting as a gas diffusion layer can also be used as the catalyst support.

Other examples taking advantage of the tiny pores of mesoporous silicon to support the catalyst exist. For instance, Hayase et al. (2004) and Liu et al. (2006a) proposed to use PSi as a buffer layer in a DHFC and a DMFC acting as the GDL and the catalyst support simultaneously. This design suppresses the extra catalyst support and diffusion layer usually necessary. Moreover, the gas can access membrane parts situated on the regions between the channels, thanks to the continuous porous layer carried out on the top of the structure (see Figure 12.8). Thus, the process is simplified. The prototypes are improved for gas access and, finally, the wet etching techniques used to perform these prototypes allow batch process. Figure 12.8 presents the process performed by Liu and co-workers (Liu et al. 2006a) for the complete micro-fuel cell. First, a silicon oxide is grown on a silicon substrate (Figure 12.8a) and patterned using photolithography (Figure 12.8b). Wet etching of the surface permits obtaining a substrate with trenches. Thereafter, the opposite side of the substrate was anodized to produce a porous layer (Figure 12.8c and d). The porous layer is used as the catalyst support and the gas diffusion layer. This support is replicated and both are sandwiching a proton exchange membrane (PEM) to carry out the fuel cell.

The other structure, proposed by Hayase et al. (2004), is presented in Figure 12.9. This figure shows the structure of the micro-fuel cell, which consists again of two silicon stacks sandwiching a PEM. PEM sheet is hot-pressed with monolithically fabricated Si electrodes. Catalyst metals are supported by PSi formed by anodization.

The different prototypes performed with mesoporous silicon show various performances during operation test. These results are summarized in the Table 12.4. This table presents, to our knowledge, all the prototypes reviewed in the literature incorporating mesoporous silicon for gas diffusion layer or catalyst support. The fuel cell type, the active surface (A_{active}), the open circuit voltage (OCV), the maximum peak power value (PP), the gas nature, and the temperature used for operation tests are also reported.

As it is seen from the results presented in Table 12.4, there is a large variation in the parameters of the PEM fuel cells. Analysis of the reasons of this variation identified several factors, which may be responsible for poor performances. They are as follows (Hayase and Saito 2005):

Lack of polymer electrolyte. It is assumed that large amounts of catalyst could not contribute to the reaction due to the lack of polymer electrolyte inside the catalyst layer.

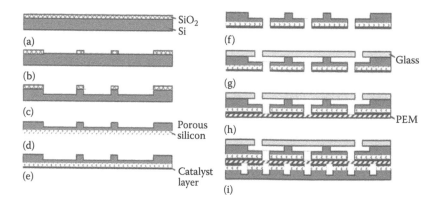

FIGURE 12.8 Double wafer with TMAH channel on one side and microporous silicon as GDL. Fabrication process of a DMFC was proposed by Liu et al. (2006b). (a) SiO₂ formation, (b) photolithography and patterning, (c) wet etching, (d) SiO₂ layer removal and anodization, (e) catalysts plating, (f) liquid channels fabrication, (g) glass process and anodic bonding, (h) PEM attachment, and (i) packaging and wiring. (From Liu X. et al., *J. Micromech. Microeng.* 16, S226, 2006. Copyright 2006: IOP. With permission.)

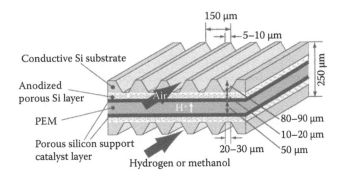

FIGURE 12.9   Schematic view of the miniature fuel cell design. A Hayase's prototype was obtained using two wafers etched with TMAH to produce channels and a microporous silicon layer acting as GDL on top of the trenches. (From Hayase M. et al., *Electrochem. Solid-State Lett.* 7, A231, 2004. Copyright 2004: the Electrochemical Society. With permission.)

TABLE 12.4   Micro-Fuel Cell Prototypes Incorporating Mesoporous Silicon in the Core System Reported in the Literature

Ref.	Fuel Cell Type	A_{active} (cm²)	OCV (V)	PP (mW)	Fuel A & K	T°
Meyers and Maynard 2002	DHFC	NR	0.9	60	H_2 & O_2	RT
Hayase et al. 2004	DHFC	0.25	0.8	1.5	H_2 & Air	80°C
	DMFC	0.25	0.3–0.45	0.045	MeOH 1 M & Air	NR
Yamazaki 2004	DHFC and RHFC	1	0.9	37	NR	40°C
	SOFC	1	1	145	NR	600°C
Liu et al. 2006a	DMFC	0.64	0.45	5	MeOH 2 M & Air	RT

Source: Gautier G. and Kouassi S., *Int. J. Energy Res.* 39, 1, 2015. Copyright 2015: John Wiley & Sons. With permission.

Note: A_{active}: the active surface, DHFC: direct hydrogen fuel cell, DMFC: direct methanol fuel cell, EtOH: ethanol, Fuel A & K: the fuel provided at anode (A) and cathode (K), MeOH: methanol, NR: when data have not been reported, OCV: the open circuit voltage of the cell, PP: the power peak during the test, RHFC: reformed hydrogen fuel cell, RT: room temperature, T°: the temperature during the test.

Poor diffusibility. The porosity and pore diameter are too small to deliver fuel to the catalyst layer.

High resistivity of PSi. Although the electrical resistivity of the Si substrate is low, resistivity of the porous layers is high and resistive loss ($I \cdot R$) might be large.

Taking into account the above, one can conclude that some drawbacks still exist using mesoporous silicon as an FC support. In fact, mesoporous silicon is an insulating material compared to the common carbon material usually used to carry out the catalyst support function. Concerning the diffusion layer function, as mesoporous silicon present very tiny pores with a fractal nature (Nychyporuk et al. 2005), the permeation time for the gas can be very important if the porous layer thickness is not well laid out (Desplobain et al. 2008). It was also found that the mesoporous layer collapses when a huge amount of catalyst metals is deposited on the PSi layer. Moreover, water removal is required when the fuel cell operates at high current. During this operating condition, the porosity can be an important limiting factor for water management in the cell because water penetrates with difficulty into a porous media if pores are too small. In fact, due to capillarity effects, water removal could be correctly ensured only if the pore size is more than tens of micrometers (Yao et al. 2006). Therefore, in order to improve the catalyst performance of the monolithically fabricated Si electrodes aimed for application in miniature fuel cells, enlargement of pore diameter and porosity of the PSi layer were proposed in many articles (Hayase and Saito 2005). For these reasons, it becomes clear that the macroporous silicon also is an interesting material for micro-fuel cell fabrication.

12.4.1.2 MACROPOROUS SILICON

In the past, many authors studied macroporous silicon for micro-fuel cell fabrication. Nevertheless, no prototype has been performed on *p*-type macroporous silicon essentially because of the brittleness of the porous structure when it comes to the process steps of fuel cell manufacturing. For the latter reason, only *n*-type macroporous silicon based fuel cells have been reported in the literature. Desplobain et al. (2009) and Kouassi et al. (2010) have worked on random initiated porous silicon (RIPS). Aravamudhan et al. (2005) and NEAH Power Company have worked on coherent porous silicon (CPS). Lee et al. (2007) have worked on RIPS and CPS and, finally, Gautier et al. (2012) have added a supplementary dimension to the substrate with macroporous silicon etching on 3D substrates. Moreover, original 3D structures have been patented by Kouassi (2006) and Desplobain and Gautier (2009). CPS technology has been performed by Aravamudhan et al. (2005) and Lee et al. (2007) to produce silicon substrate for micro-fuel cell manufacturing.

In particular, Aravamudhan et al. (2005) have highlighted the advantage given by the macropores' size. In fact, direct ethanol fuel cell prototypes (DEFC) took advantage of the capillarity effect in the macropores to drive the fuel to the reaction site when the pore size is craftily tuned. Moreover, Tang et al. (2005) highlighted the effect of a varying porosity on water management in fuel cells with a graded porosity layer improving the cell performance especially in the high current regime of the fuel cell where flooding occur. This structure could easily be performed from a macroporous silicon layer, as macropores' shape can vary from one side of the substrate to the opposite (Kouassi et al. 2010). Indeed, generally, the pore tubes present an increasing diameter when starting from anodized face to the opposite one. It can contribute to improve fuel cell performances. Nevertheless, RIPS and mostly CPS present a bulk region directly in contact with the PEM. These parts are lost surface in the fuel cell due to gas access problem and could therefore lower the fuel cell performance. A possible solution could be to perform a micro- or mesoporous layer on the macropores. Thus, the mesoporous layer will act as a gas diffusion layer. This solution has been studied by Desplobain et al. (2008) revealing that even a thin mesoporous layer has an important impact on fluid permeation. In fact, as the authors were looking for high performance miniaturized fuel cells, this 3-μm mesoporous layer generates a decrease of 2 orders of magnitude of the membrane permeation properties. Otherwise, they compared a macroporous silicon substrate with commercial substrates used for micro-fuel cell fabrication. Macroporous silicon shows permeation properties to hydrogen similar to through silicon via (TSV) substrates and superior to 0.7 and 5 μm pore diameter ceramics and graphite porous material with pores of 0.3 and 0.55 μm (Figure 12.10).

When RIPS is used to perform macroporous substrates for micro-fuel cell fabrication, other advantages can be considered compared to CPS. First, the surface roughness is more important and therefore adhesion of the fuel cell layers should be improved on this type of surface. Moreover, the process for RIPS is less expensive compared to CPS because the initiation step process is suppressed,

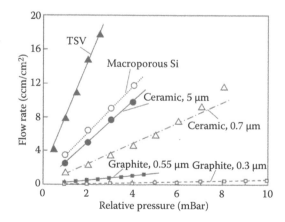

FIGURE 12.10   Hydrogen flow rates measured through different porous media of similar thicknesses used in micro-fuel cells: two graphite substrates with pores mean diameters of 0.3 and 0.55 μm, two ceramics with pores mean diameters of 0.7 and 5 μm, a TSV performed with DRIE and a macroporous silicon substrate. (Data extracted from Gautier and Kouassi 2015.)

limiting the technology cost. The first nonoptimized prototypes performed revealed performance up to 250 mW/cm² (Desplobain et al. 2009). This result clearly shows that macroporous silicon can easily be a valuable replacement to TSV obtained with DRIE. Indeed, with CPS technology as supporting substrate, PEMFC have been performed using Marsacq (2005) ink jet process initially developed on silicon substrate with TSV performed with DRIE. The performance obtained with TSV substrates from Marsacq (2005) prototypes showed a peak power of 300 mW/cm².

Macroporous silicon offers an interesting alternative to perform the fuel cell support and therefore, a more complex use of this material has been envisaged. In fact, Kouassi et al. (2010) performed a more complex structure with 3D macroporous silicon substrates (see also Gautier et al. 2012). Moreover, innovative silicon-based structures have also been patented by Kouassi (2006) and Desplobain and Gautier (2009). These substrates aim to improve the micro-fuel cell performance by increasing the total fuel cell surface with a constant footprint as illustrated in Figure 12.11. Note that Su et al. (2008) improved micro-SOFC using the same idea. The idea consists of taking advantage of the silicon bulk by micromachining techniques to improve the total effective surface of the prototype. To reach this objective, the authors have proposed to perform 3D structure with DRIE or alkaline etching on the silicon wafer surface prior to PSi fabrication. The first results were obtained with alkaline etching of one side of the silicon wafer. Afterward, subsequent PSi etching in a 30% HF mixture with acetic acid as wetting agent (in potentiostatic conditions at 3.1 V) was performed to produce macroporous silicon. It permits performing porous 3D structures, which are presented in Figure 12.12 (Kouassi et al. 2011). This picture shows a structure partially porous with 30% of the total surface presenting pores going through the silicon substrate to perform the flow channel function. This partial PSi area is due to the thickness differences on the substrate before anodization due to trenches formation.

In fact, on one hand, the pore etching rate is dependent on the holes arriving at the interface (HF is highly concentrated then and is not limiting the etching reaction). On the other hand, the substrate thickness is inhomogeneous in the etched area. Therefore, this thickness heterogeneity initiates a preferential region where most of the BSI-generated holes will be consumed during macropores etching. These regions are the bottom of the trenches where the substrate thickness is minimum leading to a shorter path for the generated holes to reach the reaction surface. When the first pores issued from the bottom of the trenches emerge at the opposite side of the wafer, a liquid leakage is created between the two electrodes of the anodization system. For this reason, the other pores still in the growth process encounter an important drop of their growth rate (due to an important part of the current flowing directly through the HF mixture). This drop brought the etching rate close to zero and limits the emergence of the pores situated in the thicker parts of the silicon substrate.

This result has been improved with a structure presenting a similar thickness in every part of the substrate. To perform this new structure, the authors carried out a double side alkaline etching of the substrate to produce a structure having a corrugated shape cross-section. This substrate

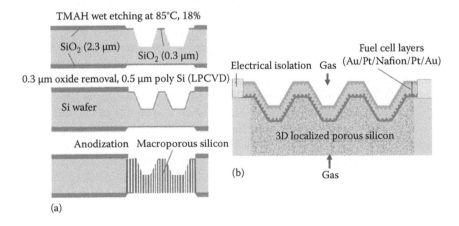

FIGURE 12.11 Cross-section schematic view of (a) the process steps performed to obtain single-side 3D porous substrates on 6-in. silicon wafers, and (b) 3D fuel cell with the fuel cell layer deposited using ink jet technology. (From Kouassi S. et al., *J. Power Sources* 216, 15, 2012. Copyright 2012: Elsevier. With permission.)

FIGURE 12.12 (a–c) SEM images of the wafer surface with macroporous areas after single side alkaline etching and an anodization under potentiostatic conditions at 3.1 V during 4 h in an HF 30%–acetic acid–water electrolyte. The white arrows in the images (a) and (b) underline the over etching occurring at the edges of the trenches when 2 silicon surfaces produce an angle inferior to 180°. The pores have been able to grow through the substrate only from the bottom of the trenches as pointed out in images (a) and (c). (From Kouassi S. et al., *Phys. Stat. Sol. (c)* 8, 1787, 2011. Copyright 2011: John Wiley & Sons. With permission.)

was consecutively anodized after a surface doping leading to the result presented in Figure 12.13. The surface doping produces a mesoporous layer that needs to be removed before permeation tests. This is performed with an alkaline etching of the substrate at room temperature.

These porous substrates have been tested on a permeation bench to compare their permeability to planar PSi substrate with macropores. After the anodization, a post-treatment is performed with an alkaline solution to free all the pores improving consequently the permeation properties of the porous substrate. The measurements revealed that the 3D structure has permeation at least similar to planar substrate. This membrane has been used to carry out micro-fuel cell with adapted techniques. In fact, as the substrate is not planar, the active layers deposition have been performed with PVD for the current collector (gold), ink jet for the catalyst layer, and spray coating for the PEM (Nafion). The first performance results obtained were around 90 mW/cm^2 with H$_2$ and air breathing (Kouassi et al. 2012). Note that the deposition techniques are not optimized for these substrates yet. Therefore, some optimization of the layer thickness, for example, could improve these first results.

Starkov et al. (2009) also proposed interesting approaches to designing electrodes for μ-FC using macroporous silicon. SEM images of PSi-based structures designed by Starkov's team, which can be used as anode and cathode in μ-FC, are shown in Figure 12.14a. The electrode comprises a layer of Pt-based catalyst (1), a gas diffusion layer (2), a channel for supplying gas (3), and the Pd/Ag contact layer (4). On the inner surface of the GDL, there is a layer of carbon nanofibers, which was deposited in order to increase the density of the three-phase boundaries (Starkov and Red'kin 2007). As it is known, the three-phase boundaries are an area where the electrochemical dissociation of hydrogen occurs. Thus, it was possible to combine in a monolithic design the functional units, which in the traditional fabrication of fuel cells are discrete and are combined into a single device using a mechanical pressing.

As can be seen from the reaction shown in Figure 12.1, a large amount of water is released at the cathode during operation of the fuel cell. According to calculations, at the production of 1 ampere-hour of electrical energy, 1.2–1.4 g of pure water is released in the cathode region. If special technological solutions are not taken, then the clogging of both the gas supply and gas

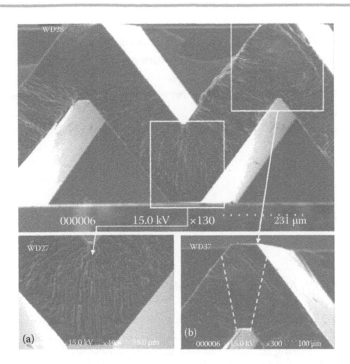

FIGURE 12.13 SEM image of the improved 3D structure of macroporous silicon for miniaturized fuel cell support application obtained after a potentiostatic etching (3.1 V) in HF–acetic acid solution during 4 h. The dotted lines (b) indicate the region where no pore growth occurred. Note that the anodized surface is on top and the illuminated one is at the bottom. (From Kouassi S. et al., *Phys. Stat. Sol. (c)* 8, 1787, 2011. Copyright 2011: John Wiley & Sons. With permission.)

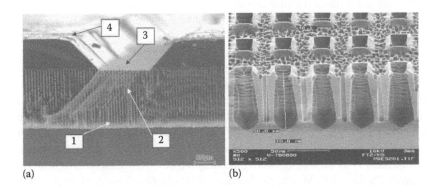

FIGURE 12.14 (a) Multifunctional μ-FC anode based on the PSi: 1—Pt catalyst layer; 2—gas diffusion layer; 3—gas flow channel; 4—Pd/Ag contact layer. (b) Porous structure of μ-FC cathode with gas flow channels (small pores) separated from catch water drains (large pores). (From Starkov V.V. and Red'kin A.N., *Phys. Stat. Sol. (a)* 204 (5), 1332, 2007. Copyright 2007: John Wiley & Sons. With permission.)

diffusion channels occurs and the efficiency of the fuel cell strongly reduces. When using mesoporous silica, it usually happens. Therefore, it is desirable to use macroporous silicon as a material for cathode fabrication. The photograph of such a cathode designed in Starkov's group (Starkov and Red'kin 2007; Starkov et al. 2009) is shown in Figure 12.14b. The peculiarity of such an electrode is that its structure contains both small and large pores.

Thus, macroporous silicon (CPS and RIPS) demonstrates interesting features for fuel cell technology such as an important permeability of the obtained substrate, low cost of the technique, and engineering possibility of the pore size and distribution. Some prototypes already exist in the literature and a company has already used this technology to develop commercial micro-fuel cells. In fact, NEAH Power is actively working on a fuel cell technology involving PSi. Prototypes integrating macroporous silicon reviewed in this chapter are reported in Table 12.5.

TABLE 12.5   Micro-Fuel Cell Prototypes Incorporating Macroporous Silicon in the Core System Reviewed from the Literature and Discussed in This Chapter

Ref.	Fuel Cell Type	A_{active} (cm²)	OCP (V)	PP (mW)	Gas A and K	T (°C)
Aravamudhan et al. 2005	DEFC	1	0.26	8	EtOH 8.5M and O_2	RT
NEAH Power	DMFC	NR	NR	100–200	MeOH and HNO_3	RT-60°C
Lee et al. 2007	DHFC	1.82	0.85	9.5	H_2 and air	RT
Desplobain et al. 2009	DHFC	1	0.9	250	H_2 and air	RT
Kouassi et al. 2012	DHFC (3D)	1	0.9	90	H_2 and air	RT

Source: Gautier G. and Kouassi S., *Int. J. Energy Res.* 39, 1, 2015. Copyright 2015: John Wiley & Sons. With permission.
Note: A_{active}: active surface area, OCP: the open circuit potential, PP: the power density peak during the test.

12.4.2 POROUS SILICON AS A PROTON EXCHANGE MEMBRANE

In the field of micro-fuel cells technologies, mesoporous and macroporous silicon can also be used as a proton exchange membrane in the MEA. Indeed, compared with Nafion®, silicon membranes can be patterned using standard photolithography processes. In addition, the problematic Nafion bonding on silicon can be avoided. Unfortunately, PSi is not naturally proton-conducting. Consequently, post-anodization processes must be performed to reach high proton conductivities. Nevertheless, the first challenge is to perform thin membranes from highly doped silicon wafers with thicknesses varying generally between 40 and 100 µm by RIE or KOH wet etching. Then, the anodization must be performed with various current densities leading to pore diameters from 5 to 40 nm. When the cell potential is measured, a slow exponential decrease can be observed when the pores reach the membrane backside (Pichonat et al. 2004). Then, a very thin (a few nanometers) nonporous silicon film, which must be removed by RIE, still remains.

In order to functionalize PSi, the first approach consists of filling the pores with sulfuric acid (Gold et al. 2004) or with diluted Nafion (Pichonat et al. 2004) after a slight oxidation of the pore surface. The second approach consists of taking advantage of the great specific area of porous materials. Then, Pichonat and Gauthier-Manuel (2006c) proposed to graft the internal surface of the pores with silane molecules after creating Si-OH bonds at the surface. Moghaddam et al. (2010) mentioned also the use of 3-mercaptopropyltrimethoxysilane to reach high proton conductivities. Starkov (2007) has shown that gel PVA/PDA (polyvinyl alcohol/phenol-2,4-disulfonic acid) electrolyte is also promising for fabrication proton exchange membranes aimed for µ-FC. According to Starkov et al. (2009), the above-mentioned electrolyte has the following advantages for application in PSi-based PEM: low gas permeability, plasticity, hydrophilicity, high ionic conductivity at ambient conditions, exceeding the conductivity of commercial membranes Nafion, and low cost. For indicated applications, Starkov and Gavrilin (2007) proposed to use gradient porous structures shown in Figure 12.15. Purposeful formation of layers with the required pore morphology, as well as the alternation of layers with different porosity, was carried out by changing the composition of the electrolyte during the anodic etching (Starkov and Gavrilin 2007). Formed thus composite membranes possessed the high proton conductivity and required mechanical strength, and provided efficient operation of µ-FC at room temperature and wide range of humidity change (Starkov et al. 2009). One should note that the encapsulation of the electrolyte into the porous structure of silicon also gives the possibility to decrease the "water swelling" phenomenon, characteristic for polymer-based membranes. Table 12.6 lists the performances of mesoporous silicon PEM membranes.

12.4.3 MICRO-MACHINED SILICON-BASED SUPPORTS FOR SMALL FUEL CELLS

In this section, we will describe silicon-based structures performed with standard MEMS technologies, such as reactive ion etching.

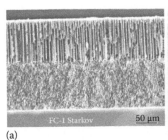
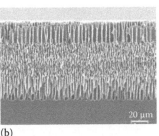

(a)   (b)

FIGURE 12.15 SEM cross-section images of PSi-based (a) two-layer and (b) three-layer gradient structures used for saturation with the aqueous solution of the PVS/PSA electrolyte. (a) First layer has 130-μm pore thickness and 5 μm average pore diameter. Morphology of the first layer of pores is direct and perpendicular to the etching plane. The second layer has 150-μm pore thickness and 1.5–2 μm average pore diameter. Morphology of these indirect pores is tangled. (b) Porosity is 70/55/70% from top to bottom layers. Porosity 55% and 70% corresponds to deep anodic etching process in HF-ethanol and HF-isopropanol-H_2O solutions. Current density was 10 mA/cm² in experiment conditions. (From Starkov V. and Gavrilin E., *Phys. Stat. Sol. (c)* 4(6), 2026, 2007. Copyright 2007: John Wiley & Sons. With permission.)

TABLE 12.6   Mesoporous Silicon Proton Exchange Membrane Performances Reported in the Literature

Ref.	Fuel Cell Type	Membrane Technology	Membrane Proton Conductivity (mS/cm)	Fuel Cell Performances (OCP – Power Density)
Pichonat et al. 2004; Pichonat and Gauthier-Manuel 2006a	DHFC	Filling by Nafion	40	0.8 V–20 mW/cm² (H_2, ambient air)
Gold et al. 2004	PEMFC	Sulfuric acid (8M) loaded	7–330	–
Nagayama et al. 2005	DHFC	Filling by Nafion	–	0.89 V–1.27 mW/cm² (H_2, ambient air)
	DHFC	Sulfuric acid (20%) loaded	–	0.97 V–12.75 mW/cm² (H_2, ambient air)
Pichonat and Gauthier-Manuel 2006b	DHFC	Pore surface grafting by silane molecules	–	0.47 V–17 mW/cm², (H_2/ambient air)
Chu et al. 2006, 2007	PEMFC	Acid loaded	–	0.65 V–94 mW/cm² (formic acid + sulfuric acid/air)
Torres-Herrero et al. 2010	–	Filling by Nafion	44	–
Wang et al. 2011	DMFC	Filling by Nafion	20–30	–
Moghaddam et al. 2010	PEMFC	Sulphonated pore surface	110	1 V–332 mW/cm² (H_2, ambient air)
Starkov et al. 2009	PEMFC	Filling by PVS/PSA	2–7	0.4 V–22 mW/cm²

Source: Data extracted from Gautier and Kouassi 2015.
Note: See Table 12.3 for abbreviations.

The silicon-based prototypes reported in the literature can be classified according to the way the silicon support has been micromachined to act as the bipolar plate and sometimes gas diffusion layer. In fact, two structures have been studied—the first one is very similar to large fuel cell bipolar plate structures as the channels are performed on one face of the silicon material (see Figure 12.16a). The second structure (Figure 12.16b) is performed using the two sides of the silicon wafer with TSV or trenches. In both cases, two techniques are separately or jointly used: DRIE and alkaline etching (KOH, TMAH). Note that the alkaline etching technique is a wet etching technique and therefore can be cheaper than DRIE as it can be extended in batch to etch

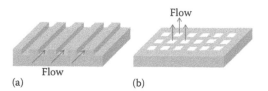

FIGURE 12.16   Different types of structures performed on silicon to act as the flow field of the micro-fabricated fuel cell. (a) Bipolar plate structure on silicon with flowfield. (b) Structure with TSV acting as flowfield. (From Gautier G. and Kouassi S., *Int. J. Energy Res.* 39, 1, 2015. Copyright 2015: John Wiley & Sons. With permission.)

many substrates simultaneously. In the first structure (Figure 12.16a), the fluid has to flow along the surface in channels on the side where the fuel cell layers are present to reach the reaction sites. In the second structure (Figure 12.16b), the fluid has to go through the substrate to reach the reaction sites.

Yen et al. (2003), Motokawa et al. (2004), or Kundu et al. (2006) and co-workers performed bipolar plate structures for micro-DMFC using DRIE to produce micro-channels for gas delivery. The typical resulting structures are presented in the example in Figure 12.17a and b, which is a magnification of the inlet part of the plate. These pictures are taken from Yen et al. (2003). Moreover, Lee et al. (2002) performed, with the same technique (i.e., DRIE) flow channels composed of pillars instead of channels like those presented in Figure 12.17c and d). This approach was meant to increase the PEM free surface for gas access as the PEM membrane is directly stacked to the silicon substrate. Thus, the silicon pillars create less masked area to gas access than trenches do. Otherwise, Yu et al. (2003) performed the same type of structure with alkaline etching of <100> oriented silicon wafer producing V shaped channels. The main advantage of this approach is that the stacks of these elementary cells can easily be carried out without complex gas delivery systems like demonstrated for large fuel cells. Nevertheless, the main drawback of this approach when miniaturized is that obstruction of the structure is likely to occur. In fact,

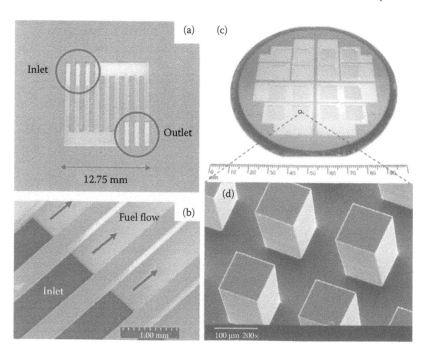

FIGURE 12.17   Silicon bipolar plate performed with DRIE: (a) global view of the bipolar plate with the channels etched on silicon and (b) a magnification of the inlet region. (From Yen T.J. et al., *Appl. Phys. Lett.* 2003; 83, 4056, 2003. Copyright 2003: the American Institute of Physics. With permission.) Bipolar plate, (c) with pillars global view and (d) magnification presenting the pillars. (From Lee S.J. et al., *J. Power Sources* 112, 410, 2002. Copyright 2002: Elsevier. With permission.)

obstruction of the channels due to flooding or membrane deformation, for example, can rapidly isolate an important part of the system from fuel delivery. This is due to gas inlets that are not in number and mostly performed on the periphery of the device. Therefore, solutions with drilled substrates have been proposed using DRIE or alkaline etching techniques. The silicon substrate can therefore be simply drilled from one face to the other like that performed by Yao et al. (2006) and by Cao et al. (2008) (for the cathode side only) using DRIE or Zhong et al. (2009) using alkaline etching. The silicon wafers can also be thinned locally using a low-cost technique like alkaline etching before DRIE, reducing the duration of the expensive DRIE process step. For instance, Kelley et al. (2002), Jankowski et al. (2002), or more recently Zhu et al. (2008), Seo and Cho (2009), and Lee et al. (2009) and co-workers mixed alkaline etching and consecutive DRIE to perform through holes on a silicon surface (on a silicon on insulator [SOI] substrate for Zhu et al. 2008).

In each study, a conductive layer of gold, platinum, silver, or nickel has been deposited to ensure the electron collection in the fuel cell. Note that Yeom et al. (2006) carried out the same type of structure using only DRIE. The typical structure previously discussed is illustrated in Figure 12.18 and an example of the process on SOI (Zhu et al. 2008) is given in Figure 12.19. There is clearly an advantage for through holes compared to channels in terms of gas access to all the reaction sites of the PEM in the fuel cell MEA. Moreover, the through holes also permit, if they are well designed, direct support of the catalyst layer of the fuel cell. Otherwise, improvements have been proposed for these types of prototypes concerning water management issues. In fact, in these structures, water has to go through the holes to be removed from the fuel cell in the opposite direction to the gas flowing to the reaction site of the cell PEM. Therefore, some authors have proposed to perform local silicon surface grafting to produce hydrophobic regions (Yao et al. 2006). This treatment allows selecting surfaces where no flooding of the holes will occur allowing constant gas feeding of the fuel cell. Another solution was also to use the capillarity effect with different hole sizes. In fact, if holes are too small, only gas can flow through and not liquid. Therefore, a substrate presenting two types of holes dimension allows selecting the gas feed function or the water removal function in different parts of the substrate.

It should be noted that in recent years in the development of μ-FC, significant progress was made in both reducing the size of these cells and increasing the power density. Results shown in Figure 12.20 can be an illustration of these achievements. Figure 12.20a shows a photograph of a μ-FC fabricated using the approach discussed above, and Figure 12.20b shows the polarization curve and current density plot of a micro-silicon fuel cell before integration. As it is seen, this microsilicon fuel cell has an open cell potential (the voltage output of the fuel cell in the zero current density limit) of 0.98 V and a peak power density of 237 mW/cm^2 (0.46 V, 520 mA/cm^2). According to Swaminathan et al. (2012), these results testify that PSi-based micro fuel cell power has excellent potential for scaling down to the microliter sizes in order to address the dual

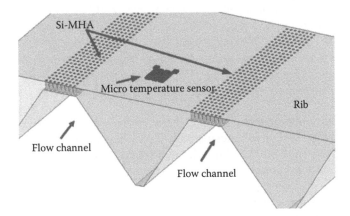

FIGURE 12.18 Schematic view of a typical substrate performed combining DRIE for silicon micro-holes arrays (Si-MHA) and KOH etching for flow channel fabrication. Note that a sensor has been integrated on this system. (From Lee C.Y. et al., *Int. J. Hydrogen Energy* 34, 6457, 2009. Copyright 2009: Elsevier. With permission.)

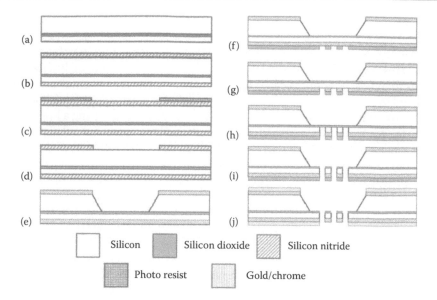

FIGURE 12.19 Example of process flow proposed on SOI for micro-fuel cell support fabrication. (a) SOI substrate, (b) SiN deposition, (c) front side photolithography, (d) SiN front side etching, (e) alkaline etching, (f) backside photolithography, (g) SiN backside etching, (h) Si DRIE, (i) oxide removal, and (j) gold/Cr deposition. (From Zhu L. et al., *J. Power Sources* 185, 1305, 2008. Copyright 2008: Elsevier. With permission.)

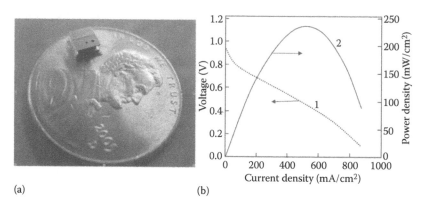

FIGURE 12.20 (a) Photograph of the MEMS hydrogen generator and integrated micro-power source with micro-silicon fuel cell, compared with US one cent coin. MEMS hydrogen generator was bonded to the micro-fuel cell separated by a porous PTFE membrane. (From Swaminathan V.V. et al., *Microfluid. Nanofluid.* 12, 735, 2012. Copyright 2012: Springer. With permission.) (b) Polarization curve and power density plot of the micro-silicon fuel cell before integration. This micro-silicon fuel cell has an open cell potential of 0.98 V and a peak power density of 237 mW/cm². Hydrogen was supplied from a hydrogen tank. (From Zhu L. et al., *J. Power Sources* 185, 1305, 2008. Copyright 2008: Elsevier. With permission.)

problem of increasing on-board energy and power density needs of microsystems. Swaminathan et al. (2012) believe that further enhancements could include reduction of MEA/control system volume so as to incorporate >50% volume fraction of fuel, improved reactor designs to combat issues of byproduct expansion and agglomeration for increased fuel utilization, and self-hydration schemes to maximize proton conductivity in the MEA. Some of these improvements could enable microscale power generation to achieve even higher energy densities and eventually drive successful miniaturization of fuel cells from the portable electronics scale into powering MEMS/NEMS. According to Swaminathan et al. (2012), the main issues in the integration of μFC in power MEMS are the effects of water transport and the hydrate byproduct swelling on hydrogen production in the microreactor.

12.5 CONCLUSION

In this chapter, we have demonstrated that PSi, compatible with clean room environment, can be a promising material to be integrated in fuel cell technology. In fact, this original material via MEMS technology permits addressing many issues in fuel cell miniaturization. In particular, as it was shown in this chapter, PSi allows performing substrates with flow channels, gas diffusion layers, and catalyst support functions. Of course, from an economic point of view, silicon is an expensive material for fuel cell fabrication even if the miniaturization could justify the technology price. Thus, additional steps in the fabrication must be as cheap as possible. For this reason, wet etching techniques are preferable to DRIE, which is more expensive. However, if one wants to address small definition or more various hole shapes and organization, alkaline etching due to its geometric restriction (Seidel et al. 1990) is not sufficient even if combined with DRIE. Therefore, PSi rapidly appeared as a promising alternative to wet etching technique, broadening the possibilities to structure the silicon for fuel cell application. Moreover, some authors have been able to demonstrate that the above-mentioned functions of PSi in FC could be achieved in less process steps than other techniques.

However, we have to note that the main issue that still needs to be addressed for this material is its use for high volume throughput. In fact, very few attempts of industrialization have been performed yet. Nevertheless, there are prerequisites that the problem of industrialization of the process of silicon porosification, which requires the development of reliable processes and equipments for large surface treatment, will be solved. This will enable a step forward from lab to fab environment, where cost and throughput are the main issues to address.

ACKNOWLEDGMENT

G.K. is grateful to the Ministry of Science, ICT and Future Planning (MSIP) of the Republic of Korea for supporting his research.

REFERENCES

Aravamudhan, S., Rahman, A., and Bhansali, S. (2005). Porous silicon based orientation independent, self-priming micro direct ethanol fuel cell. *Sens. Actuators A* **123–124**, 497–504.

Armbruster, S., Schafer, F., Lammel, G. et al. (2003). A novel micromachining process for the fabrication of monocrystalline Si-membranes using porous silicon. In: *Proceedings of 12th International Conference on Solid-State Sensors, Actuators and Microsystems*, Boston, June 8–12, pp. 246–249.

Baertsch, C.D., Jensen, K.F., Hertz, J.L. et al. (2004). Fabrication and structural characterization of self-supporting electrolyte membranes for a micro solidoxide fuel cell. *J. Mater. Res.* **19**, 2604–2615.

Barbir, F. (2006). PEM fuel cells. In: Sammes N. (Ed.) *Fuel Cell Technology*. Springer-Verlag, London, pp. 27–51.

Barbir, F., Braun, J., and Neutzler, J. (1999). Properties of molded graphite bi-polar plates for PEM fuel cells. *Int. J. New Mater. Electrochem. Syst.* **2**, 197–200.

Cao, J., Zou, Z., Huang, Q. et al. (2008). Planar air-breathing micro-direct methanol fuel cell stacks based on micro-electronic-mechanicalsystem technology. *J. Power Sources* **185**, 433–438.

Chen, X., Wu, N.J., Smith, L., and Ignatiev, A. (2004). Thin-film heterostructure solid oxide fuel cells. *Appl. Phys. Lett.* **84**, 2700–2702.

Chu, K., Shannon, M., and Masel, R. (2006). An improved miniature direct formic acid fuel cell based on nanoporous silicon portable power generation. *J. Electrochem. Soc.* **153**, A1562–A1567.

Chu, K., Shannon, M., and Masel, R. (2007). Porous silicon fuel cells for micro power generation. *J. Micromech. Microeng.* **17**, S243–S249.

D'arrigo, G. and Coffa, S. (2006). Fuel cell formed in a single layer of monocrystalline silicon and fabrication process. US patent, 11/383,088.

D'arrigo, G., Spinella, C., Arena, G., and Lorenti, S. (2003). Fabrication of miniaturised Si-based electrocatalytic membranes. *Fuel Cells Bull.* **2003**(4), 110–112.

Desplobain, S. (2009). Etude et realisation de couches de diffusion de gas en silicium poreux appliquees a la fabrication de micropiles a hydrogene. PhD Thesis, Universite Francois-Rabelais, France.

Desplobain, S. and Gautier, G. (2009). Cell holder for fuel cell. US patent, 2009/047453.

Desplobain, S., Gautier, G., Gharbage, N., Ventura, L., and Roy, M. (2008). Gas management through thick macroporous and mesoporous macroporous membranes. *Phys. Stat. Sol. (c)* **5**, 3843–3845.

Desplobain, S., Gautier, G., Ventura, L., and Bouillon, P. (2009). Macroporous silicon hydrogen diffusion layers for micro-fuel cells. *Phys. Stat. Sol. (a)* **206**, 1282–1285.

Dyer, C.K. (1990). A novel thin-film electrochemical device for energy conversion. *Nature* **343**, 547–548.

Dyer, C.K. (2002). Fuel cells for portable applications. *J. Power Sources* **106**, 31–34.

Gautier, G. and Kouassi, S. (2015). Integration of porous silicon in microfuel cells: A review. *Int. J. Energy Res.* **39**, 1–25.

Gautier, G., Kouassi, S., Desplobain, S., and Ventura, L. (2012). Macroporous silicon hydrogen diffusion layers for micro-fuel cells: From planar to 3D structures. *Micro-Electron. Eng.* **90**, 79–82.

Gold, S., Chu, K.-L., Lu, C., Shannon, M.A., and Masel, R.I. (2004). Acid loaded porous silicon as a proton exchange membrane for micro-fuel cells. *J. Power Sources* **135**, 198–203.

Hayase, M. and Saito, D. (2005). Catalyst layer formation onto meso pore porous silicon layer for miniature fuel cell. In: *Proceedings of 5th International Workshop on Micro and Nanotechnology for Power Generation and Energy Conversion Applications*, Nov. 28–30, Tokyo, Japan, pp. 128–131.

Hayase, M., Kawase, T., and Hatsuzawa, T. (2004). Miniature 250 μm thick fuel cell with monolithically fabricated silicon electrodes. *Electrochem. Solid-State Lett.* **7**, A231–A234.

Hsieh, S.S., Kuo, J.K., Hwang, C.F., and Tsai, H.H. (2004). A novel design and microfabrication for a micro PEMFC. *Microsyst. Technolog.* **10**, 121–126.

Hsieh, S.S., Huang, C.F., Kuo, J.K., Tsai, H.H., and Yang, S.H. (2005). SU-8 flow field plates for a micro PEMFC. *J. Solid State Electrochem.* **9**, 121–131.

Hsieh, S.S., Feng, C.L., and Huang, C.F. (2006). Development and performance analysis of a H_2/air micro PEM fuel cell stack. *J. Power Sources* **163**, 440–449.

Jankowski, A.F., Hayes, J.P., Graff, R.T., and Morse, J.D. (2002). Micro-fabricated thin-film fuel cells for portable power requirements. *Mater. Res. Soc. Symp. Proc.* **730**, 93–98.

Kamarudin, S.K., Daud, W.R.W., Ho, S.L., and Hasran, U.A. (2007). Overview on the challenges and developments of micro-direct methanol fuel cells (DMFC). *J. Power Sources* **163**, 743–754.

Kelley, S.C., Deluga, G.A., and Smyrl, W.H. (2002). Miniature fuel cells fabricated on silicon substrates. *AIChE J.* **48**, 1071–1082.

Kouassi, S. (2006). Fuel cell with a large exchange surface area. US patent, US8216739 B2.

Kouassi, S., Gautier, G., Desplobain, S., Coudron, L., and Ventura, L. (2010). Macroporous silicon electrochemical etching for gas diffusion layers applications: Effect of processing temperature. *Defect Diff. Forum* **297**, 887–892.

Kouassi, S., Gautier, G., Desplobain, S., and Ventura, L. (2011). Study of macroporous silicon electrochemical etching in 3D structured *n*-type silicon substrates. *Phys. Stat. Sol. (c)* **8**, 1787–1791.

Kouassi, S., Gautier, G., Thery, J. et al. (2012). Proton exchange membrane micro fuel cells on 3D porous silicon gas diffusion layers. *J. Power Sources* **216**, 15–21.

Kundu, A., Jang, J.H., Lee, H.R. et al. (2006). MEMS-based micro-fuel cell processor for application in a cell phone. *J. Power Sources* **162**, 572–578.

Kundu, A., Jang, J.H., Gil, J.H. et al. (2007). Micro-fuel cells—Current development and applications. *J. Power Sources* **170**, 67–78.

Laramie, J. and Dicks, A. (2003). *Fuel Cell Systems Explained*, 2nd ed. Wiley, Chichester, UK.

Lee, S.J., Chang-Chien, A., Cha, S.W. et al. (2002). Design and fabrication of a micro fuel cell array with "flip flop" interconnection. *J. Power Sources* **112**, 410–418.

Lee, C.Y., Lee, S.J., Huang, D.R., and Chuang, C.W. (2007). Integration of the micro-thermal sensor and porous silicon as the gas diffusion layer for microfuel cell. In: *Proc. of the 7th IEEE Conference on Nanotechnology*, August 2–5, Hong Kong, pp. 1252–1255.

Lee, C.Y., Lee, S.J., Hu, Y.C., Shih, W.P., Fan, W.Y., and Chuang, C.W. (2009). Integration of silicon micro-hole arrays as a gas diffusion layer in a micro-fuel cell. *Int. J. Hydrogen Energy* **34**, 6457–6464.

Lindroos, V., Tilli, M., Lehto, A., and Motooka, T. (2010). *Handbook of Silicon Based MEMS Materials and Technologies*. William Andrew Pub., Oxford, UK.

Litster, S. and McLean, G. (2004). PEM fuel cell electrodes. *J. Power Sources* **130**, 61–76.

Liu, X., Suo, C., Zhang, Y. et al. (2006a). Novel modification of Nafion 117 for a MEMS based micro direct methanol fuel cell (μDMFC). *J. Micromech. Microeng.* **16**, S226–S232.

Liu, X., Suo, C., Zhang, Y., Zhang, H., Chen, W., and Lu, X. (2006b). Application of MEMS technology to micro direct methanol fuel cell. In: *Proceedings of DTIP of MEMS and MOENS*, April 26–28, Stresa, Italy, TIMA edn., 6 p.

Lysenko, V., Bidault, F., Alekseev, S. et al. (2005). Study of porous structures as hydrogen reservoirs. *J. Phys. Chem. B* **109**, 19711–19718.

Marsacq, D. (2005). Les Micropiles à Combustible, une Nouvelle Génération Demicrogénérateurs Electrochimiques. CEA/Service Information-Media, Saclay/Siège, pp. 50–51.

Mathias, M.F., Roth, J., Fleming, J., and Lehnert, W. (2003). Diffusion media materials and characterization. In: Vielstich, W., Lamm, A., and Gastegier, H.A. (Eds.) *Handbook of Fuel Cells, Fundamentals, Technology and Applications, Vol. 3: Fuel Cell Technology and Applications*. John Wiley & Sons, New York, pp. 517–537.

Maynard, H.L. and Meyers, J.P. (2003). Article comprising micro fuel cell. US patent, 9/514,494.

Meyers, J.P. and Maynard, H.L. (2002). Design considerations for miniaturized PEM fuel cells. *J. Power Sources* **109**, 76–88.

Min, K.B., Tanaka, S., and Esashi, M. (2002). MEMS-based polymer electrolyte fuel cell. *Electrochemistry* **70**, 924–927.

Moghaddam, S., Pengwanga, E., Masel, R., and Shannon, M. (2008). A self-regulating hydrogen generator for micro fuel cells. *J. Power Sources* **185**, 445–450.

Moghaddam, S., Pengwang, E., Jiang, Y. et al. (2010). An inorganic–organic proton exchange membrane for fuel cells with a controlled nanoscale pore structure. *Nature Nanotechnology* **5**, 230–235.

Morse, J.D. (2007). Micro-fuel cell power sources. *Int. J. Energy Res.* **31**, 576–602.

Motokawa, S., Mohamedi, M., Momma, T., Shoji, S., and Osaka, T. (2004). MEMS-based design and fabrication of a new concept micro direct methanol fuel cell (µDMFC). *Electrochem. Commun.* **6**, 562–565.

Nagayama, G., Idera, N., Tsuruta, T., Yu, J., Takahashi, K., and Hori, M. (2005). Porous silicon as a proton exchange membrane for micro fuel cells. *Electrochemistry (Tokyo)* **73**, 939–941.

Nguyen, N.T. and Chan, S.H. (2006). Micromachined polymer electrolyte membrane and direct methanol fuel cells—A review. *J. Micromech. Microeng.* **16**, R1–R12.

Nychyporuk, T., Lysenko, V., and Barbier, D. (2005). Fractal nature of porous silicon nanocrystallites. *Phys. Rev. B* **71**, 115402.

Pichonat, T. and Gauthier-Manuel, B. (2006a). A new process for the manufacturing of reproducible mesoporous silicon membranes. *J. Membrane Sci.* **280**, 494–500.

Pichonat, T. and Gauthier-Manuel, B. (2006b). Realization of porous silicon based miniature fuel cells. *J. Power Sources* **154**, 198–201.

Pichonat, T. and Gauthier-Manuel, B. (2006c). Meosporous silicon based miniature fuel cells for nomadic and chipscale systems. *Microsyst. Technol.* **12**, 330–334.

Pichonat, T., Gauthier-Manuel, B., and Daniel Hauden, D. (2004). New proton-conducting porous silicon membrane for small fuel cells. *Fuel Cells Bull.*, August 11–14.

Presting, H., Konle, J., Starkov, V., Vyatkin, A., and König, U. (2004). Porous silicon for micro-sized fuel cell reformer units. *Mater. Sci. Eng. B* **108**, 162–165.

Schmitz, A., Tranitz, M., Wagner, S., Hahn, R., and Hebling, C. (2003). Planar self-breathing fuel cells. *J. Power Sources* **118**, 162–171.

Seidel, H., Csepregi, L., Heuberger, A., and Baumgärtel, H. (1990). Anisotropic etching of crystalline silicon in alkaline solutions: I. Orientation dependence and behavior of passivation layers. *J. Electrochem. Soc.* **137**, 3612–3626.

Seo, Y.H. and Cho, Y.H. (2009). Micro direct methanol fuel cells and their stacks using a polymer electrolyte sandwiched by multi-window microcolumn electrodes. *Sens. Actuators A* **150**, 87–96.

Shah, K., Shin, W.C., and Besser, R.S. (2003). Novel microfabrication approaches for directly patterning PEM fuel cell membranes. *J. Power Sources* **123**, 172–181.

Srikar, V.T., Turner, K.T., Andrew, T.Y., and Spearing, S.M. (2004). Structural design considerations for micromachined solid-oxide fuel cells. *J. Power Sources* **125**, 62–69.

Starkov, V.V. (2007). Multifunctional integrated fuel cells electrode on macroporous silicon. In: Veziroglu, T.N., Zaginaichenko, S.Y., Schur, D.V. et al. (Eds.) *Hydrogen Materials Science and Chemistry of Carbon Nanomaterials.* Springer, Dordrecht, Netherlands, pp. 765–771.

Starkov, V. and Gavrilin, E. (2007). Gradient-porous structure of silicon. *Phys. Stat. Sol. (c)* 4(6), 2026–2028.

Starkov, V.V. and Red'kin, A.N. (2007). Carbon nanofibers encapsulated in macropores in silicon. *Phys. Stat. Sol. (a)* **204**(5), 1332–1334.

Starkov, V., Vyatkin, A., Volkov, V., Presting, H., Konle, J., and König, U. (2005). Highly efficient to hydrogen permeability palladium membranes supported in porous silicon. *Phys. Stat. Sol. (c).* **2**(9), 3457–3460.

Starkov, V.V., Lyskov, N.V., and Dobrovolsky, Yu.A. (2009). Porous silicon for micro fuel cells. *Intern. Sci. J. Altern. Energy Ecology* **8**(76), 78–84 (in Russian).

Steele, B.C.H. and Heinzel, A. (2001). Materials for fuel-cell technologies. *Nature* **414**, 345–352.

Stewart, M.P. and Buriak, J.M. (2000). Chemical and biological applications of porous silicon technology. *Adv. Mater.* **12**, 859–869.

Su, P.C., Chao, C.C., Shim, J.H., Fasching, R., and Prinz, F.B. (2008). Solid oxide fuel cell with corrugated thin film electrolyte. *Nano Lett.* **8**, 2289–2292.

Swaminathan, V.V., Zhu, L., Gurau, B., Masel, R.I., and Shannon, M.A. (2012). Integrated micro fuel cell with on-demand hydrogen production and passive control MEMS. *Microfluid. Nanofluid.* **12**, 735–749.

Tang, Y., Stanley, K., Wu, J., Ghosh, D., and Zhang, J. (2005). Design consideration of micro thin film solid-oxide fuel cells. *J. Micromech. Microeng.* **15**, S185–S192.

Tjerkstra, R.W., Gardeniers, J.G., Kelly, J.J., and Van den Berg, A. (2009). Multi-walled microchannels: Freestanding porous silicon membranes for use in µTAS. *J. Microelectromech. Syst.* **9**, 495–501.

Torres-Herrero, N., Santander, J., Sabate, N. et al. (2010). A monolithic micro fuel cell based on a functionalized porous silicon membrane. In: *Proceedings of 8th International Conference on Advanced Semiconductor Devices & Microsystems (ASDAM)*, Oct. 25–27, Smolenice Castle, Slovakia, pp. 263–266.

Wainright, J.S., Savinell, R.F., Liu, C.C., and Litt, M. (2003). Microfabricated fuel cells. *Electrochim. Acta* **48**, 2869–2877.

Wang, X., Li, W., Chen, Z., Waje, M., and Yan, Y. (2006). Durability investigation of carbon nanotube as catalyst support for proton exchange membrane fuel cell. *J. Power Sources* **158**, 154–159.

Wang, M., Wang, X., Wu, S., Tan, Z., Liu, L., and Guo, X. (2011). Nanoporous silicon membrane with channels for micro direct methanol fuel cells. In: *Proceedings of IEEE International Conference on Nano/Micro Engineered and Molecular Systems (NEMS)*, Feb. 20–23, Kaohsiung, Taiwan, pp. 968–971.

Yamazaki, Y. (2004). Application of MEMS technology to micro fuel cells. *Electrochem. Acta* **50**, 663–666.

Yao, S.C., Tang, X., Hsieh, C.C. et al. (2006). Micro-electro-mechanical systems (MEMS)-based micro-scale direct methanol fuel cell development. *Energy* **31**, 636–649.

Yen, T.J., Fang, N., Zhang, X., Lu, G.Q., and Wang, C.Y. (2003). A micro methanol fuel cell operating at near room temperature. *Appl. Phys. Lett.* **83**, 4056–4058.

Yeom, J., Jayashree, R.S., Rastogi, C., Shannon, M.A., and Kenis, P.J.A. (2006). Passive direct formic acid microfabricated fuel cells. *J. Power Sources* **160**, 1058–1064.

Yu, J., Cheng, P., Ma, Z., and Yi, B. (2003). Fabrication of a miniature twin-fuel-cell on silicon wafer. *Electrochim. Acta* **48**, 1537–1541.

Zhao, T.S. (2009). *Micro Fuel Cells Principles and Applications*. Elsevier, London.

Zhong, Z., Chen, J., and Peng, R. (2009). Design and performance analysis of micro proton exchange membrane fuel cells. *Chinese J. Chem. Eng.* **17**, 298–303.

Zhu, L., Lin, K.Y., Morgan, R.D. et al. (2008). Integrated micro-power source based on a micro-silicon fuel cell and a micro electromechanical system hydrogen generator. *J. Power Sources* **185**, 1305–1310.

Hydrogen Generation and Storage in Porous Silicon

13

Valeriy A. Skryshevsky, Vladimir Lysenko, and Sergii Litvinenko

CONTENTS

13.1   Introduction   274

13.2   Hydrogen Storage in Solid Matrixes   274

    13.2.1   Physisorption of Hydrogen in Porous Matrixes   274

    13.2.2   Chemical Storage of Hydrogen   275

13.3   Estimation of Hydrogen Contents in PSi Using FTIR Method   276

13.4   Hydrogen Generation from Porous Silicon   280

    13.4.1   Thermally Stimulated Desorption of Hydrogen   280

    13.4.2   Chemically Stimulated Desorption of Hydrogen   281

    13.4.3   Photocatalytic Hydrogen Evolution   285

13.5   Rate of Hydrogen Generation   285

13.6   Method of Supplementary Hydrogenization   289

13.7   Hydrogen Storage and Generation in PSi Composites   290

13.8   Energetic Analysis of PSi Nanostructures   291

13.9   Conclusion   292

References   293

13.1 INTRODUCTION

In recent years, it has become increasingly clear that hydrogen appears as a new energy carrier instead of the traditional carbonaceous fuels. One of the key roadblocks to the widespread use of hydrogen as a renewable fuel is hydrogen storage.

Three main options exist for storing hydrogen (Tzimas et al. 2003):

1. As a highly compressed gas
2. As a cryogenic liquid
3. In a solid matrix

Most automakers are considering either the high-pressure gaseous hydrogen (passenger vehicles) or cryogenic liquid (also known in rockets applications). However, these first two technologies are fraught with public perception issues on safety and cannot be applied in a large field of portable devices: mobile telephones, portable computers, and other gadgets. Other issues of these technologies need to be addressed, including compression costs and safety, liquidation costs, and dormancy. Therefore, it is becoming increasingly accepted that the solid matrix method of hydrogen storage is the only option that has any hope of being efficient for hydrogen storage especially for portable device applications. For mobile applications, and as such the 2015 Department of Energy (DOE) targets for hydrogen storage are 9 wt% and 81 g/L for gravimetric and volumetric system capacities, respectively (Krishna et al. 2012; Dalebrook et al. 2013).

13.2 HYDROGEN STORAGE IN SOLID MATRIXES

Nowadays, a wide range of materials is currently being considered as potential reversible hydrogen storage media. Among them are microporous adsorbents, which physisorb molecular hydrogen at low temperatures, the interstitial hydrides, which are reactive metals reversibly absorbing dissociated atomic hydrogen into their bulk as an interstitial, the complex hydrides that bind atomic hydrogen either covalently or ionically and release it via solid state decomposition, and some alternative storage materials (Broom 2011; Krishna et al. 2012).

13.2.1 PHYSISORPTION OF HYDROGEN IN POROUS MATRIXES

One of the approaches for reversible hydrogen storage is based on physisorption of hydrogen molecules onto the specific surface of a solid matrix. Hydrogen can be stored in its molecular form by physical adsorption on the surface of a porous solid material, and typical gravimetric capacities for a range of materials are identified in Table 13.1 (Dalebrook et al. 2013).

This process is based on Van der Waals interactions of the hydrogen molecules with several atoms at the surface of a solid. Because of the weak interactions, significant physisorption is only observed at low temperatures (<273°K) and the maximum amount of adsorbed hydrogen is proportional to the specific surface area of the adsorbent solid. Materials with a large specific surface area like activated or nanostructured carbon and carbon nanotubes (CNTs) are possible substrates for physisorption. The main difference between CNTs and high surface area graphite is the curvature of the graphene sheets and the cavity inside the tube. In microporous solids with capillaries, which have a width of less than a few molecular diameters, the potential fields from opposite walls overlap so that the attractive force acting on adsorbate molecules is increased compared with that on a flat carbon surface. This phenomenon is the main motivation for the investigation of the hydrogen-CNT interaction. The theoretical maximum amount of adsorbed hydrogen is approximately 3.0 mass% for single-walled CNTs with a specific surface area of 1315 m²/g at a temperature of 77°K while the maximum experimentally measured absorption capacity of the nanostructured carbon materials is less. A large variety of different nanostructured carbon samples have been investigated using a high-pressure microbalance (Strobel et al. 1999; Nijkamp et al. 2001) at 77°K, electrochemical galvanostatic measurements at

TABLE 13.1   Properties of Selected Physical Hydrogen Storage Systems

Storage Medium	Temperature (°C)	Pressure (bar)	Capacity (wt%)
Compressed hydrogen gas	25	200	16[a]
		Up to 700	28[a]
Carbon nanotubes	27	1	0.2
	25	500	2.7
	−196	1	2.8
Graphene oxide	25	50	2.6
PIMs[b]	−196	10	2.7
HCPs[c]	−196	15	3.7
COFs[d]	−196	70	7.2
Zeolites	25	100	1.6
	−196	16	2.07
MOFs[e]	25	50	8
	−196	70	16.4
Clatrate hydrates	−3	120	4

Source: Dalebrook A.F. et al., *Chem. Commun.* **49**, 8735, 2013. Copyright 2013: Royal Society of Chemistry. With permission.
[a] Values for pure hydrogen are system gravimetric capacities, which measure storage relative to the container weight.
[b] Polymers of intrinsic microporosity.
[c] Hyper-cross-linked polymers.
[d] Covalent organic frameworks.
[e] Metal organic frameworks.

room temperature, and volumetric (mass flow) gas phase measurements at 77°K (Lee et al. 2000; Nützenadel et al. 2000).

Besides carbon nanostructures, other nanoporous materials have been investigated for hydrogen absorption. Zeolites of different pore structure and composition, for example, A, X, and Y, have been analyzed (Weitkamp et al. 1995) in the temperature range 293–573°K and at pressures of 2.5–10 MPa. The amount of absorbed hydrogen increases with temperature and absorption pressure. For example, the maximum of 1.8 mass% of absorbed hydrogen was found (Langmi et al. 2003) for a zeolite (NaY) with a specific surface area of 725 m²/g. A microporous metal-organic framework (MOF) Zn_4O (1,4-benzenedicarboxylate)$_3$ was proposed as a hydrogen storage material (Rosi et al. 2003). It was shown the MOF absorbs hydrogen proportionally to the operating pressure at 298°K. At 77°K, the amount of absorbed hydrogen can achieve 16.4 mass% (Dalebrook et al. 2013).

The big advantages of physisorption for hydrogen storage are the low operating pressure, the relatively low cost of the materials involved, and the simple design of the storage system. The rather small gravimetric and volumetric hydrogen density on carbon, together with the low temperatures necessary, are significant drawbacks.

13.2.2 CHEMICAL STORAGE OF HYDROGEN

Contrary to physical storage methods described in the preceding section, chemical storage media contain hydrogen in chemically bonded or complexed forms, or incorporated into small organic molecules. Two categories will be separately addressed where media are comprised of either solid or liquid phase components. The release of molecular hydrogen is initiated when the source material is subjected to thermal or catalytic decomposition. Particular challenges in this area involve ease of reversible charging and discharging, storage capacity and stability of the base material or catalytic systems (Dalebrook et al. 2013).

Hydrogen reacts at elevated temperatures with many transition metals and their alloys to form hydrides. Many of these compounds, (MH_n), show large deviations from ideal stochiometry (n = 1, 2, 3) and can exist as multiphase systems. The lattice structure is that of a typical metal with hydrogen atoms on the interstitial sites and for this reason, they are also called interstitial hydrides. The maximum amount of hydrogen in the hydride phase is given by the number of interstitial sites in the intermetallic compound (Switendick 1978; Westlake 1983). Metal hydrides can adsorb a large amount of hydrogen at a constant pressure. Some metal hydrides absorb and desorb hydrogen at ambient temperature and close to the atmospheric pressure.

One of the most interesting features of metallic hydrides is the extremely high volumetric density of hydrogen atoms present in the host lattice. The highest volumetric hydrogen density reported is 150 kg/m³ in Mg_2FeH_6 and $Al(BH_4)_3$. Both hydrides belong to the complex hydrides family. Metallic hydrides can reach a volumetric hydrogen density of 115 kg/m³, for example, $LaNi_5$.

Metal hydrides are very effective as storing large amounts of hydrogen in a safe and compact way. All the reversible hydrides working around ambient temperature and atmospheric pressure consist of transition metals; therefore, the gravimetric hydrogen density is limited to less than 3 mass%. Group 1, 2, and 3 light metals, for example, Li, Mg, B, and Al, give rise to a large variety of metal-hydrogen complexes. They are especially interesting because of their light weight and the number of hydrogen atoms per metal atom, which is two in many cases. The main difference between the complex and metallic hydrides is the transition to an ionic or covalent compound on hydrogen absorption.

Metallic hydrides as hydrogen storage materials began to be intensively studied since 1997, when the absorption and desorption pressure-concentration isotherms for catalyzed $NaAlH_4$ at temperatures of 180°C and 210°C were measured. The amount of the absorbed/desorbed hydrogen reached 4.2 mass% (Bogdanovich and Schwickardi 1997). The compound with the highest gravimetric hydrogen density at room temperature known today is $LiBH_4$ (18 mass%). This hydride desorbs three of the four hydrogens in the compound on melting at 280°C and decomposes into LiH and B. The desorption process can be catalyzed by adding SiO_2 and significant thermal desorption has been observed starting at 100°C (Züttel et al. 2003).

Complex hydrides constitute a relatively new field of hydrogen storage materials. The alanates have been investigated extensively during the last 10 years. The borides are especially interesting because of their very high gravimetric and volumetric hydrogen density.

From the other side, recent realizations of micro fuel cells (Mex et al. 2001; Lee et al. 2002; Meyers and Maynard 2002; Yamazaki 2004; Dzhafarov et al. 2008; Gautier 2014) conceived to be used for energy supplying in portable devices are based on low-cost Si technologies already developed and successfully used mainly in the microelectronics field throughout the world. Therefore, finding simple ways for hydrogen storage in such a type of microsystems directly compatible with silicon-based microtechnologies is one of the key issues.

13.3 ESTIMATION OF HYDROGEN CONTENTS IN PSi USING FTIR METHOD

It is known that hydrogen is present in the chemical composition of porous silicon (PSi), especially in as-prepared samples, because of etching technology (Parkhutik 1999; Föll et al. 2002). Thus, silane SiH_x (x = 1, 2, 3) groups are formed on the surface of the obtained PSi. In addition, some amount of hydrogen is incorporated into the bulk of PSi. Here, it is kept on internal Si states of different types, such as crystalline planes, inner voids and other defects of structure, as well as on impurities (especially oxygen and boron) (Ogata et al. 1995; Estreicher et al. 1999; Pritchard et al. 1999). H atoms, incorporated inside the Si lattice, can interact chemically with inner dangling silicon bonds and atoms of impurities. Thus, inner SiH_x, Si-H-B, O_xSiH, SiOH, and other complexes are formed. The peculiarities of hydrogen arrangement in the PSi lattice are similar to those in c-Si (Cerofolini et al. 1995). Moreover, hydrogen in molecular form has the possibility to be captured by the surface or to penetrate into the PSi lattice. The structure contains centers that are able to hold H_2 molecules by the forces of dipole–dipole interaction. Some of such centers are formed on the interstitial oxygen atoms. Further, hydrogen molecules can be entrapped and dissociated on defects, particularly on vacancies and self-interstitials (Estreicher et al. 1999; Pritchard et al. 1999).

Hydrogen concentration in N_H (mmol/cm³) in PSi can be estimated from absorption spectra by using the following relation used earlier for estimation of hydrogen content in amorphous Si layers (Tolstoy et al. 2003):

$$(13.1) \qquad N_H = \frac{1}{\Gamma_S} \int_{h\nu} \frac{\alpha}{h\nu} d(h\nu) = \frac{I_S}{\Gamma_S},$$

where I_S (cm⁻¹) is the integrated adsorption of the stretching band, α (cm⁻¹) is the absorption coefficient, $h\nu$ is the photon energy, and Γ_S (cm²/mmol) is the stretching oscillator strength of the Si–H bonds, which was determined from the following relation as it was proposed in Shanks et al. (1980) for amorphous silicon:

$$(13.2) \qquad \Gamma_S = 37.6 \times \frac{I_S}{I_W},$$

where I_W is experimentally measured integrated adsorption of the wagging band (near 640 cm⁻¹).

Otherwise, the hydrogen concentration value can be expressed in atomic, C_H (at.%), or mass C_M (mass%) percents as follows (Tolstoy et al. 2003):

$$(13.3) \qquad C_H = \frac{N_H m_{Si}}{1000 \rho_{Si}(1-P) + N_H m_{Si}},$$

$$(13.4) \qquad C_M = \frac{1}{1 + (m_{Si}/m_H)((1/C_H) - 1)},$$

where m_{Si} and m_H are the atomic masses of silicon and hydrogen, respectively, ρ_{Si} is the monocrystalline Si density (2.33 g/cm³), and P is the porosity of the PSi nanostructure.

Hydrogen in PSi is stored in Si-H, Si-H₂, and Si-H₃ bonds and their ratio can be estimated from experimental FTIR spectra. Typical well-known Si-H$_x$ (with x = 1, 2, 3) stretching vibration spectra of meso-PSi nanostructures are shown in Figure 13.1a. Three well-defined large bands centered at 2088, 2110, and 2137 cm⁻¹, mainly corresponding, respectively, to Si-H, Si-H₂, and Si-H₃ stretching bonds, can be clearly seen by Chazalviel and Ozanam (1997) and Bisi et al. (2000). Additionally, a fine structure of the absorption bands consisting of about seven overlapping peaks can be distinguished (especially for a low-porosity meso-PSi sample), which is probably due to the

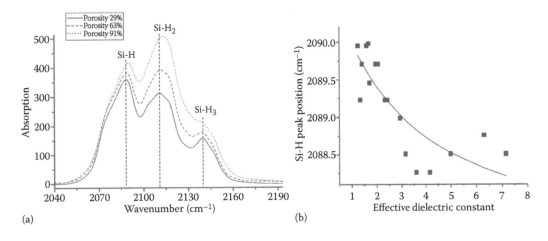

(a)   (b)

FIGURE 13.1   (a) Infrared absorption spectra of as-prepared hydrogen terminated meso-porous Si for three porosity values; (b) position of Si-H peak versus effective dielectric constant of the PSi. (From Lysenko V. et al., *Appl. Surf. Sci.* **230**, 425, 2004. Copyright 2004: Elsevier. With permission.)

specific orientations of Si-H$_x$ bonds in space induced by morphological anisotropy of Si nanocrystallites constituting the porous nanostructures. In particular, oscillation modes corresponding to Si monohydride dimmers can also be present in the range of 2075–2103 cm^{-1} (Timoshenko et al. 2003).

A slight spectral shift of the Si–H and Si–H$_2$ bands toward higher energy edge along with porosity is observed. It can be explained by influence of the porosity dependent effective dielectric constant of meso-PSi on the vibrational frequency of SiH$_x$ bonds. Indeed, experimental points corresponding to the SiH peak positions (Figure 13.1b) can be well fitted (solid curve) by the following expression proposed in (Wieder et al. 1979):

$$(13.5) \qquad \omega_{SiH} = \sqrt{\omega_0^2 - A\frac{\varepsilon_{PS}-1}{\varepsilon_{PS}+3}}$$

where ω_{SiH} is the energy of Si-H peak, ω_0 and A are fitting parameters equal to 2090 cm^{-1} and 1.2×10^4 cm^{-1}, respectively, and ε_{PS} is the porosity dependent effective dielectric constant.

Another important porosity-induced effect can also be observed from the absorption spectra. Indeed, the peak corresponding to Si–H band is higher than that of the Si–H$_2$ one for low porosity values (<63%). Thus, hydrogen is bound with Si atoms by forming mainly monohydride bonds. For high porosity values (>63%), hydrogen is mainly stored under dihydride complexes while for intermediate porosity (~63%) the peak absorption intensities of the mono- and dihydride bonds are the same. Such a remarkable progressive evolution of the Si–H$_2$ bond intensity is a sign of the porosity-dependent fractal shape of the Si nanocrystallites surface. It can be explained by increase of fractal-like roughness of the Si nanocrystallites surfaces at the atomic level favoring additional appearance of surface Si atoms with two free bonds for preferable formation of Si–H$_2$ complexes (see schematic representation in Figure 13.2). Being relatively smooth at low porosities (<63%), fractal-like surfaces of the large nanocrystallites have a low amount of Si atoms capable of binding two hydrogen atoms (Figure 13.2a). Porosity enhancement is accompanied by increase of the surface atomic roughness and therefore by the number of SiH$_2$ surface bonds (Figure 13.2b) (Lysenko et al. 2004).

Figure 13.3 shows the variation of the atomic concentration of hydrogen bound to Si atoms at the PSi-specific surface as a function of the porosity of the layer (Lysenko et al. 2005). In general, the concentration variation with porosity reflects structural evolution of the PSi-specific surface to which hydrogen is bound. Hydrogen concentration values in nano-PSi are higher than in meso-PSi layers because of the much more important specific surface to which hydrogen atoms may be bound. Indeed, as it was observed from nitrogen adsorption experiments (Herino et al. 1987) the specific surface of meso-PSi is measured to be about 200 m^2/cm^3, while the nano-PSi surface is found to be about 600 m^2/cm^3. Such an important difference of the specific surface values for the meso- and nano-PSi morphologies are explained by the difference of the nanocrystallite dimension constituting these two PSi morphological types, as can be seen in Figure 13.3b.

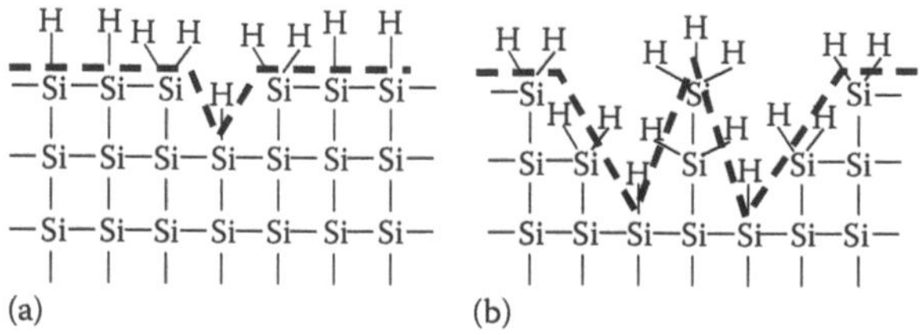

FIGURE 13.2 Schematic view of porosity dependent fractal-like roughness of Si nanocrystallites surfaces at atomic level: (a) low porosity (<60%); (b) high porosity (>60%). Dashed line points to the nanocrystallites surfaces. (From Lysenko V. et al., *Appl. Surf. Sci.* **230**, 425, 2004. Copyright 2004: Elsevier. With permission.)

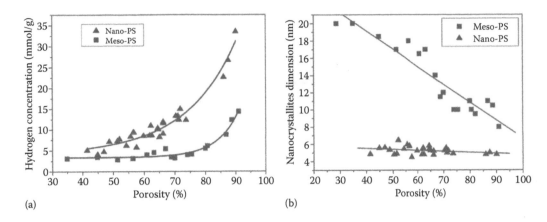

FIGURE 13.3   (a) Hydrogen concentration variation along with porosity in meso- and nano-PSi layers. Solid lines added as eye guides represent exponential fit of the experimental data. (b) Dimensions of Si nanocrystallites in meso- and nano-PSi layers versus porosity. Straight solid lines represent linear fits of the experimental data. (From Lysenko V. et al., *J. Phys. Chem. B* **109**, 19711, 2005. Copyright 2005: American Chemical Society. With permission.)

Indeed, for a given porosity value, the smaller the nanocrystallite dimension, the higher the corresponding specific surface is.

Concerning the meso-PSi samples in the 30–70% porosity range, the hydrogen concentration is very low and relatively constant (2.5–5 mmol/g) because the samples are characterized by large (Figure 13.3b) and smooth nanocrystallites. For example, the concentration of atomic hydrogen at the surface of the meso-PSi sample of 60% porosity is 3.8 mmol/g, which corresponds well to the value of the molecular hydrogen extracted from thermal desorption measurements (~2 mmol/g) (Rivolo et al. 2003). Despite the nanocrystallite decrease from 20 to 12 nm, the corresponding specific surface values remain too small (125–200 m²/g) in the meso-PSi samples with porosities <70%. The corresponding quantity of hydrogen bound at this surface is also too small to be measured with good precision. Therefore, only a slight variation of the hydrogen concentration around a constant level is found in this porosity range. Despite the almost unchangeable nanocrystallite dimensions around 10 nm, a monotonic exponential increase of the concentration (up to 13 mmol/g) in the range of high porosities (>70%) reflects an important enhancement of the PSi-specific surface induced by the extremely developed fractal-like surface roughness of the Si nanocrystallites constituting the porous layers (Lysenko et al. 2004, 2005).

The fractal nature of the PSi-specific surface roughness is completely confirmed by hydrogen concentration analysis in the nano-PSi samples. Indeed, as it can be seen in Figure 13.3a, the concentration of hydrogen increases exponentially along with porosity. An excellent exponential fit of the experimental data is represented in Figure 13.3a by a solid line. Quasi- constant values of the Si nanocrystallite dimensions along all porosity ranges means that the porosity increase is ensured only by the decrease of the number of nanocrystallites, and consequently, the observed increase of the total specific surface with porosity is only possible if an important enhancement of the fractal-like surface roughness at the nanometer scale is considered for each nanocrystallite.

Thus, the estimated content of hydrogen in PSi is quite high rather for very small nanocrystals of PSi (up to 15–35 mmol/g) and thus it can be considered quite efficient solid-state hydrogen storage material. The view of Si nanocrystals can be obtained from TEM images. Figure 13.4a shows a TEM image of a porosified grain of polycrystalline Si powder and Figure 13.4b shows spherical freestanding Si nanocrystals. As it can be seen, PSi is a sponge-like structure containing interconnected Si nanocrystals. The typical sizes of Si nanocrystals in both systems are ~3–10 nm. Immersion of nanosilicon structures in water/alcohol solutions is accompanied by evolution of hydrogen (Goller et al. 2011).

The theoretical hydrogen storage capacity of different PSi nanostructures is considered in Bunker and Smith (2011) and Song and Wu (2012).

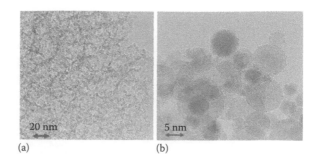

FIGURE 13.4   TEM images of nanosilicon samples: (a) Chemically porosified polycrystalline Si powder. (b) Freestanding Si nanospheres prepared from the gas phase and subsequently stain etched. The size scale is indicated by arrows. (From Goller B. et al., *Nanotechnology* **22**, 305402, 2011. Copyright 2011: IOP Publishing. With permission.)

13.4 HYDROGEN GENERATION FROM POROUS SILICON

The most widely used methods for H_2 production from PSi are

1. Heating (Gupta et al. 1988; Martin et al. 2000; Rivolo et al. 2003; Lysenko et al. 2005; He et al. 2013). This process is accompanied by thermal dissociation of SiH_x bonds and desorption of physically adsorbed atomic and molecular hydrogen.
2. Photodesorption (Collins et al. 1992; Tuyen et al. 2001; Koropecki et al. 2004; Oh et al. 2012). Absorption of photons with energy equal to or higher than corresponding binding energies of SiH_x species is accompanied by desorption of hydrogen. Usually, PSi is illuminated with ultra-violet laser (Collins et al. 1992) or visible light (Oh et al. 2012).
3. Chemical reaction with an oxidizer, for example, water (Tutov et al. 2002; Lysenko et al. 2005; Barabash et al. 2006; Bahruji et al. 2009; Tichapondwa et al. 2011; Zhan et al. 2011). The reaction between PSi and water causes desorption of hydrogen. The presence of bases, even in small amounts, drastically intensifies the reaction.

13.4.1 THERMALLY STIMULATED DESORPTION OF HYDROGEN

The desorption of hydrogen from PSi occurs when the latter is heated and it depends on how the hydrogen bound to the porous matrix as well. The convenient method to study the influence of temperature on hydrogen desorption is temperature-programmed desorption (TPD) method. Comparison between TPD curves and mass spectrum of the desorbed species shows that the desorbed phase is almost completely constituted by hydrogen molecules (Rivolo et al. 2003). A negligible contribution from desorbed silanes is found to be at least 2 orders of magnitude lower than that observed for hydrogen. Effusion curves for hydrogen desorption from fresh and aged nano-PSi samples prepared from p-Si are presented in Figure 13.5 (Lysenko et al. 2005).

Three main regions correspond to the hydrogen desorption from the fresh PSi samples: (1) a weak signal at 190°C, (2) a more intense one centered at 350°C, and (3) another one situated around 460°C. The two last intensive peaks related to the hydrogen desorption from Si-H_2 and Si-H bonds are also observed on PSi formed on *n*-type Si wafers (Martin et al. 2000). The first low temperature peak is assumed to correspond to hydrogen desorption from numerous structural defects present in the high nanoporous materials. Indeed, as indicated by the electron diffraction image (see insert in Figure 13.5), the PSi used for the thermally stimulated desorption study has amorphous structure, and therefore, atomic or even molecular hydrogen can be easily trapped in such completely disordered PSi nanostructures (Kitajima et al. 1999). TPD signal of the PSi sample after about 50 days of storage mainly preserves its general shape and is shifted by about 40°C toward the higher temperatures in comparison with the signal from the fresh samples. This shift corresponds to the storage-induced slight oxidation of the PSi samples (Benilov et al. 2007) resulting in a slight increase of binding energies of Si-H_x bonds in which the Si atoms are

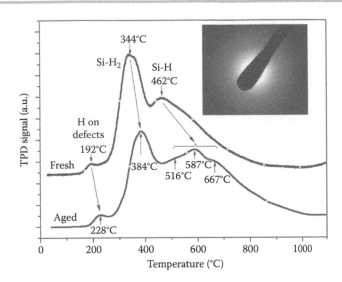

FIGURE 13.5   Effusion curves for H_2 desorption from fresh and aged nano PSi samples. Electron diffraction image given in insert reflects amorphous structure of the PSi samples. (From Lysenko V. et al., *J. Phys. Chem. B* **109**, 19711, 2005. Copyright 2005: American Chemical Society. With permission.)

back-bonded to one or more oxygen atoms, as can be deduced from IR spectroscopy (Gupta et al. 1991). Indeed, after 50 days of storage, the PSi sample has lost about 18% of the initial quantity of hydrogen. The loss of hydrogen is accompanied by an increase of silicon oxide amount due to spontaneous oxidation of PSi nanocrystallites in ambient air, as can be concluded from the growth of the IR absorption band corresponding to the Si-O vibrations (Lysenko et al. 2005).

The integration of the TPD signal corresponding to the fresh PSi sample yielded a maximal value of the total amount of desorbed molecular hydrogen (H_2) being around 33 mmol/g. It corresponds to 66 mmol/g of atomic hydrogen chemically bound at the PSi-specific surface. Taking into account that 1 g of Si contains 36 mmol of Si atoms, one can roughly conclude that each Si atom in the PSi sample is bound to two hydrogen atoms. From the fractal model point of view (Lysenko et al. 2004), it means that the fractal dimension of the nanoparticles constituting the PSi samples tends toward 2. It means that small (1–3 nm) nanocrystallites constituting the nano-PS layer can be represented as a binary solid constituted by Si and H atoms (Nychyporuk et al. 2005).

Despite the fact that the thermodesorption is a useful method to define the quantitative of hydrogen in PSi and its chemical bonding, from a practical point of view this method is rather inconvenient as an energy source for portable devices.

13.4.2 CHEMICALLY STIMULATED DESORPTION OF HYDROGEN

It is well known that Si-H bonds in the organic compounds react quite easily with water in the presence of bases (Fleming 2001). Evolution of H_2 and formation of Si-O bonds are the results as shown by the following reaction:

$$(13.6) \qquad \underset{\diagdown}{\overset{\displaystyle H}{\underset{|}{Si}}}\diagdown \;+\; H_2O \;\longrightarrow\; \underset{\diagdown}{\overset{\displaystyle OH}{\underset{|}{Si}}}\diagdown \;+\; H_2$$

Reaction 13.6 is accompanied by H_2 gas bubbling and can be used for hydrogen desorption from the surface of meso- and nano-PSi layers (Lysenko et al. 2005; Goller et al. 2011). Application of FTIR spectroscopy allows controlling Reaction 13.6 via analysis of Si-H absorption modes. As it can be seen in Figure 13.6a, the intensity of the Si-H_x wagging band near 640 cm^{-1} decreases to 25% after the sample treatment by the solution with 10^{-4} mol/L NH_3 concentration. It should be mentioned that the quantity of NH_3 molecules in such a solution is at least 50 times lower than the quantity of Si-H_x groups in the treated sample. Therefore, the ammonia acts rather as a

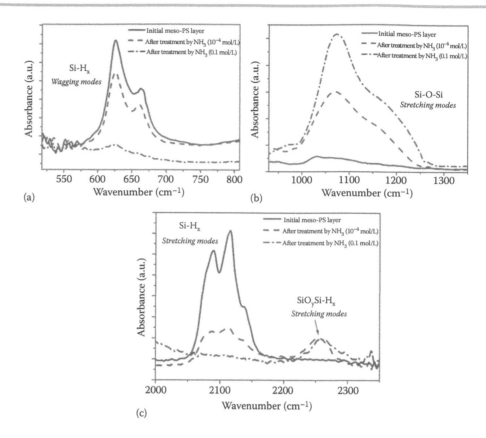

FIGURE 13.6 (a) IR absorption spectra of Si-H$_x$ wagging band corresponding to the hydrogenated meso-PSi layers treated by NH$_3$ solutions during 20 min at room temperature. IR absorption spectra of (b) Si-O-Si stretching band and (c) SiO$_y$Si-H$_x$ band appeared in the PSi samples after hydrogen desorption stimulated by NH$_3$ solutions. (From Lysenko V. et al., *J. Phys. Chem. B* **109**, 19711, 2005. Copyright 2005: American Chemical Society. With permission.)

catalyst than as a reagent in this reaction. In contrast, a treatment with 0.1 mol/L NH$_3$ solution results in complete desorption of hydrogen from the PSi surface, which is illustrated in Figure 13.6a by the disappearance of the Si-H$_x$ wagging band. Treatment of the nano-PSi samples of 80% porosity with 10^{-4} mol/L NH$_3$ solutions results in a 75% decrease of Si-H wagging band. The increase of the treatment efficiency of the nano-PSi in comparison with the meso-PSi samples is due to the much higher values of the specific surface in the nano-PSi structures.

It should also be mentioned that Reaction 13.6 is not the single one taking place during the treatment of the PSi samples with bases. Indeed, adsorption bands situated near 2260 cm^{-1} (stretching of Si-H bonds in SiO$_y$SiH$_x$ complexes) and near 1100 cm^{-1} (stretching of Si-O-Si fragments) present in the IR spectra of the PSi samples treated with NH$_3$ solutions (see Figure 13.6b,c) indicate the interaction of water with numerous highly reactive Si dangling bonds (see reaction scheme 13.7) appearing at the nanocrystallite surface after the hydrogen release induced by the initial reaction 13.6:

(13.7)

In particular, reaction of the nano-PSi samples constituted by the small Si nanoparticles (<5 nm) with 0.1 mol/LNH$_3$ results in a complete conversion of the PSi nanostructure into hydrated dioxide (SiO$_{2x}$H$_2$O), which was already observed in the IR spectra. (1) Assuming total water induced oxidation of the Si nanocrystallites from the nano-PSi samples and (2) taking into account that the oxidation of each Si atom is accompanied by the formation of two hydrogen molecules according to the following reaction:

(13.8) $$Si + 2H_2O = SiO_2 + 2H_2$$

an additional amount of the molecular hydrogen produced from the oxidation reaction is estimated to reach a value of 71 mmol/g. Thus, the maximum total amount of the molecular hydrogen issued because of Reactions 13.6 and 13.7 can be about 100 mmol per gram of the initial PSi nanostructure. Such a theoretical value for the molecular hydrogen that can be obtained from the chemical treatment of the PSi layers seems to be quite realistic, especially for the case of nano-PSi powder consisting of the smallest (2–3 nm) amorphous Si nanocrystallites (Lysenko et al. 2005).

An important issue is the ratio between the amount of hydrogen that is generated due to the rupture of silicon-hydrogen bonds in the PSi (Reaction 13.6), and hydrogen, which is released by splitting water in the presence of silicon nanoparticles (Reactions 13.7 and 13.8).

Obviously, this ratio will be determined by (1) dimension of silicon nanoparticles, (2) method of PSi production, (3) efficiency of surface coverage by hydrogen, and (4) chemical reaction conditions (temperature, catalyst type, and its concentration). Impact of these parameters on hydrogen generation for various types of PSi and Si microparticles were studied by Manilov et al. (2010a), Litvinenko et al. (2010), Goller et al. (2011), Kale et al. (2012), and Erogbogbo et al. (2013).

According to Reaction 13.8, all produced hydrogen is due to water dissociation, while the reaction of SiH$_2$ with water results:

(13.9) $$SiH_2 + 2H_2O = SiO_2 + 3H_2$$

For example, only 2/3 of the obtained hydrogen is due to the splitting of water molecules, and the other 1/3 is due to rupture of Si-H$_2$ bonds.

The conceptual illustration of a shrinking core model describing H$_2$ generation from 10 nm Si nanoparticles prepared by CO$_2$ laser pyrolysis synthesis of silicon nanoparticles is shown in Figure 13.7. At the beginning of the chemical reaction, the first 0.5 mmol of H$_2$ is generated from breaking of Si-H bonds, the following 2.0 mmol are generated from water splitting, which is accompanied by size reduction of the silicon nanocrystalline core and the silica cap grows (Erogbogbo et al. 2013).

As shown above, a large quantity of chemically bound hydrogen can be obtained from anodic etching or chemical synthesis methods. However, for high volume production of PSi at an industrial level, an electroless technology to form PSi should be preferable. Therefore, electroless stain etching of silicon powder seems to be more attractive than the electrochemical etching approach. Moreover, since the total surface area of initial silicon powder is much bigger than the area of bulk silicon substrate for the same mass, the PS production time for crystalline powder etching can be considerably reduced.

Stain etching process of silicon micro-powder was reported in Litvinenko et al. (2010). The metallurgical p-type silicon, 0.001–0.003 Ω·cm with size of particles up to several microns was used. Since size distribution in the initial powders is very large, narrower size selection of Si micro-particles constituting the powders was carried out by sedimentation technique. SEM images of the powder fractions obtained during 8 and 120 min of the sedimentation step are shown in Figure 13.8a and b, respectively. The used etching solution was H$_2$O: HF: HNO$_3$ in proportion 20:4:1 (Polisski et al. 2008). The duration of the etching process varied in the range of 10–60 min depending on the temperature and on the HNO$_3$ concentration. The etching reaction was stopped when the powder became brown. Special attention was paid to prevent dissolution of the created porous shell around the solid Si core. The reaction product in a foam form was collected, rinsed by water, ethanol and dried at room conditions. Volume of hydrogen obtained for PSi powder fabricated from different size initial silicon powder is shown in Figure 13.9 (Litvinenko et al. 2010).

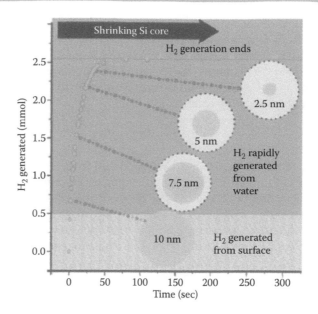

FIGURE 13.7   Shrinking core model describing H_2 generation from 10 nm Si NPs. Core spheres represent crystalline Si, and the dotted circle indicates shrinkage from the original sphere size. The light gray and dark gray areas represent hydrogen quantity generated from the hydrogenated silicon and water splitting, respectively. (From Erogbogbo F. et al., *NanoLett.* **13**, 451, 2013. Copyright 2013: American Chemical Society. With permission.)

(a)   (b)

FIGURE 13.8   SEM images of the Si powder fractions obtained during 8 (a) and 120 min (b) of the sedimentation step. (From Litvinenko S. et al., *Int. J. Hydrogen Energy* **35**, 6773, 2010. Copyright 2010: International Association of Hydrogen Energy. With permission.)

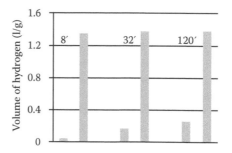

FIGURE 13.9   Volume of hydrogen produced from stain-etched silicon powder for different size fractions of the initial powders. Left (small) peaks correspond to SiH_2, right ones to H_2O for each fraction. (From Litvinenko S. et al., *Int. J. Hydrogen Energy* **35**, 6773, 2010. Copyright 2010: International Association of Hydrogen Energy. With permission.)

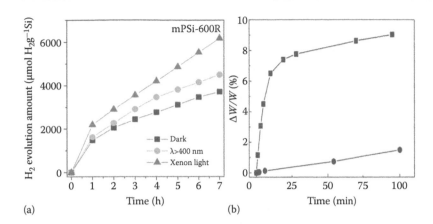

(a)   (b)

FIGURE 13.10   (a) Comparison of H_2 evolution activities of mesoporous Si under different illumination conditions. (From Dai F. et al., *Nature Commun.* 5, 3605, 2014. Copyright 2014: Nature Publishing Group. With permission.) (b) Rate of hydrogen generation in wt% versus exposure time of PSi to water/ethanol mixture (1:1) (squares) or to hydrous ethanol (circles). (From Goller B. et al., *Nanotechnology* **22**, 305402, 2011. Copyright 2011: IOP Publishing. With permission.)

13.4.3 PHOTOCATALYTIC HYDROGEN EVOLUTION

PSi is also considered a perspective candidate for hydrogen-generating materials due to photocatalytic/photoelectrochemical reactions with water (Mathews et al. 2003; Bahruji et al. 2009; Erogbogbo et al. 2013; Dai et al. 2014). Photocatalytic H_2 generation takes advantage of some of the unique abilities of semiconductors; after absorbing photons, electrons (holes) can be generated at the conduction band (valence band) of semiconductors. If the generated electrons and holes have enough energy to overcome the energetic barriers of both water reduction and oxidation and any over potentials, photocatalytic water splitting will take place (Kudo et al. 2009). Contrary to usual widegap photocatalyst materials like TiO_2, CdS, ZnS-$CuInS_2$ and so on, for which UV illumination is necessary to split water, the narrow 1.17 eV band gap of Si is quite appropriate to be photoexcited by the visible part of the solar spectrum, allowing photon capture down to the near-infrared region.

The photocatalytic activities of mesoporous silicon under different illumination conditions were studied by Dai et al. (2014). They elaborated a bottom-up synthesis of mesoporous crystalline silicon materials with high surface area and tunable primary particle/pore size via a self-templating pore formation process (these templates can be easily removed without any etchants). The advantages of these materials, such as their nanosized crystalline primary particles and high surface areas, enable increased photocatalytic hydrogen evolution rate and extended working life. Such material manifested a certain reactivity in the presence of water even under dark conditions. In addition, mesoporous crystalline silicon materials demonstrated an efficient photocatalytic reactivity under a visible light illumination (Figure 13.10a).

13.5 RATE OF HYDROGEN GENERATION

To determine hydrogen extraction kinetics from the hydrogenated PSi, it is possible to use different volumetric or gravimetric measurements (Goller et al. 2011) or recording fuel cell electrical output which is proportional to the hydrogen quantity produced during the reactions of PSi with water (Manilov et al. 2010a). The first measurement technique gives directly the volume or mass of the produced hydrogen, while the last one is much more resolved in time but it has to be calibrated. The calibration could be made, for example, by direct comparison with the first direct methods (Litvinenko et al. 2010).

The rate of hydrogen generation in water solution strongly depends on the type of silicon nanoparticles, temperature, illumination, and additives in solution. In water, freshly prepared Si nanostructures cannot be dispersed because of the hydrophobic nature of their hydrogen-terminated surface while addition of alcohol provides their efficient wetting by water (access of OH⁻ ions to the nanosilicon surface) and allows dissolution of reaction products.

Figure 13.10b shows the rate of hydrogen generation ($\Delta W/W$) in wt% with respect to the amount of PSi immersed in water/ethanol solution versus exposure time. This experiment has been performed for 2 h and the achieved $\Delta W/W$ value was about 9% (squares in Figure 13.10b). Since ethanol contains about 6% of water, an oxidation process is also possible when PSi powder is immersed in hydrous ethanol; however, the efficiency of the process is much lower (circles in Figure 13.9b). It was found that about 1:4 mixture of alcohol and water results in the highest reaction rate. Theoretical limit for the $\Delta W/W$ value, assuming that four hydrogen atoms can be generated per single Si atom, is 14%. This value has not been finally achieved because porosified polycrystalline silicon grains, according to direct observation in TEM, have a small bulk core that cannot be completely oxidized (Goller et al. 2011).

The adding of different bases (acting as catalyzers) drastically increases the generation rate. We tested three different bases: ammonia (NH_4OH), sodium hydroxide (NaOH), and potassium hydroxide (KOH), for identical concentrations of 0.21 mol/mL. This experiment was carried out to select a basis ensuring the most rapid desorption kinetics and the highest overall quantity of the generated hydrogen. The results are shown in Figure 13.11a. One can conclude that the nature of bases has a negligible influence on the amount of the generated hydrogen. In contrast, the rate of hydrogen production depends significantly on the used base. The similar result was obtained from the study of influence of base concentration on the hydrogen generation process (Figure 13.11b).

Comparing results of hydrogen generation from PSi obtained by electrochemical (anodic) etching and stain etching, one can state that H_2 is generated more rapidly with meso-PSi fabricated by electrochemical etching (Figure 13.12a). Chemically etched micropowders (Figure 13.8) treated with the H_2O:HF: HNO_3 solution are characterized by slower rates of H_2 formation, which decreases with increase of the particle size. The lowest intensities of the reaction are recorded in nonetched silicon powders. It is worth noting that a noticeable reaction of the solution containing crystalline Si is initiated at a temperature not lower than 50°C while PSi reacts already at room temperature (Manilov et al. 2010b).

The most intense generation of hydrogen from PSi is observed during the first minutes of the reaction. Such behavior of the curve is determined by the immediate breakage of Si–H_x surface groups. The developed PSi surface allows a considerable intensification of the reactions of Si–Si and Si–H bonds with water. The former effect makes a noticeable contribution to the rate of H_2 generation because silicon is less resistant to the oxidation as compared with Si-H groups, which is explained by the lower energy of the Si–Si chemical bonds (Barabash et al. 2006). PSi obtained by electrochemical etching contains the largest amount of the porous phase and, therefore, the higher number of Si–H_x groups with x ≥ 2 (Lysenko et al. 2004). The chemically etched Si powders have a less developed surface and, respectively, a smaller number of silane groups and larger dimensions

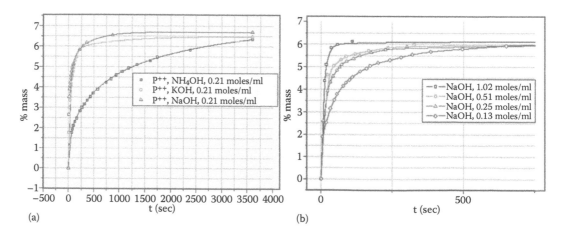

(a)

(b)

FIGURE 13.11 (a) Influence of base nature on hydrogen desorption kinetics, (b) influence of sodium hydroxide concentration on hydrogen desorption kinetics.

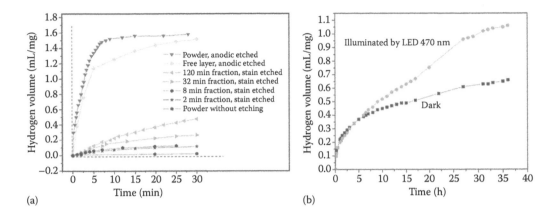

FIGURE 13.12 (a) Kinetics of hydrogen generation during 30 min of the oxidation reaction in the dark from various silicon fractions at 50°C. (b) Kinetics of hydrogen effusion from PSi fraction (32 min) in the solution under illumination with a light emitting diode (470 nm) and in the dark. (From Manilov A.I. et al., *Ukr. J. Phys.* **55**, 928, 2010. Copyright 2010: Ukrainian Journal of Physics. With permission.)

of crystalline Si cores. Moreover, such a worsening of the properties becomes more pronounced with increase in the size of particles. These considerations explain the effect of a decrease of the H_2 generation rate with increase in the size of particles in Si fractions (see Figure 13.12a). Nonetched silicon powders do not contain the porous phase and have practically no surface Si–H_x groups. In addition, the surface of nonetched silicon is partially oxidized. This results in low intensity of H_2 generation and the absence of the reaction with water at room temperature. The heating of a solution causes a rise of the penetrability of oxide layers and, therefore, stimulates the reaction. The increase in the ammonia concentration in the solution results in the intensification of the reaction. The rise of H_2 generation rate by 5–10% was recorded due to the twofold increase in the ammonia concentration. Moreover, no effect of the ammonia percentage on the total hydrogen yield was observed, which confirms the catalytic role of NH_3 (Manilov et al. 2010b).

The hydrogen generation intensifies due to the growth of the solution temperature. This fact is explained by a higher intensity of the interaction between water and silicon, as well as by an increase in the permeability of the silicon oxide layer. In addition, the only phase participating in the reaction at room temperature is the PSi phase, whereas the heating of the solution to 50°C activates the reaction of water with the crystalline phase. Respectively, the total yield of hydrogen from PSi also increases due to the heating because the c-Si residuals in the structure also get involved into the reaction. In such a way, one can separate the contributions of the crystalline and PSi phases to the total hydrogen yield by controlling temperature of the solution.

Illumination of PSi favors an increase of the hydrogen yield in the reaction, but has some peculiarities. The illumination by a red light-emitting diode had no effect on the kinetics of H_2 release. At the same time, a blue light-emitting diode (470 nm) resulted in a significant effect (Figure 13.12b). In this case, the temperature of the solution was stabilized in order to eliminate its influence on the kinetics. Similar results in the case of the ultraviolet illumination have been obtained by Bahruji et al. (2009). It is worth noting that a relative increase of the H_2 generation at the illumination (470 nm) manifests itself after a certain time of the reaction (1–5 h). At the initial moments, the kinetics of illuminated samples are similar to those of non-illuminated ones.

The maximum and the average rates of hydrogen generation for different silicon powders are shown in Figure 13.13. The maximum rate, time required to generate 1 mmol of H_2, and total produced hydrogen quantity are presented in Table 13.2. The maximum H_2 generation rate with 10 nm Si was 150 times higher than that achieved with 100 nm particles, which, in turn, was only 1.3 times that achieved with 44-μm particles. These differences cannot be accounted by differences in specific surface area. The maximum hydrogen generation from 10-nm silicon is substantially higher than for silicon, stain-etched nano-PSi, by photoactivated reaction under ultraviolet light and of the same order of magnitude as nano-Al and nano-Zn (Erogbogbo et al. 2013).

However, a highest average H_2 generation rate of 0.095 g H_2 s^{-1} g^{-1} Si (1.33 mol H_2 s^{-1} mol^{-1} Si) was obtained for mesoporous crystalline silicon materials created by the bottom-up synthesis

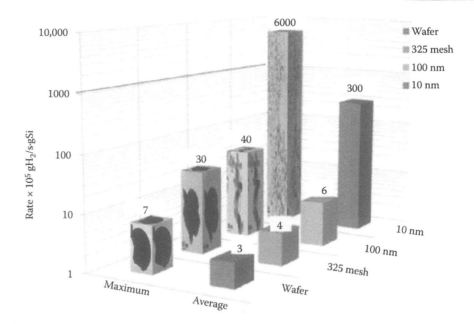

FIGURE 13.13 Comparison of hydrogen generation rates for different silicon powders. Maximum rates are in the left row of columns. Average rates are in the right row of columns. The solid line indicates the maximum reported rate for hydrogen generated from aluminum. (From Erogbogbo F. et al., *NanoLett.* **13**, 451, 2013. Copyright 2013: American Chemical Society. With permission.)

TABLE 13.2   Maximum Rate, Time Required to Generate 1 mmol of H$_2$ and Total Hydrogen Quantity mol H$_2$ per mol Si

Dimension of Particles	Maximum Rate, g(H$_2$)/s·g(Si)	Time Required to Generate 1 mmol of H$_2$, s	Total Hydrogen Produced, mol H$_2$/mol Si	Refs.
10 nm particles synthesized by laser pyrolysis	0.06	5	2.58	Erogbogbo et al. 2013
100 nm particles (Sigma Aldrich)	0.0004	811	1.25	Erogbogbo et al. 2013
<40 μm particles (Sigma Aldrich)	0.0003	3075	1.54	Erogbogbo et al. 2013
(110) Si wafer	0.00007	12.5 h	1.03	Erogbogbo et al. 2013
Stain etching particles 3–10 nm	2×10^{-8}			Goller et al. 2011
Electrochemical etched PS	2×10^{-8}			Zhan et al. 2011
Nano PS obtained by staine etching, 60°C	1.7×10^{-5}			Litvinenko et al. 2010
Photoactivated reaction under UV light	2×10^{-8}			Bahruji et al. 2009
Photocatalytic hydrogen evolution from self-templating synthesis of meso PS	0.095		1.33	Dai et al. 2014
Nano Zn and nano Al	0.01			Bunker et al. 2011; Ma and Zachariah 2010

of Si nanoparticles via a self-templating pore formation process (Dai et al. 2014). Such sample is attributed to its enlarged surface area, which provides extra contact between Si and water. This relationship between surface area and reaction rate is further demonstrated by a chemical reaction of a sample with KOH aqueous solution.

13.6 METHOD OF SUPPLEMENTARY HYDROGENIZATION

The tasks of additional hydrogen accumulation in PSi due to absorption in bulk or physical adsorption are still important. Successful experiments of enrichment of crystalline and nano-Si with hydrogen were conducted by way of restraint in molecular H_2 (Manilov et al. 2010a; Kale et al. 2012) or H_2 plasmas (Sriraman et al. 2006; Darwiche et al. 2007). Retaining of H atoms and H_2 molecules depends on a number of defects, dopants, and adsorption centers, which is increasing for more disordered surfaces. The interaction of a hydrogenated PSi surface with atomic hydrogen was theoretically and experimentally investigated (Glass et al. 1996; Manilov and Skryshevsky 2013). The hydrogen accumulation from the gaseous phase is not significant in PSi (Manilov et al. 2010a). The main reason is the initial chemically adsorbed monolayer of H atoms existing on the surface of PSi (Parkhutik 1999; Föll et al. 2002). Further, it is difficult to incorporate hydrogen inside the nanosilicon particles (Zhen-Yi et al. 2012). Therefore, only physical adsorption of hydrogen is possible, due to interaction with Si-H_x species. These processes can be modeled by methods of molecular dynamics in the field of van der Waals forces. The results of calculations show the possibility of physical hydrogen adsorption onto hydrogenated pore walls. However, sufficient desorption of physically adsorbed hydrogen from the PSi surface is obtained, which makes the long-time keeping of the adsorbate impossible. Therefore, it can explain the weak additional accumulation of hydrogen from the gaseous phase (Manilov and Skryshevsky 2013).

The maximum experimental hydrogen uptake of 2.25% at temperature of 120°C at 9.76 bar pressure was obtained by Kale et al. (2012). In this investigation, the Si nanoparticles have been synthesized by sonicating 12 μm thick and 29% porous freestanding PSi films for 4 h. Si nanoparticles have been fabricated in the range of 8–20 nm in size. Hydrogen absorption pressure composition isotherm in the pressure range of 1–10 bar and in the temperature range of 29°C–150°C is presented in Figure 13.14a. Room temperature hydrogen storage capacity of Si nanoparticles was obtained about 1.4 wt%. Hydrogen uptake by sample reduces at the temperature of 150°C. This may be attributed to the breaking of some of the Si-H_x bonds and subsequent release of hydrogen. The curves show a rising trend with increase in the pressure, meaning hydrogen uptake is directly proportional to the pressure at least in the pressure range of measurements. FTIR transmittance

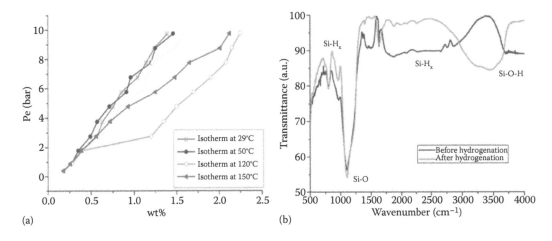

(a)   (b)

FIGURE 13.14   (a) Hydrogen absorption pressure composition isotherm of Si nanoparticles carried out in standard Seivert's type apparatus in the pressure range of 1–10 bar and in the temperature range of 29°C–150°C, (b) FTIR spectra of Si nanoparticles before hydrogenation and after hydrogenation at $P = 9.76$ bar and $T = 150$°C. (From Kale P. et al., *Int. J. Hydrogen Energy* **37**, 3741, 2012. Copyright 2010: International Association of Hydrogen Energy. With permission.)

spectrum shown in Figure 13.14b exhibits peaks around 640, 900, and 2100 cm^{-1} corresponding to wagging, stretching, and bending Si-H bonds produced during SiH$_x$ formation. Increasing sharpness of SiH$_x$ peaks in the FTIR spectra after hydrogenation was attributed to hydrogen uptake by the sample.

13.7 HYDROGEN STORAGE AND GENERATION IN PSi COMPOSITES

The additional hydrogen sorption can be realized by the creation of different PSi composites. Thus, creation of a palladium/silicon mesoporous composite is expected to modify the hydrogen generation properties of PSi. First, Pd effectively absorbs H$_2$ molecules dissociating into atoms after interaction with the metal. Therefore, using hydrogen accumulation from a gas phase, an increase of the general hydrogen content in the composite could be achieved (Rahimi and Irajizad 2006; Rather et al. 2007). The composite materials based on PSi with palladium particles incorporated inside the pores were prepared by means of Pd electroless deposition by (Manilov et al. 2010a,b) via the following schema:

(13.10) $\left[\begin{array}{c}\text{Si}-\text{H}\\\text{Si}-\text{H}\end{array}\right.$ $+\ \text{Pd}(\text{CH}_3\text{CO}_2)_2 \longrightarrow$ $\left[\begin{array}{c}\text{Si}\\\text{Si}\end{array}\right.$ $+\ \text{Pd} + 2\text{CH}_3\text{CO}_2\text{H}$

The first reaction results in decreasing of SiH$_x$ bands intensities in the spectra of PSi + Pd samples and can be easily explained by consumption of SiH$_x$ groups in the reaction of Pd^{2+} reduction:

(13.11) $\text{Si}-\text{Si} + \text{Pd}^{2+} + \text{H}_2\text{O} \longrightarrow \text{Si}^{\,\text{O}}\text{Si} + \text{Pd} + 2\text{H}^+$

The second reaction leads to appearance of silicon oxide bands and indicates the reduction of Pd^{2+} with the participation of H$_2$O in such a way that Pd has been included inside the PSi nanostructure, but is localized in a near-surface layer (approximately 50 μm thick for the PSi + 10% Pd sample). Such localization could be explained by the high rate of Pd reduction on the PSi surface in comparison with its diffusion inside the PSi layer. Furthermore, the Pd particles in the near-surface layer might block the PSi pores, preventing further Pd penetration. Apart from the Pd inside the PSi nanostructure, it is also located on the outer surface of the samples, forming relatively big (20–50 nm) nanoparticles.

Adsorption and desorption of molecular hydrogen at 50 kPa and ambient temperature as well as chemical extraction of H$_2$ during the reaction of initial PSi and PSi + Pd composites with water and base were studied (Manilov et al. 2010a,b). Reversible uptake of molecular H$_2$, probably due to the formation of palladium hydride, was detected for PSi + Pd composites, otherwise the initial PSi was found to be inactive to the H$_2$ under ambient conditions. The PSi + Pd composites react with water much slower and release less H$_2$ in comparison with initial PSi, probably due to partial pore blockage by Pd. However, the pretreatment of the composites in gaseous H$_2$ increases the volume of H$_2$ evolving in the reaction with H$_2$O at ambient temperature (Figure 13.15).

He et al. (2013) proposed to use a hydrogen storage system based on hybrid PSi single wall carbon nanotubes (SWNT). The PSi was fabricated by anodic etching of silicon wafers in alcoholic solutions of hydrogen fluoride. Direct hydrogen dosing to adsorb hydrogen on the PSi-SWNT hybrid, pure PSi, electrochemically charged PSi and SWNTs, and pure SWNTs, followed by TPD measurements, were used to quantify the hydrogen adsorption process. Before the TPD experiments, a gas mixture containing 10% H$_2$/Ar was dosed onto the sample at room temperature.

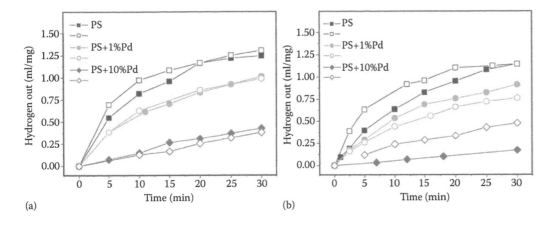

FIGURE 13.15   Hydrogen extraction kinetic from PSi and PSi+Pd in H_2O:EtOH:NH_3 (5:2:2) (a) and (5:1:1) (b) solutions. Solid symbols refer to the samples with the hydrogen gas treatment, empty symbols refer to the reference samples. (From Manilov A.I. et al., *J. Alloys Compounds* **492**, 466, 2010. Copyright 2010: Elsevier. With permission.)

The results indicate an increased hydrogen storage capacity at a lower temperature in the PSi-SWNT system relative to that from pure PSi, and pure and charged SWNTs. In addition, hydrogen adsorption in PSi-SWNT is about a factor of 2 to 6 higher than that in pure and charged PSi, and pure and charged SWNTs. In such hybrids, SWNT acts as a "catalyst" that converts inactive Si atoms to active Si atoms toward hydrogen chemisorption and as a bridge for neighbor Si atoms to allow hydrogen spillover to nearby Si atoms.

13.8 ENERGETIC ANALYSIS OF PSi NANOSTRUCTURES

Table 13.3 presents a comparative (with other hydrogen storage means) analysis of mass electrical energy that can be potentially extracted from the PSi nanostructures through a fuel cell, taking into account atomic hydrogen concentrations. Only the maximal concentration values 13, 34, and 66 mmol/g found in meso- and nano-PSi structures with enhanced porosities (>90%) are considered (Lysenko et al. 2005). Assuming quasi-identical levels for the fractal-like roughness of the specific surface for the PSi nanostructures, the observed difference in hydrogen concentrations

TABLE 13.3   Comparative Energetic Analysis of PSi Nanostructures for Their Application as Hydrogen Reservoirs in Portable Devices

Materials	Atomic Hydrogen Content (mmol/g)	Theoretical Mass Energy Density (W·h/kg)	Autonomy Operation (h)	Refs.
Meso-PSi (90%, 10 nm)	13	429	21.4	Lysenko et al. 2005
Nano-PSi (90%, 5 nm)	34	1120	56.1	Lysenko et al. 2005
Nano-PSi powder (>95%, 2–3 nm)	66	2176	108.8	Lysenko et al. 2005
$MgH_2 \rightarrow Mg + H_2$	76	2505	125.2	*MRS Bull.* 2002
$LaNi_5H_6 \rightarrow LaNi_5 + 3H_2$	14	461	23	*MRS Bull.* 2002
$(NaBH_4 + 2H_2O) \rightarrow NaBO_2 + 4H_2$	108	3560	178	*MRS Bull.* 2002
$(LiBH_4 + 4H_2O) \rightarrow LiOH + H_3BO_3 + 4H_2$	85	2802	140.1	*MRS Bull.* 2002
$NH_4BH_4 \rightarrow BN + 4H_2$	244	8043	402.1	*MRS Bull.* 2002
$NH_3BH_3 \rightarrow BN + 3H_2$	195	6428	321.4	*MRS Bull.* 2002
Li-ion batteries		150	15	Broussely et al. 1995

Note: Autonomy (h) is calculated as a time of autonomy operation of a device consuming 1 W and using 100 g of material storing hydrogen (taking into account 50% efficiency of low-temperature fuel cell).

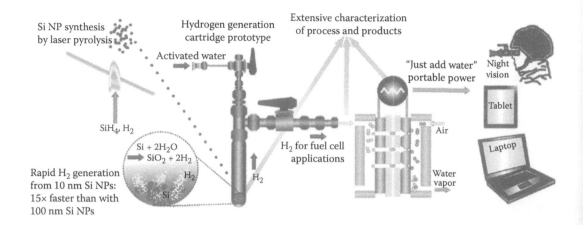

FIGURE 13.16 Schema showing CO_2 laser pyrolysis synthesis of silicon nanoparticles transferred to a custom stainless steel prototype cartridge used to generate hydrogen for fuel cell applications. (From Erogbogbo F. et al., *NanoLett.* **13**, 451, 2013. Copyright 2013: American Chemical Society. With permission.)

is explained by the similar difference in the mean dimension of the Si nanocrystallites constituting the PSi samples. The higher the hydrogen concentration, the higher the corresponding mass electrical energy is. The PSi nanostructures consisting of Si nanocrystallites with a mean dimension of about 2 nm ensure a maximal mass electrical energy value of about 2200 W·h/kg, which is quite comparable with energy values ensured by thermolysis or hydrolysis of hydrides. It is important to note that the autonomy of about 100 h could be ensured by the use of 100 g of PSi nanostructures as hydrogen reservoirs for a portable device needing 1 W of electrical power.

Application of Si nanoparticles to hydrogen generation is promising among other methods when splitting water as a hydrogen source is used. For example, efficient utilization of hydrogen generated during the reactions of nano-silicon/water in internal combustion engine has been investigated by Mehta et al. (2014). Similar to that of other nanoparticles for hydrogen generation, silicon nanoparticles are naturally expected to generate hydrogen more rapidly than bulk silicon. Figure 13.16 provides a schema showing multidisciplinary integrated approach from nanochemistry for generation of silicon nanoparticles (using the Si NP synthesis by laser pyrolisis) to the reaction with water under basic conditions allowing hydrogen generation on demand to the use of generated hydrogen in a fuel cell for portable power. Integration of nanosilicon with appropriate cartridge technologies could provide a "just add water" hydrogen-on-demand technology that would promote adoption of hydrogen fuel cells in portable power applications. However, scalable and energy-efficient processes for nanoparticle production must be implemented to expand the potential use of silicon-based H_2 generation beyond niche applications (Erogbogbo et al. 2013).

Application of Si nanoparticles as catalyst for decomposition of ammonia borane NH_3BH_3 was considered by Gangal et al. (2012). Ammonia borane has high gravimetric hydrogen storage capacity of 19.6 wt% and hydrogen release from it initiates at around 100°C and as such is compatible to meet the present-day requirements of a PEM fuel cell.

13.9 CONCLUSION

The concentration of hydrogen chemically bound to the PSi-specific surface is found to be strongly correlated to the dimension and shape of the Si nanocrystallites constituting the PSi nanostructures. Maximal values of hydrogen concentration for both meso- and nano-PSi samples are mainly due to the fractal-like shape of the nanocrystallite surface. Chemical and thermal ways for hydrogen desorption can be applied. In the last case, an additional hydrogen quantity can be generated via water splitting. Thus, the chemically induced production of the molecular hydrogen from the PSi nanostructures appears to be much more efficient than that ensured by thermal annealing. However, the chemically stimulated hydrogen desorption is completely irreversible because of the complete transformation of the PSi nanostructure into the hydrated dioxide. The PSi nanostructures consisting of Si nanocrystallites with a mean dimension of about 2 nm ensure a maximal mass electrical energy value of about 2200 W·h/kg, which is quite comparable with energy values ensured by thermolysis or hydrolysis of hydrides.

REFERENCES

Bahruji, H., Bowker, M., and Davies, P.R. (2009). Photoactivated reaction of water with silicon nanoparticles. *Int. J. Hydrogen Energy* **34**(20), 8504–8510.

Barabash, R.N., Alekseev, S.A., Zaytsev, V.N., and Barbier, D. (2006). Porous silicon modified with vinylsilanes. *Ukrainian Chem. J.* **72**(10), 78–84.

Benilov, A.I., Gavrilchenko, I.V., Benilova, I.V., Skryshevsky, V.A., and Cabrera, M. (2007). Influence of pH solution on photoluminescence of porous silicon. *Sens. Actuators A* **137**, 345–349.

Bisi, O., Ossicini, S., and Pavesi, L. (2000). Porous silicon: A quantum sponge structure for silicon based optoelectronics. *Surf. Sci. Rep.* **38**, 1–126.

Bogdanovich, B. and Schwickardi, M. (1997). Ti-doped alkali metal aluminium hydrides as potential novel reversible hydrogen storage materials *J. Alloys Compounds* **253–254**, 1–9.

Broom, D.P. (2011). *Hydrogen Storage Materials, Green Energy and Technology*. Springer-Verlag, London.

Broussely, M., Perton, F., Biensan, P. et al. (1995). Li$_x$NiO$_2$, a promising cathode for rechargeable lithium batteries. *J. Power Sources* **54**, 109–114.

Bunker, C.E. and Smith, M.J. (2011). Nanoparticles for hydrogen generation. *J. Mater. Chem.*, **21**(33), 12173–12180.

Cerofolini, G.F., Balboni, R., Bisero, D. et al. (1995). Hydrogen precipitation in highly oversaturated single-crystalline silicon. *Phys. Stat. Sol. (a)* **150**, 539–586.

Chazalviel, J.-N. and Ozanam, F. (1997). Surface modification of porous silicon. In: Canham, L. (Ed.) *Properties of Porous Silicon*. INSPEC, The Institution of Electrical Engineers, London, United Kingdom.

Collins, R.T., Tischler, M.A., and Stathis, J.H. (1992). Photoinduced hydrogen loss from porous silicon. *Appl. Phys. Lett.* **61**(14), 1649–1651.

Dai, F., Zai, J., Yi, R. et al. (2014). Bottom-up synthesis of high surface area mesoporous crystalline silicon and evaluation of its hydrogen evolution performance. *Nature Commun.* **5**, 3605, 1–11.

Dalebrook, A.F., Gan, W., Grasemann, M., Moret, S., and Laurenczy, G. (2013). Hydrogen storage: Beyond conventional methods. *Chem. Commun.* **49**, 8735–8751.

Darwiche, S., Nikravech, M., Morvan, D., Amouroux, J., and Ballutaud, D. (2007). Effects of hydrogen plasma on passivation and generation of defects in multicrystalline silicon. *Sol. Energy Mater. Sol. Cells.* **91**, 195–200.

Dzhafarov, T.D, Aydin, Y.S., and Oruc, L.C. (2008). Porous silicon based gas sensors and miniature hydrogen cells. *Jpn. J. Appl. Phys.* **47**, 8204–8207.

Erogbogbo, F., Lin, T., Tucciarone, P.M. et al. (2013). On-demand hydrogen generation using nanosilicon: Splitting water without light, heat, or electricity. *NanoLett.* **13**, 451–456.

Estreicher, S.K., Hastings, J.L., and Fedders, P.A. (1999). Hydrogen-defect interactions in Si. *Mater. Sci. Eng. B* **58**, 31–35.

Fleming, I. (2001). *Science of Synthesis: Houben-Weyl Methods of Molecular Transformations, Vol. 4: Compounds of Group 15 (As, Sb, Bi) and Silicon Compounds*. Georg Thieme Verlag, Stuttgart.

Föll, H., Christophersen, M., Carstensen, J., and Hasse, G. (2002). Formation and application of porous silicon. *Mater. Sci. Eng. B* **39**, 93–141.

Gangal, A.C., Kale, P., Edla, R., Manna, J., and Sharma, P. (2012). Study of kinetics and thermal decomposition of ammonia borane in presence of silicon nanoparticles. *Int. J. Hydrogen Energy* **37**, 6741–6748.

Gautier, G. (2014). Porous silicon and micro-fuel cells. In: Canham, L. (Ed.), *Handbook of Porous Silicon*. Springer, pp. 957–964.

Glass, J.A., Wovchko, E.A., and Yates, J.T. (1996). Reaction of atomic hydrogen with hydrogenated porous silicon—Detection of precursor to silane formation. *Surf. Sci.* **348**, 325–334.

Goller, B., Kovalev, D., and Sreseli, O. (2011). Nanosilicon in water as a source of hydrogen: Size and pH matter. *Nanotechnology* **22**, 305402, 1–4.

Gupta, P., Colvin, V.L., and George, S.M. (1988). Hydrogen desorption kinetics from monohydride and dihydride species on silicon surfaces. *Phys. Rev. B* **37**(14), 8234–8243.

Gupta, P., Dillon, A.C., Bracker, A.S., and George, S.M. (1991). FTIR studies of H$_2$O and D$_2$O decomposition on porous silicon surfaces. *Surf. Sci.* **245**(3), 360–372.

He, Z., Wang, S., Wang, X., and Iqbal, Z. (2013). Hydrogen storage in hierarchical nanoporous silicon–carbon nanotube architectures. *Int. J. Energy Res.* **37**, 754–760.

Herino, R., Bomchil, G., Barla, K., Bertrand, C., and Ginoux, J.L. (1987). Porosity and pore size distributions of porous silicon layers. *J. Electrochem. Soc.* **134**(8), 1994–2000.

Kale, P., Gangal, A.C., Edla, R., and Sharma, P. (2012). Investigation of hydrogen storage behavior of silicon nanoparticles. *Int. J. Hydrogen Energy* **37**(4), 3741–3747.

Kitajima, M., Ishioka, K., Tateishi, S. et al. (1999). Effects of crystal disorder on the molecular hydrogen formation in silicon. *Mater. Sci. Eng. B* **58**(1–2), 13–16.

Koropecki, R.R., Arce, R.D., and Schmidt, J.A. (2004). Infrared studies combined with hydrogen effusion experiments on nanostructured porous silicon. *J. Non-Cryst. Solids* **338–340**, 159–162.

Krishna, R., Titus, E., Salimian, M. et al. (2012). Hydrogen storage for energy application. In: Liu, J. (Ed.), *Hydrogen Storage*. InTech, pp. 243–266.

Kudo, A. and Miseki, Y. (2009). Heterogeneous photocatalyst materials for water splitting. *Chem. Soc. Rev.* **38**, 253–278.

Langmi, H.W., Walton, A., Al-Mamouri, M.M. et al. (2003). Hydrogen adsorption in zeolites A, X, Y and RHO. *J. Alloys Compaunds* **356–357**, 710–715.

Lee, S.M., Park, K.S., Choi, Y.C. et al. (2000). Hydrogen adsorption and storage in carbon nanotubes. *Synth. Met.* **113**(3), 209–216.

Lee, S.J., Chang-Chien, A., Cha, S.W. et al. (2002). Design and fabrication of a micro fuel cell array with "flip-flop" interconnection. *J. Power Sources* **112**(2), 410–418.

Litvinenko, S., Alekseev, S., Lysenko, V. et al. (2010). Hydrogen production from nano-porous Si powder formed by stain etching. *Int. J. Hydrogen Energy* **35**, 6773–6778.

Lysenko, V., Vitiello, J., Remaki, B., Barbier, D., and Skryshevsky, V. (2004). Nanoscale morphology dependent hydrogen coverage of meso-porous silicon. *Appl. Surf. Sci.* **230**, 425–430.

Lysenko, V., Bidault, F., Alekseev, S. et al. (2005). Study of porous silicon nanostructures as hydrogen reservoirs. *J. Phys. Chem. B* **109**, 19711–19718.

Ma, X.F. and Zachariah, M.R. (2010). Size-resolved kinetics of Zn nanocrystal hydrolysis for hydrogen generation. *Int. J. Hydrogen Energy* **35**(6), 2268–2277.

Manilov, A.I. and Skryshevsky, V.A. (2013). Hydrogen in porous silicon—A review. *Mater. Sci. Eng. B* **178**, 942–955.

Manilov, A.I., Alekseev, S.A., Skryshevsky, V.A., Litvinenko, S.V., Kuznetsov, G.V., and Lysenko, V. (2010a). Influence of palladium particles impregnation on hydrogen behavior inmeso-porous silicon. *J. Alloys Compounds* **492**, 466–472.

Manilov, A.I., Litvinenko, S.V., Alekseev, S.A., Kuznetsov, G.V, and Skryshevsky, V.A. (2010b). Use of powders and composites based on porous and crystalline silicon in the hydrogen power industry. *Ukr. J. Phys.* **55**(8), 928–935.

Martin, P., Fernandez, J.F., and Sanchez, C.R. (2000). TDS applied to investigate the hydrogen and silane desorption from porous silicon. *Phys. Stat. Sol. (a)* **182**(1), 255–260.

Mathews, N.R., Sebastian, P.J., Mathew, X., and Agarwal, V. (2003). Photoelectrochemical characterization of porous Si. *Int. J. Hydrogen Energy* **28**(6), 629–632.

Mehta, R.N., Chakraborty, M., and Parikh, P.A. (2014). Impact of hydrogen generated by splitting water with nano-silicon and nano-aluminum on diesel engine performance. *Int. J. Hydrogen Energy* **39**, 8098–8105.

Mex, L., Ponath, N., and Müller, J. (2001). Miniaturized fuel cells based on microsystem technologies. *Fuel Cells Bull.* **4**, 9–12.

Meyers, J.P. and Maynard, H.L. (2002). Design considerations for miniaturized PEM fuel cells. *J. Power Sources* **109**(1), 76–88.

MRS Bull. (2002). Special issue on hydrogen storage. **27**, 675–716.

Nijkamp, M.G., Raaymakers, J.E.M.J., van Dillen, A.J., and de Jong, K.P. (2001). Hydrogen storage using physisorption—Materials demands. *Appl. Phys. A.* **72**(5), 619–623.

Nützenadel, Ch., Züttel, A., Emmenegger, Ch., Sudan, P., and Schlapbach, L. (2000). Electrochemical storage of hydrogen in carbon single wall nanotubes. In: Tománek, D., and Enbody, R.J. (Eds.), *Science and Application of Nanotubes*. Kluwer Academic Publishing/Plenum Press, Dordrecht, pp. 205–214.

Nychyporuk, T., Lysenko, V., and Barbier, D. (2005). Fractal nature of porous silicon nanocrystallites. *Phys. Rev. B* **71**(11), 115402.

Ogata, Y., Niki, H., Sakka, T., and Iwasaki, M. (1995). Hydrogen in porous silicon: Vibrational analysis of SiH_x species. *J. Electrochem. Soc.* **142**, 195–201.

Oh, I., Kye, J., and Hwang, S. (2012). Enhanced photoelectrochemical hydrogen production from silicon nanowire array photocathode. *Nano Lett.* **12**, 298–302.

Parkhutik, V. (1999). Porous silicon—Mechanisms of growth and applications. *Solid State Electron.* **43**, 1121–1141.

Polisski, S., Lapkin, A., Goller, B., and Kovalev, D. (2008). Hybrid metal/silicon nanocomposite systems and their catalytic activity. In: *Proc. 6th Int. Conf. on Porous Semiconductors Science and Technology, PSST 2008*, Mallorca, Spain, p. 44.

Pritchard, R.E., Ashwin, M.J., Newman, R.C., and Tucker, J.H. (1999). H_2 molecules in crystalline silicon. *Mater. Sci. Eng. B* **58**, 1–5.

Rahimi, F. and Irajizad, A. (2006). Effective factors on Pd growth on porous silicon by electroless-plating: Response to hydrogen. *Sens. Actuators B* **115**(1), 164–169.

Rather, S.U., Zacharia, R., Hwang, S.W., Naik, M.U., and Nahm, K.S. (2007). Hyperstoichiometric hydrogen storage in monodispersed palladium nanoparticles. *Chem. Phys. Lett.* **438**(1–3), 78–84.

Rivolo, P., Geobaldo, F., Rocchia, M., Amato, G., Rossi, A.M., and Garrone, E. (2003). Joint FTIR and TPD study of hydrogen desorption from p^+-type porous silicon. *Phys. Stat. Sol. (a)* **197**(1), 217–221.

Rosi, N.L., Eckert, J., Eddaoudi, M. et al. (2003). Hydrogen storage in microporous metal-organic frameworks. *Science* **300**, 1127–1129.

Shanks, H., Fang, C.J., Ley, L., Cardona, M., Demond, F.J., and Kalbitzer, S. (1980). Infrared spectrum and structure of hydrogenated amorphous silicon. *Phys. Stat. Sol. (b)* **100**(1), 43–56.

Song, X. and Wu, J. (2012). Modeling of hydrogen weight storage capacity in solid porous silicon. *Adv. Mater.s Res.* **415–417**, 2322–2328.

Sriraman, S., Valipa, M.S., Aydil, E.S., and Maroudas, D. (2006). Hydrogen-induced crystallization of amorphous silicon thin films. I. Simulation and analysis of film postgrowth treatment with H_2 plasmas. *J. Appl. Phys.* **100**, art. 053514.

Strobel, R., Jörissen, L., Schliermann, T. et al. (1999). Hydrogen adsorption on carbon materials. *J. Power Sources* **84**(2), 221–224.

Switendick, A.C. (1978). The change in electronic properties on hydrogen alloying and hydride formation. In: Alefeld, G. and Völkl, J. (Eds.), *Hydrogen in Metals I. Topics in Applied Physics.* **28**. Springer, Berlin, pp. 101–129.

Tichapondwa, S.M., Focke, W.W., Del Fabbro, O., Mkhize, S., and Muller, E. (2011). Suppressing H_2 evolution by silicon powder dispersions. *J. Energetic Mater.* **29**, 326–343.

Timoshenko, V.Y., Osminkina, L.A., Efimova, A.I. et al. (2003). Anisotropy of optical absorption in birefringent porous silicon. *Phys. Rev. B* **67**(11), 113405.

Tolstoy, V.P., Chernyshova, I.V., and Skryshevsky, V.A. (2003). *Handbook of Infrared Spectroscopy of Ultrathin Films.* Wiley, New York.

Tutov, E.A., Pavlenko, M.N., Protasova, I.V., and Kashkarov, V.M. (2002). The interaction of porous silicon with water: A chemographic effect. *Tech. Phys. Lett.* **28**(9), 729–731.

Tuyen, L.T.T., Tam, N.T.T., Quang, N.H., Nghia, N.X., Khang, D.D., and Khoi, P.H. (2001). Study on hydrogen reactivity with surface chemical species of nanocrystalline porous silicon. *Mater. Sci. Eng. C* **15**(1–2), 133–135.

Tzimas, E., Filiou, C., Peteves, S.D., and Veyret, J.-B. (2003). *Hydrogen Storage: State-of-the-Art and Future, Perspective.* European Commission. DG JRC. Institute for Energy, the Netherlands.

Weitkamp, J., Fritz, M., and Ernst, S. (1995). Zeolites as media for hydrogen storage. *Int. J. Hydrogen Energy.* **20**(12), 967–1002.

Westlake, D.J. (1983). Hydrides of intermetallic compounds: A review of stabilities, stoichiometries and preferred hydrogen sites. *J. Less-Common Metals* **91**, 1–20.

Wieder, H., Cardona, M., and Guarnieri, C.R. (1979). Vibrational spectrum of hydrogenated amorphous SiC films. *Phys. Stat. Sol. (b)* **92**(1), 99–112.

Yamazaki, Y. (2004). Application of MEMS technology to micro fuel cells. *Electrochim. Acta* **50**(2–3), 663–666.

Zhan, C.Y., Chu, P.K., Ren, D. et al. (2011). Release of hydrogen during transformation from porous silicon to silicon oxide at normal temperature. *Int. J. Hydrogen Energy* **36**(7), 4513–4517.

Zhen-Yi, N.I., Xiao-Dong, P.I., and De-Ren, Y. (2012). Can hydrogen be incorporated inside silicon nanocrystals? *Chin. Phys. Lett.* **29**(7), 077801.

Züttel, A., Wenger, P., Rentsch, S., Sudan, P., Mauron, Ph., and Emmenegger, Ch. (2003). LiBH$_4$ a new hydrogen storage material. *J. Power Sources* **118**(1–2), 1–7.

PSi-Based Microreactors

Caitlin Baker and James L. Gole

CONTENTS

14.1	Introduction	298
14.2	PSi Matrix for Heterogeneous Catalysis	300
	14.2.1 PSi Supported Metal Catalysts	301
	14.2.2 PSi Supported Photocatalysis	304
14.3	PSi-Based Microreactors	307
	14.3.1 Flow-Through Membrane PSi Microreactors	308
	14.3.2 PSi Layer Enhancement of Microreactors	313
14.4	Conclusion and Outlook	316
References		316

14

14.1 INTRODUCTION

Chemical synthesis, analysis of chemical kinetics, and process development all benefit from the many advantages offered by microreactors (Jensen 2001). The submillimeter dimensions and large surface area to volume ratio of microreactors, in contrast to macroscale systems and batch reaction vessels, allow for increased heat and mass transfer for gas, fluid, and multiphase reactions (Löwe and Ehrfeld 1999; Roumanie et al. 2008). Microreactors can be operated in a continuous flow configuration, circumventing possible inconsistencies arising from batch-wise reaction processes. Precise control of reaction conditions, such as isothermally performing highly exothermic reactions, results in high selectivity, high product yield, and minimal waste (Löwe and Ehrfeld 1999; Cao et al. 2014). Furthermore, microreactors provide safe conditions for dangerous reactions such as the production of toxic, flammable, or explosive chemicals (Roumanie et al. 2008).

Scientific studies on microreactor applications have shown linear growth over the past 20 years (Seelam et al. 2013). Microreactors are particularly useful for surface-supported heterogeneous catalytic reactions, as only small volumes of catalyst are necessary, the catalyst is inherently separated from the reaction products, and essential thermal conditions are easily managed (Löwe and Ehrfeld 1999; Roumanie et al. 2008; Cao et al. 2014). These conditions allow for swift, consistent, and reliable screening of catalytic materials and optimal reaction conditions (Tiggelaar and Gardeniers 2009). Low reaction volumes are ideal not only for laboratory use but also for on-site and on-demand production of fine chemicals (Llorca et al. 2008). Examples include hydrogen fuel production through steam reforming, membrane reactor hydrogenation and dehydrogenation, and catalytic oxidation and reduction for gas purification, (Splinter et al. 2002; Ye et al. 2005; Llorca et al. 2008; Liu et al. 2014). Biosensing and glucose monitoring microreactors have been fabricated by the immobilization of enzymes or glucose oxidizing catalysts (Laurell 2002; Ensafi et al. 2014).

The geometries, such as size and architecture, and construction materials for microreactors are widely varied and chosen based on the intended reaction characteristics. Conventional microreactors are a single microchannel (Figure 14.1a), but multichannels (Figure 14.1b) are preferred as they have a lower pressure drop across the active area and are better suited for isothermal operation (Laurell 2002; Seelam et al. 2013). Membrane microreactors (Figure 14.2) have the advantage of acting as a selective barrier for separating reaction products (Seelam et al. 2013). Ideal thermal

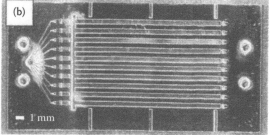

FIGURE 14.1   (a) Single channel packed bed microreactor loaded with activated carbon. The hashmarks indicate where two photomicrographs have been spliced so that the entire device may be viewed. (b) Multi-channel microreactor. (From Losey M.W. et al., *Ind. Eng. Chem. Res.* **40**(12), 2555, 2001. Copyright 2001: American Chemical Society. With permission.)

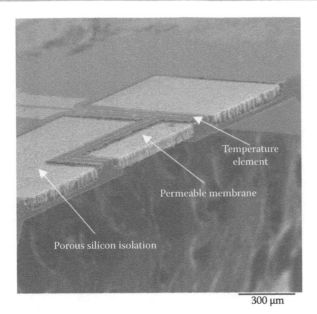

FIGURE 14.2   SEM of PSi permeable micro membrane with PSi thermal isolation and integrated heating element. (From Splinter A. et al., *Sens. Actuators B* **83**, 169, 2002. Copyright 2002: Elsevier. With permission.)

and fluid properties can be optimized through a combination of computational fluid dynamics and partial differential equations modeling mass, momentum, and energy balance. Theoretical modeling analyses take into consideration the variables of turnover rate, residence time, number of microchannels, and throughput (Kothare 2006; Seelam et al. 2013). Kim and Kwon (2006) calculated the optimal parameters of reactor volume, channel width, reactor height, and catalyst thickness for a hydrogen production microreactor combining a plug flow reactor volume equation, a power-law reaction rate expression, the energy conservation equation, and the Darcy flow equation.

Microreactors can be fabricated from polymers, ceramics, stainless steel, and silicon through microfabrication techniques such as microlithography, LIGA (a combination of lithography, electroforming, and replication), laser ablation, wet and dry chemical etching, and a variety of bonding techniques (Löwe and Ehrfeld 1999; Jensen 2001; Rebrov et al. 2001; Seelam et al. 2013). For testing and characterization, the microreactor is typically integrated into metal housing using epoxy with fluidic connections for mass flow controllers and analysis equipment (Llorca et al. 2008; Tiggelaar and Gardeniers 2009). Figure 14.3 shows the integration of a silicon microreactor into a stainless steel housing with a ceramic heater system and a schematic of the experimental setup. Connections to the microreactor are necessarily gas-tight or leak-free, making use of gaskets, sealing O-rings, and other mechanical sealants.

Silicon is a particularly favorable material for microreactors. Robust micromachining processes, due to the extensive use of silicon in the electronics and semiconductor industry, offer a high level of precision to create a variety of structures with small feature size (Pattekar and Kothare 2004). Silicon is chemically inert, thermally stable, mechanically rigid, and has beneficial electronic properties. Additionally, silicon micromachining allows for facile integration of sensors, actuators, and heating elements, illustrated in Figure 14.4 (Ye et al. 2005).

The effectiveness of heterogeneous catalysts supported on microreactor surfaces is highly dependent on the active surface area. In order to increase the catalytic performance of microreactors, various modifications have been developed to further enhance the surface area to volume ratio. This was conventionally accomplished using a fixed bed of microchannels packed with catalyst pellets (Figure 14.1) (Losey et al. 2001; Park et al. 2006), by introducing flow obstacles such as pillars (Roumanie et al. 2008), or by depositing a high-surface area catalyst coating such as zeolite (Rebrov et al. 2001). Porous silicon (PSi), formed by electrochemical anodization, is a promising material for microreactor surface-enlarging and heterogeneous catalyst support.

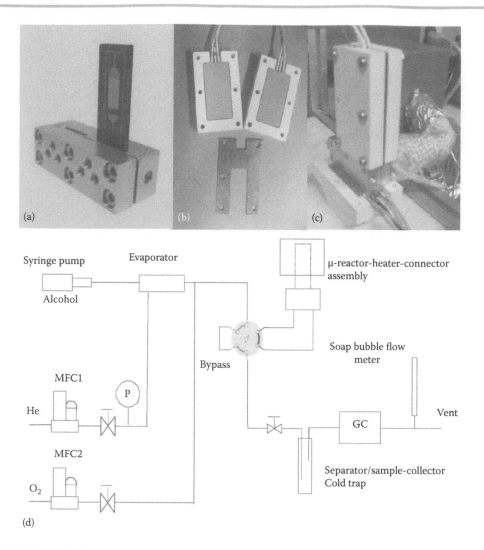

FIGURE 14.3   (a) Silicon-glass microreactor for gas-phase oxidation of rosalva inserted into a stainless steel housing and connector, (b) disassembled ceramic heater system, and (c) the complete assembled microreactor and heater system. (d) Schematic of experimental setup. (From Cao E. et al., *Processes* **2**(1), 141, 2014. Published by MDPI AG as open access.)

Freshly prepared PSi features a hydrogen-terminated surface, which can act as a reducing agent to easily form noble metal catalyst nanoparticles throughout the porous structure (Nakamura et al. 2011). Additionally, deep quantum confinement due to the presence of nanometer-sized features combined with the ability to absorb a broad spectrum of light make PSi an excellent candidate for photocatalysis (Qu et al. 2010; Su et al. 2013). As an added benefit for microreactor technology, PSi can serve as a thermal insulation barrier (Figures 14.2 and 14.4), which allows for local heating (Splinter et al. 2002; Ye et al. 2005). This chapter will focus on the design, fabrication, and advantages of PSi-based microreactors.

14.2 PSi MATRIX FOR HETEROGENEOUS CATALYSIS

Heterogeneous catalysts are those that are in a different phase than the reactants. Most commonly, the catalyst is a solid on which the gas or liquid phase reactants absorb. Once the catalyzed reaction is complete, the products then desorb from the catalyst. A great advantage of heterogeneous catalysis over homogeneous catalysis is the ease of separating the catalyst from the reaction products. Immobilization of a catalyst onto a support can facilitate this separation process in both traditional batch reactions and in the context of microreactors (Roumanie et al. 2008).

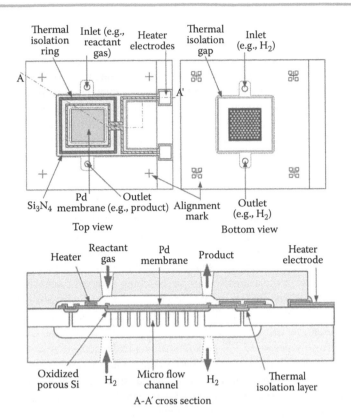

FIGURE 14.4 Schematic of Pd membrane microreactor on oxidized PSi support. (From Ye S. et al., *J. Micromech. Microeng.* **15**(11), 2011, 2005. Copyright 2005: IOP Publishing. With permission.)

Catalytic activity is often dependent on the support surface. A wide range of catalyst supports has been studied to enhance the catalytic activity. Catalysts supports are often inert, enhancing the catalytic activity by providing a high surface area per volume for dispersed catalysts and are thermally, chemically, and mechanically stable (Lokteva et al. 2008; Polisski et al. 2010). However, semiconductor supports, traditionally reducible oxides, offer the ability to electronically interact with the metal catalysts to further enhance catalytic sites. Ioannides and Verykios (1996) developed a theoretical model showing that electron transfer from the support to the metal to be the cause of observed changes in catalytic activity. Also, the transfer of electrons influences the electronic structure of the semiconductor support, creating additional catalytic sites on the support surface near the metal particles (Ioannides and Verykios 1996). PSi is a desirable choice for a catalyst support due to the high surface area porous structure and the semiconducting properties of the bulk silicon (Polisski et al. 2010). PSi can also enhance the activity of supported photocatalysts. Under irradiation, the small band gap (2.0 eV) of PSi contributes additional electron-hole pairs to the supported photocatalyst (Tank et al. 2011).

There are many methods of depositing catalysts onto the support. Metal catalyst deposition onto PSi is often accomplished by sputtering (Roumanie et al. 2008; Cao et al. 2014), galvanic reactions (Ensafi et al. 2014), wash coating, or by adding metal chlorides to the PSi etch solution (Nakamura et al. 2011). The hydrogen-terminated surface of PSi is ideal for the reduction of metal salts to form metal nanoparticle catalysts (Polisski et al. 2008; Nakamura et al. 2011; Ensafi et al. 2014; Liu et al. 2014).

14.2.1 PSi SUPPORTED METAL CATALYSTS

After an initial study confirmed the high catalytic activity of Pt loaded PSi powders, Polisski et al. (2008, 2010) systematically studied the activity of metal catalysts supported on PSi through the test reaction of CO oxidation. As a less expensive alternative to etched silicon wafers, low-cost metallurgical-grade polycrystalline Si powder was porosified through a stain-etching procedure (Polisski et al. 2008, 2010). For characterization studies, a complementary PSi layer with almost identical pore morphology was prepared via electrochemical etching on p-type silicon. Polisski et al. (2008, 2010) took advantage of the highly efficient reductive feature of the hydrogen-terminated PSi for the metal catalyst loading. The PSi powder, after being evacuated to remove etching residues, was immersed into metal salt solutions at various low temperatures and salt concentrations. As the hydrogen-terminated PSi surface is hydrophobic (Polisski et al. 2010), alcoholic solvents were used to assure wetting of the surface within the pores. The reduction reaction was monitored by measuring the pH of evolved HCl vapors. Once the reaction was complete, signaling the formation of the metal nanoparticles on the PSi, the Pt/PSi composite was carefully dried and washed in ethanol and warm water several times and then collected by centrifugation. Additionally, Pt/PSiO$_x$ samples were prepared for comparison catalytic studies by annealing the Pt/PSi composite at 1100°C for various lengths of time. Metal catalyst loading slightly lowered both the surface area and volume per gram of the PSi (160 m^2 g^{-1} and 0.3 cm^3 g^{-1}) to final values of 92 m^2 g^{-1} and 0.23 cm^3 g^{-1} for the Pt/PSi composite and 70 m^2 g^{-1} and 0.15 cm^3 g^{-1} for the Pt/PSiO$_x$ samples. Polisski et al. (2010) suggest that the lowered surface area and volume per gram are due to the metal nanoparticles forming in the mesopores and in the case of the annealed sample, blocking some of the narrower pores.

Polisski et al. (2008, 2010) demonstrated the importance of the hydrogen-terminated surface for Pt nanoparticle deposition by attempting a metal salt reduction on hydrogen effused PSi powder samples. Fourier transform infrared (FTIR) spectroscopy confirmed the presence or absence of Si-H$_x$ bonds in the freshly prepared PSi powders versus hydrogen-effused samples. The samples, prepared by heating to 400°C, did not participate in any reductive activity, confirming the importance of Si-H$_x$ groups in nanoparticle formation. Metal nanoparticle formation dependence on metal salt concentration and exposure time was investigated by *in situ* plasma resonance monitoring (Polisski et al. 2008). Metal plasma oscillation frequency response spectra were collected as the PSi layer was submerged in an Au salt solution in ethanol of various concentrations. A resonance peak at 580 nm appeared as Au metal nanoparticles were formed, distinct from the bulk resonance peak at 634 nm. Polisski et al. (2008) associated the integral area under the spectral curve with the quantity of reduced particles to create a gauge for comparing nanoparticle formation at different concentrations and exposure time. By correlating results with optical microscopy of the Au loaded PSi layers (Polisski et al. 2010), the authors determined that low temperatures and low salt concentrations allowed for a slow enough reduction rate that the salt solution filled the porous structure and then reduced for optimal homogeneity and maximum depth of the formed metal nanoparticles.

The catalytic activity was tested by flowing a CO/O$_2$ gas mixture over the Pt/PSi powder and CO$_2$ output was monitored by a gas-sampling mass spectrometer. Polisski et al. (2008, 2010) found that samples with at least a 0.9 wt% loading of Pt exhibited a sharp light-off temperature, the temperature of maximum change in reaction rate, shown in Figure 14.5. This is correlated with the

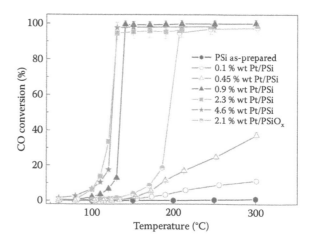

FIGURE 14.5 CO conversion as a function of temperature for different Pt/PSi and Pt/PSiO$_x$ catalysts. (From Polisski S. et al., *J. Catal.* **271**, 59, 2010. Copyright 2010: Elsevier. With permission.)

presence of metallic Pt, which is much more catalytically active than PtO, as confirmed by XPS analysis. The sharpness of the light-off curves indicates uniform metal nanoparticle size, associated with the optimal low temperature preparation described above. For a normalized comparison of the catalytic activity of the various prepared Pt/PSi composites with conventionally synthesized Pt nanoparticles, the turnover frequency (TOF) was calculated at particular CO conversion rates and temperatures. An optimal Pt loading of 2.3 wt% yielded a light-off temperature at 115°C, complete CO conversion at 120°C, and a TOF of 344 h^{-1} with 10% CO conversion. While the Pt/PSiO$_x$ sample with a similar Pt loading of 2.1 wt% had a slightly better TOF of 363 h^{-1}, the light-off temperature was much higher at 180°C. Higher Pt loading resulted in lowered catalytic activity due to lower dispersion of the Pt nanoparticles (Polisski et al. 2010). Comparatively, Pt nanoparticles, radiolitically and chemically reduced then stabilized on nonporous SiO$_2$, reported a much better TOF of 510 h^{-1}, but with an unfavorably higher light-off temperature of 150°C. The authors report that the Pt/PSi composite also performed better than Pt supported on Al$_2$O$_3$ (TOF of 180 h^{-1}). Additionally, the Pt/PSi catalysts were stable after 60 h of continuous testing. Polisski et al. (2008, 2010) suggest that the PSi supplies two key features that greatly benefit the metal assisted catalysis: the hydrogen-terminated surface allows for *in situ* synthesis of metal nanoparticles and the high surface area, even after metal nanoparticle loading, facilitates mass transport of reactants.

Pt/PSi composite powders also show promise for assisting hydrogen production from methanol combustion (Nakamura et al. 2011). A novel method to directly obtain metal-PSi composite powders was employed to create powders with room temperature catalytic activity (Nakamura et al. 2011). Beginning with metallurgical-grade polycrystalline silicon powder, PtCl$_2$ was slowly added to an HF etch solution over 10 min to varying final concentrations at an elevated temperature of 50°C. During the 60-min etch, the Pt/PSi composite powders formed as Pt layers were deposited while the Si layers dissolved (Nakamura et al. 2011). Prior to drying in air at room temperature for 24 h, the Pt/PSi composite powders were rinsed with distilled water to remove etching residue, a critical step for catalytic activity. Nakamura et al. (2011) found that omitting the rinse resulted in an unreactive material, likely due to residual HF oxidizing the metalic Pt. Catalytic activity was measured by monitoring the reaction temperature of an air/CH$_3$OH mixture passed over the Pt/PSi composite powder.

Spontaneous combustion of methanol, in the absence of external heat, was observed in a reactor loaded with Pt/PSi composite powder. Similar to the relationship between Pt loading and CO oxidation catalytic activity reported by Polisski et al. (2010), a maximum Pt loading was found for methanol combustion. A PtCl$_2$ concentration of 0.8 g/L in the etch solution produced higher maximum reactor temperatures than both lower and higher metal salt concentrations. Nakamura et al. (2011) suggest that while increasing the amount of Pt nanoparticles will enhance catalytic activity, an increase of Pt nanoparticle size will decrease the surface area, decreasing catalytic activity. As evidence for the catalytic benefits of PSi, Pt nanoparticles electrolessly plated on nonporous Si showed no enhanced catalytic activity, exhibited as a complete lack of temperature elevation in the reactor. The Pt/PSi catalysts also showed good stability, actually increasing catalytic activity with use. This may possibly be due to combustion of surface contaminants (Nakamura et al. 2011).

The hydrogen-terminated PSi surface is also advantageous for the *in situ* formation of silver nanoparticles. Catalytic reduction of nitro-aromatics has been achieved with the use of Ag nanoparticles reduced from a metal salt into the PSi layer on a silicon chip (Liu et al. 2014). Catalytic activity of the Ag/PSi chip was measured by monitoring the color change of an aqueous solution containing p-nitroaniline (p-NA) and NaBH$_4$ upon submersion of the chip. The color change, quantified by ultraviolet-visible (UV-VIS) spectroscopy, indicated a complete reduction of nitro-compounds in p-NA within 120 min. The reduction reaction did not occur in the absence of a catalyst. Although hydroxides, produced by the reduction of nitro-aromatics, are deleterious to the PSi structure, the Ag/PSi chip catalyst was rejuvenated by immersion in 30% nitric acid for several minutes followed by redeposition of the Ag nanoparticles. By utilizing a PSi layer on a chip, rather than PSi powder, the separation of catalyst and products is a simple process of just picking up the chip (Liu et al. 2014).

Conventionally facilitated through enzymatic activation, copper nanoparticles supported on PSi have shown promise in catalytically assisting glucose oxidation (Ensafi et al. 2014). Copper salts were reduced into the pores of PSi powder in a manner similar to the methods described above. The powders were incorporated into a carbon paste electrode (CPE) to function as a glucose sensor. The electrocatalytic activity of the Cu/PSi composite powders was evaluated through the use of cyclic voltammetry and chronoamperometry. Ensafi et al. (2014) reported that the glucose sensor amperometrically responded to the presence of glucose concentrations as low as 1 µmol dm^{-3} in less than 4 s. The sensor also featured long time stability, reproducibility, and selectivity to glucose in the presence of common coexisting organic substances (Ensafi et al. 2014).

14.2.2 PSi SUPPORTED PHOTOCATALYSIS

Heterogeneous photocatalysis is the process where a semiconducting catalyst absorbs photons and generates an electron-hole pair, which catalyzes redox reactions. The photocatalytic mechanism depends on various aspects of the semiconductor band structure. The band gap determines the effective range of radiation wavelengths; a photon must be absorbed and able to excite an electron from the valence to the conduction band. Reduction by the excited electron requires the base of the conduction band to be at a more negative potential than the reaction potential. Similarly, oxidation by the electron holes requires that the top of the valence band must be at a more positive potential (Abrams and Vesborg 2013). Additionally, a slow rate of electron-hole recombination ensures a higher probability for the electron-hole pair to participate in redox reactions, and thus creates a more effective photocatalyst (Tank et al. 2011). A primary application of photocatalysis is environmental remediation, which includes oxidation of organic species for cleaning wastewater, removal of volatile organic compounds and smog from air, and creating active self-cleaning surfaces (Abrams and Vesborg 2013). Photocatalysts are particularly well suited for environmental remediation because they operate at or below room temperature (Abrams and Vesborg 2013).

TiO_2 is one of the most studied photocatalysts due to its chemical inertness, high reactivity, nontoxicity, and large recombination rate (Tank et al. 2011; Abrams and Vesborg 2013; Schneider et al. 2014). Tank et al. (2011) studied the photocatalytic activity of TiO_2 nanoparticles supported on PSi and under the influence of an electric field. The PSi support, formed by electrochemically etching p-type Si wafers with a resistivity of 2 Ω·cm, served as both nanoparticle template and electrode for application of the electric field. The TiO_2 nanoparticles were synthesized in a torch operated DC arc thermal plasma reactor, using a gas phase condensation method (Tank et al. 2011). Evaporated titanium metal reacted with surrounding oxygen and the resulting metal oxide condensed on the walls of the reaction chamber. X-ray diffraction (XRD) demonstrated the crystallinity of the synthesized TiO_2 and transmission electron microscopy (TEM) determined the nanoparticles to be spherical in shape and range in size from 5 nm to 50 nm. The TiO_2 nanoparticles were deposited onto the PSi by an electrophoretic method (Tank et al. 2011). This method involved applying an electric field between the PSi and a platinum counter electrode in a solution of the TiO_2 nanoparticles dispersed in isopropyl alcohol. The samples were then dried in vacuum and heated to 100°C.

The photocatalytic activity of the PSi supported TiO_2 nanoparticles was evaluated by the degradation of methylene blue under UV radiation. In this reaction, the electrons excited to the conduction band reduce adsorbed oxygen molecules to form a superoxide and the holes reacted with water to form hydroxide radicals (Tank et al. 2011; Abrams and Vesborg 2013). To test the enhancement effect of an electric field, the samples were mounted to one of two platinum parallel plates and an electric potential ranging from −8 V to 8 V was applied. Tank et al. (2011) found that the photocatalytic degradation could be enhanced up to 18% with an applied potential of −3 V. The authors explained that the applied voltage shifted the band structure of the TiO_2 to a more negative position relative to the reduction and oxidation levels of the aqueous solution. The reducing electrons in the conduction band were more effective, being shifted to a much more negative potential than the reduction potential of the oxygen molecules. On the other hand, the ability of the holes to form hydroxide radicals was limited as the potentials were shifted closer. However, Tank et al. (2011) note that the formation of the superoxide is the dominant mechanism of the photodegradation of methylene blue, and so the overall photocatalytic activity was increased. Additionally, the photo-excited electrons in the PSi support also contributed to the reduction process (Tank et al. 2011).

Due to either deep quantum confinement or complex surface electronic states, PSi exhibits light absorption across a broad spectral range, from the UV to the visible to the near infrared, beyond that of TiO_2 (Qu et al. 2010; Su et al. 2013). Qu et al. (2010) utilized both the function as a metal catalyst support and the photocatalytic properties of PSi for the photodegradation of indigo carmine (IC) and 4-nitrophenol (4NP). The SiNWs were fabricated by metal-assisted chemical etching of *n*-type Si wafers (Figure 14.6). Pt nanoparticles, synthesized separately, were introduced to the SiNWs treated with aminopropyl-trimethoxy silane (APTMS) and dispersed in ethanol. The APTMS caused the SiNWs to have a positively charged surface, which assisted the electrostatic attraction of the Pt nanoparticles. SiNWs before and after Pt loading are shown in Figure 14.6. Qu et al. (2010) explained that the Schottky barrier formed between the metal and semiconductor (Pt and Si) can enhance the photocatalytic activity in the SiNW in addition to the isolated catalytic function of the metal. On absorption of a photon, an electron-hole pair is formed and then separated in the SiNW (Wang et al. 2011), in this case assisted by the presence of a Schottky barrier (Qu et al. 2010). In an aqueous environment, the separated electron-hole pair can react with water and dissolved O_2 to produce reactive oxygen species that can then oxidize organic pollutants (Qu et al. 2010; Wang et al. 2011). The Pt loaded SiNWs efficiently photocatalytically degraded IC and 4NP and even out-performed similarly prepared TiO_2 nanoparticles under the same experimental conditions in the presence of IR radiation (Qu et al. 2010). Although stable for up to 10 cycles of IC photodegradation, the SiNWs developed a thin layer of silicon oxide during the photodegradation of 4NP (Qu et al. 2010).

Wang et al. (2011) synthesized both porous and nonporous SiNWs by metal-assisted etching of heavily doped or lightly doped *p*- and *n*-type Si wafers. While both types of SiNWs demonstrated good photocatalytic activity for the degradation of methyl red, the heavily doped

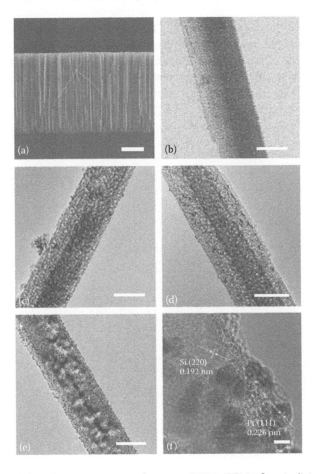

FIGURE 14.6 SEM of the (a) cross-section of porous SiNWs, TEM of an individual SiNW (b) before and (c–e) after Pt loading, and (f) HR-TEM image of Pt loaded *p*-type SiNW. (From Qu Y. et al., *J. Mater. Chem.* **20**(18), 3591, 2010. Copyright 2010: The Royal Society of Chemistry. With permission.)

porous SiNWs exhibited activity approximately two times greater than the nonporous SiNWs (Wang et al. 2011). However, hydrogen surface termination required recovery by HF treatment due to oxidation during the catalytic process, although the morphology of the SiNWs held good stability.

In an effort to improve stability in aqueous solutions, Su et al. (2013) developed a hierarchical PSi photocatalyst as an alternative to nanowires. Macroporous silicon with a nanoporous surface (NP-MPSi) was fabricated by anodic etching of *n*-type silicon in a 49% HF/ethanol solution with a volume ratio of 1:1 at a current density of 50 mA/cm^2 under irradiation of an Xe lamp. A pore size analyzer revealed that about 80% of the pore volume was due to pores with a diameter less than 20 nm, 60% of which had diameters less than 5 nm. Quantum confinement, which requires nanosilicon structures at least 5 nm or smaller, is therefore possible in the hierarchically structured NP-MPSi (Su et al. 2013). For comparison, PSi containing only macropores (MPSi) was fabricated by the same anodic etching procedure in the absence of the Xe lamp with a 49% HF/ethanol volume ratio of 1:4 (Su et al. 2013). Figure 14.7 shows the macroporous structure of MPSi and NP-MPSi and the distinct presence of nanopores only in the NP-MPSi sample.

Photocatalytic degradation of phenol was used to measure the photocatalytic oxidation ability of the NP-MPSi (Su et al. 2013). The concentration of phenol was measured by high performance liquid chromatography (HPLC) over the course of 5 h under Xe lamp irradiation. Approximately 50% of phenol was removed by both direct photolysis and photocatalytic degradation using the MPSi catalyst. In contrast, over 95% of phenol was removed in the presence of the NP-MPSi photocatalyst. Su et al. (2013) calculated that the degradation rate of phenol was almost 2.5 times greater for the hierarchically structured PSi than the MPSi. To test for visible light photocatalytic activity of the NP-MPSi, the Xe lamp was equipped with a 420-nm optical filter to remove UV light. Photolysis had no effect on the phenol degradation but 37% of phenol was removed in the presence of the NP-MPSi.

Fourier transform infrared spectroscopy (FTIR) and X-ray photoelectron spectroscopy (XPS) were performed on the MPSi and NP-MPSi before and after the photocatalysis process to investigate the stability of the photocatalysts (Su et al. 2013). FTIR revealed that oxides were formed on the surface of MPSi during the 5 h photocatalytic reaction and XPS confirmed that Si was oxidized to SiO$_2$. However, the NP-MPSi displayed negligible changes in the peaks of the FTIR and XPS spectra, indicating no oxidation of the PSi surface took place. Su et al. (2013) suggest that this is likely due to the quantum confinement effect, causing a saturation of dangling Si bonds

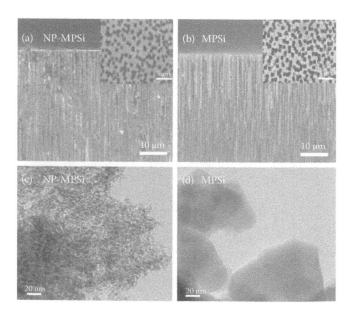

FIGURE 14.7   SEM (a, b) and TEM (c, d) images of NP-MPSi and MPSi. (From Su J. et al., *Appl. Catal. B* **138–139**, 427, 2013. Copyright 2013: Elsevier. With permission.)

with hydrogen, as the only difference between MPSi and NP-MPSi is the presence of sub-5-nm pores. Additionally, the NP-MPSi photocatalyst was used five times in successive phenol degradation experiments and held a constant efficiency above 90%. These combined results indicate that the NP-MPSi is an excellent photocatalyst, both for its ability to degrade pollutants and remain stable in aqueous solutions.

14.3 PSi-BASED MICROREACTORS

The unique attributes of PSi offer an alternative to traditional microreactor designs. Taking advantage of the channel structure of PSi, bulk silicon can be removed to expose the backside of a PSi layer to create a flow-through PSi membrane reactor, shown in Figure 14.8 (Splinter et al. 2002; Presting et al. 2004; Llorca et al. 2008). Due to the ease of etching, device fabrication, and microelectronic integration, PSi had already been recognized as an ideal candidate for devices requiring membrane functionalities. Examples include a PSi membrane filled with Nafion® (D'Arrigo et al. 2003; Pichonat and Gauthier-Manuel 2006) or sulfuric acid (Gold et al. 2004) as a proton exchange membrane fuel cell (PEMFC) or as a PEMFC integrated hydrogen gas diffusion layer (Gautier et al. 2012). As a microreactor, the PSi membranes have been evaluated as hydrocarbon steam reformers (Llorca et al. 2008) and CO preferential oxidation reactors (Splinter et al. 2002; Divins et al. 2013). However, this approach requires many steps of fabrication including predefining pore locations, etching the pores, and finally opening the backside of the pores.

Another approach of applying the advantages of PSi to microreactor technology is to form a PSi layer onto the microchannel walls of a fixed bed microreactor. Electrochemical etching of a silicon wafer with preformed reaction channels will produce a surface layer on which metal catalysts (Cao et al. 2014) or enzymes (Drott et al. 1999), the catalysts of nature, can be deposited. Roumanie et al. (2008) have reduced the processing complexity further by taking advantage of black silicon, an "inverse" of PSi, which features the same high surface area and can be fabricated using additional steps with common silicon microfabrication tools rather than a separate etching process.

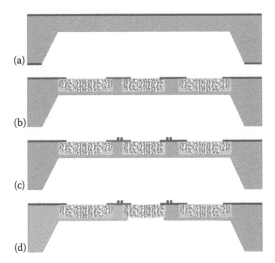

FIGURE 14.8 Schematic of a fabrication process of a micro membrane reactor. (a) A 1-μm layer of SiO_2 is used as a mask and the backside of the bulk silicon is etched by a wet etch process, (b) a low stress Si_3N_4 layer is deposited as a mask before an HF to form the PSi layer, (c) deposition of a temperature resistor heating element, and (d) the backside of the PSi membrane layer is exposed by an advanced silicon etching (ASE) process. (From Splinter A. et al., *Sens. Actuators B* **83**, 169, 2002. Copyright 2002: Elsevier. With permission.)

14.3.1 FLOW-THROUGH MEMBRANE PSi MICROREACTORS

Palladium membrane reactors are commonly used in industry for hydrogen purification or hydrogenation/dehydrogenation due to the ability of Pd to serve as both a permeable membrane and a catalyst (Ye et al. 2005). Conventional Pd membrane reactors are fabricated from a self-supported Pd layer, which must be relatively thick to maintain mechanical strength. However, as hydrogen flux through the membrane is inversely proportional to the thickness, there is much interest in producing a membrane with a thin Pd layer supported on a porous material (Seelam et al. 2013). Ye et al. (2005) have developed a thin Pd membrane microreactor supported on oxidized PSi (Figure 14.4). PSi is an ideal support material as Si wafers bond to glass reactor caps with a tight gas seal, the PSi support allows for a high-pressure differential across the Pd membrane that is required for processes such as hydrogenation, and a PSi ring provides integrated thermal isolation (Ye et al. 2005).

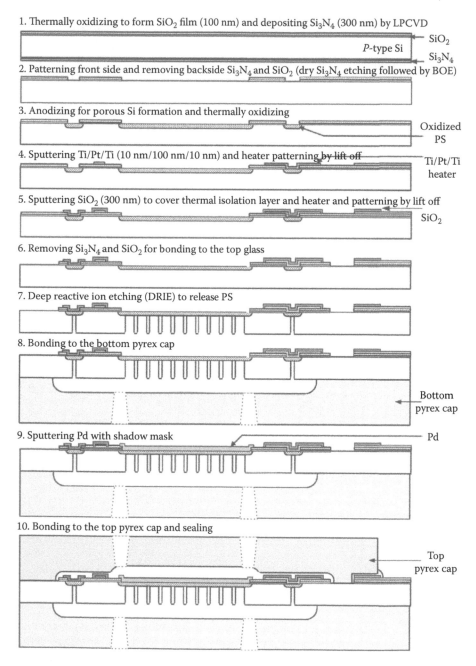

FIGURE 14.9 Fabrication process of Pd membrane on oxidized PSi microreactor. (From Ye S. et al., *J. Micromech. Microeng.* **15**(11), 2011, 2005. Copyright 2005: IOP Publishing. With permission.)

As illustrated in Figure 14.9, the porous membrane support is fabricated from a *p*-type Si wafer with a glass top and bottom cap to direct the hydrogen, reaction gas, and products. Standard film deposition and photolithography of a Si_3N_4/SiO_2 film is used to create a mask for PSi formation for the membrane and thermal isolation. A 30-μm thick layer of PSi is formed by anodization in an HF-ethanol electrolyte solution. The PSi layer is then oxidized in dry oxygen at 300°C, followed by wet oxygen atmosphere at 600°C, and annealed in nitrogen at 700°C. The oxidation of the PSi provides a stop for the DRIE, which opens up the backside of the PSi to create the porous membrane (Ye et al. 2005). An integrated Pt/Ti heater is deposited by a sputtering and lift off process and covered with SiO_2 to prevent decomposition. The Pd catalyst and membrane layer was deposited to the PSi by sputtering at 200°C, without an adhesion layer to maximize hydrogen flux.

Ye et al. (2005) studied the thermal isolation, hydrogen permeation and selectivity, and hydrogenation of 1-butene by a PSi supported Pd membrane reactor. An infrared thermal image of the microreactor (Figure 14.10) demonstrates the effectiveness of the PSi thermal isolation ring. When the oxidized PSi supported Pd membrane was heated to 90°C, the PSi ring caused an average temperature difference of 20°C between the membrane and the surrounding Si substrate. Hydrogen permeation was evaluated by supplying pressure-regulated hydrogen to the lower side of the membrane while a nitrogen stream carried any permeated hydrogen to a gas chromatography system for measurement. Helium and nitrogen permeation was tested under the same conditions to determine the membrane selectivity for hydrogen, defined as the ratio of hydrogen flux to helium or nitrogen flux. The authors found the oxidized PSi supported Pd membrane to have hydrogen selectivity comparable to the more widely studied Pd membrane sputtered on a γ-alumina support. The hydrogen permeation of the oxidized PSi supported membrane was from 40% to over 600% greater than reported values for the γ-alumina supported membrane. This was attributed to good adhesion of the Pd layer to the smooth surface of the oxidized PSi ensuring gas tightness (Ye et al. 2005).

Hydrogenation of 1-butene was carried out in the same manner as the permeability experiment to demonstrate the realization of the Pd membrane microreactor as a device for the synthesis of fine chemicals. While hydrogen was supplied through the lower Pyrex glass cap, the reactant gas of 1% 1-butene and 5% nitrogen in argon was supplied to the top of the membrane (Figure 14.10). An inline micro gas chromatograph measured the reaction products to determine conversion of 1-butene by hydrogenation. At a sufficiently low flow rate, correlating to a long residence time of 0.24 s, Ye et al. (2005) reported a 100% conversion of 1-butene. The favorable hydrogen permeability/selectivity and good mechanical strength to withstand large pressure differences suggests that the oxidized PSi supported Pd membrane microreactor has great potential for hydrogen purification and hydrogenation/dehydrogenation applications. Additionally, the PSi thermal isolation ring allows for an efficient device with lower energy consumption.

Due to the difficulties and hazards of storing and handling hydrogen for fuel cells, Llorca et al. (2008) created a macroporous silicon microreformer for onsite production of hydrogen via catalytic steam reforming of ethanol. Although ethanol requires a higher reforming temperature

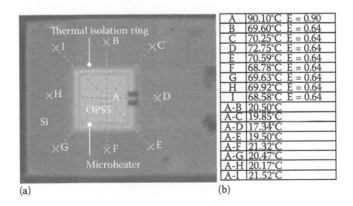

A	90.10°C	E = 0.90
B	69.60°C	E = 0.64
C	70.25°C	E = 0.64
D	72.75°C	E = 0.64
E	70.59°C	E = 0.64
F	68.78°C	E = 0.64
G	69.63°C	E = 0.64
H	69.92°C	E = 0.64
I	68.58°C	E = 0.64
A-B	20.50°C	
A-C	19.85°C	
A-D	17.34°C	
A-E	19.50°C	
A-F	21.32°C	
A-G	20.47°C	
A-H	20.17°C	
A-I	21.52°C	

(a)   (b)

FIGURE 14.10 (a) Infrared thermal image of microreactor showing thermal isolation of PSi ring and (b) the temperature table of each point indicated on the image. OPSS indicated the oxidized PSi support. (From Ye S. et al., *J. Micromech. Microeng.* **15**(11), 2011, 2005. Copyright 2005: IOP Publishing. With permission.)

than the more extensively studied methanol steam reformers, ethanol yields more hydrogen on a molar basis and is also a renewable biofuel. Macroporous silicon is used for the first time as a catalyst support structure to assist in the chemical reaction to produce hydrogen as a parallel channel flow-through microdevice (Llorca et al. 2008). This represents a good choice as silicon is stable to oxidation at the higher temperatures required for ethanol reforming (Gold et al. 2004).

The macroporous silicon microreformer was fabricated in two stages: an elegant electrochemical etching process to create the array of microchannels (Figure 14.11), followed by a deposition of Co_3O_4-ZnO catalyst onto the pore walls (Llorca et al. 2008). The square array of 220-μm deep pores was lithographically prestructured on n-type float zone silicon wafers. Predefining the position of pores is essential for PSi membrane reactors to ensure that membranes can be stacked without dead-end channels (Presting et al. 2004). An earlier study on the fabrication of PSi membranes for the purpose of hydrocarbon reforming used an SiN_x etch mask to predefine pore locations (Presting et al. 2004). Inverted pyramids, created by tetramethylammonium hydroxide (TMAH) etching, guided the subsequent electrochemical etch to create pores every 4 μm. With a solution of 5 wt% HF and 0.1 mmol Triton X-100 surfactant, the etch was carried out at 288 K for 330 min with backside illumination. The etch current was reduced as the etch progressed and the HF diffused into the growing pores. To turn the pore structure into a membrane of channels, the backside of the silicon was etched away by a TMAH solution at 358 K. The recently created front-side porous structure was protected from the TMAH etch by oxidation in O_2 at 1373 K, and the resultant oxide was removed by HF as a final step (Llorca et al. 2008).

Traditional methods of deposition were not sufficient to produce a continuous and even catalyst layer in the newly created high aspect ratio microchannels. Thus, catalyst deposition of zinc and cobalt oxides in a consistent layer approximately 100 nm thick to the walls of the macroporous silcon membrane (Figure 14.12) was completed through a novel complexation-decomposition

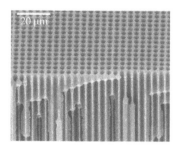

FIGURE 14.11   SEM image of macroporous silicon membrane. (From Llorca J. et al., *J. Catal.* **255**, 228, 2008. Copyright 2008: Elsevier. With permission.)

(a)   (b)

FIGURE 14.12   Microchannel (a) before and (b) after deposition of zinc and cobalt oxide catalyst layer. (From Llorca J. et al., *J. Catal.* **255**, 228, 2008. Copyright 2008: Elsevier. With permission.)

route. First, a dimethylketone solution containing Zn^{2+} and Co^{2+}, with urea as a complexing agent, was forced into the macropores using a vacuum pump until all air was removed and the channels were completely filled. Second, the filled membrane was slowly heated to 348 K for 3 h and dried at 393 K for 15 h, forming intermediate complexes, confirmed by IR spectroscopic analysis in a separate experiment. The membrane then underwent thermal decomposition and calcination at 673 K for 2 h leaving only a layer of ZnO and Co_3O_4 particles. The final chemical composition and deposition morphology were confirmed by EDX and SEM analysis. Llorca et al. (2008) estimate that each catalyst coated microchannel features an internal surface area per volume of approximately 1.3×10^6 m^2/m^3. Finally, a 1-h treatment with a helium-hydrogen mixture at 723 K activated the catalyst for ethanol steam reforming by partially reducing the cobalt oxide to the redox pair $Co^{\delta+}/Co^0$ (Llorca et al. 2008). The activated catalyst layer containing ZnO, Co_3O_4, and now metallic Co was confirmed by XRD and HRTEM.

The microreformer was constructed by gluing steel washers to the catalyst coated microporous membrane with epoxy and then integrating into a steel housing. Catalytic experiments were carried out inside a furnace as varying molar ratios of ethanol and water diluted in He were passed through the microreformer at atmospheric pressure. Reaction products were measured continuously by online mass spectrometry and compared against a noncatalyst loaded macroporous silicon microreformer. An optimal operation temperature of 773 K and steam-to-carbon ratio S/C = 3 produced a high conversion rate of ethanol and selectivity for H_2 (73.4 vol%). Only minimal amounts of CH_4, a product of ethanol thermal decomposition, were produced by the catalyst-loaded microreformer. Additionally, the residence time per microchannel was less than 5 ms. Llorca et al. (2008) suggest that the produced microreformer could theoretically produce enough H_2 to power a small device requiring several watts.

In a follow-up study, Casanovas et al. (2009) compared the microporous silicon microchannel reformer to conventional monoliths and stainless steel microreactors with channel diameters of approximately 0.9 mm and 0.7 mm. The ethanol reformers were all coated with Co_3O_4 catalyst by an *in situ* thermal decomposition of double-layered cobalt hydroxide salts in a process similar to that described earlier. The deposition method was ideal for the macroporous silicon channels as well as the ceramic and stainless steel microreactors. SEM images confirmed the consistent even catalyst layer and ultrasound exposure and mechanical vibration tests caused less than 2.5% catalyst weight loss for the ceramic and stainless steel microreactors.

Due to drastically different operating conditions, the large number of differing variables from one microreactor to the next makes a direct comparison difficult. Casanovas et al. (2009) compared the ceramic monolith, the stainless steel microreactor, and the Si-micromonolith (macroporous silicon channels) at similar ethanol conversion levels. As shown in Figure 14.13, the silicon microreactor outperforms the other microreactors. In summary, the much larger surface area per

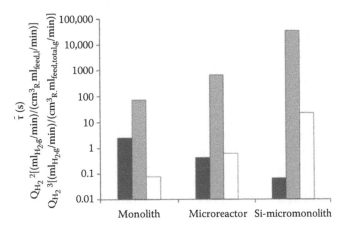

FIGURE 14.13 Comparison of catalytic performance of a ceramic monolith, stainless steel microreactor, and the macroporous Si-micromonolith microchannel reactor. Specific H_2 production rates (gray and white) and residence time (black) for an ethanol conversion level of about 33% highlight the multiple order of magnitude increase in performance of the macroporous Si microreformer. (From Casanovas A. et al., *Catal. Today* **143**, 32, 2009. Copyright 2009: Elsevier. With permission.)

unit reactor volume of the macroporous silicon monoliths creates a vastly superior microreactor. Hydrogen production rates are better not only on a per reactor volume basis but on a per unit catalyst mass basis as well. Also, the residence time is three orders of magnitude shorter for the Si-micromonolith for comparable ethanol conversion levels (Figure 14.13).

López et al. (2010) expanded on the previous study creating an experimental proof of concept ethanol reformer device featuring three microchannel membranes in series. The portable fuel cell feeding device dimensions measured $40 \times 40 \times 19$ mm. Performance was evaluated as a function of operation temperature, liquid feed load, feed composition (S/C ratio), and long-term operating conditions. The reforming reaction of ethanol was dominant over the decomposition reaction at an operating temperature of 773 K. Optimal operating conditions with a residence time between 0.4 s and 0.9 s resulted in almost complete ethanol conversion and a molar ratio of 1:6 ethanol to water produced a favorable balance of performance and energy demand. The device also suffered no performance degradation after 250 h of use. López et al. (2010) compared the results to a conventional cordierite monolith and conventional microreactor by normalizing performance by microreactor volume and injection load. The macroporous silicon microchannel outperformed the cordierite monolith with 20 times more specific hydrogen production with a residence time that is 7 times less while achieving the same selectivity to H_2 over other reaction products. In comparison to the conventional microreactor, the silicon microreactor featured an H_2 output two orders of magnitude better with a residence time 5 times lower, again at the same selectivity for H_2. The advantages of the macroporous silicon stem from the ability to form high surface areas at low volume.

The CO by-product of the ethanol steam reformer, although minimal, is still enough to poison the anode electrocatalysts of the proton exchange membrane fuel cells (PEMFC) (Divins et al. 2013). In an earlier study, for the purpose of enhancing the selectivity of resistive gas sensors to special gases in the presence of CO, Splinter et al. (2002) created a flow-through PSi micro membrane reactor coated with a palladium catalyst to convert CO into nondetectable CO_2 (Figures 14.2 and 14.8). Shown in Figure 14.2, the micro membrane reactor featured a 70-μm thick PSi layer that serves as both the membrane, opened from the backside by advanced silicon etching (ASE), and as a thermal insulator between an on-chip heating element and the bulk silicon. The on-chip heating element raised the temperature of the reactor to 140°C to assist the reaction. Palladium was introduced into the pores by submersing the entire silicon reactor into a palladium acetylacetonate solution with a toluene solvent. Catalytic activity of the micro membrane reactor was then measured as a drop in detection signal of an in-series resistive gas sensor, as the CO was converted to the nondetectable CO_2.

By functionalizing the silicon microchannels originally created for ethanol steam reforming (Llorca et al. 2008) with an Au/TiO$_2$ catalyst, Divins et al. (2011, 2013) created a CO preferential oxidation (CO-PrOx) microreactor. This CO-PrOx microreactor could then be incorporated downstream of the ethanol steam reformer as part of a complete micro-fuel cell feeding device. The catalyst deposition method involves a several hour five-step process, based on the previous complexation-decomposition method, and features the same ideal results of homogeneity of layer thickness and evenness along with good adherence. Here, the silicon micromonolith converts CO at a rate two orders of magnitude greater than that of a conventional cordierite monolith with the same catalyst coating (Divins et al. 2011, 2013).

Beginning with the macroporous silicon micromonoliths (Figure 14.11), the CO-PrOx catalysts were incorporated and activated in five steps. First, a thin layer of SiO_2 was developed on the microchannel walls by a 30-min treatment of O_2 at 1373 K. Similar to the earlier catalyst application process, titanium isopropoxide was forced into the microchannels, which then interacted with the terminal hydroxyl groups of the SiO_2 layer. Next, decomposition at 723 K for 4 h resulted in a homogeneous TiO_2 layer. Gold nanoparticles were introduced by flowing carbosilanethiol dendron protected gold nanoparticles in a toluene solution through the microchannels. The dendron encapsulation of the gold nanoparticles ensured an even distribution onto the TiO_2 layer on the microchannel walls and prevented nanoparticle aggregation, which is a common issue associated with other deposition methods (Divins et al. 2013). The even layers of catalysts are shown in Figure 14.14. A final treatment of calcining at 673 K activated the catalytic sites for CO-PrOx.

The resulting microreactor was effective at CO conversion and Divins et al. (2013) reported a 50% conversion of CO to CO_2 at around 473 K and complete conversion (CO concentrations below 5 ppm) at temperatures above 673 K. As the CO-PrOx microreactor is to be used in series

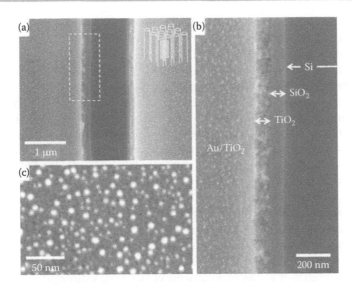

FIGURE 14.14 SEM images of cross-section of macroporous silicon channels with catalyst layers. Closeup view (b) of channel wall (a) shows separate deposited layers and gold nanoparticles (c) in Au/TiO₂ layer. (From Divins N.J., López E., Roig M. et al., *Chem. Eng. J.*, **167**, 597, 2011. Copyright 2011: Elsevier. With permission.)

with an ethanol steam reformer for the ultimate goal of H_2 production for fuel cells, selectivity of CO oxidation over the undesirable H_2 oxidation was measured as well. While an increase in reaction temperature increases the conversion of CO, the selectivity decreases. Similarly, increasing ratios of H_2/CO also enhances CO conversion yet also decreases selectivity. However, Divins et al. (2013) calculate that approximately 1 mL of the catalyst coated microchannels could convert CO at a rate required to operate in series with a hydrogen production device feeding a 3W proton exchange membrane micro fuel cell. An alternate method of selectively removing CO from H_2 would potentially be to use an advantageously sized mesh or membrane (Ye et al. 2005) to physically trap the larger CO molecules.

14.3.2 PSi LAYER ENHANCEMENT OF MICROREACTORS

To demonstrate the advantages of PSi enhanced microreactors for heterogeneously catalyzed multiphase reactions, Losey et al. (2002) fabricated and compared two multiple parallel microchannel reactors. The first device featured open channels with a packed-bed of finely divided porous particles and the second incorporated a thin PSi catalyst support layer covering microstructured channels. The PSi enhanced microstructured channels were constructed to mimic the standard microreactor structure of the packed-bed open channels while offering the ability to precisely define reactor properties such as support size, shape, arrangement, and void fraction, which is difficult to accomplish in a packed-bed design. The two contrasting microreactor designs were compared using the gas-liquid-solid hydrogenation of cyclohexene as a test reaction.

Both the open-channel packed bed and PSi coated microstructured reactors were fabricated using standard silicon microfabrication techniques. Both reactors featured 300 μm deep, 625 μm wide, and 20 mm long parallel channels formed by DRIE. A thick layer of resist was used to pattern the channels. For the PSi coated microstructure reactor, a staggered array of 50 μm diameter columns was patterned and formed during the same DRIE process step. The PSi layer was created by anodization under illumination with a silicon nitride etch mask to confine the PSi formation to the microstructured channels. A platinum catalyst was applied by flowing a Pt containing precursor solution through the reactor channels and then heating in air at 550°C for 3 h. The Pt catalyst was then reduced by flowing hydrogen at 300°C for 3 h. For the open-channel packed-bed reactor, a filter was formed at the end of the channels to contain the catalytic particles. The particles were incorporated into the channel by flowing slurry containing 50 μm to 75 μm diameter standard catalytic particles at a carefully controlled flow rate. Both microreactors were sealed by anodically bonding a glass wafer to the silicon wafer (Ye et al. 2005).

Losey et al. (2002) chose the catalytic hydrogenation of cyclohexane to test and compare the gas-liquid reaction rate and mass transfer coefficient of the two microreactors. Both the open channel packed-bed reactor and the PSi-coated microstructured reactor demonstrated a mass transfer coefficient more than 100 times larger than that of conventional laboratory reactors. Optimizing parameters that affect mass transfer, such as packing density and distribution, is facilitated by the microfabrication techniques of the microstructured reactor. Additionally, fabricating consistent optimal features across multiple devices can be more easily obtained. Losey et al. (2002) also found that the reaction rates were over 40% higher for the PSi coated microstructured reactor than the catalyst particle packed-bed reactor.

Roumanie et al. (2008) used DRIE to create a high aspect ratio nanostructuration on the silicon microreactor surface called black silicon. Black silicon, a type of PSi named for its light absorbing property, can be fabricated via DRIE, laser chemical etching, pulsed electrochemical etching, fast atom beam etching, or recently by metal-assisted chemical etching (Lu and Barron 2013). The needle-like structure of black silicon, shown in Figure 14.15, has been described as an "inverted" PSi structure (Stubenrauch et al. 2006). Roumanie et al. (2008) estimated that the planar surface area of the silicon surface increased by a factor of 15 after the formation of black silicon, greatly increasing the potential active surface area of a deposited catalyst.

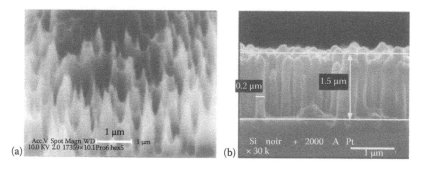

FIGURE 14.15   Black silicon (a) formed by DRIE and after Pd catalyst deposition (b). (From Roumanie M., Delattre C., Mittler F. et al., *Chem. Eng. J.* **135**, S317, 2008. Copyright 2008: Elsevier. With permission.)

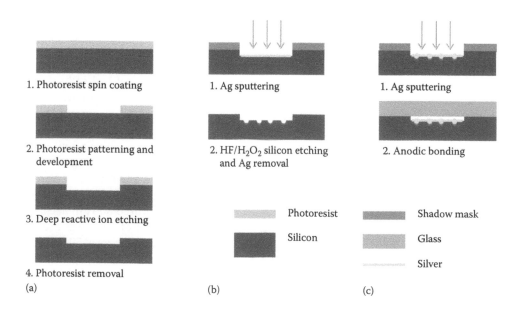

FIGURE 14.16   Schematic illustration of the microreactor fabrication process: (a) photolithography and deep reactive ion etching, (b) PSi formation, and (c) catalyst deposition and glass bonding. (From Cao E. et al., *Processes* **2**(1), 141, 2014. Published by MDPI AG as open access.)

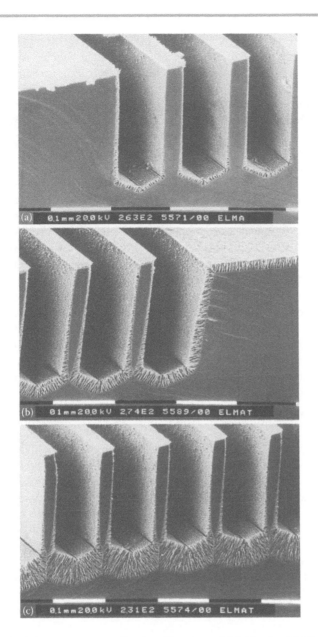

FIGURE 14.17 SEM of the reactors anodized at 10 mA cm^{-2} (a) 10 min, (b) 25 min, and (c) 50 min. (From Drott J. et al., *Microchim. Acta*, **131**, 117, 1999. Copyright 1999: Springer Science and Business Media. With permission.)

Black silicon was prepared on planar silicon, which was then diced and fitted into a fixed bed microreactor (Roumanie et al. 2008). Pt nanoparticles were deposited on both the planar silicon and black silicon (Figure 14.15b) surface by cathodic sputtering. Catalytic activity enhancement was measured by comparing the conversion of CO to CO_2 for Pt deposited surfaces with and without the black silicon nanostructuration. Roumanie et al. (2008) reported a factor of 10 increase in surface activity on the microreactors with the black silicon. For comparison, a porous alumina layer deposited on the planar silicon, requiring a several step washcoating procedure, exhibited an even greater increase in catalytic activity. However, black silicon fabrication is comparatively much simpler as it is produced by standard silicon micromachining fabrication methods (Roumanie et al. 2008).

To enhance the performance of the catalytic oxidation of rosalva to costenal within a microstructured reactor (Figure 14.3), Cao et al. (2014) supported a silver catalyst on PSi etched into the reactor surface. The overall structure of the microreactor was formed in three main steps, shown in Figure 14.16. First, a 6-mm wide reaction channel was formed by DRIE with a depth of

0.12 mm. This was followed by Ag metal-assisted HF chemical etching to form a 1- to 2-μm deep PSi layer on the reaction channel, increasing the surface area by a factor of 40,000. Finally, Ag catalyst was deposited by a sputter coater and the entire microreactor was anodically bonded with a glass wafer to contain the chemical reaction within the reactor. For comparison, a microreactor was fabricated without PSi. Catalytic activity was monitored by measuring reaction products by gas chromatography. Rosalva oxidation required lower activation energy and experienced six times more reactivity (TOF at 450°C) on the PSi supported silver catalyst versus a thin film silver catalyst (Cao et al. 2014).

In manners similar to the above-described deposition of metal catalysts on PSi for enhanced catalysis, much research is being conducted on the functionalization of PSi microreactors with enzymes such as glucose oxidase (GOx) or trypsin for glucose oxidation and other biochemical processes (Drott et al. 1999; Bengtsson et al. 2002; Palestino et al. 2008). High aspect ratio Si microreactor channels about 70 μm wide were anodically etched in an HF and ethanol solution to create a PSi surface (Drott et al. 1999; Bengtsson et al. 2002). Drott et al. (1999) created various depths of PSi layer on the microreactor surface, shown in Figure 14.17, by increasing etch duration. Enzymes were immobilized on the PSi surface in three steps: silanisation, glutaraldehyde activation, and finally coupling of the GOx or trypsin enzymes to the prepared PSi surface (Drott et al. 1999; Bengtsson et al. 2002). Compared to a similarly prepared nonporous microreactor, the PSi surface increased catalytic efficiency by 170-fold for GOx enzyme activated glucose turnover (Drott et al. 1999). In all studies, increasing the etch duration to increase the PSi layer depth and thus increase the active surface area resulted in an increase in catalytic activity. In particular, the trypsin microreactor experienced an increase in catalytic turnover of more than 200% (Bengtsson et al. 2002).

14.4 CONCLUSION AND OUTLOOK

PSi is an ideal support matrix for heterogeneous catalysts due to its high surface area, variable pore size and morphology, stability, and ease of integration into microreactor technologies. The hydrogen-terminated surface is a favorable support matrix for enzymes as well as reduction of metal salts to facilitate the creation of metal nanoparticle/PSi nanocomposites. Future study may involve the optimization of formation conditions of the PSi support to maximize catalytic activity. Advancement of the photocatalytic ability of PSi will create applications in environmental remediation, organic waste treatment, and water splitting (Qu et al. 2010; Su et al. 2013). Integration of the photocatalytic properties of PSi with PSi-enhanced microreactor design is yet to be explored. Within the context of microreactors, PSi membrane reactors show significant promise for on-site hydrogen formation (López et al. 2010). However, the selective oxidation of CO, without substantial loss of hydrogen, will be an important advancement for implementation with PEMFCs.

REFERENCES

Abrams, B.L. and Vesborg, P.C.K. (2013). Catalysts for environmental remediation—Examples in photo- and heterogeneous catalysis. In: Suib, S.L. (Ed.) *New and Future Developments in Catalysis: Catalysis for Remediation and Environmental Concerns.* Elsevier, Amsterdam, pp. 63–85. doi: 10.1016/B978 -0-444-53870-3.00004-6.

Bengtsson, M., Ekström, S., Marko-Varga, G., and Laurell, T. (2002). Improved performance in silicon enzyme microreactors obtained by homogeneous porous silicon carrier matrix. *Talanta* **56**(2), 341–353. doi: 10.1016/S0039-9140(01)00600-2.

Cao, E., Zuburtikudis, I., Al-Rifai, N., Roydhouse, M., and Gavriilidis, A. (2014). Enhanced performance of oxidation of rosalva (9-decen-1-ol) to costenal (9-decenal) on porous silicon-supported silver catalyst in a microstructured reactor. *Processes* **2**(1), 141–157. doi: 10.3390/pr2010141.

Casanovas, A., Domínguez, M., Ledesma, C., López, E., and Llorca, J. (2009). Catalytic walls and microdevices for generating hydrogen by low temperature steam reforming of ethanol. *Catal. Today* **143**, 32–37. doi: 10.1016/j.cattod.2008.08.040.

D'Arrigo, G., Spinella, C., Arena, G., and Lorenti, S. (2003). Fabrication of miniaturised Si-based electrocatalytic membranes. *Mater. Sci. Eng. C* **23**(1–2), 13–18. doi: 10.1016/S0928-4931(02)00228-X.

Divins, N.J., López, E., Roig, M. et al. (2011). A million-channel CO-PrOx microreactor on a fingertip for fuel cell application. *Chem. Eng. J.* **167**, 597–602. doi: 10.1016/j.cej.2010.07.072.

Divins, N.J., López, E., Llorca, J. et al. (2013). Macroporous silicon microreactor for the preferential oxidation of CO. In: *Proceedings of 2013 Spanish Conference on Electron Devices (CDE)*, Feb. 12–14, 2013, Valladolid, Spain, pp. 139–142. doi: 10.1109/CDE.2013.6481362.

Drott, J., Rosengren, L., Lindström, K., and Laurell, T. (1999). Porous silicon carrier matrices in micro enzyme reactors—Influence of matrix depth. *Mikrochim. Acta* **131**, 115–120. doi: 10.1007/PL00021396.

Ensafi, A.A., Abarghoui, M.M., and Rezaei, B. (2014). A new non-enzymatic glucose sensor based on copper/porous silicon nanocomposite. *Electrochim. Acta* **123**, 219–226. doi: 10.1016/j.electacta.2014.01.031.

Gautier, G., Kouassi, S., Desplobain, S., and Ventura, L. (2012). Macroporous silicon hydrogen diffusion layers for micro-fuel cells: From planar to 3D structures. *Microelectron. Eng.* **90**, 79–82. doi: 10.1016/j.mee.2011.04.003.

Gold, S., Chu, K., Lu, C., Shannon, M.A., and Masel, R.I. (2004). Acid loaded porous silicon as a proton exchange membrane for micro-fuel cells. *J. Power Sources* **135**(1–2), 198–203. doi: 10.1016/j.jpowsour.2004.03.084.

Ioannides, T. and Verykios, X.E. (1996). Charge transfer in metal catalysts supported on doped TiO_2: A theoretical approach based on metal–semiconductor contact theory. *J. Catal.* **161**(2), 560–569. doi: 10.1006/jcat.1996.0218.

Jensen, K.F. (2001). Microreaction engineering—Is small better? *Chem. Eng. Sci.* **56**(2), 293–303. doi: 10.1016/S0009-2509(00)00230-X.

Kim, T. and Kwon, S. (2006). Design, fabrication and testing of a catalytic microreactor for hydrogen production. *J. Microtech. Microeng.* **16**, 1760–1768. doi: 10.1088/0960-1317/16/9/002.

Kothare, M.V. (2006). Dynamics and control of integrated microchemical systems with application to micro-scale fuel processing. *Comput. Chem. Eng.* **30**(10–12), 1725–1734. doi: 10.1016/j.compchemeng.2006.05.026.

Laurell, T. (2002). Biocatalytic porous silicon microreactors. *Sensors Update* **10**(1), 3–32. doi: 10.1002/1616-8984(20021)10:1<3::AID-SEUP3>3.0.CO;2-Z.

Liu, X., Cheng, H., and Cui, P. (2014). Catalysis by silver nanoparticles/porous silicon for the reduction of nitroaromatics in the presence of sodium borohydride. *Appl. Surf. Sci.* **292**, 695–701. doi: 10.1016/j.apsusc.2013.12.036.

Llorca, J., Casanovas, A., Trifonov, T., Rodríguez, A., and Alcubilla, R. (2008). First use of macroporous silicon loaded with catalyst film for a chemical reaction: A microreformer for producing hydrogen from ethanol steam reforming. *J. Catal.* **255**(2), 228–233. doi: 10.1016/j.jcat.2008.02.006.

Lokteva, E.S., Rostovshchikova, T.N., Kachevskii, S.A. et al. (2008). High catalytic activity and stability of palladium nanoparticles prepared by the laser electrodispersion method in chlorobenzene hydrodechlorination. *Kinet. Catal.* **49**(5), 748–755. doi: 10.1134/S0023158408050212.

López, E., Irigoyen, A., Trifonov, T., Rodríguez, A., and Llorca, J. (2010). A million-channel reformer on a fingertip: Moving down the scale in hydrogen production. *Int. J. Hydrogen Energy* **35**(8), 3472–3479. doi: 10.1016/j.ijhydene.2010.01.146.

Losey, M.W., Schmidt, M.A., and Jensen, K.F. (2001). Microfabricated multiphase packed-bed reactors: Characterization of mass transfer and reactions. *Ind. Eng. Chem. Res.* **40**(12), 2555–2562. doi: 10.1021/ie000523f.

Losey, M.W., Jackman, R.J., Firebaugh, S.L., Schmidt, M.A., and Jensen, K.F. (2002). Design and fabrication of microfluidic devices for multiphase mixing and reaction. *J. Microelectromech. Syst.* **11**(6), 709–717. doi: 10.1109/JMEMS.2002.803416.

Löwe, H. and Ehrfeld, W. (1999). State-of-the-art in microreaction technology: Concepts, manufacturing and applications. *Electrochim. Acta* **44**(21–22), 3679–3689. doi: 10.1016/S0013-4686(99)00071-7.

Lu, Y. and Barron, A.R. (2013). Nanopore-type black silicon anti-reflection layers fabricated by a one-step silver-assisted chemical etching. *Phys. Chem. Chem. Phys.* **15**(24), 9862–9870. doi: 10.1039/C3CP51835C.

Nakamura, T., Tiwari, B., and Adachi, S. (2011). Direct synthesis and enhanced catalytic activities of platinum and porous-silicon composites by metal-assisted chemical etching. *Jpn. J. Appl. Phys.* **50**(8R), 081301. doi: 10.1143/JJAP.50.081301.

Palestino, G., Legros, R., Agarwal, V., Pérez, E., and Gergely, C. (2008). Functionalization of nanostructured porous silicon microcavities for glucose oxidase detection. *Sens. Actuators, B* **135**(1), 27–34. doi: 10.1016/j.snb.2008.07.013.

Park, H.G., Malen, J.A., Piggott, W.T. et al. (2006). Methanol steam reformer on a silicon wafer. *J. Microelectromech. Syst.* **15**(4), 976–985. doi: 10.1109/JMEMS.2006.878888.

Pattekar, A.V. and Kothare, M.V. (2004). A microreactor for hydrogen production in micro fuel cell applications. *J. Microelectromech. Syst.* **13**(1), 7–18. doi: 10.1109/JMEMS.2004.823224.

Pichonat, T. and Gauthier-Manuel, B. (2006). A new process for the manufacturing of reproducible mesoporous silicon membranes. *J. Membr. Sci.* **280**(1–2), 494–500. doi: 10.1016/j.memsci.2006.02.010.

Polisski, S., Goller, B., Lapkin, A., Fairclough, S., and Kovalev, D. (2008). Synthesis and catalytic activity of hybrid metal/silicon nanocomposites. *Phys. Stat. Sol. RRL* **2**(3), 132–134. doi: 10.1002/pssr.200802076.

Polisski, S., Goller, B., Wilson, K., Kovalev, D., Zaikowskii, V., and Lapkin, A. (2010). In situ synthesis and catalytic activity in CO oxidation of metal nanoparticles supported on porous nanocrystalline silicon. *J. Catal.* **271**(1), 59–66. doi: 0.1016/j.jcat.2010.02.002.

Presting, H., Konle, J., Starkov, V., Vyatkin, A., and König, U. (2004). Porous silicon for micro-sized fuel cell reformer units. *Mater. Sci. Eng. B* **108**(1–2), 162–165. doi: 10.1016/j.mseb.2003.10.115.

Qu, Y., Zhong, X., Li, Y., Liao, L., Huang, Y., and Duan, X. (2010). Photocatalytic properties of porous silicon nanowires. *J. Mater. Chem.* **20**(18), 3590–3594. doi: 10.1039/C0JM00493F.

Rebrov, E.V., Seijger, G.B.F., Calis, H.P.A., de Croon, M.H.J.M., van den Bleek, C.M., and Schouten, J.C. (2001). The preparation of highly ordered single layer ZSM-5 coating on prefabricated stainless steel microchannels. *Appl. Catal. A* **201**(1), 125–143. doi: 10.1016/S0926-860X(00)00594-9.

Roumanie, M., Delattre, C., Mittler, F. et al. (2008). Enhancing surface activity in silicon microreactors: Use of black silicon and alumina as catalyst supports for chemical and biological applications. *Chem. Eng. J.* **135**, S317–S326. doi: 10.1016/j.cej.2007.07.053.

Schneider, J., Matsuoka, M., Takeuchi, M. et al. (2014). Understanding TiO_2 photocatalysis: Mechanisms and materials. *Chem. Rev.* **114**, 9919–9986. doi: 10.1021/cr5001892.

Seelam, P.K., Huuhtanen, M., Keiski, R.L., and Basile, A. (2013). Microreactors and membrane microreactors: Fabrication and applications. In: Basile, A. (Ed.) *Handbook of Membrane Reactors*. Woodhead Publishing, Cambridge, pp. 188–235. doi: 10.1533/9780857097347.1.188.

Splinter, A., Stürmann, J., Bartels, O., and Benecke, W. (2002). Micro membrane reactor: A flow-through membrane for gas pre-combustion. *Sens. Actuators B* **83**(1–3), 169–174. doi: 10.1016/S0925-4005 (01)01036-X.

Stubenrauch, M., Fischer, M., Kremin, C., Stoebenau, S., Albrecht, A., and Nagel, O. (2006). Black silicon—New functionalities in microsystems. *J. Micromech. Microeng.* **16**(6), S82–S87. doi: 10.1088/0960-1317/16/6/S13.

Su, J., Yu, H., Q.X., Chen, S., and Wang, H. (2013). Hierarchically porous silicon with significantly improved photocatalytic oxidation capability for phenol degradation. *Appl. Catal. B* **138–139**, 427–433. doi: 10.1016/j.apcatb.2013.03.014.

Tank, C.M., Sakhare, Y.S., Kanhe, N.S. et al. (2011). Electric field enhanced photocatalytic properties of TiO_2 nanoparticles immobilized in porous silicon template. *Solid State Sci.* **13**(8), 1500–1504. doi: 10.1016/j .solidstatesciences.2011.05.010.

Tiggelaar, R.M. and Gardeniers, J.G.E. (2009). Silicon and glass microreactors. In: Hessel, V., Renken, A., Schouten, J.C., and Yoshida, J. (Eds.) *Micro Process Engineering*. Wiley-VCH Verlag GmbH & Co. KGaA, Weinheim, pp. 1–24. doi: 10.1002/9783527631445.ch18.

Wang, F., Yang, Q., Xu, G., Lei, N., Tsang, Y.K., Wong, N., and Ho, J.C. (2011). Highly active and enhanced photocatalytic silicon nanowire arrays. *Nanoscale* **3**(8), 3269–3276. doi: 10.1039/C1NR10266D.

Ye, S., Tanaka, S., Esashi, M., Hamakawa, S., Hanaoka, T., and Mizukami, F. (2005). Thin palladium membrane microreactors with oxidized porous silicon support and their application. *J. Micromech. Microeng.* **15**(11), 2011–2018. doi: 10.1088/0960-1317/15/11/004.

Li Batteries with PSi-Based Electrodes

Gael Gautier, François Tran-Van, and Thomas Defforge

15

CONTENTS

15.1   Introduction   320
15.2   Textured Silicon Material in Li-Battery Anode   321
     15.2.1   Principles and Characterization of Li-Batteries Silicon Anodes   321
         15.2.1.1   Lithiation/Delithiation Mechanisms   321
         15.2.1.2   SEI Formation   323
         15.2.1.3   Volume Expansion of Silicon-Based Composites   324
     15.2.2   Silicon Anode Performance Improvement   324
         15.2.2.1   Depth of Discharge   325
         15.2.2.2   Binders in the Composite   325
     15.2.3   Advantages of Silicon Texturation   326
15.3   Synthesis and Performances of Porous Silicon Anodes   327
     15.3.1   Porous Silicon Thin Films   328
         15.3.1.1   General Considerations   328
         15.3.1.2   Macroporous Silicon Anode Performances   331
         15.3.1.3   Mesoporous Silicon Anodes Performances   332
     15.3.2   Silicon and Porous Silicon Nanowires   333
     15.3.3   Porous Silicon Micro-Particles   337
15.4   Performances of Porous Silicon-Based Hybrid Structures   338
     15.4.1   Carbon Coating   338
     15.4.2   Metal Particle Decoration   339
15.5   Conclusion   339
References   340

15.1 INTRODUCTION

Since its commercialization in the early 1990s, Li-ion batteries have tried to follow the ever-increasing needs for energy storage systems, in particular for the electric/hybrid vehicles and other nomad applications (Armand and Tarascon 2008). Despite remarkable improvements over the last two decades, Li-ion batteries still suffer from limited energy density as compared to fossil energies. As every electrochemical storage device, Li-ion batteries are composed of two electrodes immersed in an electrolyte containing Li$^+$ ions. The cathode is typically made of either $LiCoO_2$, $LiNi_{1/3}Mn_{1/3}Co_{1/3}O_2$, or $LiFePO_4$ while the anode consists of graphitic carbon as illustrated in Figure 15.1 (Xu 2004). Despite reliable cycling performances, the specific capacity of graphitic carbon is limited (372 mA·h/g) compared to some other chemicals, especially silicon (see Table 15.1). Indeed, except lithium, Si presents

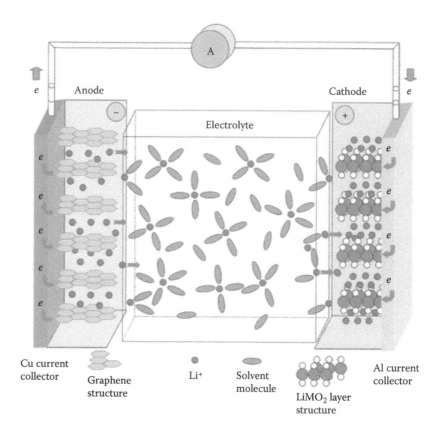

FIGURE 15.1 Schematic illustration of lithium ion batteries with graphite as anode and layered structure of metal oxide as cathode. (From Xu K. *Chem. Rev.* 104, 4303. Copyright 2004: American Chemical Society. With permission.)

TABLE 15.1 Comparison of the Theoretical Specific Capacity, Charge Density, Volume Change and Onset Potential of Various Anode Materials

Material	C	Al	Sb	Sn	Mg	Si
Density, g/cm³	2.25	2.7	6.7	7.29	1.3	2.33
Lithiated phase	LiC_6	$LiAl$	Li_3Sb	$Li_{4.4}Sn$	Li_3Mg	$Li_{4.4}Si$
Theoretical specific capacity, mA·h/g	372	993	660	994	3350	4200
Theoretical charge density, mA·h/cm³	837	2681	4422	7246	4355	9786
Volume change, %	12	96	200	260	100	320
Potential vs. Li, V	0,05	0,3	0,9	0,6	0,1	0,4

Source: Data extracted from Zhang W. *J Power Sources* 196, 13–24, 2011.

the highest specific capacity with 3579 mA·h/g at room temperature (Hatchard and Dahn 2004). However, this unique property induces several drawbacks such as a large volume expansion during Si lithiation that leads to electrode pulverization and thus electrode performance fade (Kasavajjula et al. 2007). Recently, Si nanostructures have demonstrated their ability to limit this issue.

The first part of this chapter is dedicated to the main principles of lithium-ion battery anode characterization and the main differences between Si and graphitic carbon anodes. The advantages of nanostructured silicon (including porous silicon [PSi]) compared to bulk or thin film silicon are listed. In the second part, the different synthesis paths of PSi to form Li-ion battery anode are detailed. The performances of PSi layers, particles, or nanowire-containing anodes are compared. As native Si is not stable enough to obtain acceptable cycling performances, the last part of this chapter develops PSi surface posttreatment to form a protective shell for the nanostructured Si electrode and improve its conductivity. Two kinds of materials are usually employed as protective layers: carbon coating or metal decoration.

15.2 TEXTURED SILICON MATERIAL IN Li-BATTERY ANODE

15.2.1 PRINCIPLES AND CHARACTERIZATION OF Li-BATTERIES SILICON ANODES

15.2.1.1 LITHIATION/DELITHIATION MECHANISMS

Charge-discharge profile of classical crystalline silicon-carbon based composite is described in Figure 15.2. By comparison with lithiation at elevated temperature (i.e., 415°C) in an LiCl–KCl

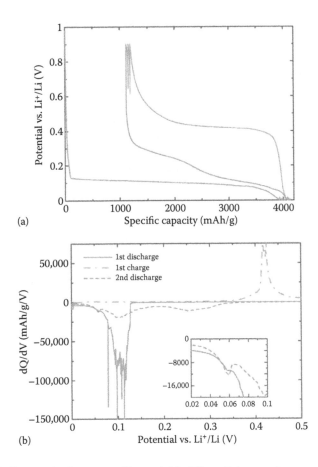

FIGURE 15.2 (a) Charge-discharge profile and (b) differential capacity versus potential for an Li half-cell cycled between 0.005 V and 0.9 V (vs. Li/Li$^+$) at 0.01 C in 1 M LiPF$_6$ ethylene carbonate/diethyl carbonate (1/2 v:v). (From Li J. and Dahn J.R., *J. Electrochem. Soc.* 154, A156, 2007. Copyright 2007: ECS – The Electrochemical Society. With permission.)

eutectic melt electrolyte, which leads to four plateaus associated with the formation of inter-metallic compounds $Li_{12}Si_7$, Li_7Si_3, $Li_{13}Si_4$, and $Li_{22}Si_5$, respectively (Wen and Huggins 1981), in agreement with the equilibrium Li–Si phase diagram, only one flat voltage region is visualized at around 0.12 V (vs. Li/Li$^+$) (Ryu et al. 2004; Li and Dahn 2007) during galvanostatic discharge at ambient temperature (Figure 15.2a). As described by Limthongkul et al. (2003), the lithiation process leads to solid-state amorphization with the formation of a two-phase structure (Obrovac and Krause 2007) comprising metastable lithiated amorphous silicon (a-Li$_y$Si) and completely unli-thiated crystalline Si particles. The large peak observed in the differential capacity curve (Figure 15.2b) is characteristic of the partial lithiation of S according to Reaction 15.1 (Li and Dahn 2007).

(15.1)
$$Si + xLi \rightarrow \left(1 - \frac{x}{y}\right) Si + \frac{x}{y} \, a\text{-}Li_y Si$$

Lithiation progression results in an increase of the thickness of the amorphous shell instead of its Li-enrichment as shown by transmission electron microscopy (TEM) (Gu et al. 2012; Liu et al. 2012), scanning electron microscopy (SEM) (Chon et al. 2011), and simulations studies (Chan et al. 2012). Recent analyses carried out by auger electron spectroscopy (AES) have confirmed the two-phase structure described previously. The composition of the Li-rich amorphous shell structure (Li$_y$Si) has been evaluated by Li and Dahn (2007) around y = 3.5 ± 0.2 from the charge recorded during the lithi-ation process. In contrast, experiments carried out on individual silicon particles by valence electron energy loss spectrometry (VEELS) (Danet et al. 2010) or AES (Radvanyi et al. 2013) have estimated the composition of the amorphous phase close to $Li_{2.9+/-0.3}Si$ and $Li_{3.1}Si$, respectively (Figure 15.3).

At the end of the first discharge, at voltage lower than 60 mV (see the peak in the inset in Figure 15.2b solid curve), $Li_{15}Si_4$ metastable phase crystallizes from the amorphous Li$_y$Si phase as described by Obrovac and Christensen (2004), and shown by the occurrence of diffraction peaks in an *in situ* X-ray diffraction (XRD) study (Hatchard and Dahn 2004; Li and Dahn 2007; Misra et al. 2012). The delithiation of the crystalline $Li_{15}Si_4$ phase is assumed to take place from both the surface and the bulk owing to the high lithium diffusivity in the lithiated phases (Key et al. 2010). The flat plateau (Figure 15.2b, dash-dotted curve) and the strong peak of the differential capacity observed around 0.41 V (vs. Li/Li$^+$) have been attributed to a two-phase region. The absence of new crystalline peaks in X-ray diffractograms and the decrease of the intensity of the $Li_{15}Si_4$ ones have evidenced that amorphous silicon is formed during the first charge. Nucleation, from small clusters, would lead to the formation of amorphous silicon matrix. The lithiated phase with iso-lated Si ions is gradually converted to an amorphous Si phase (Key et al. 2010).

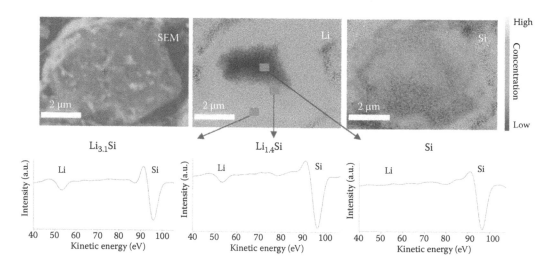

FIGURE 15.3   SEM image, AES Li and Si elemental mappings of the investigated silicon particle after 10 min of sputtering. In addition, the AES derivative spectra of Li KLL and Si LVV and the cor-responding quantification have been reported in different regions of the particle, represented by the squares on the SEM image. (From Radvanyi E. et al., *J. Mater. Chem. A* 1, 4956. Copyright 2013: RSC Publishing. With permission.)

15.2.1.2 SEI FORMATION

The first full discharge at 0.005 V (vs. Li/Li⁺) for a composite Si/C/binder realized in a binary electrolyte ethylene carbonate (EC)/dimethyl carbonate (DMC) (1:1 v/v) LiPF$_6$ 1 M has exhibited a capacity of around 4000 mA·h/g as shown in Figure 15.2. As discussed previously, the capacity associated to the richest lithiated metastable phase formed at ambient temperature corresponds to 3.75 Li inserted per atom of Si, that is, a capacity of 3579 mA·h/g according to Hatchard and Dahn (2004). The difference in capacity between the first discharge and the full lithiation of crystalline silicon was mainly attributed to the irreversible degradation of the solvent and/or salt that forms a solid electrolyte interface (SEI). The formation of this passivating layer on the Si surface has been highlighted by Fourier transform infrared spectroscopy (FTIR), X-ray photoelectron spectroscopy (XPS), photoelectron spectroscopy (PES), high resolution transmission electron microscopy (HRTEM), and secondary ion mass spectrometry (TOF-SIMS) (Wu et al. 2003; Ruffo et al. 2009; Verma et al. 2010; Philippe et al. 2012; Pereira-Nabais et al. 2014).

By quartz crystal microbalance analysis, the formation of SEI during the first charge was associated with mass gain attributed to the build-up of electrolyte reduction products on silicon surface (Ryu et al. 2008). In alkylcarbonate-based electrolytes generally utilized in such systems, the voltage limit corresponding to the reduction of the solvent is about 1 V (vs. Li/Li⁺), greatly higher than the voltage region associated with the silicon lithiation. Thus, it has been proposed by De la Hoz et al. (2013) that the reductive decomposition of the solvent would participate with the formation of the passivation layer. According to Etacheri et al. (2012) and confirmed by NMR analysis (Delpuech et al. 2013), the degradation of the solvent would be mainly responsible for irreversible capacity (rather than the salt) during the charge-discharge process.

Recently, Philippe et al. (2012) have studied the interface during the first charge-discharge in binary propylene carbonate (PC)/diethyl carbonate (DEC) (1:2 v:v) 1 M LiPF$_6$ of nano-Si based composite electrodes (Si/C/carboxymethylcellulose [CMC]) by hard and soft XPS. It was shown that the formation of the SEI starts before 0.5 V (vs. Li/Li⁺) with the formation of a thin layer of which thickness increases during the discharge process but without significant changes in composition. Chan et al. have shown by XPS and SEM analyses of the SEI formed onto Si nanowire (SiNW) a voltage dependent composition layer with a thickness higher than 10 nm (Chan et al. 2009; Ruffo et al. 2009). During the lithiation of the crystalline silicon (plateau at 100⁻¹⁰ mV vs. Li/Li⁺), the formation of Li$_2$O and Li$_x$SiO$_y$ was detected by XPS at the interface and the amount of Li$_2$O was shown to increase with the discharge process. The mechanism of SEI formation proposed by Philippe et al. (2012) is shown in Figure 15.4.

At the end of the first cycle, a large and irreversible capacity loss (more than 1100 mA·h/g in Figure 15.2) was recorded. Excepting SEI formation, irreversible process has also been partially

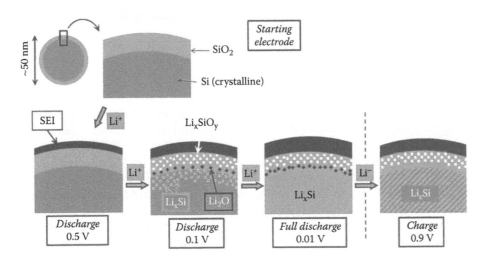

FIGURE 15.4 Schematic view of SEI formation during the first cycle depending on the discharge voltage and partial reversibility upon charge. (From Philippe B. et al., *Chem. Mater.* 24(6), 1107, 2012. Copyright 2012: American Chemical Society. With permission.)

attributed to the large volume expansion of silicon attributed to lithium insertion, which deteriorates the mechanical properties of the electrode and limits the electronic percolation inside the composite as discussed next.

15.2.1.3 VOLUME EXPANSION OF SILICON-BASED COMPOSITES

According to the phase diagram of Li-Si alloy, the most lithiated phase of silicon can contain 4.4 Li per atom of Si in a cubic phase. The crystal structure is identical to the one of the crystalline silicon although the volume per silicon atom is more than four times higher (82.4 Å^3 for $Li_{22}Si_5$ vs. 20.0 Å^3 for Si) (Boukamp et al. 1981). It is one of the highest known volume expansions among the common alloy anodes (Zhang 2011). Even if full electrochemical silicon lithiation at ambient temperature lead to the insertion of "only" 3.75 Li per Si during crystallization of $Li_{15}Si_4$ as discussed previously, the volume expansion is still striking (290% according to Gonzalez et al. 2014). These strong morphology changes in the material generate cracks and pulverization of the composite electrode and were also observed in thin amorphous silicon (a-Si) film (Maranchi et al. 2006). Ryu et al. (2004) have studied the evolution of the internal resistance of micro-Si electrode during cycling by galvanostatic intermittent titration technique. The authors have showed an increase of the electrode conductance during charge process and have explained this result by a higher conductivity of the Si-alloy compared to pure silicon (Pollak et al. 2007). On the other hand, the de-allowing, leading to the contraction of the electrode, increased its resistance imputed to less effective electronic contact between particles (mainly due to cracks in the structure also visualized at this discharge stage by AFM [Beaulieu et al. 2001; Zhang 2011]). In contrast, Kasavajjula et al. (2007) have recorded an increase of the electrode resistance during both the first charge and discharge process. In both cases, these results confirmed that, in the end, large silicon volume changes are detrimental to the cyclability and degrade the charge percolation inside the electrode and the electrical contact between active materials and current collector. As a consequence of the mechanical degradation of the electrode, a low coulombic efficiency is obtained during the first charge-discharge process (25% to 35% depending on the nature of electrode: bulk silicon, film or large particle of Si and to the nature of the electrolyte) and a rapid capacity fade occurred during the five first cycles (Ryu et al. 2004; Kim et al. 2005). In a composite electrode, the initial crack propagation would start from the surface of the silicon particles associated with their elastic deformation (Zhao et al. 2012a) or from the inner of the particles associated with their plastic deformation (Cheng and Verbrugge 2010) as illustrated in Figure 15.5a (Chan et al. 2008).

To limit mechanical degradation during volume changes upon cycling, some authors suggested decreasing the size of the silicon particles as well as to nanostructure the silicon to form nanowire or PSi, as we will discuss in the next paragraphs. The optimization of the electrode composition with different binders or by using specific additives in the electrolyte composition was also demonstrated to be efficient to improve the cyclability.

15.2.2 SILICON ANODE PERFORMANCE IMPROVEMENT

Li-Si alloys are one of the most promising anodes for lithium ion batteries due to their high storage capacity and the abundance of Si on Earth (second most abundant element) (Li et al. 1999; Kasavajjula et al. 2007; Wu et al. 2012; Chen 2013; Liang et al. 2014). Its potential capacity can be compared to theoretical capacity of other Li alloy such as tin ($Li_{4.4}Sn$: 994 mA·h/g), antimony (Li_3Sb: 660 mA·h/g), aluminum (LiAl: 993 mA·h/g) and standard graphite anode allowing a theoretical capacity of 372 mA·h/g. The capacity of Si-alloy could reach 4200 mA·h/g for the richest lithiated crystalline intermetallic compounds $Li_{22}Si_4$. The highest electrochemical lithiation of Si at room temperature produces "only" $Li_{15}Si_4$ corresponding to 3579 mA·h/g, which is still ten times higher than graphite. Nevertheless, as discussed previously, Si anode suffers from a low cyclability, which limits its industrial development compared to graphite that can be cycled over 1000 times. Different options were proposed to overcome this technological barrier and to improve the performances of Si anode in terms of electrochemical-induced aging. Several reports in the literature have shown that it is possible to adjust the electrochemical parameters as well as

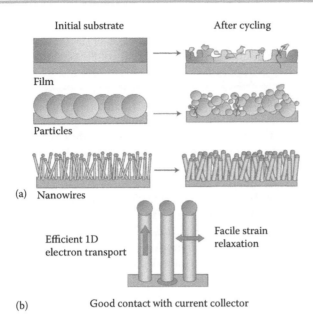

FIGURE 15.5 (a) Schematic view of morphological changes that occur in Si during cycling: The volume of silicon anodes may change up to 290% during cycling. As a result, Si films and particles tend to pulverize during repeated charge/discharge cycles. (b) However, nanostructured silicon such as NWs grown does not pulverize during cycling and exhibit efficient one-dimensional electron transport. (From Chan C.K. et al., *Nature Nanotechnol.* 3, 31. Copyright 2008: Nature Publishing Group. With permission.)

to adapt the chemical environment of the Si active material or to optimize the electrolytic media in order to tend to the optimum storage capacity without fading on cycling.

15.2.2.1 DEPTH OF DISCHARGE

Since the capacity fading is mainly induced by volume changes during charge-discharge and the expansion of the particles is almost proportional to Li incorporation, it has been suggested to limit the depth of discharge (DOD) in order to limit the volume expansion of the silicon. Obrovac and Krause (2007) studied the lithiation process during cycling and the importance to maintain a two-phase structure for better cyclability of microsized Si particle composite electrodes.

Although the limitation of DOD is efficient to improve the cyclability of the electrode, the available capacity of the anode is strongly reduced compared with theoretical storage capacity of Si. For instance, setting the cutoff voltage on discharges at 180 mV (vs. Li/Li+) limits the silicon powder anode capacity to 850–900 mA·h/g (Obrovac and Krause 2007).

15.2.2.2 BINDERS IN THE COMPOSITE

Another way to absorb the volume expansion of Si and to limit its mechanical degradation is to optimize the composition of the Si composite. Thereby, the matrix can enhance the mechanical strength of the electrode and sustain a large elastic deformation during cycling (Zhang 2011). In this case, binders can play a crucial role in the mechanical properties of composites even if their weight ratio must be limited to maintain acceptable anode specific capacity. An improvement of the anode performances (Lestriez et al. 2007; Li et al. 2007; Hochgatterer et al. 2008; Bridel et al. 2010) by using carboxymethylcellulose (CMC) instead of polyvinylidene difluoride (PVdF) has been shown. Lestriez et al. (2007) suggested that CMC would lead to a more homogeneous dispersion allowing a good contact between conductive carbon and active particles.

More recently, polysaccharide was utilized as a natural polymer in composite with Si nanoparticles (NPs). Capacity two times higher than the one obtained with NaCMC that is, 2000 mA·h/g at 1 C was reached by Kovalenko et al. (2011). Uchida et al. (2015) improved the cycle stability of microsized silicon (Si) by applying polyimide with a low breaking elongation percentage and a high tensile strength. It has been shown that the adhesion of the binder suppresses the decrepitation of the electrode and maintains a discharge capacity of 800 mA·h/g during 195 cycles.

15.2.3 ADVANTAGES OF SILICON TEXTURATION

It has been demonstrated that a decrease in particle size and/or the nano-structuration of the silicon could be an efficient method to limit the decrepitation of this material and to improve the performance stability of Si-based anodes (Chan et al. 2008; Wu et al. 2012). Previously reported onto microsized particles of different diameters (Liu et al. 2005), this proof of concept has also been confirmed using nanotubes (Park et al. 2009; Song et al. 2010; Zhou et al. 2013), nanowires (Chan et al. 2008; Lee et al. 2012), nanowalls (Wan et al. 2014), NPs (Kim et al. 2008, 2010; Liu et al. 2012), and hollow nanospheres (Yao et al. 2011; Du et al. 2014). Table 15.2 recapitulates some the silicon anode designs reported in the literature.

Chan et al. (2008) used a vapor–liquid–solid (VLS) process to grow silicon nanowires (SiNWs) directly connected to a current collector (see Figure 15.5). Thanks to Si nano-structuration as well as to the space between the adjacent nanowires, the volume changes have not been detrimental for the nanomaterial, which has preserved its mechanical stability during reversible lithiation. Moreover, the direct contact between nanowires and current collector has allowed keeping good electrical contact at the interface without electrode delamination. Using these nanostructures, the authors achieved maintaining at a discharge rate of C/20 in 1 M $LiPF_6$/ (ethylene carbonate:diethyl carbonate), a reversible capacity of 3200 mA·h/g over 10 cycles with a coulombic efficiency of 73% in the first cycle.

Liu et al. (2012) have studied by *in situ* TEM and XRD, the effects of particle size on the capacity of Si to absorb volume variations during lithiation without cracks. Figure 15.6 describes *in situ* TEM characterizations of nanosized particle during its lithiation. It was clearly demonstrated that NPs could relieve the strain during the volume expansion (with an evolution of diameter of the particle from 80 to 130 nm under full lithiation). In contrast, microparticles (diameter 940 nm) tested in the same electrochemical conditions exhibited cracks on the surface of the amorphous shell and propagated inward during the course of the discharge. Thus, for deep discharge (formation of crystalline $Li_{15}Si_4$), fracture of the particle occurred. It has been concluded that under a critical size of around 150 nm, Si particles can accommodate volume expansion without fracture (Liu et al. 2012). Similar experiments were performed with SiNWs and studied by *in situ* TEM and modelization to determine the critical diameters at which SiNWs start to pulverize (Ryu et al. 2011). Statistical studies estimated a critical diameter of cylindrical shaped

TABLE 15.2   A List of Silicon Anode Designs and Their Reported Performance

Ref.	Design	Reported Capacity (mA·h/g)	Cycles
Chan et al. 2008	Nanowires	3500	20
Maranchi et al. 2003	Thin films	3500	30
Thakur et al. 2012b	Gold coated porous silicon films	3000	50
Thakur et al. 2012c	Porous silicon particulates	3000	600
Cui et al. 2009	Carbon-silicon core-shell nanowires	2000	30
Takamura et al. 2004	50 nm thin films	2000	3000
Ge et al. 2012	Porous silicon nanowires	2000	250
Zhao et al. 2012b	Hierarchical macro/mesoporous particles	1500	50
Ge et al. 2013a	Porous silicon nanoparticles	1400	200
Cui et al. 2008	Crystalline-amorphous core-shell nanowires	1000	100
Zhang et al. 2006	Particles + carbon nanotubes	584	20
Chen et al. 2012	Hollow porous particles	2600	100
Wu et al. 2012	SiO_x covered nanotube	600	6000

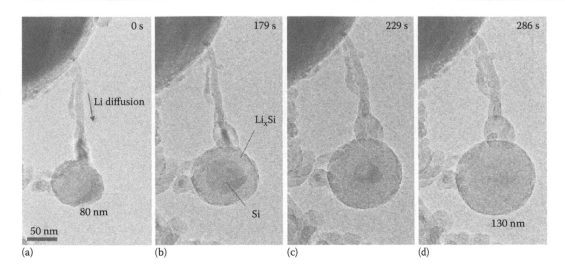

FIGURE 15.6   (a) *In situ* TEM observations of Si nanoparticles with diameter 80 nm during the lithiation process; (b, c) formation of the Si/Li$_x$Si core/shell structure under charge; (d) full lithiation of the particle with a volume expansion of around 300% without cracking. (From Liu X.H. et al., *ACS Nano.* 6(2), 1522. Copyright 2012: American Chemical Society. With permission.)

Si before fracture between 240 and 360 nm for lithiation down to 0.01 V versus Li/Li$^+$ (Lee et al. 2012). Moreover, the decrease in diameter of 1D structures ensures a rapid Li transport due to short Li conduction distances (Chan et al. 2008) that limits fracture mechanisms.

Kim et al. (2010) reported the synthesis of well-defined nanosized Si particles (from 5 to 20 nm) using reverse micelles at high pressure and temperature. With 10-nm particles size, they obtained during the first cycle a coulombic efficiency of 80% (from 1.5 to 0 V (vs. Li/Li$^+$) in 1 M LiPF$_6$–EC/DMC (1:1 v/v) at a rate of 0.2 C). Moreover, when the particles were coated with carbon, the charge capacity (3535 mA·h/g) and coulombic efficiency for the first cycle were improved up to 89% and the capacity retention was 96% after 40 cycles. Similarly, carbon-coated silicon nanowire exhibited an excellent first discharge capacity of 3344 mA·h/g with a coulombic efficiency of 84%. The influence of carbon coating will be detailed in Section 15.4. Reversible capacity of 1326 mA·h/g was retained after 40 cycles in 1 M LiPF$_6$ EC/DEC (1:1 v/v) (Huang et al. 2009). Finally, Wu et al. (2012) recently proposed preparing Si nanotubes covered by SiO$_x$ layer possessing strong mechanical properties which prevent the Si from expanding outward and thus the volume change of the electrode. A stable SEI was also formed and ensured a capacity retention of 88% after 6000 cycles. Short Li$^+$ diffusion lengths allowed fast charging/discharging rates (up to 20 C) with 1000 mA·h/g at 12 C.

Silicon porosification is also an interesting texturation alternative for Li-ion anode stability. Calculations on the different parameters defining PSi (e.g., pore size, crystallite size, and porosity) have been optimized to limit the stress during lithium insertion (Ge et al. 2012; Li et al. 2014). Ge et al. (2012) have coupled Li$^+$ diffusion and strain-induced lithium intercalation in silicon. They have concluded that the porosity was the most significant parameter to control to limit the stress in PSi. On the other hand, pore size or wall thickness does not seem to affect the stress in the porous material. Li et al. (2014) have calculated the minimal porosity of PSi particles required to limit their volume expansion during lithiation and leading to stable anode characteristics. By taking into account the porous particle dimensions, porosity higher than 80% has been assessed to minimize the volume expansion during electrode cycling (see Figure 15.7).

15.3 SYNTHESIS AND PERFORMANCES OF POROUS SILICON ANODES

Silicon nanostructures have demonstrated their ability to strongly enhance anode performances (see the previous section). Among these nanostructures, PSi seems promising as the voids in these sponge-like structures facilitate the lithium insertion and accommodate the large volume expansion during high rate charge/discharge cycling (Cho 2010; Ge et al. 2013b). We will see in

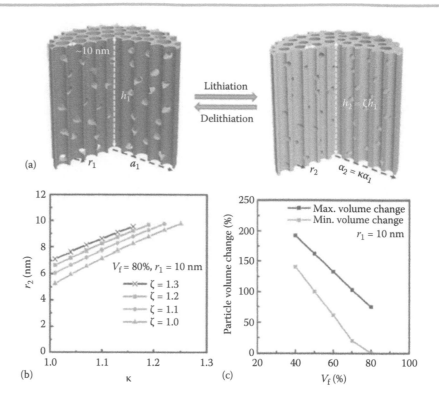

FIGURE 15.7   (a) Schematic model of mesoporous silicon particle. *r* represents the pore diameter, *a* and *h* the particle radius and height, respectively. (b) The plot illustrates the correlation of the pore radius after full lithiation (r_2), and the expansion rates of the particle along the radical direction (κ) and along the axial direction (ζ). (c) The plot illustrates the porous particle volume expansion versus its initial porosity (V_f). (From Li X. et al., *Nature Commun.* 5, 4105. Copyright 2014: Nature Publishing Group. With permission.)

the following paragraphs that these PSi anodes can be utilized as thin films, one-dimensional structures (SiNWs or porous SiNWs), or particles scattered in conductive binder.

15.3.1   POROUS SILICON THIN FILMS

As developed in the previous paragraphs, nanosized Si particles show better capacity retention and improved initial reversible capacity compared with bulk material. Nevertheless, standard anodes with Si NPs contain up to 20% of inactive binders and conductive additives. On the other hand, nano-crystalline and amorphous silicon thin films suffer from a low cycle life (Kasavajjula et al. 2007). Then, PSi thin films could present an interesting compromise between silicon NPs and bulk or thin films.

15.3.1.1   GENERAL CONSIDERATIONS

Among all the techniques known to produce PSi, electrochemical etching is the most efficient as it can provide high thickness layers with a good reproducibility. Nevertheless, some authors have mentioned the metal assisted chemical etching (MaCE) process to produce PSi anodes although this technique is restricted to thin layers (less than 1 µm) (Ivanov et al. 2014). The advantages of this method are the simplicity of the setup and the slight dependency of the final morphology on the substrate characteristics (resistivity, doping type) (Han et al. 2014). More detailed description of this method can be found in Chapter 10 in *Porous Silicon: Formation and Properties*. To obtain high thickness PSi layers, the anodization is generally preferred. PSi formation by silicon anodization is analyzed in details in *Porous Silicon: Formation and Properties*. Then, the substrate characteristics and electrolyte composition must be judiciously selected to obtain a specific morphology. Thus, to produce macropores, low doped *p*-type silicon is generally preferred as *n*-type Si requires backside illumination (Lehmann and Föll 1990) and highly doped Si is generally known to produce mesoporous silicon (Lehmann et al. 2000). To obtain large empty macropores with moderate to high porosities (between 70 and 90%), the electrolyte additives must be carefully selected. The first solution is to add a cationic surfactant such as cetyl trimethyl ammonium chloride (CTAC) (Chao et al. 2000; Defforge et al. 2013) to the aqueous electrolyte (see Figure 15.8). The second way is to use organic solvents such as dimethyl formamide (DMF) (Ponomarev and Lévy-Clément 1998).

In the case of mesoporous silicon made from p^+ or n^+ silicon, the influence of electrolyte additives is negligible and the porosity is mainly governed by the HF concentration and the applied current density. To obtain larger pores and higher porosities, very high doping levels are preferred (around 1 mΩ·cm) (Zhang 2001) (see Figure 15.9). Moreover, the pore diameter can be modulated to form multilayered mesoporous structures (Vincent 1994). These structures seem to be efficient to accommodate the silicon volume expansion during lithiation (see Section 15.3.1.3) (Cheng et al. 2014).

FIGURE 15.8   SEM cross-sectional view of macroporous silicon layer, which can be used for silicon anode batteries. In this case, the anodization was performed in an aqueous HF electrolyte mixed with CTAC as a wetting agent during 100 min at 10 mA/cm². Using a 30–50 Ω·cm *p*-type wafer, the average pore diameter is around 5 μm.

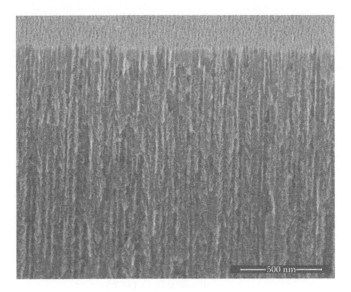

FIGURE 15.9   Example of mesoporous silicon layer, which can used for silicon anode batteries. In this case, the anodization was performed in an aqueous HF electrolyte mixed with Triton X-100® as wetting agent during 30 s at 400 mA/cm². Using a 1–3 mΩ·cm *p*-type wafer, the average pore diameter is around 60 nm.

It is generally admitted that it is preferable to remove the PSi membrane from the substrate. Indeed, the underlying silicon substrate is susceptible to be lithiated producing irreversible mechanical damages (see Section 15.2). Then, after electrochemical etching, a thin copper layer can be deposited on top of the PSi layer, for instance using a chemical process in an aqueous HF/CuSO$_4$ mixture (Quiroga-González et al. 2011). Then, a thicker electrochemically deposited metal layer can be used to lift off the membrane (see Figure 15.10). The PSi/Si interface can be previously weakened using a short electropolishing step at the end of the anodization process (Sun et al. 2012).

To increase the performance stability, some authors propose to deposit metals on the pore surfaces (Thakur et al. 2012b) or to cover it with carbon (Thakur et al. 2012a). The influence of PSi surface treatments on the anode performances will be discussed in Section 15.4.

A typical cyclic voltammogram of a PSi anode is shown in Figure 15.11. The current density and hence the charge transfer increased with cycling up to the 30th cycle. Thereafter, the current

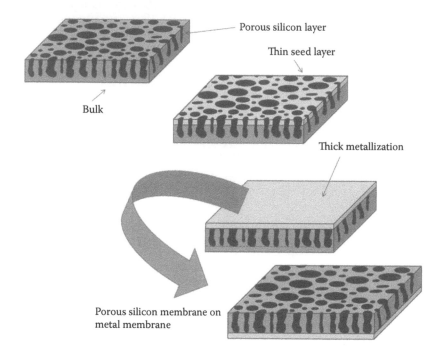

FIGURE 15.10   Schematic presentation of the structure generally adopted to study PSi membrane anode performances. After electrochemical etching, a thin metal is coated as a seed layer on top of a PSi layer. Then, a thicker metal layer can be deposited to lift off the membrane. The interface also can be previously weakened using a short electropolishing step at the end of the anodization process.

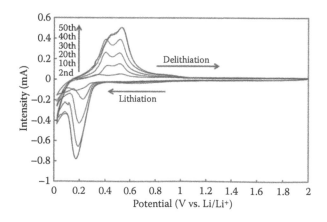

FIGURE 15.11   Typical cyclic voltammograms of a macroporous silicon/Cu anode in EC/DMC and 1 M LiPF$_6$ at a scan rate of 0.1 mV/s. (From Luais E. et al., *J. Power Sources* 242, 166. Copyright 2013: Elsevier. With permission.)

density remains constant or starts to decrease. As previously mentioned, the lithium alloying/dealloying process results in internal structural changes such as amorphization of the material. So, the activation of the silicon electrode during the first cycles can be attributed to the surface reconstruction process and the SEI formation. This behavior is also visible in the case of other nanostructured silicon anodes (see Section 15.2.3).

Then, the anodes are generally cycled between 0.01 and 1–3 V (vs. Li/Li⁺). As mentioned previously, one way to decrease stress or disintegration of thin films is to control the DOD, that is, the cutoff voltage (Section 15.2.2.1). For instance, Luais et al. (2013) showed that the capacity of a macroporous silicon electrode cycled between 0.125 and 0.7 V (vs. Li/Li⁺) remains stable after 200 cycles at 1.8 A/g. The specific capacity is then around 700 mA·h/g. This value is still fairly high compared with the theoretical capacity of graphite (372 mA·h/g) (Zhang 2011).

In the two next sections, we will show and comment on the performances of many anodes reported in the literature involving macroporous or mesoporous silicon thin films.

15.3.1.2 MACROPOROUS SILICON ANODE PERFORMANCES

The evolution of some specific capacities as a function of the cycle number is presented in Figure 15.12. The anode characteristics and characterization parameters are summarized in Table 15.3.

At first, Shin and co-workers (2002, 2005) proposed a study on macroporous silicon anodes carried forward over to copper foils. In these experiments, PSi was prepared using an acetronitrile (18 mL), TBAP (tetra-n-butylammoniumperchlorate, 1.2 g), and 49% HF (900 µL) electrolyte from *p*-type 1–20 Ω·cm wafers at 4 mA/cm² (Shin et al. 2005). The macropore average diameter is around 1.5 µm for a maximum layer thickness of 4 µm. A maximum specific capacity around 60 µA·h/cm² was demonstrated although this value continuously decreased during the 40 cycles. To the best of our knowledge, this was the first study involving a PSi layer for Li-ion battery anode application.

Kang et al. (2008) proposed a hybrid structure composed of macroporous Si covered with Si nanostructures. Note that in this article, the macropores (micrometer-sized pores according to International Union of Pure and Applied Chemistry (IUPAC)) are mentioned as "micropores" whereas the meso- or microporous film (whose pore size is below 50 nm) is described as "nanopores." To avoid any misunderstanding, we will follow the IUPAC recommendations. The structure was fabricated with an electrochemical etching process in a DMF/HF solution. The macropore diameter varied from 1 to 1.5 µm and extended up to 15 µm in depth. The walls of

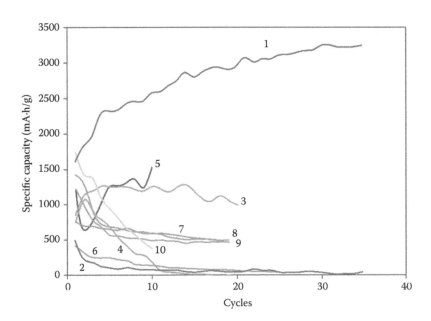

FIGURE 15.12 Galvanostatic profiles of some Li/macroporous silicon half-cells reports in the literature. PS characteristics and characterization parameters are reported in Table 15.3.

TABLE 15.3   Performances of Macroporous Silicon Anodes Reported in the Literature

Ref.	Electrolyte	J_e	ρ	Φ	P	t	PSi Treatment	C	V_{min}	V_{max}	J_c	C_i	C_f
1. Thakur et al. 2012b	HF:DMF 2:25	1.6	14–22	1	–	5.5	20 nm gold coating	35	0.1	2	0.05	3200	3200
2. Thakur et al. 2012b	HF:DMF 2:25	1.6	14–22	1	–	5.5	Bare PSi	35	0.1	2	0.05	500	40
3. Thakur et al. 2012a	HF:DMF 4:30	2	14–22	1	–	12	Pyrolyzed polyacrylonitrile	20	0.07	1.5	0.2	1270	995
4. Thakur et al. 2012a	HF:DMF 4:30	2	14–22	1	–	12	Bare PSi	20	0.07	1.5	0.2	1070	0
5. Li et al. 2013a	4% HF	6 to 18	5–15	3.1	70	190	KOH etching	10	0.01	2	0.35	1530	1530
6. Li et al. 2013a	4% HF	6 to 18	5–15	3.1	70	190	KOH etching	25	0.01	2	1.05	400	40
7. Sun et al. 2012	HF (4M), DMF	40	1–20	1.4	70	52	–	20	0.02	3	40	1430	490
8. Sun et al. 2012	HF (4M), DMF	40	1–20	1.4	70	52	–	20	0.02	3	200	760	500
9. Sun et al. 2012	HF (4M), DMF	40	1–20	1.4	70	52	–	20	0.02	3	100	1230	467
10. Luais et al. 2013	5 wt% HF, CTAC	10	30–50	5	–	50	–	10	0.05	2	200	1750	380

Note: All the wafers are *p*-type doped except in Li et al. (2013a). *S* is the substrate material under PSi layer, J_e is the etching current density in mA/cm², ρ is the wafer resistivity in ohm.cm, Φ is the average pore diameter in µm, *P* is the porosity in % before PSi treatments, *t* is the PSi layer thickness in µm, *C* is the cycle number at J_c in mA/cm² (1 to 6) or mA/g (7 to 10), V_{min} and V_{max} are the voltage cutoffs in volts, C_i and C_f are the initial and final specific capacities in mA·h/g. All the authors used 1M LiPF$_6$ electrolytes in 1:1 EC:DMC except for Li et al. (2013a) (1M LiClO$_4$ in PC: dimethoxyethane, 7:3).

these pores are covered with a nanosized structure that consists of 10 nm in diameter spherical particles. The as-prepared PSi structures show cathodic/anodic peaks for lithiation/delithiation during cyclic voltammetry with minimal destruction of the macropore wall particles even after 50 cycles. In addition, the current peak and the cumulative charge increased during these lithiation/delithiation steps.

The anodes studied by Astrova et al. (2011) or Li et al. (2013a,b) were based on a regular lattice of macropores with different periods, sizes, and shapes of the pore cross-section. The thickness of the membranes varied from 100 to 200 µm. The authors examined the capacity evolution during cycling. It was found that these anodes have an initial discharge capacity higher than the one of graphite but it decreased during the course of cyclic embedding/extraction of lithium, especially at high current densities (j = 1.05 mA/cm²) by a factor of 3 after 70 cycles. It was also shown that silicon walls gradually disintegrate depending on their initial thickness: thinner walls are less susceptible to mechanical disintegration.

Sun et al. (2012) proposed also to work on macroporous silicon membrane anodes fabricated from *p*-type silicon wafer using electrochemical anodization in an HF/DMF electrolyte. Freestanding porous membranes were achieved with high aspect ratio of 40, average pore size of 1 µm, porosity around 70%, and wall thickness of less than 0.1 µm. The maximum lithium storage was evaluated to 1425 mA·h/g (at 0.2 A/g). However, this value decreased to 500 mA·h/g after only 20 cycles. Similar final performances were obtained by Luais et al. (2013). However, it was demonstrated that macroporous anodes can be cycled between 0.125 and 0.7 V (vs. Li/Li⁺) at 1.8 A/g without damaging the structure leading to a stable capacity of around 700 mA·h/g for 200 cycles.

15.3.1.3 MESOPOROUS SILICON ANODES PERFORMANCES

A few authors have reported studies on mesoporous silicon thin films as a negative electrode in Li-ion batteries. The higher specific surface area and the thinner pore walls of mesoporous silicon layers are expected to improve the anode performances, especially its stability. The evolution of some specific capacities is presented in Figure 15.13 as a function of the cycle number. The anode characteristics and characterization parameters are summarized in Table 15.4.

The first report of a mesoporous anode carried out by electrochemistry was provided by Zhu et al. (2013a). In this work, very high porosity 80% layers were lifted off and deposited on a copper

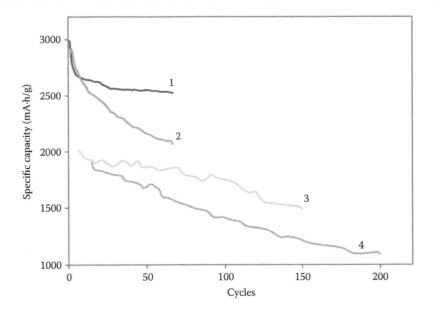

FIGURE 15.13   Galvanostatic profiles of some Li/mesoporous silicon half-cells reports in the literature. PS characteristics and characterization parameters are reported in Table 15.4.

TABLE 15.4   Performances of Mesoporous Silicon Anodes Reported in the Literature

Ref.	Electrolyte	J_e	ρ	Φ	P	t	S	C_s	V_{min}	V_{max}	J_c	C_i	C_f
1. Cheng et al. 2014	EtOH:HF (49%) 1:3	125–208	<1	10–50	51	10	Cu	70	0.05	2	0.1 C	2990	2522
2. Cheng and Verbrugge 2014	EtOH:HF (49%) 1:3	125	<1	10	–	10	Cu	70	0.05	2	0.1 C	2875	2069
3. Luais et al. 2015	30 wt.% HF Triton X-100	400	2–4	60–70	55	5	Si	150	0.1	2	300 µA/cm²	2012	1490
4. Zhu et al. 2013a	EtOH:HF (49%) 1:3	–	10–20	20	80	1	Cu	200	0.01	1	1 C	1904	1092

Note: All the wafers are *p*-type doped. See Table 15.3 captions for acronym definitions. After formation of PSi, no additional posttreatments were used. All the authors used 1 M LiPF$_6$ electrolytes in 1:1 EC:DMC.

foil. The highest reported capacity is around 2500 mA·h/g, but after 200 cycles at 1 C, this value drops to 1100 mA·h/g. Luais et al. (2015) reported higher values even with a silicon substrate. But, in this case, the porosity is lower (55%), the pores are at the edge between meso- and macroporous material (60–70 nm) and the layer is around 5 µm. The reported value evolved between 2000 and 1500 mA·h/g during 150 cycles at 300 µA/cm².

Finally, Cheng et al. (2014) reported capacities up to 2900 mA·h/g with 10-nm pores and a thick layer of 10 µm. However, this capacity clearly drops to 2000 mA·h/g after 70 cycles. The performances seem to be improved when periodic horizontal interconnected channels are performed in the pore network. These structures were carried out applying a current pulse periodically during the anodization. This nano-structuration is assumed to absorb the lattice distortion during the lithium insertion in the pore walls. Indeed, the capacity value was kept over 2500 mA·h/g but the measurements were performed only during 70 cycles.

15.3.2 SILICON AND POROUS SILICON NANOWIRES

Silicon nanowires (SiNWs) are among the first nanostructures that showed notable improvements in comparison to bulk or thin film silicon. Chan et al. (2008) showed limited capacity fading with VLS synthesized SiNWs (see Figure 15.5b). These one-dimensional structures present unique and attractive characteristics for Li-ion batteries for the reasons detailed here (Wu and Cui 2012):

- The empty space between the neighbor wires facilitates the volume expansion of Si during its lithiation.
- The one-dimensional structure enables an efficient charge transfer from the surface to the current collector.
- As already mentioned (Section 15.2), wires with tens or hundreds of nanometers in diameter are expected to limit silicon anode pulverization during repeated Li alloying/dealloying sequences (see Lee et al. 2012 for a complete study of this phenomenon).

Following this breakthrough in the field of Li-ion battery anodes, numerous experiments on SiNWs were reported in the literature. Up to now, two main SiNW growth methods are investigated: VLS method as pioneered by Chan et al. (2008) and MaCE of silicon developed by Peng et al. (2002). The latter technique, which is strongly related to PSi, is a chemical technique to produce nanostructured Si anodes. The MaCE technique is currently highly attractive due to its low cost for mass-production of either PSi or SiNWs (Hu et al. 2014). Nanowires can be obtained from Si (mono- or polycrystalline) after dipping into an HF solution mixed with noble metallic ions (Peng et al. 2002; Han et al. 2014). For Li-ion battery application, silver from a silver nitrate solution is the most widespread metal used to produce a dense carpet of high aspect ratio SiNWs. The dimensions of the nanowires (length and diameter) can be tuned by the etching conditions (Han et al. 2014). Table 15.5 recapitulates some SiNWs electrodes performances detailed in the literature.

TABLE 15.5   Performances of Monocrystalline Silicon Nanowire Electrodes as Li-Ion Batteries Reported in the Literature

Ref.	Silicon Source (Type, Orientation, Resistivity)	Posttreatment	Potential Window (V)	Minimal Capacity (mA·h/g)	Cycles	Capacity Retention (%)	Rate (C)	Coulombic Efficiency (%)
Huang et al. 2010	p-type, (100)		0.02–2	1000	30	>50	0.15 A/g	90%
Xu and Flake 2010	p-type, (100), 1–5 Ω·cm		0.01–1.5	512	15	65	0.1 mA/cm²	65
Xu et al. 2011	(100)		0.01–1.5	2348	15	71	0.1	
Quiroga-González et al. 2011	p-type, (100), 15–24 Ω·cm	Copper electrodeposition (current collector)	0–1		60			
Bang et al. 2011a	p-type, (100), 1 Ω·cm	Carbon coating CVD	0.01–1.5	530	50	>25	0.1	>90%
Nguyen et al. 2012	Si deposition (SiH₄ – MOCVD)		0–2	1420	100	44	0.15	>97%
Vlad et al. 2012	p-type, (100), 15 Ω·cm	Copper electroless	0.02–1.2	1900	50	>80	0.05	
Baek et al. 2013	p-type, (100), (111), and (110), 1–10 Ω·cm		0.01–1.5	200	15	>20	0.5	
Zhu et al. 2013b	p-type, (100), 0.3–10 Ω·cm	Graphene + annealing	0.0005–2	1335	80	77	0.05	
Kim et al. 2013	p-type, (100), 1–20 Ω·cm		0.01–2.5	1900	60	>88	1	
Kim et al. 2014	p-type, (100), 1–20 Ω·cm	Graphene + gold evaporation	0.01–3.2	1520	20	84		
Lemordant et al. 2014	p-type, (100), 30–50 Ω·cm			0.6 mA·h/cm²	50	85	0.02 mA/cm²	

Note: The binder and the electrolyte compositions may differ. All of the SiNWs were produced from MaCE technique except Quiroga et al. and Lemordant et al. (anodization + alkaline treatment).

Peng, who pioneered and developed the MaCE technique (2002), is also the first to report the use of SiNWs obtained by MaCE as an anode for Li-ion batteries (Peng et al. 2008). In this article, where a half-cell battery performance is cycled, it is hard to conclude about the efficiency of the nanowire carpet, as it is limited to a few cycles. In the following years, several experiments and improvements involving MaCE SiNWs were reported in literature (Xu et al. 2011; Nguyen et al. 2012; Zhu et al. 2013). For instance, Huang and Zhu (2010) using SiNWs grown on both sides, demonstrated that pristine SiNWs exhibit capacity retention limited to a few tens of cycles even at low current densities. The crystallographic orientation of NWs was also studied because Si is known to behave differently depending on its orientation. Baek et al. (2013) recently reported a comparative study of detached (100), (110), and (111) oriented SiNWs. NWs in a (110) crystallographic direction deliver the highest storage capacity although the poor cycling stability does not allow a conclusion in regard to the most relevant orientation or a comparison with *in situ* investigations performed on nanowires and nanorods by other groups (Lee et al. 2012; Yang et al. 2012).

To enhance the cycling stability of SiNWs, several strategies may be considered:

- First, the SiNWs can be detached from the Si substrate (e.g., by physical scratching, sonication, or using a sacrificial porous layer [Weisse et al. 2013]). The nanowires are then generally mixed with a binder and coated on the current collector, typically a copper foil. High capacity and promising stability (≈1900 mA·h/g after 60 cycles) has been demonstrated by Kim et al. (2013) using transfered SiNWs (see Figure 15.14).
- Nanowire material and surface engineering is the most efficient technique to produce high capacity and stable Si anodes. Surface engineering (carbon coating, metal decoration, etc.) is not limited to SiNWs and can also be applied to the other Si nanostructures (porous layers, particles, etc.). This will be mentioned in a separate section (Section 15.4). However, SiNWs may also be porosified to facilitate the lithium insertion and limit the cracks often observed in crystalline SiNWs.

A large number of studies have appeared in the past few years involving MaCE technique to produce mesoporous silicon nanowires (PSiNWs). This unique structure was first demonstrated in degenerated *p*-type Si (Hochbaum et al. 2009) and combines the 1D structure of NWs and the

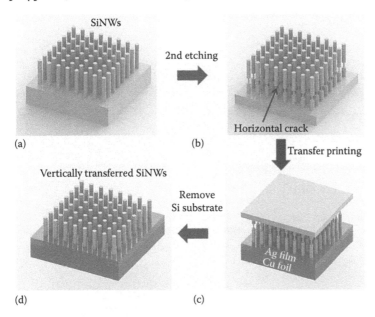

FIGURE 15.14 Schematic of a fabrication process to obtain vertically aligned detached silicon nanowires connected with Ag/Cu current collector. (a, b) Dual-step metal-assisted etching to weaken the wires by forming horizontal cracks during the second step. (c) Contact printing of SiNW onto an Ag-coated Cu electrode under high temperature and pressure. (d) Silicon substrate removal to finalize transferring of SiNWs. (From Kim H.J. et al., *ChemSusChem* 6(11), 2144. Copyright 2013: Wiley. With permission.)

high surface area of the mesoporous framework. Table 15.6 recapitulates some PSiNW electrode performances detailed in the literature. Wang and Han (2010) presented the first application of PSiNWs as an anode for Li-ion batteries. The results seemed promising especially for graphene-wrapped PSiNWs (capacity >2000 mA·h/g at 0.2 A/g current density after 20 cycles showing capacity retention of 87%). Unfortunately, in this article, no comparison between crystalline and porous SiNWs was proposed. This comparison was done by Bang et al. (2011a). Despite better performances during the initial cycles, the carbon-coated SiNW anode capacity dropped rapidly (capacity retention <25% after 50 cycles at 0.1 C) whereas with PSiNWs, almost stable performances were obtained during the whole test with a capacity retention higher than 95% after 50 cycles at 0.1 C (capacity ≈ 1450 mA·h/g). Recently, Ge et al. (2012) revealed remarkable performances using PSiNWs. Anodes partially made of PSiNWs were subjected to long tests (up to 2000 cycles) and delivered stable and high capacity performances (>1000 mA·h/g at 1 C) as illustrated in Figure 15.15. Moreover, to define the limits of this configuration, high current density charge/discharge cycles were applied (up to 4.5°C) showing no significant capacity drop. To the best of our knowledge, this configuration still shows the best performances using SiNWs as a battery anode.

TABLE 15.6   Performances of Mesoporous Silicon Nanowire Electrodes as Li-Ion Batteries Reported in the Literature

Ref.	Silicon Source	Posttreatment	Potential Window (V)	Minimal Capacity (mA·h/g)	Cycles	Capacity Retention (%)	Rate (C)	Coulombic Efficiency (%)
Liu et al. 2010	Silicon powder			1200	>150	>95		
Bang et al. 2011b	Silicon powder	Carbon coating, CVD	0.02–1.2	2400	70	95	0.2	
Wang and Han 2010	p-type, (100)	Graphene	0.005–0.8	2041	20	82	0.05	>96
Bang et al. 2011a	p-type, (100), 0.008 Ω·cm	Carbon coating, CVD	0.01–1.5	1450	50	95	0.1	>98
Ge et al. 2012	p-type, (100), <0.005 Ω·cm		0.01–2	1000	2000	95	1	
Wang et al. 2013	p-type, (100), <0.005 Ω·cm	Carbon coating, CVD	0.02–2	1500	200	75	0.84 A/g	>97
Yang et al. 2013b	p-type, (100), 0.005–0.02 Ω·cm	Carbon, MPCVD	0.01–1	1650	500	82	1.3 A/g	>98

Note: The binder and the electrolyte compositions may differ.

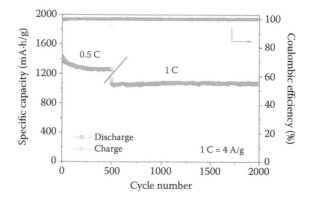

FIGURE 15.15   Plot of charge/discharge capacity of MaCE PSiNW electrode during 2000 cycles at current rate 0.5 C (for 500 cycles) and then 1 C. During the test, the electrolyte was composed of LiPF$_6$ (1 M) mixed with EC:DEC (1:1 vol.). The voltage window was set between 0.01 and 2 V (vs. Li/Li$^+$). (From Ge M. et al., *Nano Lett.* 12, 2318. Copyright 2012: American Chemical Society. With permission.)

Due to their competitive and stable performances as Li-ion batteries, SiNWs obtained by MaCE are on the way toward industrial production. Nexeon® already produces partially porous SiNWs through MaCE of metallurgical-grade Si particles (Loveridge et al. 2010; Liu et al. 2010; Armstrong et al. 2014). Metallurgic-grade Si is a low-cost source that can be textured by MaCE (the same way as monocrystalline Si) to produce partially or fully porous SiNWs (Li et al. 2013c). MaCE treatment of low-grade Si presents two main advantages. As previously mentioned, it increases the surface area of the particles (Bang et al. 2011b) and it limits the metallic contamination (Li et al. 2013c).

Except for the MaCE technique, other chemical ways to produce SiNWs have recently been reported in the literature. Föll et al. (2010) developed a dual-step etching technique to produce long (a few hundred micrometers) Si micro/nanowires. First, high-aspect-ratio ordered Si macropore arrays were etched in low-doped p-type Si. Then, Si wires were revealed by chemical etching: either anisotropic (KOH) or isotropic (HF-HNO$_3$-acetic acid) etching (Föll et al. 2011). Finally, copper was electroplated on one tip and used as a current collector. Despite a complex process flow, stability for over 60 cycles was reached using this electrode configuration. Recently, another nanostructure morphology, which is similar to SiNWs, has been developed for lithium-ion battery purposes: Si nano-ribbons. These two-dimensional structures have been carried out using a dual-step chemical etching procedure (Gautier et al. 2011). First, low-doped p-type Si was made porous in HF–acetic acid mixture, leading to macroporous silicon filled with micro-/mesoporous material. An additional alkaline etching step selectively dissolved the mesoporous material releasing macroporous silicon framework similar to nano-ribbons. These nanostructures have recently been studied as anode for Li-ion batteries and yielded interesting cycle stability (Lemordant et al. 2014; Santos-Peña et al. 2014).

15.3.3 POROUS SILICON MICRO-PARTICLES

In addition to SiNWs and PSiNWs, 3D PSi particles have also been intensively studied mainly thanks to their low-production costs. It has been shown that particles can be obtained through many techniques, although most of the porosification methods are similar to NW fabrication (anodization or MaCE). Another technique named magnesiothermic reduction also enables the formation of PSi particles using silica or porous silica as a source of Si (Bao et al. 2007). This technique has been widely studied for Li-ion battery application in the past few years (Liu et al. 2013; Bok et al. 2014). Nevertheless, only PSi particles obtained by chemical (MaCE) or electrochemical etching are related in this section.

Anodization of Si usually enables the formation of a homogeneous PSi layer (as developed in Section 15.3.1.3) instead of PSi particles. However, a succession of high and low current densities during the electrochemical etching process leads to the formation of PSi with modulated porosity. Indeed, applying periodical high current pulses locally increases the porosity of the porous film, leading to alternating high and low-porosity layers (called "perforated PSi") (Qin et al. 2014). During the ultrasonication of these perforated PSi layers, fractures occur along the weaker point (i.e., the perforated region). This leads to the formation of high surface area PSi NPs with a homogeneous thickness and one-directional pore orientation. These particles, usually used for biomedical applications (cf. Anglin et al. 2008), have been recently tested as anodes (Li et al. 2014). The porosity as well as the size of the silicon crystallites was controlled to limit the volume expansion. As for the electrode performances, despite a moderate capacity ($\approx 500-600$ mA·h/g at 1 A/g), the carbon-coated particles exhibited high capacity retention during long cycling test (>80% capacity retention after 1000 cycles). Macroporous particles obtained by electrochemical etching followed by ultrasonication step have also been studied by Thakur et al. (2012c). These structures also exhibited interesting performances with stable capacity for up to 600 cycles under optimized conditions.

Depending on the solution composition, the MaCE technique may produce SiNWs of a PSi layer (Kolasinski 2005). PSi particles produced by MaCE have also been widely studied by the community (Liu et al. 2011; Lin et al. 2012; Zhao et al. 2012b; Ge et al. 2013a; He et al. 2013). For instance, Ge et al. (2014) developed a low-cost way to produce PSi particles for Li-ion battery anodes. Milled low-grade Si powder was employed as a PSi precursor. Then, the particles were immersed in the etching solution composed of $Fe(NO_3)_3$ instead of the usual $AgNO_3$. Their "low-cost" PSi particles exhibited 550 mA·h/g after 600 cycles at 1 C.

15.4 PERFORMANCES OF POROUS SILICON-BASED HYBRID STRUCTURES

Although serious efforts carried out on Si porosification and nano-structuration have considerably enhanced the performances of lithium battery anodes, capacity fade after a few tens or hundreds of cycles is still hardly avoidable. Consequently, PSi material postengineering is essential to avoid early or late Si pulverization and thus reach high stability battery anodes. It is important to note that most of the studies reported in the literature attempt to optimize Si surface. Indeed, only a few articles deal with bare PSi or SiNWs. Among the different options to stabilize the porous framework, carbon coating is the most considered. Besides, metal decoration has also been intensively tested in the last few years.

15.4.1 CARBON COATING

Since carbon shows higher charge/discharge stability performances as compared to silicon, Si/C composites have thus been widely studied for two decades (Wilson and Dahn 1995; Kasavajjula et al. 2007). Several comparisons were reported on the stability of PSi particle anodes with and without carbon coating. In particular, Figure 15.16 gives a comparison of the cycling stability of the various samples of nanoporous silicon/carbon composites (Zheng et al. 2007).

Carbon-coated materials present higher capacity retention although a lower initial capacity is usually obtained. The first reason is the volume expansion limitation during lithiation/delithiation steps attributed to the carbon membrane. Moreover, carbon-coated Si nanostructures present a higher electrical conductivity. There are several reports in the literature of EIS showing a lower interface resistance of carbon-coated layer compared to pristine Si surface (Tao et al. 2012; Wang et al. 2012; Chen et al. 2014).

Carbon may be coated on PSi through many ways:

■ Liquid-phase infiltration of organic species such as *p*-phenylenediamine (Yoo et al. 2013, 2014), dopamine (Ru et al. 2014), polyacrylonitrile (Thakur et al. 2012a,c), sucrose, or glucose (He et al. 2013; Yang et al. 2013a) followed by a high temperature annealing (700–900°C) under neutral atmosphere.

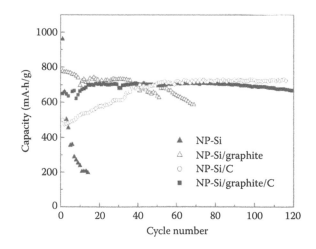

FIGURE 15.16   Plots representing cycling performances of various samples based on PSi NPs-carbon graphite composites. (From Zheng Y. et al., *Electrochim. Acta* 52, 5863, 2007. Copyright 2007: Elsevier. With permission.)

■ Gas-phase carbonization technique was also widely employed. A precursor gas is inserted in the oven and mixed with Ar. Most of the time, acetylene is utilized as a carbon precursor (Kim and Cho 2008; Bang et al. 2011b, 2012; Bok et al. 2014).

■ Recently, Zhang et al. (2014) combined both PSi formation and carbon coating thanks to the Rochow reaction. Metallurgical-grade Si particles decorated with copper reacted with CH_3Cl to form carbon-coated PSi particles. This interesting PSi formation process led to stable anode performances (800 mA·h/g during 100 cycles at 0.05 A/g).

15.4.2 METAL PARTICLE DECORATION

Metal NPs are known to enhance silicon-based anode performances. Copper (Vlad et al. 2012; Bok et al. 2014), gold (Thakur et al. 2012b; Kim et al. 2014), or silver (Yu et al. 2010; Chen et al. 2012; Du et al. 2013; Bok et al. 2014) were used to decorate PSi nanostructures (SiNWs or PSi particles). Bok et al. (2014) showed that silver and copper particles enabled a stable PSi anode performance during 50 cycles by enhancing the electrical properties of the porous framework (1220 mA·h/g and 1410 mA·h/g for 50 cycles at 0.2 C rate using Ag- and Cu-loaded PSi particles, respectively). Electrochemical impedance spectroscopy (EIS) evidenced that the charge transfer resistance was divided by 10 using a silver-coated electrode. In comparison with bare Si, metal particle decoration usually leads to both better capacity performance and retention. For instance, Yu et al. (2010) observed that silver-coated macroporous particles presented better capacity than pristine macroporous particles even during the first cycle.

A direct comparison between carbon coating and metal particle decoration to determine which strategy is the most efficient seems difficult since the mass additive (carbon or metal) is rarely mentioned in the literature. Nevertheless, one can argue that the cost of the surface functionalization by noble metal decoration is its main drawback and could limit its industrial application.

15.5 CONCLUSION

For the last decades, silicon in Li-ion battery anodes has met growing interest of the scientific community thanks to its unique performances (highest known gravimetric capacity with 3579 mA·h/g at room temperature). However, cycling performances of such material are still disappointing. Indeed, the large volume expansion (up to 290% for fully lithiated silicon at ambient temperature) is responsible for pulverization of silicon electrode and thus severe performance decay after a few cycles. Structuring silicon to the micro- and nanoscale is one way to overcome these issues and maintain silicon anode performances to an acceptable range. In this chapter, we highlighted the advantages of PSi structures for Li-ion battery anodes. The empty space of PSi structures is known to reduce the stress induced by the volume expansion and thus prevents early electrode pulverization (Li et al. 2014). Moreover, PSi electrode provides a large surface area accessible to the electrolyte. This limits the diffusion length of Li^+ ions from the electrolyte to the silicon and thus enables higher rate capacities.

The different morphologies of PSi (macro- or mesoporous silicon, silicon and PSi nanowires) and their various efficiencies have been detailed in this chapter and summarized in different tables. As acceptable performance stability (up to a few thousand cycles) has recently been demonstrated in the literature (Ge et al. 2012; Li et al. 2014), two trends are emerging. The first one is the continuous improvement of silicon nano-engineering to combine high capacity and lossless anode performance. For this purpose, complex nanostructures involving silicon but also metallic or carbon-based species are currently under investigation. The second trend is the cost-efficient and environmentally friendly production of PSi to meet industrial needs, while maintaining acceptable electrode performances. Low cost chemicals (Ge et al. 2014), and unexpected silicon sources such as beach sand (Favors et al. 2014) or rice husks (Liu et al. 2013) are thus experimented.

REFERENCES

Anglin, E.J., Cheng, L., Freeman, W.R., and Sailor, M.J. (2008). Porous silicon in drug delivery devices and materials. *Adv. Drug Delivery Rev.* **60**(11), 1266–1277.

Armand, M. and Tarascon, J.M. (2008). Building better batteries. *Nature* **451**, 652–657.

Armstrong, M.J., O'Dwyer, C., Macklin, W.J., and Holmes, J.D. (2014). Evaluating the performance of nano-structured materials as lithium-ion battery electrodes. *Nano Res.* **7**(1), 1–62.

Astrova, E.V., Fedulova, G.V., Smirnova, I.A., Remenyuk, A.D., Kulova, T.L., and Skundin, A.M. (2011). Porous silicon based negative electrodes for lithium ion batteries. *Techn. Phys. Lett.* **37**(8), 731–734.

Baek, S.H., Park, J.S., Bae, E.J., Jeong, Y.I., Noh, B.Y., and Kim, J.H. (2013). Influence of the crystallographic orientation of silicon nanowires in a carbon matrix on electrochemical performance as negative electrode materials for lithium-ion batteries. *J. Power Sources* **244**, 515–520.

Bang, B.M., Kim, H., Lee, J.P., Cho, J., and Park, S. (2011a). Mass production of uniform-sized nanoporous silicon nanowire anodes via block copolymer lithography. *Energy Environ. Sci.* **4**(9), 3395–3399.

Bang, B.M., Kim, H., Song, H.K., Cho, J., and Park, S. (2011b). Scalable approach to multi-dimensional bulk Si anodes via metal-assisted chemical etching. *Energy Environ. Sci.* **4**(12), 5013–5019.

Bang, B.M., Lee, J.I., Kim, H., Cho, J., and Park, S. (2012). High-performance macroporous bulk silicon anodes synthesized by template-free chemical etching. *Adv. Energy Mater.* **2**(7), 878–883.

Bao, Z., Weatherspoon, M.R., Shian, S. et al. (2007). Chemical reduction of three-dimensional silica micro-assemblies into microporous silicon replicas. *Nature* **446**, 172–175.

Beaulieu, L.Y., Eberman, K.W., Turner, R.L., Krause, L.J., and Dahn, J.R. (2001). Colossal reversible volume changes in lithium alloys. *Electrochem. Solid-State Lett.* **4**(9), A137–A140.

Bok, T., Choi, S., Lee, J., and Park, S. (2014). Effective strategies for improving the electrochemical properties of highly porous Si foam anodes in lithium-ion batteries. *J. Mater. Chem. A*, **2**(34), 14195–14200.

Boukamp, B.A., Lesh, G.C., and Huggins, R.A. (1981). All-solid lithium electrodes with mixed-conductor matrix. *J. Electrochem. Soc.* **128**(4), 725–729.

Bridel, J.S., Azais, T., Morcrette, M., Tarascon, J.M., and Larcher, D. (2010). Key parameters governing the reversibility of Si/carbon/CMC electrodes for Li-ion batteries. *Chem. Mater.* **22**(3), 1229–1241.

Chan, C.K., Peng, H., Liu, G. et al. (2008). High-performance lithium battery anodes using silicon nanowires. *Nat. Nanotechnol.* **3**(1), 31–35.

Chan, C.K., Ruffo, R., Hong, S.S., and Cui, Y. (2009). Surface chemistry and morphology of the solid electrolyte interphase on silicon nanowire lithium-ion battery anodes. *J. Power Sources* **189**(2), 1132–1140.

Chan, M.K., Wolverton, C., and Greeley, J.P. (2012). First principles simulations of the electrochemical lithiation and delithiation of faceted crystalline silicon. *J. Am. Chem. Soc.* **134**(35), 14362–14374.

Chao, K.J., Kao, S.C., Yang, C.M., Hseu, M.S., and Tsai, T.G. (2000). Formation of high aspect ratio macropore array on p-type silicon. *Electrochem. Solid-State Lett.* **3**(10), 489–492.

Chen, J. (2013). Recent progress in advanced materials for lithium ion batteries. *Materials* **6**(1), 156–183.

Chen, D., Mei, X., Ji, G. et al. (2012). Reversible lithium-ion storage in silver-treated nanoscale hollow porous silicon particles. *Angew. Chem., Int. Ed.* **51**(10), 2409–2413.

Chen, S., Bao, P., Huang, X., Sun, B., and Wang, G. (2014). Hierarchical 3D mesoporous silicon@graphene nanoarchitectures for lithium ion batteries with superior performance. *Nano Res.* **7**(1), 85–94.

Cheng, Y.T. and Verbrugge, M.W. (2010). Diffusion-induced stress, interfacial charge transfer, and criteria for avoiding crack initiation of electrode particles. *J. Electrochem. Soc.* **157**(4), A508–A516.

Cheng, H., Xiao, R., Bian, H., Li, Z., Zhan, Y., Tsang, C.K., Lu, C.Z., and Li, Y.Y. (2014). Periodic porous silicon thin films with interconnected channels as durable anode materials for lithium ion batteries. *Mater. Chem. Phys.* **144**, 25–30.

Cho, J. (2010). Porous Si anode materials for lithium rechargeable batteries. *J. Mater. Chem.* **20**, 4009–4014.

Chon, M.J., Sethuraman, V.A., McCormick, A., Srinivasan, V., and Guduru, P.R. (2011). Real-time measurement of stress and damage evolution during initial lithiation of crystalline silicon. *Phys. Rev. Lett.* **107**(4), 045503.

Cui, L.F., Ruffo, R., Chan, C.K., Peng, H., and Cui, Y. (2008). Crystalline-amorphous core–shell silicon nanowires for high capacity and high current battery electrodes. *Nano Lett.*, **9**(1), 491–495.

Cui, L.-F., Yang, Y., Hsu, C.-M., and Cui, Y. (2009). Carbon–silicon core–shell nanowires as high capacity electrode for lithium ion batteries. *Nano Lett.* **9**, 3370–3374.

Danet, J., Brousse, T., Rasim, K., Guyomard, D., and Moreau, P. (2010). Valence electron energy-loss spectroscopy of silicon negative electrodes for lithium batteries. *Phys. Chem. Chem. Phys.* **12**(1), 220–226.

Defforge, T., Diatta, M., Valente, D., Tran-Van, F., and Gautier, G. (2013). Role of electrolyte additives during electrochemical etching of macropore arrays in low-doped silicon. *J. Electrochem. Soc.* **160**(4), H247–H251.

De la Hoz, J.M.M., Leung, K., and Balbuena, P.B. (2013). Reduction mechanisms of ethylene carbonate on Si anodes of lithium-ion batteries: Effects of degree of lithiation and nature of exposed surface. *Appl. Mater. Interfaces* **5**, 13457–13465.

Delpuech, N., Dupré, N., Mazouzi, D. et al. (2013). Correlation between irreversible capacity and electrolyte solvents degradation probed by NMR in Si-based negative electrode of Li-ion cell. *Electrochem. Comm.* **33**, 72–75.

Du, F.H., Wang, K.X., Fu, W. et al. (2013). A graphene-wrapped silver–porous silicon composite with enhanced electrochemical performance for lithium-ion batteries. *J. Mater. Chem. A* **1**(43), 13648–13654.

Du, F.H., Li, B., Fu, W., Xiong, Y.J., Wang, K.X., and Chen, J.S. (2014). Surface binding of polypyrrole on porous silicon hollow nanospheres for Li-ion battery anodes with high structure stability. *Adv. Mater.* **26**(35), 6145–6150.

Etacheri, V., Geiger, U., Gofer, Y. et al. (2012). Exceptional electrochemical performance of Si-nanowires in 1, 3-dioxolane solutions: A surface chemical investigation. *Langmuir* **28**(14), 6175–6184.

Favors, Z., Wang, W., Bay, H.H. et al. (2014). Scalable synthesis of nano-silicon from beach sand for long cycle life Li-ion batteries. *Sci. Rep.* **4**, 5623.

Föll, H., Hartz, H., Ossei-Wusu, E., Carstensen, J., and Riemenschneider, O. (2010). Si nanowire arrays as anodes in Li ion batteries. *Phys. Stat. Sol. RRL* **4**(1–2), 4–6.

Föll, H., Carstensen, J., Ossei-Wusu, E., Cojocaru, A., Quiroga-González, E., and Neumann, G. (2011). Optimized Cu-contacted Si nanowire anodes for Li ion batteries made in a production near process. *J. Electrochem. Soc.* **158**(5), A580–A584.

Gautier, G., Defforge, T., Kouassi, S., and Coudron, L. (2011). Metal-free disordered vertical sub-micron silicon wires produced from electrochemical p-type porous silicon layers. *Electrochem. Solid-State Lett.* **14**(8), D81–D83.

Ge, M., Rong, J., Fang, X., and Zhou, C. (2012). Porous doped silicon nanowires for lithium ion battery anode with long cycle life. *Nano Lett.* **12**(5), 2318–2323.

Ge, M., Rong, J., Fang, X., Zhang, A., Lu, Y., and Zhou, C. (2013a). Scalable preparation of porous silicon nanoparticles and their application for lithium-ion battery anodes. *Nano Res.* **6**(3), 174–181.

Ge, M., Fang, X., Rong, J., and Zhou, C. (2013b). Review of porous silicon preparation and its application for lithium-ion battery anodes. *Nanotechnol.* **24**(42), 422001.

Ge, M., Lu, Y., Ercius, P., Rong, J., Fang, X., Mecklenburg, M., and Zhou, C. (2014). Large-scale fabrication, 3D tomography, and lithium-ion battery application of porous silicon. *Nano Lett.* **14**(1), 261–268.

Gonzalez, J., Sun, K., Huang, M., Lambros, J., Dillon, S., and Chasiotis, I. (2014). Three dimensional studies of particle failure in silicon based composite electrodes for lithium ion batteries. *J. Power Sources* **269**, 334–343.

Graetz, J., Ahn, C.C., Yazami, R., and Fultz, B. (2003). Highly reversible lithium storage in nanostructured silicon. *Electrochem. Solid-State Lett.* **6**, A194–A197.

Gu, M., Li, Y., Li, X., Hu, S., Zhang, X., Xu, W., Thevuthasan, S., Baer, D.R., Zhang, J., Liu, J., and Wang, C. (2012). In situ TEM study of lithiation behavior of silicon nanoparticles attached to and embedded in a carbon matrix. *ACS Nano* **6**(9), 8439–8447.

Han, H., Huang, Z., and Lee, W. (2014). Metal-assisted chemical etching of silicon and nanotechnology applications. *Nano Today* **9**, 271–304.

Hatchard, T.D. and Dahn, J.R. (2004). In situ XRD and electrochemical study of the reaction of lithium with amorphous silicon. *J. Electrochem. Soc.* **151**(6), A838–A842.

He, M., Sa, Q., Liu, G., and Wang, Y. (2013). Caramel popcorn shaped silicon particle with carbon coating as a high performance anode material for Li-ion batteries. *ACS Appl. Mater. Interfaces* **5**(21), 11152–11158.

Hochbaum, A.I., Gargas, D., Hwang, Y.J., and Yang, P. (2009). Single crystalline mesoporous silicon nanowires. *Nano Lett.* **9**(10), 3550–3554.

Hochgatterer, N.S., Schweiger, M.R., Koller, S. et al. (2008). Silicon/graphite composite electrodes for high-capacity anodes: Influence of binder chemistry on cycling stability. *Electrochem. Solid-State Lett.* **11**(5), A76–A80.

Hu, Y., Peng, K.Q., Liu, L. et al. (2014). Continuous-flow mass production of silicon nanowires via substrate-enhanced metal-catalyzed electroless etching of silicon with dissolved oxygen as an oxidant. *Sci. Rep.* **4**, 3667.

Huang, R., Fan, X., Shen, W., and Zhu, J. (2009). Carbon-coated silicon nanowire array films for high-performance lithium-ion battery anodes. *Appl. Phys. Lett.* **95**(13), 133119.

Ivanov, S., Vlaic, C.A., Du, S., Wang, D., Schaaf, P., and Bund, A. (2014). Electrochemical performance of nanoporous Si as anode for lithium ion batteries in alkyl carbonate and ionic liquid-based electrolytes. *J Appl. Electrochem.* **44**, 159–168.

Kang, D., Corno, J.A., Gole, J.L., and Shin, H. (2008). Microstructured nanopore-walled porous silicon as an anode material for rechargeable lithium batteries. *J. Electrochem. Soc.* **155**(4), A276–A281.

Kasavajjula, U., Wang, C., and Appleby, A.J. (2007). Nano- and bulk-silicon-based insertion anodes for lithium-ion secondary cells. *J. Power Sources* **163**(2), 1003–1039.

Key, B., Morcrette, M., Tarascon, J.M., and Grey, C.P. (2010). Pair distribution function analysis and solid state NMR studies of silicon electrodes for lithium ion batteries: Understanding the (de)lithiation mechanisms. *J. Am. Chem. Soc.* **133**(3), 503–512.

Kim, H. and Cho, J. (2008). Superior lithium electroactive mesoporous Si@carbon core–shell nanowires for lithium battery anode material. *Nano Lett.* **8**(11), 3688–3691.

Kim, J.W., Ryu, J.H., Lee, K.T., and Oh, S.M. (2005). Improvement of silicon powder negative electrodes by copper electroless deposition for lithium secondary batteries. *J. Power Sources* **147**(1), 227–233.

Kim, H., Han, B., Choo, J., and Cho, J. (2008). Three-dimensional porous silicon particles for use in high-performance lithium secondary batteries. *Angew. Chem.* **120**(52), 10305–10308.

Kim, H., Seo, M., Park, M.H., and Cho, J. (2010). A critical size of silicon nano-anodes for lithium rechargeable batteries. *Angew. Chem. Int. Ed.* **49**(12), 2146–2149.

Kim, H.J., Lee, J., Lee, S.E. et al. (2013). Polymer-free vertical transfer of silicon nanowires and their application to energy storage. *ChemSusChem* **6**(11), 2144–2148.

Kim, H.J., Lee, S.E., Lee, J. et al. (2014). Gold-coated silicon nanowire–graphene core–shell composite film as a polymer binder-free anode for rechargeable lithium-ion batteries. *Physica E* **61**, 204–209.

Kolasinski, K.W. (2005). Silicon nanostructures from electroless electrochemical etching. *Curr. Opin. Solid State Mater. Sci.* **9**(1), 73–83.

Kovalenko, I., Zdyrko, B., Magasinski, A. et al. (2011). A major constituent of brown algae for use in high-capacity Li-ion batteries. *Science* **334**, 75–79.

Lee, S.W., McDowell, M.T., Berla, L.A., Nix, W.D., and Cui, Y. (2012). Fracture of crystalline silicon nanopillars during electrochemical lithium insertion. *Proc. Natl. Acad. Sci.* **109**(11), 4080–4085.

Lehmann, V. and Föll, H. (1990). Formation mechanism and properties of electrochemically etched trenches in n-type silicon. *J. Electrochem. Soc.* **137**(2), 653–659.

Lehmann, V., Stengl, R., and Luigart, A. (2000). On the morphology and the electrochemical formation mechanism of mesoporous silicon. *Mater. Sci. Eng. B* **69**, 11–22.

Lemordant, D., Ghamouss, F., Tran-Van, F. et al. (2014). Sub-micron silicon wires produced by electrochemical etching of a silicon wafer as possible anodes for lithium micro-batteries. In: *Proc. of the 14th Ulm Electrochemical Talks*, June 23–26, 2014, Ulm, Germany.

Lestriez, B., Bahri, S., Sandu, I., Roué, L., and Guyomard, D. (2007). On the binding mechanism of CMC in Si negative electrodes for Li-ion batteries. *Electrochem. Commun.* **9**(12), 2801–2806.

Li, J. and Dahn, J.R. (2007). An in situ X-ray diffraction study of the reaction of Li with crystalline Si. *J. Electrochem. Soc.* 154, A156–A161.

Li, H., Huang, X., Chen, L., Wu, Z., and Liang, Y. (1999). A high capacity nano Si composite anode material for lithium rechargeable batteries. *Electrochem. Solid-State Lett.* **2**(11), 547–549.

Li, J., Lewis, R.B., and Dahn, J.R. (2007). Sodium carboxymethyl cellulose a potential binder for Si negative electrodes for Li-ion batteries. *Electrochem. Solid-State Lett.* **10**(2), A17–A20.

Li, G.V., Kulova, T.L., Tolmachev, V.A. et al. (2013a). Structural transformation of macroporous silicon anodes as a result of cyclic lithiation processes. *Semiconductors* **47**(9), 1275–1281.

Li, X., Gu, M., Kennard, R. et al. (2013b). Macroporous silicon as highly stable anodes for lithium rechargeable batteries. In: *Proc. of 224th ECS Meeting*, Oct. 27–Nov. 1, 2013, San Francisco, p. 984.

Li, X., Xiao, Y., Bang, J.H. et al. (2013c). Upgraded silicon nanowires by metal-assisted etching of metallurgical silicon: A new route to nanostructured solar-grade silicon. *Adv. Mater.* **25**(23), 3187–3191.

Li, X., Gu, M., Hu, S. et al. (2014). Mesoporous silicon sponge as an anti-pulverization structure for high-performance lithium-ion battery anodes. *Nat. Commun.* **5**, 4105.

Liang, B., Liu, Y., and Xu, Y. (2014). Silicon-based materials as high capacity anodes for next generation lithium ion batteries. *J. Power Sources* **267**, 469–490.

Limthongkul, P., Jang, Y.I., Dudney, N.J., and Chiang, Y.M. (2003). Electrochemically-driven solid-state amorphization in lithium-silicon alloys and implications for lithium storage. *Acta Mater.* **51**(4), 1103–1113.

Lin, C.C., Yen, Y.C., Wu, H.C., and Wu, N.L. (2012). Synthesis of porous Si particles by metal-assisted chemical etching for Li-ion battery application. *J. Chin. Chem. Soc.* **59**(10), 1226–1232.

Liu, W.R., Guo, Z.Z., Young, W.S. et al. (2005). Effect of electrode structure on performance of Si anode in Li-ion batteries: Si particle size and conductive additive. *J. Power Sources* **140**(1), 139–144.

Liu, Y., Chen, B., Cao, F., Chan, H.L., Zhao, X. and Yuan, J. (2011). One-pot synthesis of three-dimensional silver-embedded porous silicon microparticles for lithium-ion batteries. *J. Mater. Chem.* **21**(43), 17083–17086.

Liu, X.H., Zhong, L., Huang, S., Mao, S.X., Zhu, T., and Huang, J.Y. (2012). Size-dependent fracture of silicon nanoparticles during lithiation. *ACS Nano* **6**(2), 1522–1531.

Liu, N., Huo, K., McDowell, M.T., Zhao, J., and Cui, Y. (2013). Rice husks as a sustainable source of nanostructured silicon for high performance Li-ion battery anodes. *Sci. Rep.* **3**, 1919.

Loveridge, M., Lain, M., Liu, F., Coowar, F., Macklin, B., and Green, M. (2010). High performance silicon anode materials for next generation lithium-ion batteries. In: *The 15th International ECS Meeting on Lithium Batteries, Meeting Abstracts*, Abstract # 12.

Luais, E., Sakai, J., Desplobain, S., Gautier, G., Tran-Van, F., and Ghamouss, F. (2013). Thin and flexible silicon anode based on integrated macroporous silicon film onto electrodeposited copper current collector. *J. Power Sources* **242**, 166–170.

Luais, E., Ghamouss, F., Wolfman, J. et al. (2015). Mesoporous silicon anode for thin film lithium-ion microbatteries. *J. Power Sources* **274**, 693–700.

Maranchi, J.P., Hepp, A.F., and Kumta, P.N. (2003). High capacity, reversible silicon thin-film anodes for lithium-ion batteries. *Electrochem. Solid-State Lett.* **6**, A198–A201.

Maranchi, J.P., Hepp, A.F., Evans, A.G., Nuhfer, N.T., and Kumta, P.N. (2006). Interfacial properties of the a-Si/Cu: Active–inactive thin-film anode system for lithium-ion batteries. *J. Electrochem. Soc.* **153**(6), A1246–A1253.

Misra, S., Liu, N., Nelson, J., Hong, S.S., Cui, Y., and Toney, M.F. (2012). In situ x-ray diffraction studies of (de) lithiation mechanism in silicon nanowire anodes. *ACS Nano* **6**(6), 5465–5473.

Nguyen, S.H., Lim, J.C., and Lee, J.K. (2012). Electrochemical characteristics of bundle-type silicon nanorods as an anode material for lithium ion batteries. *Electrochim. Acta* **74**, 53–58.

Obrovac, M.N. and Christensen, L. (2004). Structural changes in silicon anodes during lithium insertion/extraction. *Electrochem. Solid-State Lett.* **7**(5), A93–A96.

Obrovac, M.N. and Krause, L.J. (2007). Reversible cycling of crystalline silicon powder. *J. Electrochem. Soc.* **154**(2), A103–A108.

Park, M.H., Kim, M.G., Joo, J. et al. (2009). Silicon nanotube battery anodes. *Nano Lett.* **9**(11), 3844–3847.

Peng, K.Q., Yan, Y.J., Gao, S.P., and Zhu, J. (2002). Synthesis of large-area silicon nanowire arrays via self-assembling nanoelectrochemistry. *Adv. Mater.* **14**(16), 1164–1167.

Peng, K., Jie, J., Zhang, W., and Lee, S.T. (2008). Silicon nanowires for rechargeable lithium-ion battery anodes. *Appl. Phys. Lett.* **93**(3), 033105.

Pereira-Nabais, C., Światowska, J., Rosso, M. et al. (2014). Effect of lithiation potential and cycling on chemical and morphological evolution of Si thin film electrode studied by ToF-SIMS. *ACS Appl. Mater. Interfaces* **6**(15), 13023–13033.

Philippe, B., Dedryvère, R., Allouche, J. et al. (2012). Nanosilicon electrodes for lithium-ion batteries: Interfacial mechanisms studied by hard and soft X-ray photoelectron spectroscopy. *Chem. Mater.* **24**(6), 1107–1115.

Pollak, E., Salitra, G., Baranchugov, V., and Aurbach, D. (2007). In situ conductivity, impedance spectroscopy, and ex situ raman spectra of amorphous silicon during the Insertion/Extraction of lithium. *J. Phys. Chem. C* **111**(30), 11437–11444.

Ponomarev, E.A. and Lévy-Clément, C. (1998). Macropore formation on p-type Si in fluoride containing organic electrolytes. *Electrochem. Solid-State Lett.* **1**(1), 42–45.

Qin, Z., Joo, J., Gu, L., and Sailor, M.J. (2014). Size control of porous silicon nanoparticles by electrochemical perforation etching. *Part. Part. Syst. Charact.* **31**(2), 252–256.

Quiroga-González, E., Ossei-Wusu, E., Carstensen, J., and Föll, H. (2011). How to make optimized arrays of Si wires suitable as superior anode for Li-ion batteries. *J. Electrochem. Soc.* **158**(11), E119–E123.

Radvanyi, E., De Vito, E., Porcher, W. et al. (2013). Study of lithiation mechanisms in silicon electrodes by Auger electron spectroscopy. *J. Mater. Chem. A* **1**(16), 4956–4965.

Ru, Y., Evans, D.G., Zhu, H., and Yang, W. (2014). Facile fabrication of yolk–shell structured porous Si–C microspheres as effective anode materials for Li-ion batteries. *RSC Adv.* **4**(1), 71–75.

Ruffo, R., Hong, S.S., Chan, C.K., Huggins, R.A., and Cui, Y. (2009). Impedance analysis of silicon nanowire lithium ion battery anodes. *J. Phys. Chem. C* **113**(26), 11390–11398.

Ryu, J.H., Kim, J.W., Sung, Y.E., and Oh, S.M. (2004). Failure modes of silicon powder negative electrode in lithium secondary batteries. *Electrochem. Solid-State Lett.* **7**(10), A306–A309.

Ryu, Y.G., Lee, S., Mah, S., Lee, D.J., Kwon, K., Hwang, S., and Doo, S. (2008). Electrochemical behaviors of silicon electrode in lithium salt solution containing alkoxy silane additives. *J. Electrochem. Soc.* **155**(8), A583–A589.

Ryu, I., Choi, J.W., Cui, Y., and Nix, W.D. (2011). Size-dependent fracture of Si nanowire battery anodes. *J. Mech. Phys. Solids* **59**, 1717–1730.

Santos-Peña, J., Lemordant, D., Ghamouss, F. et al. (2014). A study of the Si/electrolyte interfaces in lithium half-cells with submicron silicon wires showing enhanced capacity retention. In: *Proc. of the 65th International Society of Electrochemistry*, Aug. 31–Sept. 5, Lausanne, Switzerland.

Shin, H.-C., Shi, Z., Seals, L.T., Gole, J.L., and Liu, M. (2002). Porous silicon-based electrodes for lithium batteries. In: *Proc. of 202nd ECS Meeting*, Oct. 20–25, 2002, Salt Lake City, UT, p. 799.

Shin, H., Corno, J.A., Gole, J.L., and Liu, M. (2005). Porous silicon negative electrodes for rechargeable lithium batteries. *J. Power Sources* **139**, 314–320.

Song, T., Xia, J., Lee, J.H. et al. (2010). Arrays of sealed silicon nanotubes as anodes for lithium ion batteries. *Nano Lett.* **10**(5), 1710–1716.

Sun, X., Huang, H., Chu, K., and Zhuang, Y. (2012). Anodized macroporous silicon anode for integration of lithium-ion batteries on chips. *J. Electron. Mater.* **41**(9), 2369–2375.

Takamura, T., Ohara, S., Uehara, M., Suzuki, J., and Sekine, K. (2004). A vacuum deposited Si film having a Li extraction capacity over 2000 mA·h/g with a long cycle life. *J. Power Sources* **129**, 96–100.

Tao, H.C., Fan, L.Z., and Qu, X. (2012). Facile synthesis of ordered porous Si@C nanorods as anode materials for Li-ion batteries. *Electrochim. Acta* **71**, 194–200.

Thakur, M., Pernites, R.B., Nitta, N. et al. (2012a). Freestanding macroporous silicon and pyrolyzed polyacrylonitrile as a composite anode for lithium ion batteries. *Chem. Mater.* **24**, 2998–3003.

Thakur, M., Isaacson, M., Sinsabaugh, S.L., Wong, M.S., and Biswal, S.L. (2012b). Gold-coated porous silicon films as anodes for lithium ion batteries. *J. Power Sources* **205**, 426–432.

Thakur, M., Sinsabaugh, S.L., Isaacson, M.J., Wong, M.S., and Biswal, S.L. (2012c). Inexpensive method for producing macroporous silicon particulates (MPSPs) with pyrolyzed polyacrylonitrile for lithium ion batteries. *Sci. Rep.* **2**, 795.

Uchida, S., Mihashi, M., Yamagata, M., and Ishikawa, M. (2015). Electrochemical properties of non-nano-silicon negative electrodes prepared with a polyimide binder. *J. Power Sources* **273**, 118–122.

Verma, P., Maire, P., and Novák, P. (2010). A review of the features and analyses of the solid electrolyte interphase in Li-ion batteries. *Electrochim. Acta* **55**(22), 6332–6341.

Vincent, G. (1994). Optical properties of porous silicon superlattices. *Appl. Phys. Lett.* **64**(18), 2367–2369.

Vlad, A., Reddy, A.L.M., Ajayan, A. et al. (2012). Roll up nanowire battery from silicon chips. *Proc. Natl. Acad. Sci.* **109**(38), 15168–15173.

Wan, J., Kaplan, A. F., Zheng, J. et al. (2014). Two dimensional silicon nanowalls for lithium ion batteries. *J. Mater. Chem. A* **2**(17), 6051–6057.

Wang, X.L. and Han, W.Q. (2010). Graphene enhances Li storage capacity of porous single-crystalline silicon nanowires. *ACS Appl. Mater. Interfaces* **2**(12), 3709–3713.

Wang, M.S., Fan, L.Z., Huang, M., Li, J., and Qu, X. (2012). Conversion of diatomite to porous Si/C composites as promising anode materials for lithium-ion batteries. *J. Power Sources* **219**, 29–35.

Wang, B., Li, X., Qiu, T. et al. (2013). High volumetric capacity silicon-based lithium battery anodes by nanoscale system engineering. *Nano Lett.* **13**(11), 5578–5584.

Weisse, J.M., Lee, C.H., Kim, D.R., Cai, L., Rao, P.M., and Zheng, X. (2013). Electroassisted transfer of vertical silicon wire arrays using a sacrificial porous silicon layer. *Nano Lett.* **13**(9), 4362–4368.

Wen, C.J. and Huggins, R.A. (1981). Chemical diffusion in intermediate phases in the lithium-silicon system. *J. Solid State Chem.* **37**(3), 271–278.

Wilson, A.M. and Dahn, J.R. (1995). Lithium insertion in carbons containing nanodispersed silicon. *J. Electrochem. Soc.* **142**(2), 326–332.

Wu, H. and Cui, Y. (2012). Designing nanostructured Si anodes for high energy lithium ion batteries. *Nano Today* **7**(5), 414–429.

Wu, X., Wang, Z., Chen, L., and Huang, X. (2003). Ag-enhanced SEI formation on Si particles for lithium batteries. *Electrochem. Comm.* **5**(11), 935–939.

Wu, H., Chan, G., Choi, J.W. et al. (2012). Stable cycling of double-walled silicon nanotube battery anodes through solid–electrolyte interphase control. *Nat. Nanotechnol.* **7**, 310–315.

Xu, K. (2004). Nonaqueous liquid electrolytes for lithium-based rechargeable batteries. *Chem. Rev.* **104**(10), 4303–4418.

Xu, W. and Flake, J.C. (2010). Composite silicon nanowire anodes for secondary lithium-ion cells. *J. Electrochem. Soc.* **157**(1), A41–A45.

Xu, W., Vegunta, S.S.S., and Flake, J.C. (2011). Surface-modified silicon nanowire anodes for lithium-ion batteries. *J. Power Sources* **196**(20), 8583–8589.

Yang, H., Huang, S., Huang, X. et al. (2012). Orientation-dependent interfacial mobility governs the anisotropic swelling in lithiated silicon nanowires. *Nano Lett.* **12**(4), 1953–1958.

Yang, X., Shi, C., Zhang, L., Liang, G., Ni, S., and Wen, Z. (2013a). Preparation of three dimensional porous silicon with fluoride-free method and its application in lithium ion batteries. *Electrochem. Solid-State Lett.* **2**(11), M53–M56.

Yang, Y., Ren, J.G., Wang, X. et al. (2013b). Graphene encapsulated and SiC reinforced silicon nanowires as an anode material for lithium ion batteries. *Nanoscale* **5**(18), 8689–8694.

Yao, Y., McDowell, M.T., Ryu, I. et al. (2011). Interconnected silicon hollow nanospheres for lithium-ion battery anodes with long cycle life. *Nano Lett.* **11**(7), 2949–2954.

Yoo, J.K., Kim, J., Lee, H. et al. (2013). Porous silicon nanowires for lithium rechargeable batteries. *Nanotechnology* **24**(42), 424008.

Yoo, J.K., Kim, J., Choi, M.J. et al. (2014). Extremely high yield conversion from low-cost sand to high-capacity Si electrodes for Li-ion batteries. *Adv. Energy Mater.* **4**, 1400622.

Yu, Y., Gu, L., Zhu, C., Tsukimoto, S., van Aken, P.A., and Maier, J. (2010). Reversible storage of lithium in silver-coated three-dimensional macroporous silicon. *Adv. Mater.* **22**(20), 2247–2250.

Zhang, X. (2001). Electrochemistry of Silicon and Its Oxide. Kluwer Academic Pub, New York.

Zhang, W. (2011). A review of the electrochemical performance of alloy anodes for lithium-ion batteries. *J. Power Sources* **196**, 13–24.

Zhang, Z., Wang, Y., Ren, W. et al. (2014). Scalable synthesis of interconnected porous silicon/carbon composites by the rochow reaction as high-performance anodes of lithium ion batteries. *Angew. Chem., Int. Ed.* **53**(20), 5165–5169.

Zhang, Y., Zhang, X.G., Zhang, H.L. et al. (2006). Composite anode material of silicon/graphite/carbon nanotubes for Li-ion batteries. *Electrochim. Acta* **51**(23), 4994–5000.

Zhao, K., Pharr, M., Wan, Q. et al. (2012a). Concurrent reaction and plasticity during initial lithiation of crystalline silicon in lithium-ion batteries. *J. Electrochem. Soc.* **159**(3), A238–A243.

Zhao, Y., Liu, X., Li, H., Zhai, T., and Zhou, H. (2012b). Hierarchical micro/nanoporous silicon Li-ion battery anodes. *Chem. Commun.* **48**(42), 5079–5081.

Zheng, Y., Yang, J., Wang, J., and Li, Y.N. (2007). Nano-porous Si/C composites for anode material of lithium-ion batteries. *Electrochim. Acta* **52**(2007), 5863–5867.

Zhou, X., Wan, L.J., and Guo, Y.G. (2013). Electrospun silicon nanoparticle/porous carbon hybrid nanofibers for lithium-ion batteries. *Small* **9**(16), 2684–2688.

Zhu, J., Gladden, C., Liu, N., Cui, Y., and Zhang, X. (2013a). Nanoporous silicon networks as anodes for lithium ion batteries. *Phys. Chem. Chem. Phys.* **15**, 440–443.

Zhu, Y., Liu, W., Zhang, X. et al. (2013b). Directing silicon–graphene self-assembly as a core/shell anode for high-performance lithium-ion batteries. *Langmuir* **29**(2), 744–749.

PSi-Based Supercapacitors

16

Diana Golodnitsky, Ela Strauss, and Tania Ripenbein

CONTENTS

16.1   Introduction   348

16.2   Supercapacitors—Basic Definitions   349

16.3   Electrode Materials for Supercapacitors   355

16.4   Porous-Silicon Supercapacitors   356

    16.4.1   Fabrication of Porous-Silicon Supercapacitors   357

        16.4.1.1   Integrated on-Perforated-Silicon-Chip Micro-Supercapacitors   357

        16.4.1.2   Supercapacitors with PSi Electrodes   360

16.5   Electrochemical Performance of Porous Silicon-Based Supercapacitors   364

    16.5.1   Integrated on-Perforated-Si-Chip Supercapacitors   364

    16.5.2   Supercapacitors with Porous Silicon Electrodes   368

16.6   Summary   371

References   372

16.1 INTRODUCTION

Environmental protection and new energy development have recently become growing industries, in which energy-storage devices are highlighted as crucial components. The electrochemical or double-layer capacitors constitute an important transient-energy-storage technology for rapid-charge/discharge applications. Electrochemical capacitors (double-layer capacitors [EDLC] and pseudo-capacitors) are intrinsically high-power devices (between 5 and 15 kW/kg) of limited energy-storage capability and long cycle life. Compared with batteries, supercapacitors have higher power density with shorter charge/discharge time, longer shelf and cycle life, but relatively lower energy density. In some aspects, they bridge the gap between conventional capacitors and batteries in that they display energy density higher than that of conventional capacitors and higher power density than batteries (Ye et al. 2005; Huang et al. 2008; Largeot et al. 2008; Simon & Gogotsi 2008; Sharma & Bhatti 2010; Gileadi 2011; Bittner et al. 2012; Choi et al. 2012; Shukla et al. 2012; M. Li et al. 2013; Oakes et al. 2013; A. Yu et al. 2013; G. Yu et al. 2013; Cheng et al. 2014; Gu & Yushin 2014; Kang et al. 2014) (Figure 16.1).

However, insufficient energy density of electrochemical capacitors limits the optimal discharge time to less than a minute, while many applications clearly need more. The first electrochemical capacitor based on the electric-double-layer theories of Helmholtz (1853), Gouy-Chapman, Grahame, and Stern (Gileadi 2011; A. Yu et al. 2013) was patented by the General Electric Company in 1957 (Sharma & Bhatti 2010). In the same year, the Standard Oil Company, Cleveland, Ohio (SOHIO) patented a device that stored energy in the double-layer interface (Sharma & Bhatti 2010). The Pinnacle Research Institute (PRI) introduced the term "ultracapacitor" (1982) (Sharma & Bhatti 2010). The Nippon Electric Company (NEC) of Japan licensed the technology. "Supercapacitor" was the trade name of their first commercial device. Whatever the trade name, it generally refers to an electrochemical double-layer capacitor (EDLC) which stores electrical energy in the interface between an electrolyte and a solid electrode (Sharma & Bhatti 2010).

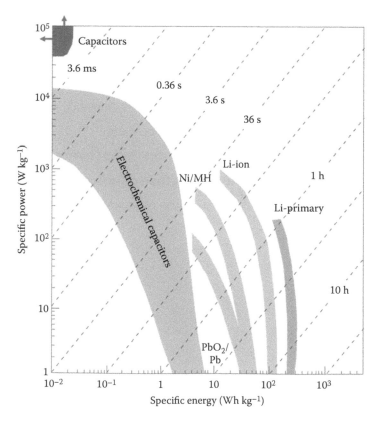

FIGURE 16.1 Ragone plot for various electrical energy storage devices. (From Simon P. and Gogotsi Y., *Nat. Mater.* **7**, 845, 2008. Copyright 2008: Macmillan Publishers Limited. With permission.)

Since the early days of the development of electrochemical capacitors in the late 1950s, there has not been a good strategy for increasing the energy density. Only incremental improvements in performance were achieved from the 1960s to 1990s. The impressive increase in performance that has been demonstrated in the last year or so is due to the discovery of new electrode materials and improved understanding of ion behavior in small pores, as well as the design of new hybrid systems combining faradaic and capacitive electrodes. Much of the research on electrochemical capacitors is concerned with increasing their energy density with the least sacrifice in power capability and cycle life for deep discharges. Fundamentally, the requirements for thin-electrode design with the use of materials having nanoscale characteristics and minimum resistance present a materials science and manufacturing challenge.

Carbon has been used as a high-surface-area electrode material ever since development of the electrochemical capacitor began. Porous activated carbon (AC), carbon nanotubes, metal oxides, and conducting polymers are currently the most widely known electrode materials.

Since the surface area of activated carbon is about 1000 m²/g of material, this enables the creation of a capacitor cell with a specific capacitance of 100 F/g and the possibility of operating devices rated at many thousands of farads (Simon & Gogotsi 2008; Sharma & Bhatti 2010; Choi et al. 2012; Shukla et al. 2012; Gu & Yushin 2014; Kang et al. 2014).

Silicon materials have not been used for supercapacitors because of the extreme reactivity of silicon with electrolytes. Until recently, there have been only a few investigations of silicon materials in capacitors (Ye et al. 2005; Largeot et al. 2008; Cheng et al. 2014; Kang et al. 2014), and these have involved device configurations with specific capacitances orders of magnitude lower (5 mF/g) than those of carbon materials for on-chip micro-supercapacitors (Oakes et al. 2013). This is not the case for porous silicon (PSi), one of the advantages of which is the possibility of assembling porous, high-surface-area templates that can maintain electrical interconnection and be controllably produced for optimization of devices planned. In addition, for integrated on-chip devices, such as micro-electromechanical systems (MEMS) or micro-robots, volumetric performance is often overlooked. In recent years, the development of new silicon-based supercapacitors known as micro-supercapacitors or micro-ultracapacitors, which could supply micropower sources for energy harvesting, has aroused considerable attention.

This review presents the routes of preparation of PSi substrates and PSi electrodes. The electrochemical performance characteristics of PSi-EDLCs and pseudo-capacitors are also discussed.

16.2 SUPERCAPACITORS—BASIC DEFINITIONS

The basic differences between the electrostatic capacitors and supercapacitors are briefly addressed here.

Capacitors store charge/energy on the surfaces of metal or metallized-plastic-film electrodes with a dielectric medium in between to achieve physical charge separation. Therefore, capacitance is a function of the properties of the dielectric and the surface areas of the electrodes, which are critical since the opposing charges are in close proximity. Most configurations contain a layered arrangement with a separation distance on the scale of micrometers, which is volumetrically inefficient. An electrolytic capacitor is similar in construction to an electrostatic capacitor but has an ion-conducting electrolyte in direct contact with the metal electrodes. Aluminum electrolytic capacitors, for example, consist of two aluminum-conducting foils (coated with an insulating oxide layer) and a paper spacer soaked in electrolyte (Sharma & Bhatti 2010). The oxide layer serves as the dielectric and is very thin. This results in capacitance per unit volume that is higher than that for electrostatic capacitors.

Supercapacitors have very high surface area that is obtained by the use of a molecular-thin layer of electrolyte, rather than a manufactured sheet of dielectric material, to separate the charges. The supercapacitor resembles an electrostatic capacitor, but differs in that it offers very high capacitance in a small package. Supercapacitors rely on the separation of charge at an electrified interface that is measured in fractions of a nanometer. The operating voltage is limited by the breakdown voltages of the solvents with aqueous electrolytes (0.9 to 1.1 V) and organic electrolytes (2.5 to 3 V). The lifetime of supercapacitors is virtually indefinite and their energy efficiency rarely falls below 95% when they are kept within their design limits. Their power density is

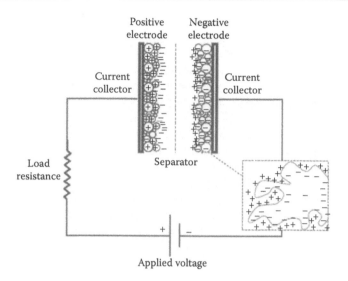

FIGURE 16.2   Schematic sketch of an electrical double-layer capacitor. (From Shukla A.K. et al., *Electrochim. Acta* **84**, 165, 2012. Copyright 2012: Elsevier. With permission.)

higher than that of batteries while their energy density is generally lower. However, unlike batteries, almost all of this energy is available in a reversible process. The energy density of capacitors is very low—less than 1% in the case of supercapacitors—but the power density is often higher because of the high dielectric-breakdown voltage.

An electrochemical double-layer capacitor is fabricated from two electrodes that are separated by an ion-permeable membrane (separator) and an electrolyte. The double layer is formed at the interface between each electrode and the electrolyte (Gileadi 2011; Shukla et al. 2012; A. Yu et al. 2013; Kang et al. 2014). Figure 16.2 shows a schematic diagram of a double-layer capacitor (Shukla et al. 2012).

When external power is applied, the electrodes are equally electrically charged, one electrode positively and the second negatively. The positive electrode is formed by depletion of electrons in the electrode, which results in an excess of positive charge that attracts an equal number of negative charges in the electrolyte near the electrode. Since the system is electrically neutral, a similar phenomenon occurs at the other electrode. Accumulated electrons in the negative electrode attract positive ions from the adjacent electrolyte, forming another double layer. Hence, a complete double-layer capacitor has two electrical double layers, one at each electrode-electrolyte interface. The properties of these two double layers determine the performance of the electrochemical capacitor.

A diffuse double layer, which is also called the Gouy–Chapman diffuse layer, is formed because of thermal fluctuation in the solution (Figure 16.3) (Kang et al. 2014). Such a diffuse double layer is formed by a gradient of ions with a high concentration near the electrode surface that gradually decreases with the distance from the electrode to the solution. The thickness of the diffuse layer depends on the temperature, the concentration of the electrolyte, the number of charge carriers (ions), and the dielectric constant of the electrolyte. When the diffuse layer is very thin, a compact array of ions and solvent molecules forms near the electrode surface and its thickness is effectively equal to the diameter of an ion/solvent molecule. This leads to the formation of a compact Helmholtz electrical double-layer. The Helmholtz layer can be divided into two layers—the inner Helmholtz plane (IHP) and the outer Helmholtz plane (OHP). The differential capacitance of the IHP is significantly affected by the specific adsorption of ions on the electrode surface. Figure 16.3 (Kang et al. 2014) shows the coexistence of the two layers that are described and fully explained by the Stern–Grahame model.

The potential drop across the double layer (V) is expressed by Equation 16.1 for the case that the concentration of the electrolyte is high enough to form the Helmholtz layer (A. Yu et al. 2013, p. 56):

(16.1)
$$V = \frac{q}{C_{dl}}$$

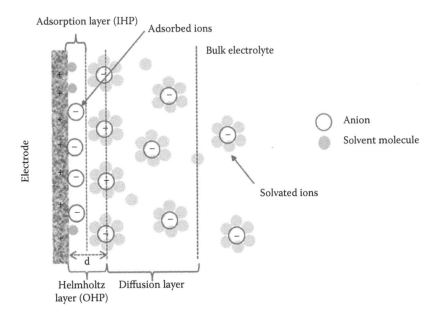

FIGURE 16.3 Schematics of an electrochemical double layer and its electrode/electrolyte interface model. (Reprinted with permission from Kang J. et al., *Electrochim. Acta* **115**, 587, 2014. Copyright 2014: Elsevier. With permission.)

The potential range over which the electrode double layer can be charged or discharged without interference from electrochemical oxidation/reduction reactions is called the "double-layer range/window" (A. Yu et al. 2013). When no electrode reactions occur over a certain potential range, the electrode is defined as an ideal polarizable electrode, that is, the electrode behaves as a capacitor. Only capacitive current flows following a change in potential over this range.

The double-layer potential range strongly depends on the electrode and electrolyte (solvent and salt) materials. Therefore, in order to obtain an ideal polarized electrochemical capacitor it is important to choose these materials in a way that enables the widest stable electrochemical-potential window.

The most practical electrode materials in supercapacitors are carbon-based materials that have an almost ideal polarizable potential window in different electrolyte solutions. However, their surface reversible redox reactions may produce pseudocapacitance that is actually beneficial because it contributes to the total capacitance (will be discussed later).

The electrolyte of a supercapacitor consists of a salt dissolved in a liquid, gel, or in a solid polymer matrix. Electrolytes used in EDLCs may be divided into three classes: (1) aqueous (solutions of acids, bases, and salts), (2) organic (for pure EDLCs—most commonly solutions of tetraethyl-ammonium tetrafluoroborate [TEATFB] salt either in anhydrous acetonitrile [AN] or propylene carbonate [PC] solvents), and (3) ionic liquids (IL). Aqueous electrolytes are typically stable over the range of 0.6–1.4 V in symmetric EDLCs, organic electrolytes are stable up to 2.2–3.0 V, and ionic liquids up to 2.6–4.0 V (Gu & Yushin 2014).

The charge density on an electrode of electrochemical capacitor (or any other capacitor) is expressed in Equation 16.2. Q is the charge accumulated on the electrode side (equal to the charge in the electrolyte-solution side) and A is the effective electrode surface area that is in contact with the electrolyte.

(16.2)
$$q = \frac{Q}{A}$$

The differential capacitance (or the capacitance density) is defined as the capacitance divided by the effective surface area. The units of capacitance density are farads per square meter (F/m²), farads per gram (F/g), or microfarads per square centimeter (μF/cm²).

Electrochemical-impedance spectroscopy (EIS) is a very powerful technique for characterizing electrochemical phenomena in the analysis of double-layer capacitors. The magnitude of the AC impedance of a supercapacitor cell, $\left(Z_{cell}^{j}\right)$, is shown in Equation 16.3.

$$(16.3) \qquad \left|Z_{cell}^{j}\right| = \sqrt{(R_{esr})^2 + \left(\frac{1}{2\pi f C_{dl}^{T}}\right)^2}$$

R_{esr} is the equivalent series resistance, f is the AC frequency in hertz, and C_{dl}^{T} is the total double-layer capacitance. The values of R_{esr} and C_{dl} are calculated from Equation 16.3. It is important to emphasize that the value of R_{esr} should be as small as possible because a high value of R_{esr} limits the charge/discharge current (rate), which can be delivered at a given voltage. In addition, high R_{esr} causes the accumulation of heat in the supercapacitor device, which is followed by rapid degradation of the EDLC components and performance characteristics.

In general, the equivalent-circuit model of an EDLC is composed of one or more pure capacitors (C) coupled with their equivalent resistances (R) which are arranged in series or in parallel. This model structure is useful for providing quantitative information on parameter variations with ease of interpretation and simple simulation. The simplified equivalent circuit of a double layer consists of two capacitors—C_H, the differential capacitance of the Helmholtz layer, and C_{diff}, the differential capacitance of the diffusion layer—coupled in series (A. Yu et al. 2013).

The reciprocal of the overall double-layer differential capacitance [$\mu F/cm^2$] is given by Equation 16.4 (A. Yu et al. 2013).

$$(16.4) \qquad \frac{1}{C_{dl}} = \frac{1}{C_H} + \frac{1}{C_{diff}}$$

Since a double layer is formed at each of the electrodes in an electrochemical cell, the overall capacitance of a supercapacitor is the equivalent of two differential capacitances connected in series. The capacitance for the positive electrode is given by Equation 16.5:

$$(16.5) \qquad C_{dl,p} = \left(\frac{C_{H,p} C_{diff,p}}{C_{H,p} + C_{diff,p}}\right)$$

If the supercapacitor is symmetric, that is, the two electrodes are identical, then the total equivalent capacitance is equal to half of the individual electrode capacitance.

$$(16.6) \qquad C_{dl}^{T} = 0.5_{dl,p}$$

If the supercapacitor is asymmetric, the electrode with the smaller capacitance dominates the total capacitance.

The equivalent circuit for an ideal electrochemical supercapacitor consists of a resistor connected in series with a capacitor (the double-layer capacitor) as described in Figure 16.4.

The ohmic component is the equivalent series resistance, R_{esr}, which originates from the contact resistance between the current collector and the active material of the electrode, the resistance of the active material that forms the electrode on the current collector, the resistance of external contacts, and the resistance of the electrolyte. C_{dl}^{T} is the total equivalent capacitance of the double layer.

It is worth noting that an ideal EDLC is independent of the working frequency or applied voltage when its capacitance and internal resistance are evaluated. In practice, however, dependence of the capacitance and resistance on frequency and voltage is commonly observed (Kang et al. 2014).

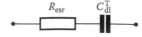

FIGURE 16.4 Equivalent circuit of supercapacitor. R_{esr} is the equivalent ohmic resistance and C_{dl}^{T} is the total capacitance.

In addition, practical EDLC devices suffer from charge leakage (defined as self-discharge), which results from potential-dependent charge-transfer reactions. These deviations from ideal capacitive behavior of the EDLC are attributed mainly to ionic chemical/physical adsorption and diffusion impedance, incomplete polarization of the porous electrode, and faradaic charge-transfer resistance caused by the voltage differential across the electrode/electrolyte interface. From the evaluation of impedance data, the electrochemical interface behavior described by simple electrical elements (R, L, and C) in an equivalent circuit does not take into account many details of ion transport in the electrolyte and multiple physical processes that occur simultaneously in the system. These include specific adsorption of ions into pore sites and diffusion phenomena taking place in the interfacial region and in the bulk electrolyte. Therefore, the physical interpretation of equivalent circuits of an EDLC is still doubtful and unclear. An advanced equivalent-circuit model of supercapacitors was recently developed by Kang et al. (2014). Three major aspects were taken into account by the authors. First, based on the theory of the interfacial layer in the EDLC, the total capacitance was approximated by a combination of three separate capacitances, those of the adsorption layer (inner Helmholtz layer), the compact double layer (outer Helmholtz layer), and the diffusion layer in the electrolyte. It was represented by the combination of Helmholtz-layer capacitance (C_H) in series with the diffusion-layer contribution (W), and adsorption capacitance (C_{ads}) placed in parallel. Therefore, the equivalent circuit of an EDLC was modeled by three capacitive elements. The second aspect is the definition of the interfacial resistance by the combination of charge-transfer resistance (R_{ct}) (in the case of a pseudo-capacitor) and adsorption resistance (R_{ads}) in the Helmholtz layer (Figure 16.5a). Third, the circuit was modified by considering the bulk-medium processes to determine the impedance of the practical device over the whole range of frequencies (Figure 16.5b). As a result, the complete equivalent circuit for the EDLC is given in Figure 16.5c. This circuit model, on its validation by the EIS data was successfully applied to the characterization of practical EDLC devices in aqueous 1M H_2SO_4 and organic 1M Et_4NBF_4/PC electrolytes.

The resulting expression for the total impedance of the cell, including interfacial double-layer, adsorption, and bulk phenomena is shown in Equation 16.7 (Kang et al. 2014).

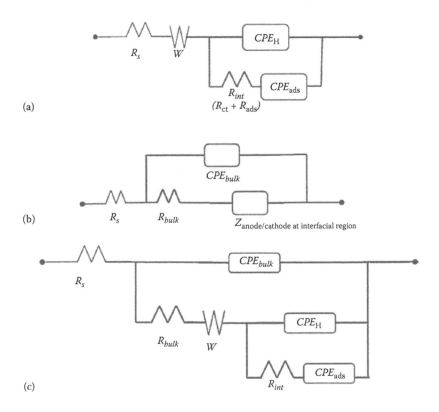

FIGURE 16.5 Equivalent circuits modelling of (a) interfacial processes at the double layer, (b) considering bulk processes, and (c) the complete circuit. (From Kang J. et al., *Electrochim. Acta* **115**, 587, 2014. Copyright 2014: Elsevier. With permission.)

(16.7)
$$Z = R_S + \left[\frac{1}{Z_{CPE_{bulk}}} + \cfrac{1}{R_{bulk} + Z_W + \left[\cfrac{1}{Z_{CPE_H}} + \cfrac{1}{R_{int} + Z_{CPE_{ads}}} \right]^{-1}} \right]^{-1}$$

In practice, capacitance values are determined by cyclic-voltammetry (CV) techniques, by chronopotentiometry (CP) in a three-electrode configuration, and by charge/discharge of the two-electrode-cell configuration (Ye et al. 2005; Simon & Gogotsi 2008; Sharma & Bhatti 2010; Gileadi 2011; Choi et al. 2012; Shukla et al. 2012; A. Yu et al. 2013; Cheng et al. 2014; Gu & Yushin 2014; Kang et al. 2014).

Plots in Figure 16.6 (Kang et al. 2014) show the cyclic-voltammetry curves of coin-type cells, measured in aqueous (1M H_2SO_4) and organic (1M Et_4NBF_4/PC) electrolytes. Since the two electrodes are identical and ideal activated carbon is a non-faradaic material, the current response of supercapacitor is symmetrical and similar to that of an ideal EDLC. Although the aqueous electrolyte shows some distortion of the ideal CV shape near 1 V, as a result of the decomposition of the electrolyte, it is clear that the organic electrolyte showed no distortion up to 2.7 V (Kang et al. 2014).

The specific capacitance ($\mu F/cm^2$) is calculated from Equation 16.8

(16.8)
$$C_{dl} = i \left(\frac{dt}{dV} \right)$$

where i is the current density, (A/cm^2), and dV/dt is the scan rate (V/sec).

There are three options for charging and discharging supercapacitors: (1) charge/discharge at constant voltage, (2) charge/discharge at constant current, (3) discharge at constant power. Important performance characteristics for the evaluation of an electrochemical capacitor are energy and power density. The energy density can be expressed as

(16.9)
$$E = \int_0^q V_{sc} \, dq = \frac{1}{2} C_{dl}^T V_{sc}^2$$

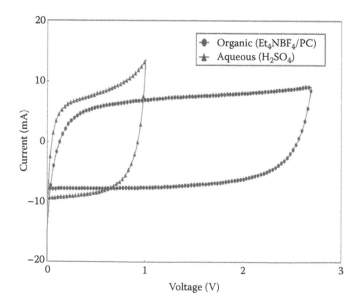

FIGURE 16.6 Cyclic voltammetry for organic and aqueous electrolytes at the scan rate 20 mVs^{-1}. (From Kang J. et al., *Electrochim. Acta* **115**, 587, 2014. Copyright 2014: Elsevier. With permission.)

Here, V_{sc} is supercapacitor voltage, q (coulomb/cm^2) is the total charge stored in the supercapacitor, and C_{dl}^T (F/cm^2) is the double-layer capacitance of the cell.

In electrochemical power devices (including supercapcitors), the specific energy density is commonly expressed as in Equation 16.8, where C_{dl}^T is replaced by C_{sp}, the specific capacitance (Equation 16.11). C_m is the measured capacitance and m(g) is the mass of the electrode material.

(16.10)
$$E_m = \frac{1}{2}\frac{C_m}{m}V_{sc}^2 = \frac{1}{2}C_{sp}V_{sc}^2$$

(16.11)
$$C_{sp} = \frac{C_m}{m}$$

The specific energy density depends on the conductivity and electrochemical stability of electrolyte, type of material, morphology, surface area and porosity of electrodes, and type of current collector. It also depends on the interaction (adsorption or redox reactions in the case of pseudo-capacitor) between the ions in the electrolyte and the active material of the electrode. Power is defined as the amount of energy released per time unit. In the MKS unit system, the unit measure of power is the watt (joule per second [J/s]). Power can be normalized to mass and volume. The definition of the specific power density (P_m) is defined as

(16.12)
$$P_m = \frac{I_{cell}V_{cell}}{m}$$

where P_m is the power density (measured in W/kg), V_{cell} is the cell voltage, I_{cell} is the current density in the cell (measured in A/cm^2), and m is the mass loading (kg/cm^2).

As shown in Ragone plots (see Figure 16.1), supercapacitors are characterized by higher power densities and lower energy densities than other devices such as batteries and fuel cells. Therefore, overcoming the disadvantage of low energy density is currently the major focus and challenge in supercapacitor research and development.

16.3 ELECTRODE MATERIALS FOR SUPERCAPACITORS

The electrode is the key part of the electrochemical capacitor (EC); hence, the electrode materials are the most important factors in determining the properties of ECs. Since charge separation in the electrical double layer (DL) occurs over a few angstroms, the DL follows the topography of the electrode surface (Figure 16.2). For both the EDLC and the pseudo-capacitor, the most effective approach to improving the specific capacitance is the use of nanomaterials, which have suitable pore size and distribution, accessible surface area, and good interconnected channels (Ye et al. 2005; Largeot et al. 2008; Simon & Gogotsi 2008; Sharma & Bhatti 2010; Choi et al. 2012; Shukla et al. 2012; A. Yu et al. 2013; Cheng et al. 2014; Gu & Yushin 2014; Kang et al. 2014). This results in high specific surface areas and short ion-diffusion paths. It was observed, however, that the double-layer type of capacitance was not directly proportional to the surface area because the pores are too small for the electrolyte to enter. It seems obvious that the large pores can be easily and quickly accessed, while the accessibility of smaller pores is much more limited kinetically. On the other hand, Largeot et al. (2008) showed that subnanometer pores promote desolvation of ions in the vicinity of the electrode surface and contribute to a significant increase in gravimetric capacitance, which arises from a smaller distance of charge separation. From a fundamental point of view, there is still a lack of understanding of the double-layer charging in the confined space of micropores, where there is insufficient room for the formation of the Helmholtz layer and where the diffuse layer strongly depends on the shape and size of the pores (Largeot et al. 2008; Simon & Gogotsi 2008). For mesoporous materials (pores larger than 2 nm), the traditional model describing the charge of the double layer is used (Huang et al. 2008). For pores smaller than 1 nm, it is assumed that ions enter a cylindrical pore and lineup, thus forming the "electric-wire-in-cylinder" model of a capacitor (Largeot et al. 2008).

In general, electrode materials in supercapacitors can be divided into the following types: carbon materials, transition-metal oxides, and conducting polymers. EDLCs, the most common devices at present, use high-surface-area carbon-based active materials. For symmetric EDLCs, various chemistries and technologies being developed are focused on carbons with specific capacitance (F/g) significantly greater than the present values of 150–200 F/g in aqueous electrolytes and 80–120 F/g in organic electrolytes. In the current EDLC technology, activated-carbon porous electrodes are used. Novel carbon materials (carbide-derived carbon, zeolite-templated carbon, carbon aerogels, carbon nanotubes, onion-like carbon and graphene), which have high surface area, high electrical conductivity, as well as a range of shapes, sizes, and pore-size distributions, are constantly being developed and tested as potential supercapacitor electrodes (Gu & Yushin 2014). Increase in pore volume, while a desired electrode-material property, is associated with reduction in pore-wall thickness, which renders capacitance enhancement difficult. Nanoporous carbons are more expensive than activated carbon materials by an order of magnitude and present experimental difficulties when manufactured with tailored nanostructures and when designed to match the electrolyte characteristics of size, charge and solvation/desolvation of ions. Activated carbon in electrochemical double-layer electrodes was found to undergo slow aging. This effect becomes apparent only when capacitors are polarized to voltages at or above specified limits, or at high temperatures, for at least several months. A combination of "postmortem" analysis methods (Bittner et al. 2012) shows that pores are clogged with organic species formed by decomposition of the electrolyte, and carbon suffers from reduction/oxidation.

Typical active pseudocapacitive materials include transition-metal oxides and electronically conducting redox polymers. The majority of publications focused on metal oxides deals with ruthenium oxide (RuO_2). This oxide is widely studied because it is conductive and has three distinct oxidation states accessible over a 1.2 V voltage range. The pseudocapacitive behavior of RuO_2 in acidic solutions is ascribed to rapid reversible electron transfer together with electrosorption of protons on the surface of the RuO_2 particles. While providing high specific capacitance (760 F/g), ruthenium oxide is expensive. Hence, other, less expensive electrode materials with good capacitive characteristics such as Fe_3O_4, NiO, MnO_2, Co_3O_4, $Ni(OH)_2$, and $Co(OH)_2$ have been investigated (Simon & Gogotsi 2008; Shukla et al. 2012; M. Li et al. 2013; G. Yu et al. 2013; Gu & Yushin 2014). The mechanism of charge storage on metal oxides is based on surface adsorption of electrolyte cations (K^+, Na^+) and incorporation of protons to form, for example, $MnOOK_xH_y$.

Electroactive polymers include polyaniline, polypyrrole, polythiophene, and their derivatives. Among them, polypyrrole (PPy) is considered one of the most promising electrode materials for supercapacitors because of the advantages of facile synthesis, low cost, environmental friendliness, and high capacitance. The capacitance of the PPy electrode is affected by the type, size, and valence of the ions used in preparing solutions and testing electrolytes.

16.4 POROUS-SILICON SUPERCAPACITORS

Silicon is the second-most abundant element on the surface of the planet and has had a revolutionary impact on the electronics and solar industries. These industries have driven the price of silicon raw materials to a range of \$2–\$30 per kilogram for metallurgical grade to electronic grade, respectively (Oakes et al. 2013). Recently, PSi has been studied as an alternative to carbon in EDLCs; however, until now there have been only a small number of investigations in the field. PSi has also been proposed for use in integrated on-Si-chip microcapacitors and some of the examples are addressed in this chapter. PSi offers extreme flexibility of the material in terms of processing (easily tailorable microstructure), chemical inertness, as well as customizable surface area and conductivity. The inherent resistivity of the PSi can be reduced using subsequent doping and metallization or carbonization processes of the macropore surface. For example, plating with nickel followed by heat treatment produces NiSi, the resistivity of which is lower by orders of magnitude than that of activated carbon and graphite. An inert material is especially helpful in enabling wide electrochemical voltage windows. Because of the higher density of silicon than that of carbon, the Si-based EDLSs would have higher volumetric energy density.

16.4.1 FABRICATION OF POROUS-SILICON SUPERCAPACITORS

Since the energy that can be stored in a capacitor is proportional to the interfacial area, it is obvious that an effective way to increase the amount of charge is to enlarge the effective surface area. Three-dimensional (3D) architectures enable better exploitation of space in the energy-storage device resulting in a smaller footprint while retaining high power and energy density due to the high surface-to-volume ratio. This strategy has been successfully adopted by making use of the fact that silicon can be highly structured (Van den Meerakker et al. 2000; Laermer & Schilp 2003). Several techniques, such as deep reactive-ion etching (DRIE), electrochemical etching of silicon wafers, decomposition of silanes and disilanes and other methods have been tested for the preparation of PSi for supercapacitor applications. Some of the methods of fabrication of PSi supercapacitors are briefly described next. However, as found for the carbon-based supercapacitors, increasing of the pores' volume alone does not scale linearly with the capacity improvement. A variety of experiments showed that the size of the pores should be smaller than the solvated ion size. This is in order to enable partial or complete removal of the solvation shell of cation and anion allowing the ions access the pores (Simon & Gogotsi, 2008). While there is still insufficient systematic data for the PSi electrodes, the major requirements for the carbon-based supercapacitors seem valid for the PSi with the multilevel size of the micro-, meso-, and macropores in PSi electrodes fitted to the type and the composition of electrolyte.

16.4.1.1 INTEGRATED ON-PERFORATED-SILICON-CHIP MICRO-SUPERCAPACITORS

Notable work has been done on microsupercapacitors, the main goal of which has been to improve the performance in a limited area by using high-capacity active materials and 3D structures. Many studies are currently dedicated to the design of micro-supercapacitors with different types of carbons or pseudo-capacitive materials (RuO_2, MnO_2). However, their integration in microelectronic circuits is still a challenge. Intricate silicon-based micro-supercapacitors should facilitate integration. Moreover, such devices could directly be manufactured on chips.

The basic principles in the design of the microsupercapacitor are as follows: electrode loading should be sufficient to offer high capacity, the electrodes should be porous in order to provide high surface area and, in addition, the electrodes must be narrow and very close to each other, so that the migration distance of ions can be short and result in a rapid charge/discharge rate (Shen et al. 2011). It is established that the interdigital structure of electrodes allows more materials to be loaded per unit area with thicker electrodes, as compared to the loading of rolled and planar (sandwich) microcapacitors (Chmiola et al. 2010). The supercapacitor architectures proposed in the literature are either vertical trench-like (Roozeboom et al. 2009) or interdigital in-plane structures (Shen et al. 2011).

The practical fabrication of the interdigital silicon structures is based on the Bosch DRIE process (Laermer & Schilp 2003; Notten et al. 2007; Shen et al. 2011), which involves repeated alternating exposure of a photoresist-masked silicon wafer to etchant plasma and passivant plasma. The fluorine radicals from the etching gas (sulfur hexafluoride, SF_6) attack and etch silicon spontaneously, forming volatile silicon fluorides like SiF_4. A passivating gas, octafluorocyclobutane (C_4F_8), builds up teflon-like polymer films on the treated silicon wafer. Since the polymer dissolves very slowly, it is deposited on the sidewalls and protects them from etching. These etch/deposit steps alternate many times over, and result in a series of very small isotropic etch steps taking place only at the bottom of the etched pits. In Ripenbein (2010), the DRIE process for the high-aspect-ratio interdigital Si-on-chip architectures was developed according to the following requirements: homogeneous etch depth and width of the microchannel or microcontainer and low side-wall roughness. To etch, for example, a 0.5-mm-thick silicon wafer 100–1000 etch/deposit steps are needed. Short cycles yield smoother walls, and long cycles yield a higher etch rate.

Trench on-Si-chip microcapacitors were developed by Philips and NXP (Murray et al. 2007; Gautier et al. 2008; Roozeboom et al. 2009). Figure 16.7 (Roozeboom et al. 2009) shows an on-Si-chip microcapacitor with a vertical DRIE-etched 1.5–2.0 μm diameter and 30 μm depth trenches, for which surface-area enhancement by a factor of 20–25 is achieved. The pore walls of the etched wafers are made highly conductive by P-indiffusion. The pores are then filled with a

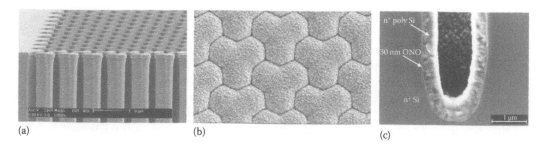

(a)   (b)   (c)

FIGURE 16.7   (a) Side view of 3D trench-array after RIE-etching; (b) top view after filling with a capacitors stack and (c) cross-sectional view of a single capacitor unit. (From Roozeboom F. et al., *Procedia Chem.* **1**, 1435, 2009. Copyright 2009: Elsevier, and Notten P.H.L. et al., *Adv. Mater.* **19**, 4564, 2007. Copyright 2007: WILEY-VCH Verlag. With permission.)

30-nm dielectric layer stack consisting of 5-nm thermal oxide, 20-nm LPCVD nitride, and 5-nm oxide ("ONO") (Notten et al. 2007). Next, 0.7-μm-thick *n*-type doped poly-Si is deposited by low-pressure chemical vapor deposition (LPCVD). Metal electrode contacts are made by the deposition of an aluminum layer and patterning.

The high-density capacitor based on macroporous silicon has been recently presented by Vega et al. (2014). The macroporous silicon obtained by multistep light-assisted electrochemical etching forms one of the electrodes of the capacitor. The porous substrate of 13.3 mm in diameter has about of 8.7 × 10⁶ pores with footprint area enhancement factor (AEF) of about 83. The limit in pore depth is set by wafer thickness and etching time. Practical sizes are around 300 μm for 4-μm pitch macroporous silicon, which incurs in an etching time of about 8 h. This results in a high surface-to-volume ratio around 0.6 m²/cm³. The insulating 15-nm-thick layer of silicon dioxide is thermally grown on the electrode. Finally, nickel electrode is electrodeposited inside the pores.

In-plane interdigital PSi structures are prepared by the same DRIE method as that used for the generation of vertical trenches. Figure 16.8 (Shen et al. 2011) schematically shows the process of the fabrication of an on-Si in-plane interdigital microsupercapacitor. High-aspect-ratio interdigital channels (105 μm wide and 90 μm deep) separated by a 15-μm wall are formed by inductive plasma etching. The channels are coated by a metal current collector and an insulation layer to prevent current leakage through the substrate. Finally, the electrode material is injected into the channels; the wall is etched and filled by electrolyte.

The electrode material used in in-plane interdigital solid EDLC (Figures 16.9 and 16.10) (Shen et al. 2011) is composed of 1000 m²/g activated carbon and PVDF. Figure 16.9 shows photographs of a cell at different stages of the preparation process: (a) top view of the cell with interdigital channels separated by a silicon wall and a gold current collector at the bottom of the channels; (b) micrograph of a certain part of the cell without electrode materials. Figure 16.10 shows SEM micrographs of the cell after the silicon walls between electrodes are etched: (a) the top view of the interdigital electrodes; (b) the cross-section view. The white bar in this SEM graph is 100 μm long.

Electrochemical microcapacitors with carbon electrodes of in-plane interdigital configuration were developed by Pech et al. (2010a,b). Carbon-based electrodes containing activated-carbon powder with a PTFE polymer binder were introduced to the channels by inkjet-printing

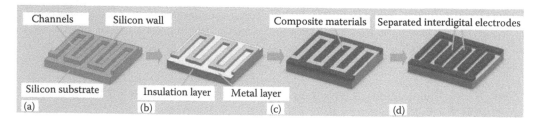

(a)   (b)   (c)   (d)

FIGURE 16.8   Scheme of fabrication of interdigital on-Si microcapacitor. (From Shen C. et al., *J. Power Sources* **196**, 10465, 2011. Copyright 2011: Elsevier. With permission.)

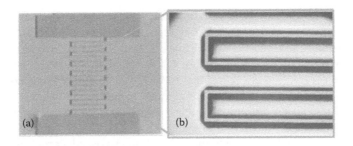

FIGURE 16.9   Photographs of a cell in different procedures: (a) top view of the cell with interdigitated channels separated by a silicon wall, a gold layer at the bottom of the channels serves as the current collector; (b) micrograph of a certain part of the cell without any electrode materials. (From Shen C. et al., *J. Power Sources* **196**, 10465, 2011. Copyright 2011: Elsevier. With permission.)

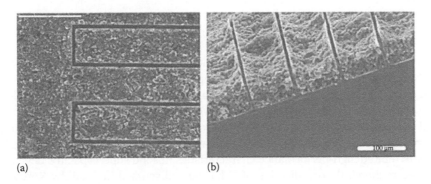

FIGURE 16.10   SEM graphs of the cell after the silicon wall between electrodes are etched: (a) the top view of the interdigitated electrodes; (b) the cross-section view, and the white bar in this SEM graph is 100 μm long. (From Shen C. et al., *J. Power Sources* **196**, 10465, 2011. Copyright 2011: Elsevier. With permission.)

technology (Pech et al. 2010a). In Pech et al. (2010b), a layer of nanostructured carbon onions several micrometers thick was electrophoretically deposited on 3D interdigital on-Si substrates.

DRIE procedure combined with LIGA technology was utilized to prepare an on-perforated-Si-chip interdigital 3D pseudo-capacitor with an aspect ratio of up to 100:1 (Sun & Chen 2009; Sun et al. 2010). A 3D redox symmetrical supercapacitor has a 3D "through-structure," which consists of two unconnected periodic 100-μm-wide "beams" coated by PPy films as the electrodes. For the 525-μm-thick silicon substrate of 1 cm² area the "interdigital-beams" architecture provides more than one order of magnitude greater effective surface area than a similar planar structure of the same footprint.

Perforated silicon microchannel plates (Si-MCPs) were tested as substrates for miniature supercapacitors in (M. Li et al. 2013) and (Miao et al. 2009) (Figure 16.11). In Chen et al. (2008), Yuan et

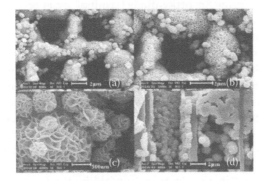

FIGURE 16.11   SEM images of the Co(OH)₂/Ni/Si-MCP structure: (a) top-view; (b) amplification of a single micropore; (c) magnified image of (b); (d) cross-sectional view. (From Li M. et al., *J. Mater. Chem. A* **1**, 532, 2013. Copyright 2014: Royal Society of Chemistry. With permission.)

al. (2009), Yu et al. (2010), and M. Li et al. (2013), the standard microelectronics-fabrication steps were combined with photo-assisted anodic dissolution (the method is described in Section 16.3.1.2) to form ordered-silicon high-aspect-ratio MCP as a substrate for a pseudo-microcapacitor. The substrate was coated by a thin nickel layer to reduce silicon resistivity. Nanometer-size $Co(OH)_2$ flakes electrodeposited on ordered three-dimensional Ni/Si-MCPs. $Co(OH)_2$ has a complex anisotropic morphology with flakes and particles on the nanometer scale and a network-like structure including interconnected nanoflakes and nanoparticles. Continuous covering of the microchannel by the flakes and nanoparticles is achieved. There are large quantities of nanoparticles on the surface, which increase the specific surface of the Si-MCPs micro pseudocapacitor. The nanoparticles deposited on the surface of nanoflakes adhere firmly to the surface.

In another example, the 3D ordered nickel oxide/silicon (NiO/Si-MCP) electrode materials are used for microchannel plate pseudo-supercapacitors. The active nickel oxide electrode material is synthesized by electroless plating of nickel on the surface of the Si-MCP followed by annealing under oxygen.

16.4.1.2 SUPERCAPACITORS WITH PSi ELECTRODES

In addition to perforated silicon substrates, electroless and anodic dissolution methods are widely used for the preparation of PSi-electrodes for supercapacitors. Macroporous silicon has been used as an electrode material in a parallel-plate capacitor with an oxide-nitride-oxide dielectric sandwich (Hummel & Chang 1992; Romstad & Veje 1997; Tada et al. 1997). Both ordered and disordered PSi structures were prepared by wet-chemical or electrochemical etching of p- or n-type silicon wafers. PSi nanowires prepared by metal-assisted etching and modified by a graphene layer exhibited high-performance characteristics in solid supercapacitor devices (Alper et al. 2012). Electrochemical porosification of Si has been discussed in detail by Korotcenkov and Cho (2010) and in Chapters 2–4 of this book. The influence of electrolyte composition and operating parameters, wafer doping and orientation, lighting, magnetic field, and ultrasonic agitation on the structure of PSi are addressed. Fabrication of certain PSi supercapacitors by electrochemical methods is briefly described next.

16.4.1.2.1 Preparation of PSi Electrodes by Anodic Dissolution   Anodic dissolution of silicon is typically used to thin or polish silicon wafers and to manufacture thick PSi layers. Several factors govern the process: the electrolyte, the design of the experimental system, and operating conditions (Lehmann & Gösele 1991; Lang 1996; Bisi et al. 2000; Lehmann 2002). The etching electrolyte is usually a dilute aqueous solution of HF mixed with ethanol. The mixtures of HF and organic solvents are also used in PSi technology. The current-voltage (I-V) characteristic of a silicon electrolyte contact is determined by both the semiconductive nature of the electrode and by the composition of the electrolyte. A Schottky diode describes the silicon-electrolyte interface as long as the charge transfer is limited by charge supply from the electrode. Under cathodic polarization, both n- and p-type silicon are stable. The major cathodic reaction is the reduction of water at the Si/HF interface, with formation of hydrogen gas. This usually occurs only at high cathodic overpotentials, or, using Schottky-diode terminology, at reverse breakdown (the voltage that can be applied without causing an exponential increase in the current).

Under anodic polarization, p-type silicon dissolves. At high current densities, the process is controlled by the diffusion of the fluoride ions in the electrolyte near the electrode surface. At low anodic-current densities, the process is controlled by charge transfer at the surface of the silicon electrode. The average pore size of PSi structures varies over four orders of magnitude, from nanometers to tens of micrometers. PSi structures of a pore width $d < 2$ nm, are considered to be mainly microporous; for 2 nm $< d <$ 50 nm mainly mesoporous; and macroporous if $d > 50$ nm. All three kinds of pores have their origin in different physical effects and can be obtained under a variety of conditions, and with widely differing morphologies. Key parameters are the HF concentration, the doping type and level of the silicon (n, n^+, p, p^+), and in some cases, the state of illumination (rear or frontal illumination) (Steiner & Lang 1995; Föll et al. 2002).

For integrated devices, the formation of PSi is carried out on a lithographically patterned surface. DRIE may precede the wet process in order to complete the desired design of a device.

16.4.1.2.2 Preparation of PSi Electrodes by Metal-Assisted Etching   Metal-assisted etching combines the advantages of anodic dissolution and electroless methods. In this technique, the current is generated by a micro-galvanic element formed between silicon and a noble metal (Au, Pt, Ag, Pd, etc.) particle at its surface in an HF: ethanol oxidizing-agent mixture. Metal-assisted etching is easy to handle; it does not need an external current source and complicated etching cell. High etch rates and thick PSi layers are achievable by this method in a manner similar to anodic dissolution. The pore size, porosity, morphology of the etched surface, and rate of pore formation are affected by several factors: the type and thickness of the precious metal, the doping type and the level of the silicon substrate, and the composition of the etching solution. The electrolyte composition is a very important parameter for the preparation of PSi by metal-assisted etching. The pore dimensions and porosity change with the change in concentration of HF, ethanol, and the oxidizing agent in the etching mixture. Lowering the concentration of hydrofluoric acid is typically followed by increase in pore sizes and porosity. The oxidizing agent decomposes to form holes, which participate in the dissolution reaction and partially oxidize the silicon. The most popular oxidizers are hydrogen peroxide (H_2O_2), nitric acid (HNO_3), sodium persulfate ($Na_2S_2O_8$), potassium permanganate ($KMnO_4$), and potassium dichromate ($K_2Cr_2O_7$). If the concentration of the oxidizer is increased, the pore's dimensions, porosity, and the etching rate increase as well. However, if the rate of injection of holes increases, silicon etching becomes isotropic and occurs in the chemical polishing regime. The third component of the electrolyte is ethanol, which wets the surface and reduces the surface tension of the solution. On increasing the ethanol concentration, finer pores and a better uniformity of the porous layer are typically achieved. Inasmuch as the generation of holes occurs by decomposition of the oxidizing agent, PSi can be formed in the dark regardless of the doping type and level. However, the difference in doping level influences the etch depth away from metal-coated areas.

The metal-assisted formation of PSi can be very useful in producing micropores on highly resistive silicon, a goal difficult to achieve by the electrochemical method (Hadjersi et al. 2005). In addition, the process is very selective because the layers of porous silicon are formed only in the regions where the metal was deposited. Furthermore, there is no direct correspondence of the pore size with the size or spacing of the deposited metal islands, that is, the metal coating does not act as an etch mask. PSi layers with different morphologies and properties can be produced by varying only the type of metal, the dopant type, and level. The simplicity of the process makes it very attractive for mass production. In Ripenbein et al. (2010) it was found that the rate of etching is higher and the pores larger in low-doped, p-type silicon (14.22 Ohm·cm, $\sim 9 \times 10^{14}$ cm^{-3}) as compared to the high-doped material (3–5 $\times$ 10^{-4} Ohm·cm, $\sim 1 \times 10^{20}$ cm^{-3}). Figure 16.12a–d shows cross-sectional HRSEM images of the silicon sample, etched in HF/C_2H_5OH/H_2O_2 solutions of different compositions. Interconnected pores propagating anisotropically are seen in the micrographs. It was found that an increase in the HF concentration is followed by a decrease in pore size; hydrogen peroxide increases pore dimensions; increasing the concentration of ethanol does not influence the size of the pores, but promotes their orientation and the uniformity of the PSi layer.

Recently, a magnesiothermic reduction method based on the reaction of metallic Mg with a SiO$_2$ precursor was developed for the preparation of PSi. Three-dimensional microporous

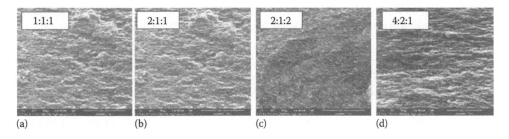

FIGURE 16.12 Cross-sectional SEM images of a wall between two containers etched in: (a) HF:C_2H_5OH:H_2O_2 (1:1:1) for 9 min; (b) HF:C_2H_5OH:H_2O_2 (2:1:1) for 9.5 min; (c) HF:C_2H_5OH:H_2O_2 (2:1:2) for 10 min; (d) HF:C_2H_5OH:H_2O_2 (4:2:1) for 11 min. The scale bar indicates 500 nm. Pore size, nm: (a) 10.2; (b) 8.6; (c) 10.0; (d) 7.0. (From Ripenbein T. et al., *Electrochim. Acta* **56**, 37, 2010. Copyright 2010: Elsevier. With permission.)

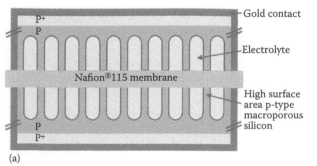
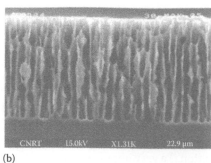

(a)   (b)

FIGURE 16.13   (a) Schematic of the symmetrical electrochemical capacitor made of two identical PSi electrodes filled with an electrolyte and separated by a Nafion115 membrane. (b) Cross-sectional SEM image of the electrode. (From Desplobain S. et al., *Phys. Stat. Sol. C* **4**(6) 2180, 2007. Copyright 2007: WILEY-VCH Verlag. With permission.)

nanocrystalline silicon and ordered mesoporous silicon films have been fabricated by such a reduction method (Bao et al. 2007; Richman et al. 2008). However, this reduction reaction is usually conducted at temperatures above 630°C, and some of the byproducts, such as magnesium silicates, generated during the reduction process are hard to remove. The PSi materials prepared have relatively low specific surface areas, and this limits their application in supercapacitors. In addition, the silicon sources are usually limited to pure silica species, such as mesoporous SBA-15 MCM-48 and microporous silicate (Chen et al. 2011; Guo et al. 2011; Zhu et al. 2011a,b).

A simple sodiothermic reduction method has been developed by L. Li et al. (2009) and Wang et al. (2013) for the preparation of PSi with the use of aluminosilicate zeolite NaY as a precursor. The presence of amorphous PSi with a surface area of 571 m^2/g was shown by XRD, TEM, and BET tests.

The symmetrical structure of an electrochemical capacitor with macroporous silicon-based electrodes, prepared by metal-assisted etching, is described by Desplobain et al. (2007) (Figure 16.13). Nafion 115 membrane is used as the ion-permeable, electron-insulating separator. The macroporous silicon was prepared from a *p*-type silicon wafer with 30–50 Ω·cm resistivity and (100) crystalline orientation by electrochemical etching in HF-H_2O solution with a TritonX-100 surfactant. PSi is unstable in contact with many organic electrolytes. It is well known that it undergoes intercalation of cations followed by extreme volumetric changes. This, in turn, may cause breakup of the electrode and electrical isolation of the active material. Volume expansion/contraction of silicon subjects its surface to a continuous reduction reaction of liquid electrolyte followed by the formation of a fresh solid electrolyte interphase (SEI). Oakes et al. (2013) proposed to protect the PSi surface by graphene-coating, which in addition has high electron conductivity. Highly doped silicon wafers were anodically etched in a 50% HF and ethanol solution. The etching parameters were optimized to yield 75% porosity 4-μm-thick PSi layer. After etching, the samples were treated with C_2H_2/H_2/Ar gas mixtures at 850°C to form a stable, passive coating of graphene. A schematic representation of both a pristine and graphene-carbon-coated PSi supercapacitor device is shown in Figure 16.14. Cross-sectional SEM images show that the porous structure is not altered by the graphene layer (Oakes et al. 2013).

16.4.1.2.3 PSi Nanowires as Supercapacitor Electrodes   In recent years, silicon nanowires (SiNW) have been the subject of considerable interest due to their importance in the field of functional electronic devices and as anode material for Li-ion batteries. SiNWs can be prepared by chemical and physical deposition, laser ablation, thermal evaporation, and other methods. In Peng et al. (2002) and Zahedinejad et al. (2011) homogeneous arrays of PSi nanowires were proposed for application to supercapacitors. PSi-NWs with an average pore diameter of about 10 nm are produced by self-assembly synthesis based on a low-temperature metal–assisted electroless etch process. However, these porous nanowires were found to be highly reactive and dissolved rapidly when exposed to electrolytes containing low concentrations of salts (Chiappini et al. 2010). Thin layers of silicon carbide (SiC) were used by Alper et al. (2012) to protect the PSi nanowires, yielding SiC/PSi-NWs. While SiC coatings were nanometers-thick and successfully

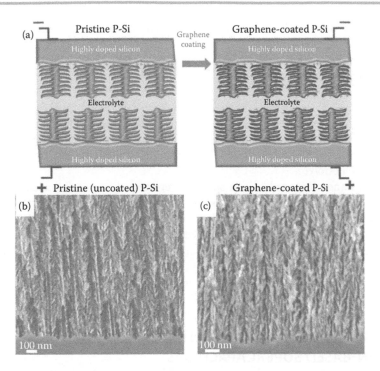

FIGURE 16.14   (a) Scheme of the effect of coating P-Si on the capacitive charge storage properties. SEM cross-sectional images of PSi showing the interface between the etched PSi and the silicon wafer for the case of (b) uncoated, pristine PSi and (c) graphene coated PSi. (From Oakes L. et al., *Sci. Rep.* **3**, Article number: 3020, 2013. Published by Nature Publishing Group as open access. With permission.)

mitigated silicon degradation in aqueous electrolytes, pore blockage and loss of capacitance were observed during electrochemical cycling. In Alper et al. (2014) it was found that when PSi-NW arrays are exposed to the CH$_4$-Ar gas mixture at about 900°C, a deposit of an ultrathin carbon sheath forms over the wires. This protective layer enables electrolyte access to the porous surface area while still mitigating silicon degradation. Scanning-electron micrographs of the nanowires (Figure 16.15, Alper et al. 2014) revealed that their vertical orientation is maintained after carbonization and no significant structural damage is observed. Scanning-transmission electron microscopy (STEM) analysis indicated that the nanoscale pore structure of the PSiNWs is preserved, a key factor in realizing the full potential of energy storage of these materials.

Thissandier et al. (2012, 2013) synthesized highly doped silicon nanowires by chemical vapor deposition with gold catalysts. Both *n*-type and *p*-type SiNWs have a diameter of about 100 nm and a length of about 10–11 µm. Microcapacitors with SiNW electrodes were investigated as electrode materials in a standard organic-based electrolyte. In a very recent publication, Thissandier et al. (2014) used nanoporous anodic alumina as a template for silicon nanowire growth. The nanowire density was about 8·10^9 nanowires/cm^2, that is, 160 times the one of SiNWs grown from 100 nm gold colloids (Thissandier et al. 2012).

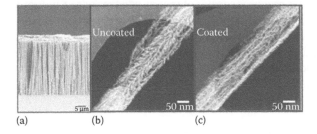

FIGURE 16.15   (a) SEM micrograph of PSi nanowire array post carbonization. (b) Representative STEM image of a PSiNW. (c) Representative STEM image of a C/PSiNW. (From Alper J.P. et al., *Nano Lett.* **14**(4), 1843, 2014. Copyright 2014: American Chemical Society. With permission.)

In Aradilla et al. (2014) SiNW electrodes with a length of nanowires approximately 15 μm and a diameter of 50 nm were grown in a CVD reactor by the vapor–liquid–solid (VLS) method with the use of a gold catalyst on a highly doped *n*-Si (111) substrate. Poly(3,4-ethylenedioxythiophene) (PEDOT) films were electrochemically deposited on the nanowires and tested in a symmetric pseudo-microsupercapacitor.

Highly ordered NiO-coated Si nanowire arrays were fabricated by Lu et al. (2011) as electrode materials for electrochemical pseudo-capacitors via depositing Ni on electroless-etched Si nanowires and subsequently annealing. The 30 to 300 nm diameter Si nanowires were found to be uniformly wrapped by a rough layer of 13 to 14 nm NiO nanoparticles. The electrode has been tested in alkaline electrolyte.

Very recently, nanostructured silicon spheres were tested as possible electrode material for supercapacitors. The spheres were synthesized by hydrolyzed tetraethylorthosilicate (TEOS) with cetyltrimethylammonium bromide (CTAB) as a template (Liu et al. 2012). The silica spheres obtained were converted to MgO/Si replicas with gaseous Mg as the reducing agent. The hollow silicon spheres exhibit a high BET surface area of 928 m^2/g and an average Barrett–Joyner–Halenda (BJH) pore size of 3.2 nm. The thickness of the spherical shell was estimated to be about 30 nm. EDS and XRD tests show that the spheres are pure crystalline silicon.

16.5 ELECTROCHEMICAL PERFORMANCE OF POROUS SILICON-BASED SUPERCAPACITORS

Table 16.1 presents some electrochemical characteristics of integrated and MCP on-Si-chip microsupercapacitors, in which PSi is used as a substrate. Activated carbon, carbon nanotubes, and polypyrrole (PPy) are the major electrode materials.

In Table 16.2, the data are collected for the EDLCs and pseudo-capacitors, in which PSi was used as an active electrode material.

In most publications dealing with the development of integrated on-Si-chip microcapacitors the areal specific capacitance values are presented because the footprint of the device is crucial for microelectronic applications. The authors wish to mention that precise comparison of the electrochemical performance of versatile PSi supercapacitors was not always possible because of the absence in the articles of some experimental details, like area or thickness of the electrode, loading, and so on. In agreement with Gogotsi and Simon (2011), we state that reporting the energy and power densities per unit weight of active material alone may not give a realistic picture of the performance of the assembled device because the weight of the other device components must also be taken into account. Electrochemical capacitors contain current collectors, electrolyte, separator, binder, connectors, and packaging, in addition to carbon, PSi or other active material-based electrodes. Since the weight of the active materials accounts for about 30% of the total mass of the packaged commercial EDLC, a factor of 3 to 4 is frequently used to extrapolate the energy or power of the device from the performance of the material. Thus, the energy density of 20 Wh/kg of carbon, for example, will reflect about 5 Wh/kg of the packaged cell. However, this extrapolation is valid only for electrodes with thicknesses and densities similar to those of commercial electrodes (100 to 200 μm or about 10 mg/cm^2 active material loading). An electrode of the same material that is about one order of magnitude thinner or lighter will further reduce the energy density three to fourfold (from 5 down to 1.5 Wh/kg on the basis of cell weight). Much of the uncertainty stems from reporting gravimetric, rather than volumetric energy and power densities of materials and devices; this is particularly valid for porous materials and microcapacitors. Therefore, in the majority of cases, the data presented in the review are those actually published in the articles with some possible estimation of the uncertainty.

16.5.1 INTEGRATED ON-PERFORATED-Si-CHIP SUPERCAPACITORS

Integrated vertical trench on-perforated Si-chip electrostatic microcapacitors with silicon oxide/nitride/oxide dielectric-layer stack deliver 8 μF/cm^2 of geometric footprint (Murray et al. 2007).

TABLE 16.1   Electrochemical Performance of Supercapacitors Fabricated on Interdigital, Trench, and MCP-Based PSi Substrates

Structure/Matrix	Electrode Material/Current Collector	Electrolyte	Capacitance		Specific Power, mW/cm²	Specific Energy µWh/cm²	Ch/dis Cycles Number	Ref.
			µF/cm²	F/g				
3D interdigital pseudo-capacitor	PPy doped by ClO₄/Ni/Si	1M NaCl	30	N/A	N/A	N/A	N/A	Sun & Chen 2009
3D interdigital pseudo-capacitor	PPy/Ti/SiO₂/Si	0.5M NaCl	56	N/A	0.56	N/A	N/A	Sun et al. 2010
3D interdigital EDLC	TiC-CDC[1] film/Si	1M Et₄N BF₄[2] in PC	1500 (35 F/cm³)	N/A	84.0	0.8		Huang et al. 2013
Planar PS integrated EDLC	Graphene/PS	EMIBF₄[3]	N/A	N/A	1 kW/kg 1000 kW/m³	4.8–4.9 Wh/kg[a] 3.2 kWh/cm³	5000	Oakes et al. 2013
Si MCP	NiO	2M KOH	N/A	586.4	N/A	N/A	500	Miao et al. 2009
Si MCP	Nano Ni flakes/Co(OH)2	2M KOH	6900	559.15	N/A	N/A	1000	Li et al. 2013
Interdigital EDLC	Activated carbon/Au/SiO₂/Si	1M Et₄NBF₄ in PC	1600–2700 µ/cm³	N/A	45.0	1.8	N/A	Pech et al. 2010b
Interdigital EDLC	Nanostructured carbon onions/Si	1M Et₄NBF₄ in PC	900 µF/cm² 1300 µF/cm³	N/A	1 kW/cm³	N/A	N/A	Pech et al. 2010a
Interdigital EDLC	Nanostructured carbon onions	PIP₁₃FSI:PYR₁₄FSI 1:1	1100	N/A	99.0	4.0	N/A	Huang et al. 2013
Interdigital nanoporous self-supported 3D EDLC	Activated carbon/pressed nickel foam	1M NaNO₃	90.7–97.2	96.0	51.5 (12.7 W/g)	N/A	100	Shen et al. 2011
3D-trench PS capacitor	n-type doped poly-Si/Cu	LPCVD nitride and 5 nm oxide	8.0	N/A	N/A	0.11	N/A	Notten et al. 2007; Roozeboom et al. 2009
3D-trench PS capacitor	Al₂O₃/HfO₂/Al₂O₃ stacks	ONO	23	N/A	N/A	N/A	N/A	Roozeboom et al. 2007
3-D patterned silicon	Macroporous doped Si	SiO₂	11	N/A	N/A	N/A	N/A	Vega et al. 2014

Note: 1: TiC-CDC—titanium carbide derived carbon; 2: Et₄NBF₄—Tetraethylammonium tetrafluoroborate; 3: EMIBF₄—1-ethyl-3-methylimidazolium tetrafluoroborate (ionic liquid); 4: PIP₁₃FSI:PYR₁₄FSI.

[a] Packaging factor ~50% gives 2.5–3.5 Wh/kg.

With the use of a voltage of 10 V, it has been calculated that about 0.40 mJ/cm² (0.11 µWh/cm²) of energy can be stored in these 3D-integrated capacitors, making these devices very attractive for low-loss decoupling to suppress cross-talk in high-frequency circuits (Roozeboom et al. 2009). A 3D patterned silicon supercapacitor (Vega et al. 2014), which consists of cylindrical pores etched in the silicon has AEF ranging from 83 to 142. The strong doping of the structure creates a deep degenerate layer with metal-like behavior and low ohmic losses. Electroplated nickel fills the pores completely without voids and provides low ESR of about 1 ohm. The achieved specific capacitance is about 11 µF/cm². The fabricated devices are able to work in an extended frequency range of up to 50 MHz.

As mentioned above, electrochemical capacitors deliver much higher capacitance values because of the use of electrode materials with very high surface areas and a molecule-thin layer of electrolyte.

Electrochemical microcapacitors with inkjet-printed in-plane interdigital carbon electrodes (Figure 16.16, Pech et al. 2010a) displayed symmetrical capacitive behavior with the typical double-layer rectangular shape and good chemical stability on cycling over a wide potential range of 2.5 V in 1M Et₄NBF₄ propylene carbonate electrolyte. The capacitance of the EDLC was 2.1 mF/cm² and its power density 44.9 mW/cm².

TABLE 16.2   Electrochemical Performance of EDLCs and Pseudo-Capacitors with Porous Silicon or Silicon Nanowires Electrodes

Type of Capacitor	Electrode Material	Electrolyte	Capacitance		Specific Power	Specific Energy	Cycles Number	Ref.
			μF/cm²	F/g				
EDLC	Gold/PSi	0.25M Et$_4$NBF$_4$/PC	200	0.005	N/A	N/A	N/A	Rowlands et al. 1999
EDLC	PSi-p⁺	1M PC:DMC (1:3) LiClO$_4$	40	N/A	N/A	N/A	N/A	Desplobain et al. 2007
EDLC	Au/Si-p⁺/Si-p/PSi-p⁺/Au	20% v/v H$_2$SO$_4$	320	N/A	N/A	N/A	N/A	Desplobain et al. 2007
EDLC	Graphene/PSi	PEO: EMIBF$_4$	N/A	N/A	1–8 kW/kg	2–10 Wh/kg	1000	Westover et al. 2014
EDLC	Graphene/PSi	EMIBF$_4$	N/A	N/A	1 kW/kg	3.2 mWh/cm³ 4.9 Wh/kg	5000	Oakes et al. 2013
EDLC	p/n SiNWs	1M Et$_4$NBF$_4^2$/PC	34–46	N/A	1.6 mW/cm²	1 nWh/cm²	200	Thissandier et al. 2012
EDLC	Ultra-dense and highly doped SiNWs	EMI-TFSI	16.5–34.6	N/A	16–400 mW/cm²	80 nWh/cm²	400-300,000	Thissandier et al. 2014
EDLC	C/PSi-NWs	3.5M KCl	325	N/A	5–0.1 mW/cm²	0.6–300 µWh/cm²	5000	Alper et al. 2014
Pseudo EDLC	PEDOT⁴/SiNWs	PYR$_{13}$ TFSI⁵	9000	32–36	85 kW/kg	10 Wh/kg 2.5 µWh/cm²	3500	Aradilla et al. 2014
Pseudo EDLC	Nanostructured silicon hollow spheres	0.5M Na$_2$SO$_4$	N/A	145–166	N/A	N/A	300	Liu et al. 2012
Pseudo EDLC	NiO/SiNWs	2M KOH	N/A	787.5	N/A	N/A	500	Lu et al. 2011

Note: The highest capacitance, specific power, and energy density values, calculated either from CV or cycleability tests of the full devices, are presented: 1: PEO—polyethylene oxide; EMIBF$_4$—1-ethyl-3-methylimidazolium tetrafluoroborate (ionic liquid); 2: Et$_4$NBF$_4$—Tetraethylammonium tetrafluoroborate; 3: EMIM-TFSI—1-ethyl-3-methylimidazolium bis-(trifluoromethylsulfonyl) imide; 4: PEDOT—Poly(3,4-ethylenedioxythiophene); 5: PYR$_{13}$ TFSI—N-methyl-N-propylpyrrolidinium bis(trifuoromethylsulfonyl)-imide (ionic liquid).

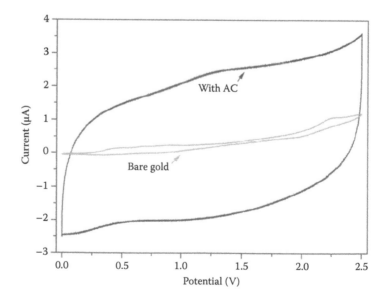

FIGURE 16.16 Cyclic voltammogramms of a 3D-on-Si micro-supercapacitor uncoated and coated with activated carbon electrodes in 1M Et$_4$NBF$_4$ propylene carbonate electrolyte at a scan rate of 100 mV/s. (From Pech D. et al., *J. Power Sources* **195**, 1266, 2010. Copyright 2010: Elsevier. With permission.)

The specific capacitance of the in-plane interdigital on-perforated-Si micro EDLC with onion-like electrophoretically deposited carbon (OLC) electrodes (Pech et al. 2010b) was lower than the specific capacitance of interdigital activated-carbon (AC) electrodes (Pech et al. 2010a). However, at a scan rate of 100 V/s the onion-like carbon EDLC exhibited 0.9 mF/cm^2, which is much higher than that of AC-EDLC and is comparable to the values (0.4–2 mF/cm^2) reported in Pech et al. (2010a) and In et al. (2006) at much lower scan rates. The high scan rate, indicative of high instantaneous power, is thus concomitant with high specific capacitance, which is explained by the significant (7 μm) thickness of the OLC nanosized active-electrode-material film. At a power of 20 to 100 W/cm^3, the OLC interdigital on-Si EDLC demonstrated 2–0.5 mWh/cm^3 volumetric specific energy. The combination of carbon-onion morphology providing a fully accessible surface area, with a binder-free electroporetic deposition (EPD) technique and a micro-interdigital-device design led to this high-power energy performance. Similarly to Pech et al. (2010a), Huang et al. (2013) fabricated a microsupercapacitor onto PSi substrate by electrophoretic deposition of onion-like-carbon as electrode material. Using a eutectic mixture of ionic liquids (PIP$_{13}$FSI:PYR$_{14}$FSI 1:1) as the electrolyte enabled successful operation of the supercapacitor in a wide range of temperatures from −50 up to 80°C. The micro-devices showed good performance at room temperature with a 1.1 mF/cm^2 capacitance per footprint area of device at 200 mV/s at 3.7 V, a specific energy of 4 μWh/cm^2 and a power density of 99 mW/cm^2. The authors demonstrated that despite the leakage current, low-temperature performance of micro-supercapacitors is better than macroscopic devices using the same electrode material and the same electrolyte.

The geometric capacitance of a 3D-on-Si symmetric pseudocapacitor with interdigital 2 μm-thick PPyClO$_4$ electrodes, calculated from CV at a scan rate of 100 mV/s and from galvanostatic discharge at a current density of 1 mA/cm^2, is more than one order of magnitude higher (about 30 mF/cm^2) (Sun & Chen 2009; Sun et al. 2010) than the capacitance of an EDLC with 7-μm-thick OLC electrodes (Pech et al. 2010b). The 3D supercapacitor on silicon micromachined by DRIE with PPy electrodes polymerized for 20 min (Sun & Chen 2009) shows specific capacitance of 56 mF/cm^2 and specific power of 0.56 mW/cm^2 at a scan rate of 20 mV/s.

Among devices with interdigital electrodes, a 3D microsupercapacitor comprising interdigital self-supporting AC-PVDF composite electrode materials and NaNO$_3$ aqueous electrolyte shows 90.7 mF/cm^2, high energy storage (about 12.7 kW/kg), and power-delivery (51.5 mW/cm^2) properties (Shen et al. 2011). These properties are due to the combined factors of the narrowest microgap between electrodes incorporated in the three-dimensional silicon structure, a heavy load of the nanoporous electrode material per unit area, and the relatively high specific capacitance of the material.

Miniature pseudo-supercapacitors with electrodes of electrochemically deposited nanostructured Co(OH)$_2$ flakes on three-dimensional ordered nickel-coated silicon microchannel plates were tested in 2M KOH by M. Li et al. (2013). CV curves of the Co(OH)$_2$ electrode in 2M KOH were received at various sweep rates over the potential range of 0.4 to 0.6 V (Figure 16.17; M. Li

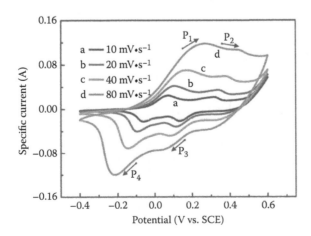

FIGURE 16.17 CVs of 3D-on Si MCP pseudocapacitor with Co(OH)$_2$ flake electrodes. (From Li M. et al., *J. Mater. Chem. A* **1**, 532, 2013. Copyright 2014: Royal Society of Chemistry. With permission.)

et al. 2013). The CV recorded for pseudo-capacitive $Co(OH)_2$ electrodes differs from that of the rectangular double-layered capacitors. It is suggested that instead of a pure electrical double-layer capacitance, the measured pseudo-capacitance is dominated by a redox mechanism. The high specific capacitance of 6.90 F/cm^2 at a discharge current density of 10 mA/cm^2 and good electrochemical stability up to 1000 cycles were ascribed to the synergistic effect of regular porous structure of the high-surface-area silicon substrate combined with a mesoporous structure of $Co(OH)_2$ and rapid ion diffusion (M. Li et al. 2013).

The results of electrochemical testing of the NiO/Si-MCP nanocomposite electrode in a 2M KOH solution reveal typical electrochemical capacitive behavior over the potential range of 0.6 to 1.0 V. The specific capacitance of approximately 586.4 F/g decreases slightly with 4.8% loss after 500 cycles (Miao et al. 2009).

16.5.2   SUPERCAPACITORS WITH POROUS SILICON ELECTRODES

Supercapacitors with PSi and silicon nanowires as active electrode materials were tested in aqueous, organic, and ionic-liquid electrolytes. Selected data are summarized in Table 16.2. Silicon materials have not been used for supercapacitors because of the extreme reactivity of silicon with electrolytes. Until 2013, there were only a few investigations of silicon materials in electrochemical environments. Therefore, there are almost no data for uncoated PSi.

EDLCs comprising gold-coated PSi electrodes and a 0.25 M Et_4NBF_4/PC electrolyte showed capacitance of approximately 0.2 mF/cm^2 (corresponding to a specific capacitance of 5 mF/g) (Rowlands et al. 1999). This compares favorably with a value of 0.18 mF/cm^2 for cells with indium tin oxide (ITO) on glass electrodes.

The inherent surface electrical conductivity of the in-silicon macropores is intrinsically low. In order to take advantage of the high-surface-area PSi, the conductivity was efficiently increased by combining doping and metallization processes (Desplobain et al. 2007). These PSi supercapacitors, pristine and coated by a semi-transparent gold layer, were tested in dilute sulfuric acid electrolyte and in 1M PC:DMC (1:3) $LiClO_4$ with Nafion separator. It was found that utilization of PSi electrodes instead of continuous plates, results in a significant drop of the equivalent series resistance (ESR) from 55 to 7 Ω/cm^2. Increase of the pore depth from 8 to 80 μm is followed by a capacitance increase from 0.058 to 0.178 mF/cm^2. While organic electrolytes enable charge/discharge at higher voltage than aqueous electrolytes, the capacitance in 1M PC:DMC (1:3) $LiClO_4$ was 2.5 times lower. The highest capacitance value—0.320 mF/cm^2—was obtained with B-doped and gold-metallized p-type macroporous silicon electrodes in 20% v/v H_2SO_4.

EDLCs with PSi electrodes modified by graphene were tested in 1-ethyl-3-methylimidazolium tetrafluoroborate($EMIBF_4$) ionic-liquid electrolytes, vacuum-infiltrated into PSi and PEO:$EMIBF_4$ electrolyte with different mass ratios of polymer to ionic liquid (Oakes et al. 2013; Chatterjee et al. 2014; Westover et al. 2014). The best results were observed for the plasticized-polymer electrolyte 1:1 and 1:3 PEO:$EMIBF_4$ ratios. Galvanostatic charge–discharge cycling of solid-state devices at a rate of 1.5 A/g (Figure 16.18a,b) (Westover et al. 2014) indicates stable, triangular charge–discharge curves representative of good device performance, especially for the 1:3 combination. Coulombic efficiencies of about 98% are measured for these devices, emphasizing applicable reversibility. These discharge characteristics hold true for discharge currents ranging from 1 to 4 A/g in both electrolytes (Figure 16.18c,d). Cyclic-voltammetry curves at 100 mV/s scan rates (Figure 16.18e) indicate stable double-layer-energy storage up to voltages above 3.6 V for the 1:1 electrolyte and up to about 3 V for the 1:3 electrolyte. The first case is comparable to the performance of state-of-the-art graphene-based ultracapacitors (Liu et al. 2010; Gogotsi & Simon 2011) and close to the potential window of $EMIBF_4$. The EDLCs with graphene-coated PSi-$EMIBF_4$ ionic-liquid electrolyte yield up to 4.9 Wh/kg (3.2 kWh/m^3) at power densities close to1 kW/kg (1000 kW/m^3) (Oakes et al. 2013). These results confirm the fact that electrode morphology, pore structure, and electrochemical stability are necessary factors in the design of electrochemical supercapacitors. Graphene coating simultaneously passivates surface-charge traps and provides an ideal electrode-electrolyte electrochemical interface. This leads to an improvement in energy density by a factor of 10–40, and an electrochemical window that is twice as wide, compared to identically structured pristine PSi. From the CV scans of the graphene-coated PSi–EDLC in

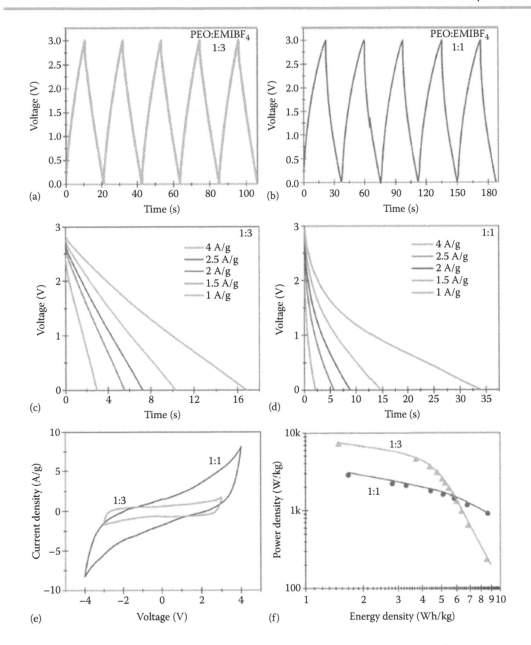

FIGURE 16.18 (a and b) Five consecutive galvanostatic charge–discharge curves for load-bearing devices with different PEO:EMIBF$_4$ mass ratios of (a) PEO:EMIBF$_4$ 1:3 and (b) PEO:EMIBF$_4$ 1:1. (c and d) Discharge curves at different discharge currents ranging from 1 to 4 A/g for (c) PEO:EMIBF$_4$ 1:3 and (d) PEO:EMIBF$_4$ 1:1. (e) Cyclic voltammetry measurements taken at scan rates of 100 mV/s for both 1:3 and 1:1 ratios of PEO:EMIBF$_4$ and (f) Ragone plot illustrating the energy–power performance characteristics of the load-bearing supercapacitor devices with 1:3 and 1:1 PEO:EMIBF$_4$ mass ratios. (From Westover A.S. et al., *Nano Lett.* **14**(6), 3197–3202, 2014. Copyright 2014: American Chemical Society. With permission.)

plasticized solid 1:3 IL electrolyte (Westover et al. 2014), energy density of 2 to 10 Wh/kg and power density of about 8 kW/kg were obtained.

Super-microcapacitors and pseudo-capacitors composed of PSi electrodes provide electrochemical performance characteristics higher than those, with silicon-nanowire electrodes and much higher than that of bulk silicon electrodes, which may be caused by low active surface area of the latter. Porosification of SiNWs significantly improves performance of micro-EDLCs.

EDLCs with highly doped *n/p*-SiNW electrodes show quasi-ideal double-layer-capacitive behavior in 1M Et$_4$NBF$_4$-PC organic electrolyte over a narrow potential window between –1.3 and

−0.5 V versus Ag⁺/Ag. The capacitances of highly doped SiNW-electrode capacitors calculated from the CV curves are roughly similar to those for the n-SiNW electrode (34 μF/cm²) and for the p-SiNW electrode (46 μF/cm²) (Thissandier et al. 2012). For comparison, n- and p-type bulk silicon show capacitances of about 5 μF/cm², which is lower by a factor of about 7 than those of n/p-SiNW electrodes. The capacitance increases with the length of the SiNWs and has been improved with the use of 20 μm-long SiNW electrodes. Micro-devices with several doping-type combinations exhibit less than 0.5% capacitance loss over 200 cycles with a maximum power density of 1.6 mW/cm². Recently developed micro-supercapacitors with ultra-dense and highly doped SiNW electrodes grown in the alumina template and ionic liquid as electrolyte demonstrated a very promising cycling stability featuring less than 1% of losses after 300,000 cycles (Thissandier et al. 2014). The maximal energy density and specific power values of 0.08 mWh/cm² and 400 mW/cm² have been achieved for these devices.

Alper et al. (2014) synthesized PSi nanowires via a lithography-compatible low-temperature wet etch and encapsulated in an ultrathin graphitic carbon sheath for EDLC electrodes. Values of specific capacitance reaching 325 mF/cm² are achieved; these represent the highest specific EDLC capacitance for planar microsupercapacitor electrode materials obtained up to now. A Ragone plot comparing the performances of different planar microsupercapacitor materials (Alper et al. 2014) is presented in Figure 16.19a. The C/PSiNW device exhibits energy densities in ionic liquids that are the highest published to date for planar supercapacitor electrode materials, while maintaining power densities that are comparable to those of reduced graphene oxide carbon-nanotube (RGO–CNT) composites. This moderate power limitation is a result of the total ESR of the electrodes, which is 170 Ω for about 0.8 cm² total device area, calculated from impedance-spectroscopy results. The C/PSiNW lifetime behavior in EMIM-TFSI shows that the materials are robust with 83% capacitance retention after 5000 cycles at a charge–discharge current of 3 mA/cm².

A novel pseudo-microsupercapacitor based on highly n-doped silicon nanowires coated with poly(3,4-ethylenedioxythiophene) (PEDOT) has been recently demonstrated by Aradilla et al. (2014). A two-electrode pseudo-capacitor with a Whatman glass-fiber paper separator, soaked in an ionic liquid (N-methyl-N-propylpyrrolidinium bis(trifluoromethylsulfonyl)imide) electrolyte was able to deliver a specific capacity of 9 mF/cm² (36 F/g). The ESR was calculated to be about 26 Ω·cm² with a specific energy of 10 Wh/kg and a maximal power density of 85 kW/kg at a cell voltage of 1.5 V. The device exhibited a long lifetime and outstanding electrochemical stability, retaining 80% of the initial capacitance after 3500 galvanostatic charge–discharge cycles at a high current density of 1 mA/cm². The improvement of the capacitive properties compared with

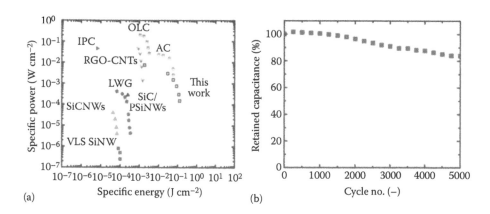

(a) (b)

FIGURE 16.19 (a) Ragone plot comparing the energy and power density of some C/PSiNW device to other planar microsupercapacitor electrode materials: activated carbon (AC), inkjet printed carbon (IPC), laser-written graphene (LWG), onion-like carbon (OLC), reduced graphene oxide–carbon nanotube hybrid (RGO–CNT), silicon carbide-coated PSi nanowires (SiC/PSiNW), silicon carbide nanowires (SiCNWs), and vapor–liquid–solid grown silicon nanowires (VLS-SiNW). (b) Results of extended cycling C/PSiNW device at charge–discharge current of 3 mA cm⁻². (From Alper J.P. *et al.*, *Nano Lett.* **14**(4), 1843, 2014. Copyright 2014: American Chemical Society. With permission.)

the bare SiNWs was attributed to the pseudo-capacitive behavior induced by the conducting polymer coating.

The electrochemical capacitive properties of hollow silicon spheres were investigated by cyclic voltammetry (CV) and chronopotentiometry in 0.5 M Na_2SO_4 aqueous electrolyte (Liu et al. 2012). The CV curves of the silicon hollow-sphere electrode exhibited a widely opened loop between 0.4–0.8 V, implying good capacitive-energy storage ability. In comparison, commercial silicon (Sigma-Aldrich, 325 mesh) showed a very narrow loop, even narrower than that of the conducting additive acetylene blacks under the same test conditions. Close examination of the CV curves of the silicon hollow-sphere electrode revealed that the loops are not rectangular as usually shown for carbon capacitors. There exist two redox peaks—at 0.45 and 0.65 V—at a scanning rate of 5 mV/s; this indicates that the capacity of the silicon hollow-sphere electrode probably arises from pseudo-capacitance behavior resulting from the redox mechanism. Because of the redox reactions, the charge–discharge curves deviate from ideal linear plots as well. The capacitances calculated from the chronopotentiograms are 145, 156, 166 F/g at current densities of 1.25, 0.75, and 0.5 A/g, respectively.

Comparison of the presented electrochemical data shows that the values of specific capacitance of 2D pseudo-microsupercapacitors based on PEDOT-coated SiNWs (Aradilla et al. 2014) are of the same order of magnitude as those of PPy symmetric redox pseudo-capacitors utilizing 3D inter-digital PSi substrates prepared by MEMS technologies (56 mF/cm²) (Sun & Chen 2009), but is only one-third that of 3D in-plane pseudo-capacitors with AC-PVDF composite electrodes (90.7 mF) (Shen et al. 2011). On the other hand, it is worth noting that SiNW-PEDOT redox symmetric micro-pseudo-capacitors afford better results in terms of specific capacitance (8–9 mF/cm²) than interdigital on-perforated-chip micro-supercapacitors comprising onion-like carbon electrodes prepared by electrophoretic deposition (0.9 mF/cm²) (Pech et al. 2010b).

The results presented highlight the fact that PSi-NWs coated by carbon, graphene, and nickel oxide materials are quite promising as an extremely high-energy-density planar microsupercapacitor-electrode material, which may enable smaller footprint-sized microsensors and microsystem devices (Alper et al. 2014; Lu et al. 2011). These results are also interesting as they indicate that the entirety of the porous structure is not only conformally coated and protected, but is electrochemically active. Because of their high surface area, electrochemical stability, and the ultrathin nature of the coating, these materials may enable a variety of applications that benefit from high surface area, nanostructured Si-based materials. The facile nature of the preparation process should boost further efforts of development of versatile conducting and protecting coatings of PSi-NWs materials for high-energy density applications in supercapacitors.

16.6 SUMMARY

Recently, PSi has been studied as an alternative to carbon in supercapacitors. In a brief overview of state-of-the art EDLC and pseudo-microsupercapacitor technology, in which PSi is utilized as a substrate or active electrode material, we have compared preparation routes, microstructural features, the fabrication, and electrochemical-performance characteristics of these energy-storage devices. While in the view of the authors, carbon powders will likely continue to dominate mainstream EDLC manufacturing, PSi electrodes and substrates may serve specialized applications (e.g., integrated silicon-on-chip microcapacitors) focused on high-energy-density operation at high rates and competing with electrolytic capacitors. The specific use of silicon as an abundant material, capable of being transformed into an electrochemical capacitor with the use of known silicon-chip-fabrication methods, opens new avenues to integrated-device applications. Novel approaches, such as tailoring the architecture of electrodes and coating of PSi by graphene, which protects the silicon from undesirable reactions with the electrolyte, show great promise. The use of processes already developed in the electronics industry for forming on-chip supercapacitors with controllable porosity, thickness, and nanoparticle or nanowire silicon morphology, makes possible the combination of efficient energy storage with existent silicon-based technology platforms in diverse technologies such as solar cells, sensors, and microelectronics. We would like to emphasize the fact that the specific capacitance of PSi and nanowires per footprint is of the same order of magnitude as that of high-surface-area porous carbon materials, which promises widespread application of PSi supercapacitors for the future. It is worth noting that the hybrid redox symmetric micro-supercapacitors have shown better results in terms of specific capacitance (mF/cm²) and energy density (Wh/kg) characteristics than EDLCs.

Considering the electrolytes for PSi-EDLCs, ionic liquids (IL) and their mixtures with organic solvents seem to be proper candidates if the IL cost could be significantly reduced. High conducting ion gels, and solid and quasi-solid composite electrolytes offer the advantage of improved safety, being free of liquid leakage, and decreased self-discharge. However, despite the growing number of publications in the last decade, systematic experimental and theoretical studies on charge-storage and ion-transport mechanisms are essential for the design of prospective PSi supercapacitors.

REFERENCES

Alper, J.P., Vincent, M., Carraro, C., and Maboudian, R. (2012). Silicon carbide coated silicon nanowires as robust electrode material for aqueous micro-supercapacitor. *Appl. Phys. Lett.* **100**(16), 163901.

Alper, J.P., Wang, S., Rossi, F. et al. (2014). Selective ultrathin carbon sheath on porous silicon nanowires: Materials for extremely high energy density planar micro-supercapacitors. *Nano Lett.* **14**(4), 1843–1847.

Aradilla, D., Bidan, G., Gentile, P. et al. (2014). Novel hybrid micro-supercapacitor based on conducting polymer coated silicon nanowires for electrochemical energy storage. *RSC Adv.* **4**, 26462–26467.

Bao, Z., Weatherspoon, M.R., Shian, S. et al. (2007). Chemical reduction of three-dimensional silica microassemblies into microporous silicon replicas. *Nature* **446**, 172–175.

Bisi, O., Ossicini, S., and Pavesi, L. (2000). Porous silicon: A quantum sponge structure for silicon based optoelectronics. *Surf. Sci. Rep.* **38**(1–3), 1–126.

Bittner, A.M., Zhu, M., Yang, Y. et al. (2012). Ageing of electrochemical double layer capacitors. *J. Power Sources* **203**, 262–273.

Chatterjee, S., Carter, R., Oakes, L. et al. (2014). Electrochemical and corrosion stability of nanostructured silicon by graphene coatings: Toward high power porous silicon supercapacitors. *J. Phys. Chem. C* **118**, 10893–10902.

Chen, X., Lin, J., Yuan, D. et al. (2008). Obtaining of high area-ratio free-standing silicon microchannel plate via modified electrochemical procedure. *J. Micromech. Microeng.* **18**(3), 037003.

Chen, W., Fan, Z., Dhanabalan, A., Chen, C., and Wang, C. (2011). Mesoporous silicon anodes prepared by magnesiothermic reduction for lithium ion batteries. *J. Electrochem. Soc.* **158**(9), A1055–A1059.

Cheng, J.P., Zhang, J., and Liu, F. (2014). Recent development of metal hydroxides as electrode material of electrochemical capacitors. *RSC Adv.* **4**, 38893–38917.

Chiappini, C., Liu, X., Fakhoury, J.R., and Ferrari, M. (2010). Biodegradable porous silicon barcode nanowires with defined geometry. *Adv. Funct. Mater.* **20**(14), 2231–2239.

Chmiola, J., Largeot, C., Taberna, P.L., Simon, P., and Gogotsi, Y. (2010). Monolithic carbide-derived carbon films for micro-supercapacitors. *Science* **328**, 480–483.

Choi, N.S., Chen, Z., Freunberger, S.A. et al. (2012). Challenges facing lithium batteries and electrical double-layer capacitors. *Angew. Chem. Int. Ed.* **51**, 9994–10024.

Desplobain, S., Gautier, G., Semai, J., Ventura, L., and Roy, M. (2007). Investigations on porous silicon as electrode material in electrochemical capacitors. *Phys. Stat. Sol. (c)* **4**(6), 2180–2184.

Föll, H., Christophersen, M., Carstensen, J., and Hasse, G. (2002). Formation and application of porous silicon. *Mater. Sci. Eng. R* **39**(4), 93–141.

Gautier, C., Ledain, S., Jacqueline, S. et al. (2008). Silicon based system in package: Improvement of passive integration process to avoid TBMS failure. *Microelectron. Reliab.* **48**, 1258–1262.

Gileadi, E. (2011). *Physical Electrochemistry: Fundamentals, Techniques and Applications.* Wiley-VCH, Weinheim.

Gogotsi, Y. and Simon, P. (2011). True performance metrics in electrochemical energy storage. *Science* **334**, 917–918.

Gu, W. and Yushin, G. (2014). Review of nanostructured carbon materials for electrochemical capacitor applications: Advantages and limitations of activated carbon, carbide-derived carbon, zeolite-templated carbon, carbon aerogels, carbon nanotubes, onion-like carbon, and graphene. *WIREs Energy Environ.* **3**(5), 409–533.

Guo, M., Zou, X., Ren, H., Muhammad, F., Huang, C., Qiu, S., and Zhu, G. (2011). Fabrication of high surface area mesoporous silicon via magnesiothermic reduction for drug delivery. *Microporous Mesoporous Mater.* **142**(1), 194–201.

Hadjersi, T., Gabouze, N., Ababou, A. et. al. (2005). Metal-assisted chemical etching of multicrystalline silicon in HF/ Na2S2O8 produces porous silicon. *Mater. Sci. Forum* **480–481**, 139–144.

Huang, J., Sumpter, B.G., and Meunier, V. (2008). Theoretical model for nanoporous carbon supercapacitors. *Angew. Chem. Int. Ed.*, **47**, 520–524.

Huang, P., Pech, D., Lin, R. et al. (2013). On-chip micro-supercapacitors for operation in a wide temperature range. *Electrochem. Commun.* **36**, 53–56.

Hummel, R.E. and Chang, S.S. (1992). Novel technique for preparing porous silicon. *Appl. Phys. Lett.* **61**, 1965–1967.

In, H.J., Kumar, S., Shao-Horn, Y., and Barbastathis, G. (2006). Origami fabrication of nanostructured, three-dimensional devices: Electrochemical capacitors with carbon electrodes. *Appl. Phys. Lett.* **88**, 0831041.

Kang, J., Wen, J., Jayaram, S.H., Yu, A., and Wang, X. (2014). Development of an equivalent circuit model for electrochemical double layer capacitors (EDLCs) with distinct electrolytes. *Electrochim. Acta* **115**, 587–598.

Korotcenkov, G. and Cho, B.K. (2010). Silicon porosification: State of the art. *Crit. Rev. Solid State Mater. Sci.* **35**(3), 153–260.

Laermer, F. and Schilp, A. (2003). Method of anisotropic etching silicon. US Patent 5501893.

Lang, W. (1996). Silicon microstructuring technology. *Mater. Sci. Eng. R.* **17**(1), 1–55.

Largeot, C., Portet, C., Chmiola, J., Taberna, P.L., Gogotsi, Y., and Simon, P. (2008). Relation between the ion size and pore size for an electric double-layer capacitor. *J. Am. Chem. Soc.* **130**(9), 2730–2731.

Lehmann, V. (2002). Electrochemistry of Silicon: Instrumentation, Science, Materials and Applications. Wiley-VCH Verlag GmbH, Weinheim.

Lehmann, V. and Gösele, U. (1991). Porous silicon formation: A quantum wire effect. *Appl. Phys. Let.* **58**(8), 856–858.

Li, L., Zhou, X.S., Li, G.D., Pan, X.L., and Chen, J.S. (2009). Unambiguous observation of electron transfer from a zeolite framework to organic molecules. *Angew. Chem. Int. Ed.* **48**(36), 6678–6682.

Li, M., Xu, Sh., Liu, T. et al. (2013). Electrochemically-deposited nanostructured $Co(OH)_2$ flakes on three-dimensional ordered nickel/silicon microchannel plates for miniature supercapacitors. *J. Mater. Chem. A* **1**, 532–540.

Liu, C.G., Yu, Z., Neff, D., Zhamu, A., and Jang, B.Z. (2010). Graphene-based supercapacitor with an ultra-high energy density. *NanoLett.* **10**(12), 4863–4868.

Liu, M.P., Li, C.H., Du, H.B., and You, X.Z. (2012). Facile preparation of silicon hollow spheres and their use in electrochemical capacitive energy storage. *Chem. Commun.* **48**, 4950–4952.

Lu, F., Qiu, M., Qi, X., Yang, L. et al. (2011). Electrochemical properties of high-power supercapacitors using ordered NiO coated Si nanowire array electrodes. *Appl. Phys. A* **104**, 545–550.

Miao, F., Tao, B., Ci, P., Shi, J., Wang, L., and Chu, P.K. (2009). 3D ordered NiO/silicon MCP array electrode materials for electrochemical supercapacitors. *Mater. Res. Bull.* **44**, 1920–1925.

Murray, F., LeCornec, F., Bardy, S. et al. (2007). Silicon-based system-in-package: Breakthroughs in miniaturization and nano-integration supported by very high quality passives and system level design tools. *Mater. Res. Soc. Symp. Proc.* **969**, 27–35.

Notten, P.H.L., Roozeboom, F., Niessen, R.A.H., and Baggetto, V. (2007). 3-D integrated all-solid-state rechargeable batteries. *Adv. Mater.* **19**(24), 4564–4567.

Oakes, L., Westover, A., Mares, J.W. et al. (2013). Surface engineered porous silicon for stable, high performance electrochemical supercapacitors. *Sci. Rep.* **3**, 3020.

Pech, D., Brunet, M., Durou, H. et. al. (2010a). Ultrahigh-power micrometer-sized supercapacitors based on onion-like carbon. *Nat. Nanotechnol.* **5**, 651–654.

Pech, D., Brunet, M., Taberna, P.L. et. al. (2010b). Elaboration of a microstructured inkjet-printed carbon electrochemical capacitor. *J. Power Sources* **195**, 1266–1269.

Peng, K.Q., Yan, Y.J., Gao, S.P., and Zhu, J. (2002). Synthesis of large area silicon nano-wires arrays via self-Assembling nano-electrochemistry. *Adv. Mater.* **14**(16), 1164–1167.

Richman, E.K., Kang, C.B., Brezesinski, T. et al. (2008). Ordered mesoporous milicon through magnesium reduction of polymer templated silica thin films. *Nano Lett.* **8**(9), 3075–3079.

Ripenbein, T. (2010). Novel Technological Processes for the Development and Preparation of Li-Ion 3D Microbattery. PhD Thesis, Tel Aviv University.

Ripenbein, T., Golodnitsky, D., Nathan, M., and Peled, E. (2010). Novel porous-silicon structures for 3D-microbatteries. *Electrochim. Acta* **56**(1), 37–41.

Romstad, F.P. and Veje, E. (1997). Experimental determination of the electrical band-gap energy of porous silicon and the band offsets at the porous silicon/crystalline silicon heterojunction. *Phys. Rev. B* **55**(8), 5220–5225.

Roozeboom, F., Klootwijk, J.H., Verhoeven, J.F.C. et al. (2007). ALD options for Si-integrated ultrahigh-density decoupling capacitors in pore and trench designs. *ECS Trans.* **3**(15), 173–181.

Roozeboom, F., Bergveld, H.J., Nowak, K. et al. (2009). Ultrahigh-density trench capacitors in silicon and their application to integrated DC-DC conversion. *Procedia Chem.* **1**, 1435–1438.

Rowlands, S.E., Latham, R.J., and Schlindwein, W.S. (1999). Supercapacitor devices using porous silicon electrodes. *Ionics* **5**, 144–149.

Sharma, P. and Bhatti, T.S. (2010). A review on electrochemical double-layer capacitors. *Energy Convers. Manage.* **51**, 2901–2912.

Shen, C., Wang, X., Zhang, W., and Kang, F. (2011). A high-performance three-dimensional micro supercapacitor based on self-supporting composite materials. *J. Power Sources* **196**, 10,465–10,471.

Shukla, A.K., Banerjee, A., Ravikumar, M.K., and Jalajakshi, A. (2012). Electrochemical capacitors: Technical challenges and prognosis for future markets. *Electrochim. Acta* **84**, 165–173.

Simon, P. and Gogotsi, Y. (2008). Materials for electrochemical capacitors. *Nature Mater.* **7**, 845–854.

Steiner, P. and Lang, W. (1995). Micromachining applications of porous silicon. *Thin Solid Films* **255**, 52–58.

Sun, W. and Chen, X. (2009). Preparation and characterization of polypyrrole films for three-dimensional MEMS supercapacitor. *J. Power Sources* **193**, 924–929.

Sun, W., Zheng, R., and Chen, X. (2010). Symmetric redox supercapacitor based on micro-fabrication with three-dimensional polypyrrole electrodes. *J. Power Sources* **195**, 7120–7125.

Tada, T., Hamoudi, A., Kanayama, T., and Koga, K. (1997). Spontaneous production of 10 nm Si structures by plasma etching using self-formed masks. *Appl. Phys. Let.* **70**, 2538–2540.

Thissandier, F., Le Comte, A., Crosnier, O. et al. (2012). Highly doped silicon nanowires based electrodes for micro-electrochemical capacitor applications. *Electrochem. Commun.* **25**, 109–111.

Thissandier, F., Pauc, N., Brousse, T. et al. (2013). Micro-ultracapacitors with highly doped silicon nanowires electrodes. *Nanoscale Research Letters*, **8**, 38.

Thissandier, F., Dupré, L., Gentile, P. et al. (2014). Ultra-dense and highly doped SiNWs for micro-supercapacitors electrodes. *Electroch. Acta* **117**,159–163.

Van den Meerakker, J.E.A.M., Elfrink, R.J.G., Roozeboom, F., and Verhoeven, J.F.C.M. (2000). Etching of deep macropores in 6 in. Si wafers. *J. Electrochem. Soc.* **147**, 2757–2761.

Vega, D., Reina, J., Pavón, R., and Rodríguez, A. (2014). High-density capacitor devices based on macro-porous silicon and metal electroplating. *IEEE Trans. Electron Dev.* **61**(1), 116–122.

Wang, J.F., Wang, K.X., Du, F.H., Guo, X.X., Jiang, Y.M., and Chen, J.S. (2013). Amorphous silicon with high specific surface area prepared by a sodiothermic reduction method for supercapacitors. *Chem. Commun.* **49**, 5007–5009.

Westover, A.S., Tian, J.W., Bernath, S. et al. (2014). A multifunctional load-bearing solid-state supercapacitor. *NanoLett.* **14**(6), 3197–3202.

Ye, C., Lin, Z.M., and Hui, S.Z. (2005). Electrochemical and vapacitance properties of rod-shaped MnO_2 for supercapacitor. *J. Electrochem. Soc.* **152**(6), 1272–1278.

Yu, Z.J., Dai, Y., and Chen, W. (2010). Electrochemical deposited nanoflakes $Co(OH)_2$ porous films for electrochemical capacitors. *J. Chin. Chem. Soc.* **57**(3A), 423–428.

Yu, A., Chabot, V., and Zhang, J. (2013). *Electrochemical Supercapacitors for Energy Storage and Delivery: Fundamentals and Applications*. CRC (Taylor & Francis Group), Boca Raton, FL.

Yu, G., Xie, X., Pan, L., Bao, Z., and Cui, Y. (2013). Hybrid nanostructured materials for high-performance electrochemical capacitors. *Nano Energy* **2**, 213–234.

Yuan, D., Ci, P., Tian, F. et al. (2009). Large size *p*-type silicon microchannel plates prepared by photo-electro chemical etching. *J. Micro/Nanolith. MEMS MOEMS* **8**(3), 033012.

Zahedinejad, M., Khaje, M., Erfanian, A., and Raissi, F. (2011). Successful definition of nanowire and porous Si regions of different porosity levels by regular positive photoresist using metal-assisted chemical etching. *J. Micromech. Microeng.* **21**, 065006.

Zhu, J., Liu, R.B., Xu, J., and Meng, C.G. (2011a). Preparation and characterization of mesoporous silicon spheres directly from MCM-48 and their response to ammonia. *J. Mater. Sci.* **46**(22), 7223–7227.

Zhu, J., Liu, R.B., Xu, J., and Meng, C.G. (2011b). Synthesis of mesoporous silicon directly from silicalite-1 single crystals and its response to thermal diffusion of ZnO clusters. *J. Mater. Sci.* **46**(11), 3840–3845.

Porous Silicon as a Material for Thermoelectric Devices

17

Androula G. Nassiopoulou

CONTENTS

17.1 Introduction 376

17.2 General Considerations 376

17.3 Porous Si as a Thermoelectric Material 378

    17.3.1 Thermal Conductivity and Other Thermoelectric Parameters
of Si Nanowires 378

    17.3.2 Thermal Conductivity and Thermoelectric Properties of PSi 379

17.4 Thermoelectric Generators Using Porous Silicon 382

17.5 Summary and Future Trends 383

References 384

17.1 INTRODUCTION

Thermal management is currently one of the key issues in several applications. In nanoelectronics, for example, one of the major limiting factors in complementary metal oxide semiconductor (CMOS) devices and integrated circuits (ICs) scaling down is heat dissipation. Efficient cooling is needed in order to assure electronic chip reliability. Thermoelectric (TE) devices constitute a promising technique to cool ambience or convert wasted thermal energies into useful power. Thermoelectric generators are reliable devices and do not need any special maintenance. Although they have a poor yield, they are already used in different applications, as for example, in electronics (see koolatron.com), portable refrigeration (see mcleancoolingtech.com), cool/heat car seats, cooling of diodes and so on. A huge amount of wasted heat from automotive, different industry applications, and human activities can be converted into useful energy using TE devices.

In order to compete with traditional techniques of converting heat into electrical power (vapor compression techniques, for example), the figure of merit, ZT, which characterizes a thermoelectric material, should reach values well above 3 for the current ~1.5 to 2. A renewed interest in thermoelectrics started in the 1990s, when new strategies were proposed to enhance ZT (Kanatzidis 2010; Lan et al. 2010; Alam and Ramakrishna 2013; Martín-González et al. 2013; Leadley et al. 2014; Mouis et al. 2014). One method to enhance ZT is to reduce lattice thermal conductivity. This is achieved by using different approaches, such as phonon scattering using mass fluctuations in mixed crystals, and grain boundary scattering or interface scattering in thin films or multilayers. Low dimensional materials show advantages towards enhancing ZT. They are used either to enhance the power factor or to reduce the lattice thermal conductivity. Low-dimensional materials currently investigated are thin films and superlattices, as well as semiconductor nanowires. (Hicks and Dresselhaus 1993; Venkatasubramanian et al. 2001; Boukai et al. 2008; Hochbaum et al. 2008; Donadio and Galli 2010; Chen et al. 2012; Neophytou et al. 2013).

In low-dimensional materials, the ZT factor is increased by two means: the first is quantum confinement, which increases the power factor, and the second is the additional scattering of phonons due to a large number of internal interfaces. Phonon confinement affects the entire phonon relaxation rate and this influences the thermal conductivity of nanostructures, which is lower compared to that of their bulk counterparts.

One of the most investigated low-dimensional systems is porous silicon (PSi), which results from the electrochemical dissolution of bulk crystalline Si. PSi is a material composed of interconnected silicon nanostructures (nanowires, nanocrystallites), which give to the material very different properties than those of bulk Si. Both its thermal and electrical conductivities are very different from those of bulk Si. At high porosity, the material shows high resistivity (Ben-Chorin et al. 1995; Theodoropoulou et al. 2003) and very low thermal conductivity (Nassiopoulou and Kaltsas 2000; Lysenko et al. 2002), this last being lower than that of the best Si-based thermal insulators (Valalaki and Nassiopoulou 2013, 2014a).

A lot of research was devoted to the determination of PSi thermal conductivity (Nassiopoulou 2014). However, the interest on its thermoelectric properties has been raised only recently. Few theoretical works were devoted to these properties (Yu et al. 2010; Guo and Huang 2014; Lee 2014) and pointed out the potential of PSi as an efficient low dimensional thermoelectric material. In this review, we first discuss the interest of low dimensional materials in thermoelectrics and second the potential of using PSi in thermoelectric generators, either as a thermal isolation platform or as an active thermoelectric material.

17.2 GENERAL CONSIDERATIONS

Thermoelectric (TE) conversion takes advantage of the voltage difference, which is generated when a temperature gradient exists between two contacts of a thermoelectric material (Seebeck effect, see Figure 17.1). The Seebeck effect is mainly related to the difference in the energy distribution of free carriers between the two contacts. The Seebeck coefficient is defined as

$$S = \frac{\Delta T}{\Delta V}$$

(17.1)

Hot end
(T_H, V_H)

Cold end
(T_c, V_c)

Electron flux

FIGURE 17.1   Seebeck effect, showing an electron flux between a hot and a cold end of a thermo-electric material, which induces voltage difference between these two ends.

where ΔV is the voltage difference between two ends with temperature difference ΔT, see Figure 17.1.

The performance of a thermoelectric material is characterized by its figure of merit ZT, which is given by

(17.2)
$$ZT = \frac{S^2 \cdot \sigma \cdot T}{k}$$

In Equation 17.2, S is the Seebeck coefficient of the material (V/K), σ is its electrical conductivity ($\Omega \cdot m$), T is the absolute temperature, and k is the material thermal conductivity (W/m·K). In a semiconductor, heat conduction is mediated both by free carriers (electrons or holes) and by lattice vibrations (phonons), with respective contribution to thermal conductivity, so $k = k_e + k_{ph}$, where k_e is the thermal conductivity due to charge carrier transport (electrons, holes) and k_{ph} is the lattice thermal conductivity due to phonons. ZT determines the fraction of the Carnot efficiency that can be theoretically attained by a thermoelectric material (Wood 1984). The quantity $S^2 \cdot \sigma$ is the so-called power factor. A good thermoelectric material is characterized by a large value of ZT, attained by a large Seebeck coefficient S, a large electron conductivity to reduce Joule heating, and a low thermal conductivity to keep the temperature difference across the material. In a thermoelectric generator (TEG), the thermoelectric materials are connected in series from an electrical point of view and in parallel from a thermal point of view. A TEG cell is shown schematically in Figure 17.2. In such a device, the maximum efficiency n_{max} cannot exceed the Carnot efficiency limit, n_C:

(17.3)
$$n_C = \frac{T_H - T_C}{T_H}$$

where T_H is the temperature at the hot side (K), and T_C is the temperature at the cold side (K).

The first generation of thermoelectric devices was based on bulk thermoelectric materials with ZT ~ 0.8–1 (Chen et al. 2012; Claudio et al. 2012; Martín-González et al. 2013). The corresponding devices operated at 5–6% conversion efficiencies. The second generation of bulk thermoelectric materials, with almost two times higher ZT (1.3–1.7), were using nanoscale inclusions and compositional inhomogeneities to dramatically reduce the lattice thermal conductivity, thus reaching conversion efficiencies of the corresponding devices as high as 11–15%. Recent approaches in this direction are expected to raise ZT by a factor of 2 and the power efficiency above 20%, depending on ΔT (Kanatzidis 2010). Bi_2Te_3 is currently the best thermoelectric material and the Bi_2Te_3–Sb_2Te_3 superlattice, together with PbTe-PbSe, were the first to show extremely low thermal conductivities (Venkatasubramanian et al. 2001; Harman et al. 2002).

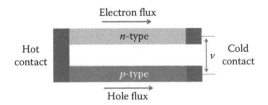

Electron flux

n-type

Hot contact

Cold contact

v

p-type

Hole flux

FIGURE 17.2   Thermoelectric generator using an *n*-type and a *p*-type material in series.

Bulk crystalline Si is a poor thermoelectric material (ZT ~ 0.01 at 300K) (Weber and Gmelin 1991) due to its high thermal conductivity (~150 W/m·K at room temperature). Nanoscale Si can have much lower thermal conductivity, comparable to or even lower than that of bulk amorphous Si (~1 W/m·K at RT). The characteristic length scales can be comparable to the dominant values of the phonon mean free path Λ and the phonon wavelength λ (for bulk crystalline Si $\Lambda \sim 300$ nm and $\lambda \sim 1$–2 nm at RT). At length scales smaller than Λ, there is increased boundary scattering that leads to reduction in k.

17.3 POROUS Si AS A THERMOELECTRIC MATERIAL

Depending on the porosity, PSi is composed of interconnected nanowires and nanodots of different sizes. It is so relevant to start with discussing the thermal conductivity and other thermoelectric properties of Si nanowires, which can help to understand the corresponding PSi thermoelectric properties.

17.3.1 THERMAL CONDUCTIVITY AND OTHER THERMOELECTRIC PARAMETERS OF Si NANOWIRES

Heavily doped semiconductors are considered as the most efficient thermoelectric materials because heat-carrying electrons are restricted from being close to the Fermi energy for metals (MacDonald 1962). For a metal or highly doped semiconductor, the Seebeck coefficient S is proportional to the energy derivative of electronic states. In low-dimensional semiconductors, the density of electronic states has sharp peaks and theoretically high thermopower (Hicks and Dresselhaus 1993). However, what has been proven to be more effective in getting high ZT values in nanostructured systems is the reduction of heat transport through enhancement of phonon scattering. This is achieved by reducing one or more dimensions of the material to values smaller than the mean free path of phonons and larger than that of electrons and holes, which reduces k without decreasing σ.

Experimental results on Si nanowires with cross-sectional area of 10 nm × 20 nm and 20 nm × 20 nm showed that they exhibit a ZT ~1 at 200K, attributed to phonon effects (Boukai et al. 2008). Surface roughness is very important for the reduction of nanowire thermal conductivity. Si nanowires fabricated by metal-assisted chemical etching and having a rough surface yielded ZT ~0.6 at room temperature (Hochbaum et al. 2008). Figure 17.3 illustrates the power factor at 300K calculated as $S^2/\rho \sim 3.3 \cdot 10^{-3}$ W/m·K², and the ZT values of an individual Si nanowire. Both ZT and the power factor decrease with decreasing temperature.

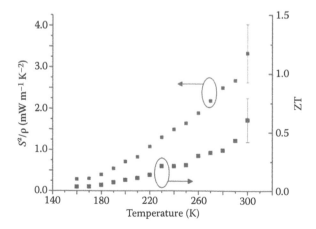

FIGURE 17.3   Single nanowire power factor and calculated ZT values as a function of temperature derived from measurements of the thermal conductivity k of 52 nm nanowires and ρ and S measurements of 48 nm nanowires. (From Hochbaum A.I. et al., *Nature* 451(7175), 163, 2015. Copyright 2015: Macmillan Publishers Limited. With permission.)

A theoretical work using molecular dynamics (Donadio and Galli 2010) in thin crystalline and core-shell Si nanowires demonstrated that in crystalline Si nanowires the presence of a crystalline/amorphous interface greatly influences the vibrational spectrum, turning vibrational propagating modes into diffusive modes. Vibrational modes are the majority in crystalline nanowires, while diffusive modes dominate in amorphous Si (a-Si). The combined contribution of propagating and diffusive modes is at the origin of the peculiar temperature dependence of the thermal conductivity of Si nanowires. It was found experimentally that k shows nearly linear dependence of $k(T)$ at low T and a broad plateau above ~150K, which extends beyond room temperature (Li et al. 2003; Donadio and Galli 2010). This is different from the behavior of bulk semiconductors, where $k(T)$ ~ T^3 for $T \rightarrow$ 0K and $k(T)$ ~ $1/T$ at higher temperatures, after a maximum is reached.

17.3.2 THERMAL CONDUCTIVITY AND THERMOELECTRIC PROPERTIES OF PSi

The thermal conductivity of PSi has been extensively investigated at room temperature (see review in Nassiopoulou 2014). Its temperature dependence was investigated in Gesele et al. (1997) and De Boor et al. (2011) down to 40K, while recently a full investigation of highly porous Si thermal conductivity in the temperature range 5–350K was reported (Valalaki and Nassiopoulou 2013, 2014a,b). The main results are as follows: PSi thermal conductivity is much lower than that of bulk crystalline Si and it does not show any maximum as a function of temperature (see Figure 17.4).

A plateau-like behavior is observed in the temperature range from 5 to 20K, explained by the "fractons" interpretation of the vibrational excitations in the fractal porous Si skeleton. In the temperature range 20–350K, a monotonic increase of k with temperature is observed, fitted with simplified classical models. One major result is that the thermal conductivity of PSi can be made even lower than that of any conventional low thermal conductivity material (a-Si and different Si oxides and nitrides) if the porosity is high (above 60%). This is illustrated in Figure 17.5 for a 40-μm thick PSi layer with a porosity of 63%.

In a recent paper, Guo and Huang (2014) used nonequilibrium and equilibrium molecular dynamics simulations, together with lattice dynamics calculations, to investigate the thermal conductivity of nanoporous Si with anisotropic pore pitches. They found a remarkable tunable anisotropy in phonon transport, mainly due to the difference in phonon relaxation times by the channel confinement. The significant decrease in thermal conductivity in the nanoporous material was attributed to both the zone folding and the suppression of relaxation times. The importance of junction effects to an effective thermal conductivity, considerably lower than that predicted by only considering diffusive boundary scattering, was pointed out and evaluated.

For its use as a thermoelectric material, PSi should show low thermal conductivity, obtained at high porosity, but at the same time, the electrical conductivity should be kept sufficiently high. There is a tradeoff between these two parameters, depending on the porosity. By increasing the porosity, the thermal conductivity of the material decreases, while its electrical resistivity

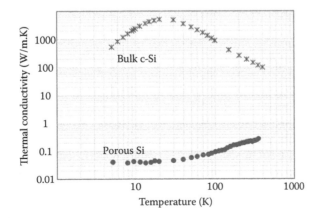

FIGURE 17.4   Comparison of PSi thermal conductivity as a function of temperature with that of bulk crystalline Si. (Data extracted from Valalaki and Nassiopoulou 2013, 2014a.)

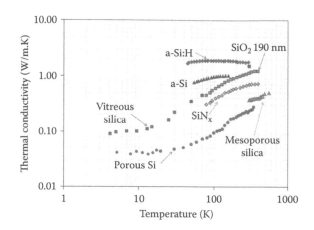

FIGURE 17.5 Comparison of PSi thermal conductivity (Valalaki and Nassiopoulou 2013, 2014a) as a function of temperature with that of other low thermal conductivity materials (results reproduced from: a-Si and a-Si :H (Cahill et al. 1989), mesoporous silica (Shin et al. 2009), vitreous silica (Smith et al. 1978) and silicon nitride and 190 nm SiO$_2$ film (Lee and Cahill 1997).

increases. Therefore, the porosity of the material should be kept moderate and an optimum should be determined for optimum thermoelectric performance. Moreover, the Si nanowires composing the PSi skeleton should be highly doped for a higher power factor ($S^2 \cdot \sigma$). For optimized thermoelectric properties, it is thus necessary to design and fabricate the appropriate PSi material by tuning the fabrication conditions.

PSi has been recently proposed as a potential thermoelectric material by different groups (Lee et al. 2008; Tang et al. 2010; Lee 2014). Theoretical work was published by Lee et al. (2008) using classical molecular dynamics and ab initio density functional theory on model PSi structures with periodically arranged nanometer-sized pores on n-type crystalline Si (Figure 1 of Lee 2008). The incentive for the above work was the large ZT values obtained experimentally from Si nanowires (SiNWs) by different authors (Boukai et al. 2008; Hochbaum et al. 2008). A value of ZT equal to 0.6 for 50 nm diameter Si wires versus 0.01 for bulk crystalline Si was reported (Hochbaum et al. 2008). Boukai et al. (2008) also reported high ZT values from SiNWs with cross-sectional areas of 10 nm × 10 nm and 20 nm × 20 nm. A 100-fold increase in ZT compared to that of the bulk was obtained in a large temperature range from 100K to 350K, including ZT ~ 1 at 200K. The above results provided strong motivation to study PSi as a thermoelectric material, based on its structure, composed of Si nanowires and nanodots. The theoretical results by Lee et al. (2008) were obtained at room temperature. They calculated the thermal conductivity of the material and the electrical conductivity of the porous structure, which was found to decrease by a factor of 2–4 compared to the bulk, the Seebeck coefficient exhibiting a twofold increase for carrier concentrations less than 2×10^{19} cm^{-3}. From these results, they predicted the figure of merit ZT to increase by 2 orders of magnitude over that of bulk Si. This important enhancement of ZT was attributed to the greatly reduced thermal conductivity of the material, together with the moderate reduction of the power factor, attributed to pore ordering. Their results are depicted in Figure 17.6.

In a recent work published in 2014, Joo-Hyang Lee (2014) reported on the thermoelectric performance of strained nanoporous Si containing cylindrical pores in a periodic arrangement. They used density functional theory and the Boltzman transport equation. They found that the electrical conductivity σ and the Seebeck coefficient S of the nanoporous material are not changed by strain under biaxial or shear strain. On the other hand, orthorhombic strain increases significantly σ and S by 68% and 110%, respectively, compared to the unstrained material. Combined with the thermal conductivity, a maximum value for ZT as high as 0.8 was predicted (Lee 2014).

Experimental results concerning the thermoelectric properties of a PSi structure were obtained by Tang et al. (2010) for a nanoporous Si material called "holey silicon." It contained holes, produced on an Si ribbon, fabricated by either nanosphere or block–copolymer lithography. Material porosity was 35%. They reported ZT values as a function of temperature, ranging from 0.05 at 150K to 0.4 at room temperature. These results are presented in Figure 17.7 (ZT = f(T)).

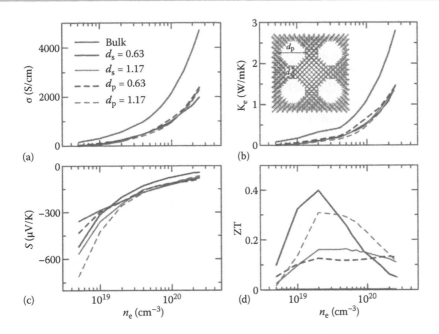

FIGURE 17.6   Theoretical predictions of the thermoelectric parameters (σ [a], k [b], S [c], and ZT [d]) of PSi with periodically arranged nanometer-size circular pores on n-type crystalline Si for different pore sizes d and spacing d_s as a function of the doping concentration. Inset of (b): circular nanoporous structure used in the calculations. Pores were along [001] crystallographic direction. (From Lee J.-H. et al., *Nano Lett.* 8(11), 3750, 2008. Copyright 2008: American Chemical Society. With permission.)

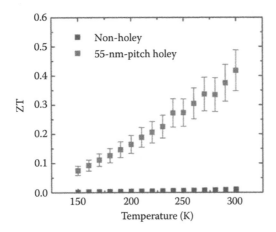

FIGURE 17.7   Experimental results of ZT values as a function of temperature of a non-holey (bulk) and a 55-nm-pitch holey Si. (From Tang J. et al., *Nano Lett.* 10(10), 4279, 2010. Copyright 2010: American Chemical Society. With permission.)

The thermoelectric properties of electrochemically etched porous Si with 60% porosity were studied by De Boor et al. (2012). Free standing PSi membranes, 70-90 µm thick, were used in this respect. After etching, the samples were doped using spin-on-dopant, in order to increase PSi conductivity. The doped material had a conductivity exceeding 1000 S/cm and an effective thermal conductivity at room temperature of ~ 7.6 W/m·K. The values of the measured figure of merit, ZT, exceeded 0.02 at room temperature, compared to the 0.01 of bulk crystalline Si. This value is, however, still low, much lower than that obtained with "holy Si." The main bottleneck is the high resistivity of highly porous Si. Further studies, investigating also PSi with lower porosity, are needed in order to fully exploit the possibilities offered by PSi in thermoelectrics.

17.4 THERMOELECTRIC GENERATORS USING POROUS SILICON

A thermoelectric generator is composed of three main parts. The first part is the thermoelectric material in the form of thermocouples, which convert a temperature difference into electrical power. The second part is a thermal isolation platform placed under the "hot" contact of the thermocouples and used to keep the temperature locally high. The third part is the encapsulation of the device, designed to have sufficient thermal isolation between the hot and cold contacts of the thermocouples. This is necessary in order to effectively capture heat changes either on the hot or on the cold part of the device.

The interest of using PSi as an active thermoelectric material is very new and no TEG devices based on it were reported thus far. On the other hand, its use as a platform on the Si wafer for local thermal isolation on bulk crystalline Si has been used in several thermal devices, including gas flow sensors (Kaltsas and Nassiopoulou 1999; Kaltsas et al. 2002), accelerometers (Goustouridis et al. 2007), and chemical gas sensors (Tsamis et al. 2003). In this respect, a thick PSi layer is locally grown on the Si wafer, on which the hot parts of the active elements of the device are integrated (Nassiopoulou et al. 2000). Recently, the same thermal isolation method was applied to an Si TEG device (Hourdakis and Nassiopoulou 2013). It was composed of a large number of Si integrated thermocouples, made of a *p*-doped polycrystalline silicon in contact with Al. Their "hot" contacts were lying on arrays of PSi areas, while their cold contacts were on bulk crystalline Si. A schematic view of a small number of thermocouples, connected in series, is illustrated in

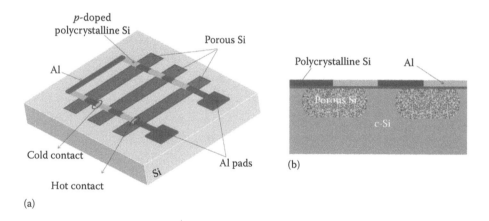

(a)   (b)

FIGURE 17.8   (a) A schematic three-dimensional view of a small number of thermocouples, connected in series and (b) a cross-sectional view of a "hot" thermocouple contact on PSi and a "cold" contact on bulk crystalline Si. (From Hourdakis E. and Nassiopoulou A.G., *Sensors* 13(10), 13596, 2013. Published by MDPI as open access.)

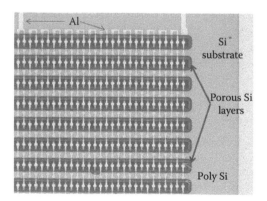

FIGURE 17.9   Top view optical image of a part of the TEG device. (From Hourdakis E. and Nassiopoulou A.G., *Sensors* 13(10), 13596, 2013. Published by MDPI as open access.)

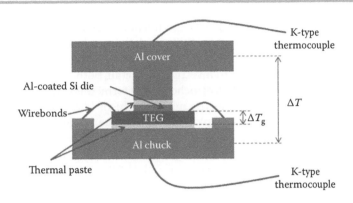

FIGURE 17.10   Schematic representation (cross-sectional view) of the encapsulated TEG cell, illustrating the Al chuck and the Al cover, used to transfer macroscopic temperature differentials to the TEG. (From Hourdakis E. and Nassiopoulou A.G., *Sensors* 13(10), 13596, 2013. Published by MDPI as open access.)

Figure 17.8a. In Figure 17.8b, we see a cross-sectional view of a "hot" thermocouple contact on PSi and a "cold" contact on bulk crystalline Si.

A plan view optical microscope image of the TEG top area is shown in Figure 17.9. The TEG was encapsulated using an Al plate with an arm reaching the TEG surface, the "cold" parts being protected with a thick photoresist and the "hot" parts being in contact with the Al plate through a patterned Al layer. The back plate of the encapsulated device was an Al chuck.

Figure 17.10 shows a cross-section of the TEG in its housing. With the above-described TEG, an output voltage of 16.55 mV/K, an output power of 0.39 μW/cm², and a power efficiency factor $\Phi = 0.0039$ μW/cm²·K² were achieved with a 50-μm thick PSi thermal isolation layer and a macroscopic temperature differential of 10K.

This result is competitive to that obtained with TEG devices using complicated and fragile freestanding micromechanical structures on the Si wafer (Venkatasubramanian et al. 2001; Kao et al. 2010; Xie et al. 2010; Li et al. 2011; Curtin et al. 2012; Perez-Marín et al. 2014).

17.5 SUMMARY AND FUTURE TRENDS

PSi is an interesting material for use in thermoelectric devices due to its structure and morphology, composed of a silicon skeleton of interconnected nanowires and nanodots, separated by voids (pores). It can be used either as a low thermal isolation platform on the Si wafer for the integration of a silicon-based thermoelectric generator, or as an active thermoelectric material. Both the structure of the PSi skeleton and pore size are tunable by the type and resistivity of the Si wafer, on which the PSi layer is formed, and the electrochemical conditions used for its formation.

For the application of PSi as a local thermal isolation platform on the Si wafer, a highly porous material is needed, which shows very low thermal conductivity, namely two orders of magnitude lower than that of bulk crystalline Si at room temperature and more than four orders of magnitude lower than that of Si at cryogenic temperatures. Compared to other low thermal conductivity materials, highly porous Si shows one of the lowest thermal conductivities, especially at low temperatures. This is attributed to phonon confinement in its low dimensional structure. In addition to the low thermal conductivity, PSi constitutes a material that is fully compatible with Si processing and is easily integrated with Si devices and integrated circuits on a single chip. A first Si-based thermoelectric generator using PSi local thermal isolation has been recently demonstrated by Hourdakis and Nassiopoulou (2013). This TEG shows competitive performance compared to other more complicated ones.

For the use of PSi as an active thermoelectric material, a different structure and porosity than above is needed. In this case, a compromise between low thermal conductivity and sufficiently high electrical conductivity is needed in order to achieve a thermoelectric material with high figure of merit. Thus, one has to choose a moderate porosity material and find the optimum material structure and morphology in this respect. Recent theoretical calculations have shown the interest of using PSi as a thermoelectric material. An experimental work was made on a model PSi structure (holy Si), fabricated using patterning techniques. A thermoelectric figure of merit as high as 0.4 at room temperature was demonstrated for that material, having a porosity of 35%.

Based on the above, it is interesting to investigate in detail, both theoretically and experimentally, the thermoelectric properties of electrochemically formed PSi, in order to assess the possibility of using it as an active thermoelectric material. This is both challenging and interesting because the material structure and morphology are not unique, but are tunable by the wafer type and resistivity and the electrochemical conditions used for PSi formation. Either a sponge-like structure could be investigated, or an anisotropic structure with vertical pores, similar to the case of the patterned holy Si investigated by Tang et al. (2010).

REFERENCES

Alam, H. and Ramakrishna, S. (2013). A review on the enhancement of figure of merit from bulk to nano-thermoelectric materials. *Nano Energy* 2(2), 190–212.

Ben-Chorin, M., Möller, F., Koch, F., Schirmcher, W., and Eberhard, M. (1995). Hopping transport on a fractal: AC conductivity of porous silicon. *Phys. Rev. B* 51(4), 2199–2213.

De Boor, J., Kim, D.S., Ao, X. et al. (2011). Temperature and structure size dependence of the thermal conductivity of porous silicon. *EPL (Eur. Lett.)* 96(1), 16001.

De Boor, J., Kim, D.S., Ao, X., Becker, M., Hinsche, N.F., Mertig, J., Zahn, P., and Schmidt, V. (2012). Thermoelectric properties of porous Si. *Appl. Phys. A* doi 10.1007/s 00339-012-6879-5.

Boukai, A.I., Bunimovich, Y., Tahir-Kheli, J., Yu, J.-K., Goddard, W.A., and Heath, J.R. (2008). Silicon nanowires as efficient thermoelectric materials. *Nature* 451(January), 168–171.

Cahill, D.G., Fischer, H.E., Klitsner, T., Swartz, E.T., and Pohl, R.O. (1989). Thermal conductivity of thin films: Measurements and understanding. *J. Vac. Sci. Technol. A* 7(3), 1259–1266.

Chen, Z.-G., Han, G., Yang, L., Cheng, L., and Zou, J. (2012). Nanostructured thermoelectric materials: Current research and future challenge. *Prog. Natural Sci.: Mater. Intern.* 22(6), 535–549.

Claudio, T., Schierning, G., Theissmann, R. et al. (2012). Effects of impurities on the lattice dynamics of nanocrystalline silicon for thermoelectric application. *J. Mater. Sci.* 48(7), 2836–2845.

Curtin, B.M., Fang, E.W., and Bowers, J.E. (2012). Highly ordered vertical silicon nanowire array composite thin films for thermoelectric devices. *J. Electron. Mater.* 41(5), 887–894.

Donadio, D. and Galli, G. (2010). Temperature dependence of the thermal conductivity of thin silicon nanowires. *Nano Lett.* 10(3), 847–851.

Gesele, G., Linsmeier, J., Drach, V., Fricke, J., and Arens-Fischer, R. (1997). Temperature-dependent thermal conductivity of porous silicon. *J. Phys. D: Appl. Phys.* 30(21), 2911–2916.

Goustouridis, D., Kaltsas, G., and Nassiopoulou, A.G. (2007). A silicon thermal accelerometer without solid proof mass using porous silicon thermal isolation. *IEEE Sensors J.* 7(7), 983–989.

Guo, R. and Huang, B. (2014). Thermal transport in nanoporous Si: Anisotropy and junction effects. *Int. J. Heat Mass Transfer* 77, 131–139.

Harman, T.C., Taylor, P.J., Walsh, M.P., and LaForge, B.E. (2002). Quantum dot superlattice thermoelectric materials and devices. *Science* 297(5590), 2229–2232.

Hicks, L.D. and Dresselhaus, M.S. (1993). Thermoelectric figure of merit of a one-dimensional conductor. *Phys. Rev. B* 47(24), 16631–16634.

Hochbaum, A.I., Chen, R., Delgado, R.D. et al. (2008). Enhanced thermoelectric performance of rough silicon nanowires. *Nature* 451(7175), 163–167.

Hourdakis, E. and Nassiopoulou, A.G. (2013). A thermoelectric generator using porous Si thermal isolation. *Sensors* 13(10), 13596–13608.

http://www.koolatron.com (accessed July 19, 2012).

http://www.mcleancoolingtech.com (accessed July 13, 2012).

Kaltsas, G. and Nassiopoulou, A. (1999). Novel C-MOS compatible monolithic silicon gas flow sensor with porous silicon thermal isolation. *Sens. Actuators A* 76(1–3), 133–138.

Kaltsas, G., Nassiopoulos, A.A., and Nassiopoulou, A.G. (2002). Characterization of a silicon thermal gas-flow sensor with porous silicon thermal isolation. *IEEE Sensors J.* 2(5), 463–475.

Kanatzidis, M.G. (2010). Nanostructured thermoelectrics: The new paradigm? *Chem. Mater.* 22, 648–659.

Kao, P.-H., Shih, P.-J., Dai, C.-L., and Liu, M.-C. (2010). Fabrication and characterization of CMOS-MEMS thermoelectric micro generators. *Sensors* 10(2), 1315–1325.

Lan, Y., Minnich, A.J., Chen, G., and Ren, Z. (2010). Enhancement of thermoelectric figure-of-merit by a bulk nanostructuring approach. *Adv. Funct. Mater.* 20(3), 357–376.

Laser Cooling for TO Packages using Embedded Thin-Film Thermoelectric Coolers-Nextreme Thermal Solutions, Inc., http://www.nextreme.com/media/pdf/Nextreme_Laser_%0ADiode_Cooling_Test_Report_Jan10.pdf.

Leadley, D., Shah, V., Ahopelto, J. et al. (2014). Thermal isolation through nanostructuring. In: Balestra F. (Ed.) *Beyond CMOS Nanodevices 1.* Wiley-ISTE, Hoboken, NJ, pp. 331–364.

Lee, J.-H. (2014). Significant enhancement in the thermoelectric performance of strained nanoporous Si. *Phys. Chem. Chem. Phys.* 16(6), 2425–2429.

Lee, S.-M. and Cahill, D.G. (1997). Heat transport in thin dielectric films. *J. Appl. Phy.* 81(6), 2590.

Lee, J.-H., Galli, G.A., and Grossman, J.C. (2008). Nanoporous Si as an efficient thermoelectric material. *Nano Lett.* 8(11), 3750–3754.

Li, D., Wu, Y., Kim, P., Shi, L., Yang, P., and Majumdar, A. (2003). Thermal conductivity of individual silicon nanowires. *Appl. Phys. Lett.* 83(14), 2934.

Li, Y., Buddharaju, K., Singh, N., Lo, G.Q., and Lee, S.J. (2011). Chip-level thermoelectric power generators based on high-density silicon nanowire array prepared with top-down CMOS technology. *IEEE Electron Dev. Lett.* 32(5), 674–676.

Lysenko, V., Périchon, S., Remaki, B., and Barbier, D. (2002). Thermal isolation in microsystems with porous silicon. *Sens. Actuators A* 99(1–2), 13–24.

MacDonald, D.K.C. (1962). *Thermoelectricity: An Introduction to the Principles.* Wiley & Sons, New York.

Martín-González, M., Caballero-Calero, O., and Díaz-Chao, P. (2013). Nanoengineering thermoelectrics for 21st century: Energy harvesting and other trends in the field. *Renewable Sustainable Energy Rev.* 24, 288–305.

Mouis, M., Chávez-Ángel, E., Sotomayor-Torres, C. et al. (2014). Thermal energy harvesting. In: Balestra F. (Ed.) *Beyond CMOS Nanodevices 1,* Wiley-ISTE, Hoboken, NJ, pp. 135–220.

Nassiopoulou, A.G. (2014). Thermal isolation of porous Si. In: Canham L. (Ed.) *Handbook of Porous Si.* Springer, Berlin.

Nassiopoulou, A.G. and Kaltsas, G. (2000). Porous silicon as an effective material for thermal isolation on bulk crystalline Silicon. *Phys. Stat. Sol. (a)* 182(1), 307–311.

Neophytou, N., Zianni, X., Kosina, H., Frabboni, S., Lorenzi, B., and Narducci, D. (2013). Simultaneous increase in electrical conductivity and Seebeck coefficient in highly boron-doped nanocrystalline Si. *Nanotechnology* 24(20), 205402.

Perez-Marín, A.P., Lopeandía, A.F., Abad, L. et al. (2014). Micropower thermoelectric generator from thin Si membranes. *Nano Energy* 4, 73–80.

Shin, S., Ha, T.-J., Park, H.-H., and Cho, H.H. (2009). Thermal conductivity of BCC-ordered mesoporous silica films. *J. Phys. D: Appl. Phys.* 42(12), 125404.

Smith, T., Anthony, P., and Anderson, A. (1978). Effect of neutron irradiation on the density of low-energy excitations in vitreous silica. *Phys. Rev. B* 17(12), 4997–5008.

Tang, J., Wang, H.-T., Lee, D.H. et al. (2010). Holey silicon as an efficient thermoelectric material. *Nano Lett.* 10(10), 4279–4283.

Theodoropoulou, M., Krontiras, C.A., Xanthopoulos, N. et al. (2003). Transient and AC electrical conductivity of porous silicon thin films. *Phys. Stat. Sol. (a)* 197(1), 279–283.

Tsamis, C., Tserepi, A., and Nassiopoulou, A.G. (2003). Fabrication of suspended porous silicon micro-hotplates for thermal sensor applications. *Phys. Stat. Sol. (a)* 197(2), 539–543.

Valalaki, K. and Nassiopoulou, A.G. (2013). Low thermal conductivity porous Si at cryogenic temperatures for cooling applications. *J. Phys. D: Appl. Phys.* 46(29), 295101.

Valalaki, K. and Nassiopoulou, A.G. (2014a). Thermal conductivity of highly porous Si in the temperature range 4.2 to 20 K. *Nanoscale Res. Lett.* 9(1), 318.

Valalaki, K. and Nassiopoulou, A.G. (2014b). Porous Silicon as an efficient local thermal isolation platform on the Si wafer in the temparature range 5–350K. In: *Proceedings of 11th International Workshop on Low Temperature Electronics (WOLTE),* July 7–9, Grenoble, France, pp. 61–64.

Venkatasubramanian, R., Siivola, E., Colpitts, T., and O'Quinn, B. (2001). Thin-film thermoelectric devices with high room-temperature figures of merit. *Nature* 413, 597–602.

Weber, L. and Gmelin, E. (1991). Transport properties of silicon. *Appl. Phys. A Mater. Sci. Process.* 53, 136–140.

Wood, C. (1984). High-temperature thermoelectric energy conversion—II. Materials survey. *Energy Convers. Management* 24(4), 341–343.

Xie, J., Lee, C., Wang, M.-F., and Feng, H. (2010). Wafer-level vacuum sealing and encapsulation for fabrication of CMOS MEMS thermoelectric power generators. In: *Proceedings of 2010 IEEE 23rd International Conference on Micro Electro Mechanical Systems (MEMS).* Jan. 24–28, Hong Kong, pp. 1175–1178.

Yu, J.-K., Mitrovic, S., Tham, D., Varghese, J., and Heath, J.R. (2010). Reduction of thermal conductivity in phononic nanomesh structures. *Nature Nanotechnol.* 5(10), 718–721.

Porous Silicon–Based Explosive Devices

18

Monuko du Plessis

CONTENTS

18.1 Introduction   388

18.2 History of Technology   388

18.3 Porous Silicon Device Manufacture   389

     18.3.1 Porous Silicon Properties   389

     18.3.2 Internal Surface Stabilization   391

     18.3.3 Choice of Oxidants   391

     18.3.4 Pore Impregnation   393

18.4 Ignition of the Oxidant   394

18.5 Porous Silicon Explosive Properties   395

     18.5.1 Energy Yield   395

     18.5.2 Propagation Velocity   397

     18.5.3 Mechanical and Electrostatic Sensitivity   398

18.6 Applications of Porous Silicon Explosive Devices   398

18.7 Conclusions   399

References   400

18.1 INTRODUCTION

It is relatively easy in today's silicon semiconductor integrated circuit technology to manufacture more than 1 billion devices onto one single chip. This mature technology has followed the well-known Moore's law for decades, but further scaling down of device dimensions may require a "more than Moore" rather than "more of Moore" approach (Kent and Prasad 2008). The "more than Moore" approach entails the integration of additional nonelectronic functionality onto the chips. These functionalities may include photonic, mechanical, and microelectromechanical systems (MEMS) devices, as well as biological interfaces, for example, micro fluidic structures. For these new functionalities to be accepted in the microelectronics environment, the technological impact on the current well-established microchip fabrication technology should be as low as possible. Several decades after the discovery of the porous silicon (PSi) electrochemical etch technique (Uhlir 1956), new possibilities arose in terms of innovative compatible functionalities, one of which is the utilization of PSi as an energetic material.

Energetic materials store chemical energy and can be classified into three classes, namely propellants, pyrotechnics, and explosives (Rossi et al. 2007). Explosives release their energy in fast detonation processes, while propellants and pyrotechnics processes undergo slower deflagration processes. Traditional energetic materials can be produced by the mixing of oxidizer and fuel constituents into one molecule to produce monomolecular energetic materials, for example, trinitrotoluene (TNT), or the mixing of oxidizer powders (e.g., nitrates or perchlorates) and fuel powders (e.g., carbon) to produce composite energetic materials. These composites exhibit high energy density, but their energy release rates are slower than monomolecular material because the mass transport rate is limited by the granulometry of the reactants. Reducing the particle size to the nanoscale would increase the burning rates due to the reduction of the mass-transport rate, thus making nanoscale energetic materials attractive alternatives to monomolecular structures.

Nanoporous silicon filled with an oxidizer is a promising structure to realize an energetic material that is compatible with mainstream silicon process technology (Rossi et al. 2007). The potential yield of energy of the exothermic reaction of silicon and oxygen is higher than that of the most common carbon-based explosives. Since the spacing between silicon and oxidizing atoms is at the atomic scale in PSi (Clément et al. 2005), there is a significant increase of the oxidation reaction rates which are crucial for explosive reactions.

18.2 HISTORY OF TECHNOLOGY

The first reference to PSi explosive behavior was in 1992 (McCord et al. 1992) when it was reported that a "flash of light and an audible pop" could be observed when concentrated HNO_3 was dropped on the surface of dry PSi. The same reaction could not be induced using concentrated H_2SO_4, and only PSi prepared by anodization showed this violent reaction. This was a first indication that the type of oxidant, as well as the morphology and chemical structure of the PSi, will determine the explosive behavior of the material.

Almost a decade later, the next report on the explosive properties of PSi was made (Kovalev et al. 2001) that nano-explosions could be accomplished at cryogenic temperatures. When freshly prepared, the internal surface of PSi is almost completely covered by hydrogen (Grosman and Ortega 1997) and the hydrogen-terminated PSi serves as a reservoir of hydrogen. Over the temperature range 4.2–90 K the immersion of the porous layer in liquid oxygen resulted in a strong explosion (Kovalev et al. 2001). Furthermore, the explosive reaction was absent if Si nanocrystals assembling the porous layers were completely oxygen terminated (the oxidation was performed at 900°C and was controlled by infrared optical absorption measurements). Therefore, both the filling of pores by oxygen and the termination of the surface of Si nanocrystals by hydrogen were crucial for the explosive reaction.

The major breakthrough was reported in 2002 describing the accidental discovery of a nano-explosion in PSi filled with a solid state oxidant at room temperature (Mikulec et al. 2002). Addition of the oxidizing agent as a dilute nitrate salt solution (instead of the liquid agents used previously) allowed the preparation of a solid material that could be detonated in a more controlled fashion. Samples that contained a large amount of surface oxide (as determined by Fourier transform infrared spectroscopy, FTIR) were usually not explosive.

18.3 POROUS SILICON DEVICE MANUFACTURE

PSi is prepared using an electrochemical etch technique, usually using an HF:ethanol electrolyte. The pores of the porous region are filled with an oxidant, and when heat is applied to the device, a strong exothermic reaction occurs (Becker et al. 2009).

18.3.1 POROUS SILICON PROPERTIES

Table 18.1 gives typical PSi etch parameters to realize explosive devices for a selected number of references. It is evident that a wide range of etch parameters can be used, with silicon resistivity ranging from 1 mΩ·cm to 20 Ω·cm, and current densities ranging from 20 mA/cm^2 to 100 mA/cm^2. It was reported (Mikulec et al. 2002) that regardless of the silicon impurity doping density and conduction type (*n*- or *p*-type), most samples could be exploded. In general, almost all PSi layers used for explosive devices make use of p-type material because the electrochemical etch can then be done without an illumination source. However, a recent study investigated in more detail the effect of doping density and doping type on the explosive properties (Parimi et al. 2014). It was found that the dopant atoms have minimal effect on the interaction between PSi and complex oxidizer molecules such as perchlorate salts. In contrast, the internal PSi structure was found to have a strong effect on the reactivity of elemental sulfur as oxidizer because the substrate doping properties (doping density and conductivity type) influence the internal structure of the PSi formed during etching.

An innovative method well suited for batch processing is the galvanic corrosion technique in which thick (up to 150 μm) films can be prepared without an external power supply (Becker et al. 2010). With a chemical oxidant like H$_2$O$_2$ present in the HF:ethanol solution, a noble metal such as platinum or gold deposited on the silicon serves as a cathode to form a galvanic cell without an external current source. The morphology of the PSi can be controlled by the choice of electrolyte composition, oxidizing agent, noble metal, silicon resistivity, and silicon dopant type (*p*- or *n*-type) in the galvanic set-up.

Table 18.2 lists typical PSi properties used in explosive devices. It can be observed that for best explosive performance the porosity should be in the range of 70%, pore size in the order of 3 to 4 nm, and porous layer thickness ranging from 20 to 80 μm.

TABLE 18.1 The Electrochemical Etch Parameters of Selected Explosive Devices

Ref.	Resistivity Ω·cm	Current Density mA/cm^2	Electrolyte HF:Ethanol
Kovalev et al. 2001	10^{-3}–10	20–70	1:1 (50% HF)
Mikulec et al. 2002	4	50	1:1 (49% HF)
Clément et al. 2004	10^{-3}–1	20–87	1:1 (50% HF)
du Plessis 2007	0.2–0.3	76.5	1.5:1 (50% HF)
Churaman et al. 2008a	1–10	20	1:1 (25% HF)
Plummer et al. 2008	3–6	22.5	1:1 (40% HF)
Mason et al. 2009	$1–5 \times 10^{-3}$	60.1–73.4	1:1 (49% HF)
Becker et al. 2011	1–20	Galvanic	3:1 (49% HF) 2.4% vol of 30% H$_2$O$_2$ added
Wang et al. 2012	0.1–0.3	100	3:1 (40% HF)
Ohkura et al. 2013	0.1–0.9	50	1:1 (48% HF)

Source: du Plessis M., *Propellants, Explos., Pyrotech.* 39, 348, 2014. Copyright 2014: John Wiley & Sons. With permission.

TABLE 18.2   The Porous Silicon Properties of Selected Explosive Devices

Ref.	Porosity (%)	Pore Size (nm)	Layer Thickness (µm)
Kovalev et al. 2001	40–70	2–10	
du Plessis 2007	65	3.4	57
Churaman et al. 2008a		4	40
Plummer et al. 2008	55.9	9.7–14.7	15
Currano and Churaman 2009	67	3.4	33
Becker et al. 2011	65–83	2.4–2.9	65–95
Plummer et al. 2011	52.3–72.2	2.9–6.4	4.2–72
Wang et al. 2012		2–7	80.6–86.1
Piekiel et al. 2014	70–72	3.14–3.32	20

Source: du Plessis M., *Propellants, Explos., Pyrotech.* 39, 348, 2014. Copyright 2014: John Wiley & Sons. With permission.

It was reported (Clément et al. 2005) that the highest energy yield of the explosive reaction was assured where the stoichiometric ratio of SiX_2, where X is oxygen, for example, could be realized. This condition was reached for porous layer porosities in the range of 70% for most oxidizers. It was also demonstrated that a pore size in the range 3 nm to 4 nm was optimal for devices using sodium perchlorate as oxidant (du Plessis and Conradie 2006). Another study concluded that a pore size of 11.8 nm was optimal for the oxidant aluminium nitrate (Plummer et al. 2008). A later study (Piekiel et al. 2013) of propagation speeds reported that the fastest combustion using sodium perchlorate as oxidizer was observed for PSi with a pore size of 3.32 nm, a porosity of 72% and a specific surface area of 900 m^2/g. In the case of the monohydrate oxidizer $NaClO_4$, the chemical reaction is be given by

(18.1) $$2\ Si + NaClO_4 \cdot H_2O \rightarrow 2\ SiO_2 + NaCl + H_2O$$

Equation 18.1 (Fradkin and Gany 2012) requires a molar ratio of 0.5 sodium perchlorate to silicon for complete oxidation of all the PSi. Assuming that the oxidizer completely fills all of the pore volume and that all of the sodium perchlorate will react with all of the silicon, the molar ratio of oxidizer (sodium perchlorate) to fuel (silicon) was computed using the densities and molecular weights of oxidizer and fuel. The oxidizer to fuel molar ratio as a function of porosity is plotted in Figure 18.1, where it is shown that the minimum porosity required achieving the required 0.5 molar ratio for complete combustion is 70% for the monohydrate sodium perchlorate. Experimental molar ratios normally lie below these values because the porous structure is not completely filled with the oxidizer. In the best case yet reported (Becker et al. 2010) a pore

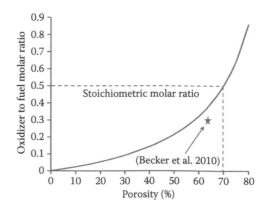

FIGURE 18.1   The oxidizer to fuel molar ratio of PSi for monohydrate $NaClO_4$ as a function of PSi porosity assuming all void space is filled with the oxidizer. The stoichiometric molar ratio of 0.5 is also shown, as well as an experimental value achieved.

filling factor of about 75% could be achieved for an oxidizer to fuel molar ratio of 0.3 at 63% porosity. This experimental value is also indicated in Figure 18.1.

Generally, the aim would be to achieve the highest explosive power in the smallest volume by making the porous layer as thick as possible. The filling of the pores with an oxidant from the front surface, however, will be more difficult using thick porous layers because the oxidant will only penetrate a certain distance into the porous structure. Therefore, too thick a porous layer will not be advisable.

Drying of the samples after the PSi formation is achieved by rinsing the samples in ethanol or methanol, followed by a rinse in the drying agent hexane, and after rinsing drying it subsequently in a gentle stream of nitrogen (Clements et al. 2008). The samples can also be rinsed using the drying agent pentane to ensure the structural stability of the pores (Churaman et al. 2010a). Pentane has a much lower surface tension than either methanol or ethanol, and has been shown to reduce PSi cracking during drying, especially for high porosity material (Becker et al. 2009).

18.3.2 INTERNAL SURFACE STABILIZATION

PSi layers have a large internal specific surface area (up to 1000 m^2/cm^3), and therefore the spacing between the fuel (silicon) and the oxidizing atom is at the atomic scale (Clément et al. 2005). It has been determined that about 20% of the silicon atoms are located at the surface of the nanocrystals (Grosman and Ortega 1997). The surface of freshly etched PSi is covered with Si-H_x groups, and during storage the Si-H_x groups are replaced with Si-O_x groups; that is, the surface is dehydrogenated (Lazaruk et al. 2007). A relationship was established between the decrease in the energy intensity of the PSi layer explosions after storage and the subsequent loss of hydrogen in the course of storage (Lazaruk et al. 2007).

To achieve long-term stability, it is necessary for the PSi internal surfaces to be stabilized. This can be achieved by thermal annealing of thePSi sample in an oxygen atmosphere below 250°C (Clément et al. 2005), when the oxygen is mainly back bonded to surface silicon atoms, while the hydrogen atoms covering the surface silicon atoms remain unaffected (Milewski et al. 1994) and the PSi surface remains organophilic. At higher temperatures, effusion of hydrogen atoms occurs and the oxygen will be bonded directly to the surface, making it more hydrophilic (Clément et al. 2004). This annealing at temperatures below 250°C did not influence the reactivity of freshly etched samples, but annealing at higher temperatures decreased the efficiency of the explosive reaction.

The activation energy of the PSi/oxygen reaction can be tailored by controlling the native layer before the oxidizer loading (Rossi et al. 2007). The backbonded oxygen layer resulted in significant surface stability against aging. The reactivity of the explosive device decreased as the level of oxidation increased and could be used to tailor it (Subramanian et al. 2008). Partial or complete oxidation of the surface could also render PSi/oxidizer mixtures less sensitive to friction and impact. A method to ensure the complete coverage of hydrogen atoms at the silicon surface was described as dipping the freshly electrochemically etched PSi (Koch and Clément 2007) in a solution of HF and ethanol. This removed the thin oxygen layer formed after exposure to air and resaturated the dangling bonds at the surface with hydrogen. The closed hydrogen cover was stable and the oxygen back-bonded during the low-temperature oxygen anneal.

18.3.3 CHOICE OF OXIDANTS

Several types of oxidants have been used in nano-explosive devices. In the most detailed investigation into the properties of a range of oxidants (Clément et al. 2004) it was found that the perchlorates were much more efficient for energetic explosions than the nitrates. A summary of the most important oxidizer properties investigated are shown in Table 18.3.

Although more efficient, perchlorates contain crystal water and are therefore hygroscopic, a distinct disadvantage. There are a number of nonhygroscopic perchlorates, like $KClO_4$ and NH_4ClO_4, but they are not solvable in common solvents used for filling the pores with the oxidant. The crystal water and the hygroscopic nature of the perchlorates are disadvantages in terms

TABLE 18.3   Properties of Some Common Oxidizers

Oxidizer	Solvent (Solubility)	Remarks
Perchlorates		
$Ca(ClO_4)_2 \cdot 4H_2O$	Me (237 g/100 g)	Strongly hygroscopic, but very efficient
	Et (166 g/100 g)	
NH_4ClO_4	Me (6 g/100 g)	Does not stay in pores; weaker reaction, lesser oxygen yield
	Ac (>6 g/100 g)	due to water production
$NaClO_4 \cdot 1H_2O$	Me (~181 g/100 g)	Less hygroscopic and stay inside the pores
	Et (<181 g/100 g)	
$KClO_4$	Me (<1.7 g/100 g)	Not solvable in any common solvent
	Et (<1.7 g/100 g)	
	Ac (<1.7 g/100 g)	
Nitrates		
$Ca(NO_3)_2 \cdot 4H_2O$	Me (>54 g/100 g)	Strongly hygroscopic
	Et (54 g/100 g)	
NH_4NO_3	Me (17 g/100 g)	Does not stay inside the pores
	Et (4 g/100 g)	
KNO_3		Bad solubility, therefore no reaction
Others		
Sulfur	CS_2 (good)	The only non-hygroscopic material that stays inside the pores

Source: Clément D. et al., *Phys. Stat. Sol. (a)* **202**, 1357, 2005. Copyright 2005: John Wiley & Sons. With permission.
Note: Ac = Acetone; Et = Ethanol; Me = Methanol.

of long-term stability, but it was reported (Clément et al. 2004) that the presence of crystal water was necessary for keeping the oxidizer salt inside the pores. Nonhygroscopic perchlorates with no crystal water (e.g., NH_4ClO_4) tend to creep out of the pores after the solvents have evaporated.

Most oxidizers are available in a powder format, and filling the pores from a solution containing the oxidizer is the preferred technique to impregnate the pores of the PSi with an oxidizer. Due to the organophilic surface nature of the PSi, water does not penetrate the pores, and other solvents have to be used (Clément et al. 2004). Methanol and ethanol are frequently used as solvents, but acetone may also be used in some instances. However, oxidizers solvable in organic liquids are mostly hygroscopic. As was reported (Clément et al. 2005), sulfur has a rather low melting point, around 113°C, which is below its ignition temperature. Since molten sulfur can wet the PSi surface, the pores can actually be filled using melted sulfur.

Although KNO_3 has poor solubility in solvents, impregnation of pores could be achieved with a 10% aqueous solution of KNO_3 and subsequent rotary drying (Lazarouk et al. 2005). Gadolinium nitrate has also been used, and a 0.2 M solution of $Gd(NO_3)_3 \cdot 6H_2O$ in ethanol was used in an explosive device (Mikulec et al. 2002) for emission spectroscopy experiments.

Aluminum nitrate as an oxidizer was reported (Plummer et al. 2008), where a 0.2 M solution in ethanol was used for pore impregnation. The results indicated that aluminum nitrate was not as effective an explosive material as either gadolinium nitrate or sodium perchlorate. In the same study, a high-explosive material RDX (cyclo-1,3,5-trimethylene-2,4,6-trinitramine) was added (0.2 M in acetone) to the PSi pores. The results were not encouraging, possibly because RDX is significantly deficient in oxidizing species.

In summary, the most popular oxidizers are calcium perchlorate and sodium perchlorate, not only due to the energetic reactions, but also because they tend to stay inside the pores (Churaman et al. 2008b) and seem to be more stable. It has been observed that some oxidizers tend to creep out of the pores, and it was experimentally shown that the migration of the oxidizer might be dependent on the content of water molecules in the pores. Using anhydrous ethanol versus methanol to dissolve the salt minimizes the water content in the pores. Due to the larger concentration of water in the methanol, the drying process appears to be causing the dislocation of the salt as the water inside the porous network begins to evaporate. This process tends to push the salt crystals out of the pores.

The relative energies released by three different oxidants are illustrated in Figure 18.2 (du Plessis 2007) with the sodium perchlorate explosion the most energetic.

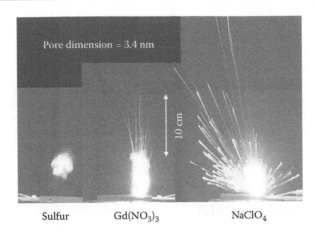

Pore dimension = 3.4 nm

10 cm

Sulfur   Gd(NO$_3$)$_3$ NaClO$_4$

FIGURE 18.2   Optical emissions from three different oxidizer nano-explosions. (From du Plessis M., *Sens. Actuators, A* **135**, 666, 2007. Copyright 2006: Elsevier. With permission.)

18.3.4 PORE IMPREGNATION

Due to the organophilic surface of PSi, water does not penetrate into the pores effectively and ethanol or methanol is a better solvent to use due to its better wettability. The very first solid-state PSi explosive samples (Mikulec et al. 2002) were processed by rinsing the samples after porous etching in ethanol and hexane, and then drying the samples under a stream of nitrogen. The dried porous surface area was then covered with 10 μL of a 0.2 M solution of Gd(NO$_3$)$_3$ · 6H$_2$O in ethanol, and then allowed to dry in air for at least 1 h. Later, a sequential filling method to achieve the highest possible mass of oxidizer confined in the pores and a stoichiometric ratio of oxygen to silicon was proposed (Clément et al. 2005). By gravimetric measurements, it was found that filling of the pores three times is sufficient for a 30% Ca(ClO$_4$)$_2$ solution in methanol to reach stoichiometric filling. Additional fillings did not further increase the mass of the oxidizer. For saturated solutions of NaClO$_4$, it was sufficient to fill the pores two times. The highest filling factors achieved were ~50% of total pore volume. This is enough to realize the stoichiometric ratio of SiO$_2$ for Ca(ClO$_4$)$_2$.

Using gravimetric techniques, the effective filling factor of the pores was determined (du Plessis and Conradie 2006) as a function of pore size. The solutions used to perform the experiment were 1 g of S in 5 mL of CS$_2$, 1 g of NaClO$_4$·1H$_2$O in 5 mL of methanol, and 1 g of Gd(NO$_3$)$_3$·6H$_2$O in 5 mL of ethanol. The effective filling factor of the pores was defined as the ratio of pore volume filled by the oxidant after sequential filling, relative to the total pore volume. Results showed that the filling factors increased with pore size, as expected, and reached a maximum plateau value at around 5 to 8 nm pore size. It is also noted that NaClO$_4$·1H$_2$O had the highest filling factor of the three oxidants investigated, approaching a maximum filling factor of about 75%. The gadolinium nitrate and sulfur reached maximum filling factors of about 50%, correlating well with the results reported earlier (Clément et al. 2005) that the highest filling factors achieved were in the order of 50% for most oxidizers.

A typical procedure to fill the pores with an oxidant was described (Becker et al. 2009) as follows: The oxidizer (preferably NaClO$_4$) was dissolved in either methanol or anhydrous ethanol. The oxidizers were applied (the pores were filled) by dropping the oxidizer solution with a pipette directly on the PSi surface. In most cases, the pores were filled with two drops of solution. This was enough solution to completely cover the PSi region. If a multi-step filling procedure was necessary, the first two drops were allowed to visually dry before applying additional drops. The solution was dried in a humidity-controlled box flooded with N$_2$ gas to reduce the humidity to about 30%. It was found experimentally (du Plessis 2014) that the maximum depth of penetration of the oxidizer solution into the PSi layer is between 50 and 100 μm. In general, the pore size and porosity in relatively thick PSi layers (≈20 μm) will vary as a function of distance from the front surface (Herino et al. 1987), resulting in the oxidizer impregnation of the pores to be not homogenous, with more oxidizer near the surface.

The immersion method of impregnation by immersing the sample in the methanoic or ethanoic solution, followed by sample drying and evaporation of the solvent, shows limitations in the amounts of material that can be deposited into the porous structure and in the size of molecules that must diffuse into the pores (Herino 1997). The immersion technique was used (Mason et al. 2009) to fill PSi samples with $NaClO_4 \cdot 1H_2O$ and $Ca(ClO_4)_2 \cdot 4H_2O$ by immersing the sample in a 1:1 oxidizer/methanol solution. The sample and solution were placed under a rough vacuum at 0.165 atm for a prescribed filling time of 5 or 15 min. The samples were then dried at 38°C overnight. In general, it was found that in some cases large patches of the sample did not ignite. The conclusion was made that this behavior was likely caused by poor oxidizer loading, an effect that has not been reported for the sequential pipette filling technique.

18.4 IGNITION OF THE OXIDANT

The first PSi explosive prototypes were ignited by scratching the surface with a diamond scribe, or by using a small electric spark (Mikulec et al. 2002). Subsequent developments showed that the oxidants can be ignited thermally, electrically, or optically (Clément et al. 2005). Thermal ignition was performed using a hot plate, and it was experienced that below a certain heating rate the oxidizer decomposes and does not explode. Experimentally determined initiation temperatures for some common oxidizers are approximately 200°C for $Ca(ClO_4)_2$, 320°C for $NaClO_4$, and 255°C for S (Clément et al. 2004). The oxidizer $Gd(NO_3)_3$ could be initiated at approximately 240°C (du Plessis and Conradie 2006).

Electrical ignition makes use of heating elements like hotwires. Hotwire finite element simulations were performed (Churaman et al. 2008b) to design hotwires deposited directly on the PSi, making use of the much lower thermal conductance of the PSi, and thus ensuring more efficient hotwire initiation. The hotwire stack was a 20/100/380 nm Ti/Pt/Au sandwich of layers. Further optimization of the hotwire stack led to width/length dimensions of 15 μm/75 μm and 15 μm/125 μm, and ignition times with a 5 V supply were 120 μs and 88 μs, respectively, with peak currents less than 100 mA (Churaman et al. 2010a). The electrical initiation energy was in the range of 0.1 mJ. The use of Cr micro bridges as reliable hotwire heating elements was also demonstrated, and initiation energies varied from 0.15 to 0.29 mJ to ignite PSi devices with $NaClO_4$ as oxidizer, with ignition delay times varying from 80 to 110 μs (Wang et al. 2012). The driving voltage in this case was, however, quite high at around 100V.

From the above studies, it is evident that the reaction time of the oxidizer ignition process after the onset of the hotwire voltage is a function of the hotwire geometry. However, it has been shown that with approximately 0.1 mJ of electrical energy, a PSi device with sodium perchlorate as oxidizer can be ignited within 100 μs using a hotwire element.

The optical ignition of PSi explosive devices by direct laser illumination has been reported at a variety of wavelengths and laser power densities. Optical ignition with a single pulse from a YAG laser could also initiate the explosive reaction (Clément et al. 2004), although the power density and pulse duration were not reported. Laser ignition using a 514 nm laser at 37.7 mW and a power density of 2.7 kW/cm² at a standoff distance of 23 cm was also reported (Churaman et al. 2010b). In another laser initiation application, a Q-switched Nd:YAG laser of 532 nm wavelength was used (Wang et al. 2013). The laser had a 15-ns pulse duration and the igniting energy was 264 mJ. No information regarding the time response of the optical ignition processes was given. A problem area with direct laser illumination is the fact that protecting PSi layers from atmospheric humidity using a thin polymer coating prevents direct laser ignition even with long exposure times to the laser (Churaman et al. 2010b).

To overcome the above drawback, two other possibilities were investigated (Plummer et al. 2014). The first technique used direct laser ignition by an infrared laser pulse at 1064 nm wavelength with illumination from the rear side of the wafer toward the PSi at the front surface. It was determined that the threshold energy for ignition was 650 mJ in a 15-ns pulse width. The second technique investigated ignition of PSi devices by a weak stress wave propagating through the silicon substrate. This shock wave was produced by a laser-generated stress wave created at the backside of the wafer, traveling through the bulk silicon material and interacting with the PSi film on the other side of the wafer. Wave velocities were somewhat lower than the speed of sound through silicon. Using photon doppler velocimetry, it was determined that these waves are weak stress waves with a threshold intensity of 131 MPa in the silicon substrate to ignite the PSi with sodium perchlorate as oxidizer. These ignition techniques lead to the possibility that PSi devices may be hermetically sealed from the environment. This would prevent degradation of the material on oxidation in air and accelerated in moist environments, a problem often encountered with this material (Plummer et al. 2014).

Optical ignition with a light source containing a broad band of wavelengths, such as a xenon (Xe) flash lamp, has also been reported (Ohkura et al. 2013). Flash ignition is practically convenient because it is nonintrusive, low-cost, and can achieve distributed ignition of energetic materials to enhance their energy release rates. It was found that the minimum flash ignition energy decreases with decreasing the film thickness and increasing the film porosity. In general, it was demonstrated that freestanding PSi films could be optically ignited in ambient air by a low power Xe flash with energy density < 1 J/cm^2.

A very interesting initiation technique combines optical and mechanical means (Morris et al. 2012). Energy can be stored as residual stress in a deposited thin film. This novel actuator powered by residual thin film mechanical stress absorbed 25 W/cm^2 of optical power from a 532-nm visible laser, heated, and released up to 22 nJ of mechanical energy, sufficient to release almost 10 kJ/g of chemical energy from the sodium perchlorate impregnated PSi. The irradiation level needed was nearly 80 times less than previous direct optical initiation via laser. Since the actuation process is the result of a thermal process, the time response is relatively slow. Irradiation with 25 W/cm^2 of optical power resulted in an explosion within 30 ms.

Another novel development combined carbon nanotubes and PSi. This nanostructured energetic material combined multi-walled carbon nanotubes, mixed with ferrocene, with PSi impregnated with sodium perchlorate (Malec et al. 2010). It was demonstrated that the PSi could be exploded using the carbon nanotubes as photosensitive initiators using a camera flash. As the temperature of the carbon nanotubes is raised indirectly by the camera flash, which induces photo-acoustic vibrations, some of the carbon is oxidized releasing the metal nanoparticles into the surroundings and in the presence of oxygen the carbon will combust with a self-sustaining flame. The PSi explosion is ignited using the light induced multi-walled carbon nanotubes ignition.

18.5 POROUS SILICON EXPLOSIVE PROPERTIES

The efficiency of the explosive reactions can be described by two parameters, namely the release of energy and the reaction rate (Kovalev et al. 2001). Other characteristics of relevance are the gas released by the explosion, as well as the sensitivity of the explosive device to mechanical shock and electrostatic discharge.

18.5.1 ENERGY YIELD

To estimate the energy yield of the explosion, the binding energies of the Si-H, Si-Si, O-O, and Si-O bonds, as well as the energy of water formation, were taken into account (Kovalev et al. 2001). If only the surface silicon atoms were supposed to be converted into SiO$_2$, and assuming that all hydrogen atoms were removed from the surface and interact with oxygen, the energy yield of, however, was estimated to be 12 kJ/g. This is significantly higher than the energy yield of TNT (4.2 kJ/g). The typical density of the porous layer was assumed to be 1 g/cm^3 in this estimation.

It is known that hydrogen is present in the chemical composition of freshly etched PSi (Manilov and Skryshevsky 2013), and that the explosive properties of PSi depend strongly on the role of Si-H bonds (Kovalev et al. 2001). The fundamental explosion is the energy released under the interaction of oxygen molecules with dangling Si bonds acting as free radicals. The three principal steps in the PSi explosion are as follows: (1) The explosion is ignited by rupturing the surface Si-H bonds and creating dangling Si bonds. Hydrogen atoms covering PSi are the buffer between Si atoms and molecular oxygen, which prevents oxygen and silicon interaction, followed by (2) hydrogen is removed from the surface via an exothermic reaction between oxygen and hydrogen, which forms water or OH radicals, and (3) it initializes the interaction of surface Si atoms directly with oxygen, and the oxidation of the silicon nanostructure is achieved. Although hydrogen plays an important role in the chain reaction, it does not get liberated in the form of hydrogen gas H_2, but rather reacts with the fuel and oxidizer to form Si-H and O-H radicals. The majority of the gas generation released by the explosion is probably in the form of water vapor (Churaman et al. 2010a).

One of the first energy yields of PSi explosive devices was measured using a calorimetric bomb-test for PSi filled with calcium perchlorate as oxidizer (Clément et al. 2004, 2005). The energy yield was measured as 7.3 kJ/g, which is higher than the yield of TNT, and compared well with the most powerful explosions known at that stage; all with energy yields less than 8 kJ/g.

The influence of pore size on the energetic properties of PSi was determined experimentally using differential scanning calorimetry (DSC) techniques (Plummer et al. 2008). The highest energy output was measured for the oxidant sodium perchlorate, almost 9 kJ/g, which compared well with the theoretical predicted energy yield of 10.35 kJ/g. Other bomb calorimetry experiments (Subramanian et al. 2008) using low-cost PSi prepared from the etching of silicon powders, reported an energy yield of 8.1 kJ/g for sodium perchlorate.

The ability to tune the energy release rate and propagation speed, as well as the gas generated during the explosion, has been demonstrated (Churaman et al. 2010a). The proposed reaction of the PSi/sodium perchlorate system is

$$(18.2) \qquad n\,SiH_2 + (1\text{-}n)\,Si + (0.5 + 0.25n)\,NaClO_4 \rightarrow SiO_2 + NaCl + n\,H_2O + heat$$

where SiH_2 arises from the hydrogen surface termination of the pore surface. The coefficient n represents the ratio of silicon atoms on the surface to total silicon atoms in the PSi volume. In some samples being studied, about 38% of the silicon atoms are estimated to be on the surface ($n = 0.38$). This corresponds to a ratio of hydrogen to silicon of up to 0.76 (Currano et al. 2009). By changing the quantity of hydrogen at the surface via thermal annealing of the freshly etched PSi at low temperatures, the amount of gas generated could be tuned. It was demonstrated (Churaman et al. 2010a) that low temperature annealing in an oxygen atmosphere had a dramatic effect on the quantity of gas generated, while the energy output was constant at about 5.5 kJ/g for all devices.

The effect of pore size on the explosive energy is well illustrated in Figure 18.3. The energy released by the oxidizer gadolinium nitrate with a pore size of 3.4 nm is significantly more than those devices with 2.3 nm and 8.2 nm pore sizes. Similar responses were observed for other oxidizers. An approximate figure of merit (FOM) was defined based on the sound of the explosion, physical damage done to the sample, the distance that particles were ejected from the sample, and the time integral of the optical signal (du Plessis 2007). Using this FOM, the relative strengths of explosions as a function of pore size for different oxidizers were estimated (du Plessis and Conradie 2006). A pore size in the region of 3 to 4 nm was optimal for devices using the oxidizers shown in Figure 18.2 (du Plessis 2008).

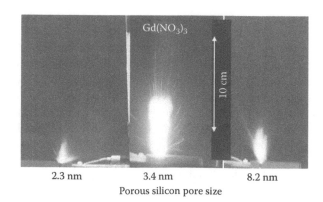

FIGURE 18.3   Optical emissions from different pore size nano-explosions using $Gd(NO_3)_3$ as oxidizer. (From du Plessis M., *Mater. Sci. Eng. B* **147**, 226, 2008. Copyright 2007: Elsevier. With permission.)

For propulsion applications, the amount of gas generated during the reaction is of importance. A calorimeter bomb chamber was used to conduct pressure measurements to determine the amount of gas generated by the $NaClO_4$ reaction (Currano et al. 2009). The measured gas generation per gram of active mass was 0.0189 mol/g. This is sufficient for use in some propellant applications. For comparison, black powder produces 0.0125 mol/g and the best modern propellants produce in the order of 0.045 mol/g. By changing the quantity of hydrogen at the surface via thermal annealing of the freshly etched PSi at low temperatures, the amount of gas generated could be tuned. It was demonstrated that low temperature annealing in an oxygen atmosphere had a dramatic effect on the quantity of gas generated in the $NaClO_4$ reaction, with almost no effect on the net energy output (Churaman et al. 2010a). The gas generation per gram of active mass was 0.0129 mol/g for devices not annealed at low temperature. When annealed at 250°C for 1 min in an oxygen environment, the gas generation was only 0.004 mol/g, a factor three smaller. This means that the removal of hydrogen from the surface of the PSi when low temperature annealing takes place in oxygen will decrease the amount of gas generation without influencing the energy released.

18.5.2 PROPAGATION VELOCITY

The measurement of the propagation velocity is problematic because there may be an erratic movement of the wave front that complicates the interpretation of the results. Certain trends, however, were identified. Propagation measurements indicated that lower porosity samples exhibit faster propagation speeds (Churaman et al. 2010a). The increased reaction velocity of the lower porosity samples is attributed to an increased speed of sound relative to the higher porosity cases (Fan et al. 2002).

Using PSi devices impregnated with sodium perchlorate, the explosive properties were investigated in detail as a function of porous layer thickness, pore size, and porosity (Plummer et al. 2011). The sample burning rates were investigated using fiber-optic velocity probes. The variation of burning rates with sample porosity was clearly demonstrated. The different porosities were prepared by varying the HF concentrations in the PSi etch between 10 and 40%. Each sample was etched at a current density of 22.5 mA/cm². Using data from the study (du Plessis 2014), the reaction velocity as a function of pore size can be plotted and is shown in Figure 18.4, where a maximum reaction velocity is achieved for a pore size of approximately 4 nm. The reaction velocity is also plotted as a function of porosity in Figure 18.5. In the confined samples, the reaction velocity peaked at a porosity of approximately 70%, but dropped sharply above this point. This correlates

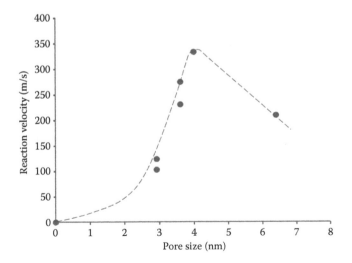

FIGURE 18.4 Reaction velocity of PSi explosive device as a function of pore size with $NaClO_4$ as oxidant. The dashed line is added to guide the eye. (From du Plessis M., *Propellants, Explos., Pyrotech.* **39**, 348, 2014. Copyright 2014: John Wiley & Sons. With permission.)

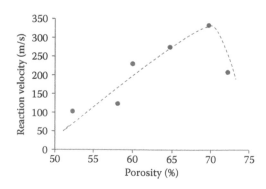

FIGURE 18.5   Reaction velocity of PSi explosive device as a function of porosity with NaClO₄ as oxidant. The dashed line was added to guide the eye.

well with a previous study (du Plessis 2008) where the reaction energy released was shown to be a strong function of pore size. It was also shown that the reaction velocity increased with porous layer thickness, reaching a plateau for layers thicker than 30 μm. Using data extracted from the study (Plummer et al. 2011), the confined reaction velocity as a function of porous layer thickness was plotted and shown to increase with layer thickness to reach a maximum for layers thicker than 30 μm (du Plessis 2014).

An interesting development demonstrated that the propagation velocity can actually be controlled by structural modifications, for example, manufacturing pillars of PSi on the surface (Parimi et al. 2012). The propagation speed for PSi with Mg(ClO₄)₂ as oxidizer could be varied from 1 to 500 m/s, depending on the structural surface profiling used. Following on this work, the effect on the reaction speed by nanoporous silicon containing microchannels was studied (Piekiel et al. 2014). The conclusion was made that the implementation of microchannels into a rapidly reacting PSi system produced significantly enhanced combustion rates. The enhanced reaction rate observed in the channeled PSi structures was thought to be a different mechanism than the change from conductive to convective heat transfer postulated earlier (Parimi et al. 2012). Evidence has been presented to strongly suggest that acoustic waves initiate and propagate combustion and that acoustic waves traveling through the PSi film carry enough energy to ignite the sensitive PSi channel structures and propagate the reaction (Piekiel et al. 2014).

18.5.3 MECHANICAL AND ELECTROSTATIC SENSITIVITY

A series of tests were performed to quantify the sensitivity of energetic nanoporous silicon filled with sodium perchlorate to mechanical shock (Churaman et al. 2008b). Beginning with half-sine shocks of 53 g for 10 ms, and increasing the shocks up to 5 131 g with duration of 0.2 ms, no spontaneous ignitions could be initiated, indicating a low sensitivity to shock.

Electrostatic discharge (ESD) sensitivity tests were carried out (Subramanian et al. 2008) on PSi explosive devices manufactured using micron-sized silicon powders and forming a network of PSi (4–5 nm nanocrystals) on the particle surface through chemical etching. NaClO₄ was used as oxidizer and the ESD testing was done according to military standards. The material was tested both as a powder and as a thin film. The discharge energy was increased up to 25 mJ with no ignitions. In another study (Thiruvengadathan et al. 2012) it was found that nanoporous silicon with NaClO₄ as oxidant pass ESD testing with at least 16 mJ energy.

18.6 APPLICATIONS OF POROUS SILICON EXPLOSIVE DEVICES

Several applications of nanostructured energetic composites have been proposed (Zhou et al. 2014) and PSi explosive devices can be used in many of these applications. One advantage of nanostructured energetic composites is its easy integration with MEMS. Table 18.4 classifies various applications according to different attributes of the explosive.

TABLE 18.4   Various Applications of Nano-Energetic Composites

Attribute	Applications
Heat release	Reactive bonding
	Micro ignition and rapid initiation
	Materials processing
	Micro power generation
Gas evolution	Micro actuation/propulsion
	Pressure mediated molecular delivery
Reaction products	Material synthesis
	Inactivation of biological agent
Others	Propellant additives
	Hydrogen production
	Nanocharges

Source: Zhou X. et al., *Appl. Mater. Interfaces* **6**, 3058, 2014. Copyright 2014: the American Chemical Society. With permission.

A number of applications for the still emerging technology of PSi explosives have been proposed (Föll et al. 2006). The first application to be considered was that of using gadolinium nitrate PSi devices to act as the excitation source, but also as the matrix for the analyte, in atomic emission spectroscopy analysis (Mikulec et al. 2002). The emission spectrum captured from a typical PSi/gadolinium nitrate explosion did not have any significant atomic emission lines over the wavelength range studied (400 to 900 nm), except for a sharp peak at 589 nm due to sodium impurity, presumably from airborne contaminants. The material functioned well as an alternative excitation source for atomic emission spectroscopy. The energy associated with the explosion was sufficient to excite atomic emission from the alkali metals and Pb, in addition to molecular emission from Ba and Sr.

An airbag initiator for the automotive industry was also developed (Clément et al. 2004, 2005) where the PSi layer filled with an oxidant was placed in direct thermal contact with a heating bridge, allowing the initiation of a standard booster charge of an industrial airbag. Fully integrated electronic detonators in the mining industry were the aim of research (du Plessis and Conradie 2006) to eventually develop PSi nano-explosive devices to be used as a primer for secondary explosives. It was shown (du Plessis 2008) that using $NaClO_4$ as oxidant, and an aluminium thin film hot wire, that the nano-explosion can be initiated within 100 µs using less than 1 mJ of electrical energy.

Nano-explosive devices can be used in self-destructive silicon chips and in the integrated circuit technology to divide silicon wafers into chips (Lazarouk et al. 2005), as well as a source of energy for silicon micro actuators (Lazaruk et al. 2007). The integration of nano-explosive PSi devices with MEMS sensor technology has already been demonstrated (Currano and Churaman 2009), as well as the first measurements of thrust generation using nanoporous energetic silicon. Energetic materials are of interest in the microelectromechanical systems (MEMS) field because of the possibility for vastly denser energy storage and the added functionality of achieving burning or detonation with on-chip integration of a MEMS device, which is not currently available with traditional MEMS.

The realization of the first nano-explosive device to propel miniature mechanical devices in the form of microrobots was achieved recently (Churaman et al. 2012). Instead of using a mechanical spring to store energy, nanoporous energetic silicon was chosen to store and release energy chemically. Microrobot mobility was demonstrated in the form of vertical thrust as stored chemical energy was rapidly converted to mechanical energy. To ignite the nanoporous energetic silicon, 180 µJ of electrical energy was consumed to release approximately 1.9 J of chemical energy.

18.7 CONCLUSIONS

It has been shown that PSi can be utilized as a nano-energetic material. Since the discovery of the first solid-state PSi explosive device in 2002, significant progress has been made to gain further insights into this emerging technology.

REFERENCES

Becker, C., Currano, L., and Churaman, W. (2009). Characterization and improvements to porous silicon processing for nanoenergetics. *U.S. Army Research Laboratory Report* ARL-TR-4717, US Army Research Laboratory, Adelphi, MD.

Becker, C.R., Currano, L.J., Churaman, W.A., and Stoldt, C.R. (2010). Thermal analysis of the exothermic reaction between galvanic porous silicon and sodium perchlorate. *Appl. Mater. Interfaces* **2**(11), 2998–3003.

Becker, C.R., Apperson, S., Morris, C.J., Gangopadhyay, S., Currano, L.J., Churaman, W.A., and Stoldt, C.R. (2011). Galvanic porous silicon composites for high-velocity nanoenergetics. *Nano Lett.* **11**, 803–807.

Churaman, W., Currano, L., Singh, A.K., Rai, U.S., Dubey, M., Amirtharaj, P., and Ray, P.C. (2008a). Understanding the high energetic behavior of nano-energetic porous silicon. *Chem. Phys. Lett.* **464**, 198–201.

Churaman, W., Currano, L., Dubey, M., and Becker, C. (2008b). Fabrication and characterization of nanoporous energetic silicon. In: *Proc. 26th Army Sci. Conf.*, December 1–4, 2008, Orlando, FL, MP-03(1–6).

Churaman, W., Currano, L., and Becker, C. (2010a). Initiation and reaction tuning of nanoporous energetic silicon. *J. Phys. Chem. Solids* **71**, 69–74.

Churaman, W.A., Becker, C.R., Metcalfe, G.D., Hanrahan, B.M., Currano, L.J., and Stoldt, C.R. (2010b). Optical initiation of nanoporous energetic silicon for safing and arming technologies. *Proc. SPIE* 7795, 779506.

Churaman, W.A., Currano, L.J., Morris, C.J., Rajkowski, J.E., and Bergbreiter, S. (2012). The first launch of an autonomous thrust-driven microrobot using nanoporous energetic silicon. *J. Microelectromech. Syst.* **21**(1), 198–205.

Clément, D., Diener, J., and Kovalev, D. (2004). Explosive porous silicon: From laboratory accident to industrial application. In: *Proc. 35th Int. Conf. ICT*, June 29–July 2, 2004, Karlsruhe, Germany, 5-1–5-11.

Clément, D., Diener, J., Gross, E., Künzner, N., Timoshenko, V.Yu., and Kovalev, D. (2005). Highly explosive nanosilicon-based composite materials. *Phys. Status. Solidi A* **202**(8), 1357–1364.

Clements, L., Puskar, L., Tobin, M.J., Harding, F., Thissen, H., and Voelcker, N.H. (2008). Preparation of chemical gradients on porous silicon by a dip coating method. *Proc. SPIE* 7267, 72670Q.

Currano, L.J. and Churaman, W.A. (2009). Energetic nanoporous silicon devices. *J. Microelectromech. Syst.* **18**(4), 799–807.

Currano, L., Churaman, W., and Becker, C. (2009). Nanoporous silicon as a bulk energetic material. In: *Proc. IEEE Transducers 2009 Conference*, Denver, CO, June 21–25, 2009, pp. 2172–2175.

du Plessis, M. (2007). Properties of porous silicon nano-explosive devices. *Sens. Actuators, A* **135**, 666–674.

du Plessis, M. (2008). Nanoporous silicon explosive devices. *Mater. Sci. Eng. B* **147**(2–3), 226–229.

du Plessis, M. (2014). A decade of porous silicon as nano-explosive material. *Propellants Explos. Pyrotech.* **39**(3), 348–364.

du Plessis, M. and Conradie, C. (2006). Nano-explosions in porous silicon. *Proc. SPIE* 6037, 60370X-1–60370X-10.

Fan, H.J., Kuok, M.H., Ng, S.C., Boukherroub, R., Baribeau, J.-M., Fraser, J.W., and Lockwood, D.J. (2002). Brillouin spectroscopy of acoustic modes in porous silicon films. *Phys. Rev. B* **65**, 165330(1–8).

Föll, H., Carstensen, J., and Frey, S. (2006). Porous and nanoporous semiconductors and emerging applications. *J. Nanomater.* **91635**, 1–10.

Fradkin, E. and Gany, A. (2012). Investigation of nanoporous silicon-based energetic materials. *Int. J. Energ. Mater. Chem. Propul.* **11**(2), 135–148.

Grosman, A. and Ortega, C. (1997). Chemical composition of "fresh" porous silicon. In: Canham L. (Ed.) *Properties of Porous Silicon.* INSPEC, London, pp. 145–153.

Herino, R. (1997). Impregnation of porous silicon. In: Canham L. (Ed.) *Properties of Porous Silicon.* INSPEC, London, pp. 66–76.

Herino, R., Bomchil, G., Barla, K., Bertrand, C., and Ginoux, J.L. (1987). Porosity and pore size distributions of porous silicon layers. *J. Electrochem. Soc.* **134**(8), 1994–2000.

Kent, J.P. and Prasad, J. (2008). Microelectronics for the real world: "Moore" versus "More than Moore." In: *Proc. IEEE Custom Integrated Circuits Conference*, September 21–24, 2008, San Jose, CA, pp. 395–402.

Koch, E.-C. and Clément, D. (2007). Special materials in pyrotechnics: VI. Silicon—An old fuel with new perspectives. *Propellants Explos. Pyrotech.* **32**(3), 205–212.

Kovalev, D., Timoshenko, V.Y., Künzner, N., Gross, E., and Koch, F. (2001). Strong explosive interaction of hydrogenated porous silicon with oxygen at cryogenic temperatures. *Phys. Rev. Lett.* **87**(6), 68301/1–68301/4.

Lazarouk, S.K., Dolbik, A.V., Jaguiro, P.V., Labunov, V.A., and Borisenko, V.E. (2005). Fast exothermic processes in porous silicon. *Semicond.* **39**(8), 881–883.

Lazaruk, S.K., Dolbik, A.V., Labunov, V.A., and Borisenko, V.E. (2007). Combustion and explosion of nanostructured silicon in microsystem devices. *Semicond.* **41**(9), 1113–1116.

Malec, C.D., Voelcker, N.H., Shapter, J.G., and Ellis, A.V. (2010). Carbon nanotubes initiate the explosion of porous silicon. *Mater. Lett.* **64**, 2517–2519.

Manilov, A.I. and Skryshevsky, V.A. (2013). Hydrogen in porous silicon—A review. *Mater. Sci. Eng., B* **178**, 942–955.

Mason, B.A., Son, S.F., Cho, K.Y., Yetter, R.A., and Asay, B.W. (2009). Combustion performance of porous silicon-based energetic composites. In: *Proc. 45th AIAA/ASME/SAE/ASEE Joint Propul. Conf.*, August 2–5, 2009, Denver, CO.

McCord, P., Yau, S.-L., and Bard, A.J. (1992). Chemiluminescence of anodized and etched silicon: Evidence for a luminescent siloxene-like layer on porous silicon. *Science* **257**, 68–69.

Mikulec, F.V., Kirtland, J.D., and Sailor, M.J. (2002). Explosive nanocrystalline porous silicon and its use in atomic emission spectroscopy. *Adv. Mater.* **14**, 38–41.

Milewski, P.D., Lichtenwalner, D.J., Mehta, P., Kingon, A.I., Zhang, D., and Kolbas, R.M. (1994). Light emission from crystalline silicon and amorphous silicon oxide (SiO$_x$) nanoparticles. *J. Electron. Mater.* **23**(1), 57–62.

Morris, C.J., Laflin, K.E., Churaman, W.A., Becker, C.R., Currano, L.J., and Gracias, D.H. (2012). Initiation of nanoporous energetic silicon by optical triggered, residual stress powered microactuators. In: *Proc. IEEE 25th Int. Conf. Micro Electro Mech. Syst.* January 29–February 2, 2012, Paris, France, pp. 1245–1248.

Ohkura, Y., Weisse, J.M., Cai, L., and Zheng, X. (2013). Flash ignition of freestanding porous silicon films: Effects of film thickness and porosity. *Nano Lett.* **13**, 5528–5533.

Parimi, V.S., Tadigadapa, S.A., and Yetter, R.A. (2012). Control of nanoenergetics through organized microstructures. *J. Micromech. Microeng.* **22**, 055011.

Parimi, V.S., Tadigadapa, S.A., and Yetter, R.A. (2014). Effect of substrate doping on microstructure and reactivity of porous silicon. *Chem. Phys. Lett.* **609**, 129–133.

Piekiel, N.W., Churaman, W.A., Morris, J., and Currano, L.J. (2013). Combustion and material characterization of porous silicon nanoenergetics. In: *Proc. IEEE 26th Int. Conf. Micro Electro Mech. Syst.* January 20–24, 2013, Taipei, Taiwan, pp. 449–452.

Piekiel, N.W., Morris, C.J., Currano, L.J., Lunking, D.M., Isaacson, B., and Churaman, W.A. (2014). Enhancement of on-chip combustion via nanoporous silicon microchannels. *Combust. Flame* **161**, 1417–1424.

Plummer, A., Cao, H., Dawson, R., Lowe, R., Shapter, J., and Voelcker, N.H. (2008). The influence of pore size and oxidising agent on the energetic properties of porous silicon. *Proc. SPIE 7276*, 72670P.

Plummer, A., Kuznetsov, V., Joyner, T., Shapter, J., and Voelcker, N.H. (2011). The burning rate of energetic films of nanostructured porous silicon. *Small* **7**(23), 3392–3398.

Plummer, A., Kuznetsov, V.A., Gascooke, J., Shapter, J., and Voelcker, N.H. (2014). Laser shock ignition of porous silicon based nano-energetic films. *J. Appl. Phys.* **116**, 054912.

Rossi, C., Zhang, K., Estève, D., Alphonse, P., Tailhades, P., and Vahlas, C. (2007). Nanoenergetic materials for MEMS: A review. *J. Microelectromech. Syst.* **16**(4), 919–931.

Subramanian, S., Tiegs, T., Limaye, S., Kapoor, D., and Redner, P. (2008). Nanoporous silicon based energetic materials. In: *Proc. 26th Army Sci. Conf.* December 1–4, 2008, Orlando, FL, MP-08(1–7).

Thiruvengadathan, R., Belarde, G.M., Bezmelnitsyn, A., Shub, M., Balas-Hummers, W., Gangopadhyay, K., and Gangopadhyay S. (2012). Combustion characteristics of silicon-based nanoenergetic formulations with reduced electrostatic discharge sensitivity. *Propellants Explos. Pyrotech.* **37**, 359–372.

Uhlir, A. (1956). Electrolytic shaping of germanium and silicon. *Bell Syst. Tech. J.* **35**, 333–347.

Wang, S., Shen, R., Ye, Y., and Hu, Y. (2012). An investigation into the fabrication and combustion performance of porous silicon nanoenergetic array chips. *Nanotech.* **23**, 435701.

Wang, S., Shen, R., Yang, C., Ye, Y., Hu, Y., and Li, C. (2013). Fabrication, characterization, and application in nanoenergetic materials of uncracked nano porous silicon thick films. *Appl. Surf. Sci.* **265**, 4–9.

Zhou, X., Torabi, M., Lu, J., Shen, R., and Zhang, K. (2014). Nanostructured energetic composites: Synthesis, ignition/combustion modeling, and applications. *Appl. Mater. Interfaces* **6**, 3058–3074.

Index

Page numbers followed by f and t indicate figures and tables, respectively.

A

Accelerometers, 253
AC conductivity, of PSi, 122–123
Acoustic holes, for dynamic response, 133
Activated-carbon porous electrodes, 356, 364
Active device isolation, 110–112, 111f, 112f
  FIPOS, 110
  IPOS, 110, 111f
  n-type process, 110, 111f
  p-type process, 110
Active devices, hybrid substrates for, 119–120, 120f
Active PSi and top electrode
  compact PSi between, 23, 23f
  other materials between, 24–25, 24f–25f
Advanced porous silicon membrane (APSM)
  Bosch process, 138–139, 138f
Advanced silicon etching (ASE), 312
AFM microscopy, 95, 324
Ag/PSi chip, catalytic activity of, 303
Airbag igniters, 253
Al/Au contact, EL emission, 22, 23f
Alcaline texturization *vs.* macroporous texturization, of Si wafers, 214–216, 214f–217f
ALD, *see* Atomic layer deposition (ALD)
Alkaline etching technique, 265–266
Alkaline fuel cells (AFC), 252t
All-optical chip, 71
"All-Si" tandem solar cells, 232, 232f
Alpha particles, 245
Alumina wafers, 255
Aluminum (Al), as top contact, 22, 22f
Aluminum electrolytic capacitors, 349
Aluminum nitrate, 392
3-aminopropyltriethoxysilane, 94
Amino propyltriethoxysilane (APTES), 95
Aminopropyl-trimethoxy silane (APTMS), 305
Ammonia borane, 292
Annealing, high-pressure water vapor (HWA), 18, 18f, 19f
Anode(s), 250, 251f
  Li-battery, PSi-based
    micro-particles, 337–338
    SiNWs, 333–337, 334t, 335f–336f, 336t
    synthesis and performances of, 327–338
    thin films, 328–333
  Li-battery, textured Si-material in, 321–327
    advantages, 326–327, 326t, 327f
    binders in composite, 325–326
    charge-discharge profile, 321–322, 321f
    depth of discharge, 325
    designs and reported performance, 326t

    lithiation/de-lithiation mechanisms, 321–322, 322f
    performance improvement, 324–326
    principles and characterization, 321–324
    SEI formation, 323–324, 323f
    volume expansion of Si-based composites, 324, 325f
Anodic dissolution, PSi electrodes by, 360
Anodic oxidation, 6–8, 7f, 8f
Anodic polarization, of PSi, 6–8
  chemical reaction, 8
  hole injection, 6–8
Antireflecting covering, silicon photodetectors with PSi-based, 43–45, 44f–45f
Antireflection, 90
Antireflection coating (ARC), 89–92, 90f–91f, 212
  improvement in, 92
  lifetime of, 91
  nano PSi as
    advantages, 218
    background, 218
    disadvantages, 220
    influence on solar cell performance, 219–221, 219f–220f
    PSi emission and quantum efficiency of solar cells, 221
    requirements for, on Si solar cells, 217–218
    SiNx:H layer, 220–221
    technology of, 218–219, 219f
  properties, 91
  screen-printed multicrystalline silicon solar cell with, 92
Anti-resonant reflecting optical waveguide (ARROW), 98
APCVD (atmospheric pressure chemical vapor deposition), 147
APTES (amino propyltriethoxysilane), 95
ARC, *see* Antireflection coating (ARC)
As-formed PSi, devices including, 13, 13t
Atmospheric pressure chemical vapor deposition (APCVD), 147
Atomic layer deposition (ALD), 184
Atomic-resolution electron holography, 166
Auger electron spectroscopy (AES), 322
Au/TiO$_2$ catalyst, 312

B

Ballistic electron surface-emitting display (BSD), 172–173, 173f
Band diagram, PSi-based cold cathodes, 169f
Batteries
  classical chemical, 241

  Li-ion, with PSi-based electrodes, 319–339, 320f; *see also* Li-ion batteries, with PSi-based electrodes
    hybrid structures performances, 338–339
    overview, 320–321
    synthesis and performances of anodes, 327–338
    textured silicon material in anode, 321–327
  micro lithium, 241
  radioactive, 240, 240f
  vs. supercapacitors, 348
Beta emitters, 245, 245t
Betavoltaic effect, 240–241
Betavoltaics/betavoltaic devices, 239–246
  advantages, 240–241
  applications, 241
  background, 240, 241
  3D design, 242–243
  design, 241
  disadvantages, 241
  efficiency *vs.* semiconductor bandgap, 242, 242f
  functioning, 240
  microbatteries, 240–241, 240f, 243f
    electron flux density, 241
    PSi-based, 242–244, 243f–244f
    radioisotopes for PSi-based, 245–246, 245t
  optimization via PSi, 242–245, 243f–245f
  overview, 240–241
  principle of, 240
  radioactive material used in, 241
  radioisotopes for, 245–246, 245t
Binders, in Si-based composites, 325–326
Biosensing, microreactors, 298
Biosensors, diffraction-based, 93–94, 94f
Bipolar plates, fuel cells, 255
Black silicon, 314–315, 314f
Boltzman transport equation, 380
Boron-doped silicon wafers, 81
Boron-silica-glass cantilevers, 132–133, 133f
Bottom up approach, SiNWs, 229
Bragg grating waveguide, fabrication, 96–97, 97f
Bragg mirror, 225, 225f
  reflection coefficient, 225–226, 226f
Bragg reflector, 54, 60, 100f
  application, 59
    for Si solar cells, 225–226, 225f–226f
  concave PSi-based, 89
  reflectance spectra of, 55, 56f
  in solar cells, PSi as, 147
Bruggeman approximation, 121

Bulk micromachining technique, 89, 130; *see also* Surface micromachining (SMM)
Bulk 2D photonic crystals, 60–63, 61f–64f
Buried oxide (BOX) layer, 144, 145

C

Canon Inc., 144
Cantilevers
  applications, 131
  boron-silica-glass, 132–133, 133f
  fabrication, 131–132, 132f
  SMM and, 131–133, 132f–133f
Capacitors, 119; *see also* Supercapacitors
  double-layer, 348; *see also* Electrochemical double-layer capacitor (EDLC)
  "electric-wire-in-cylinder" model, 356
  electrochemical, 348; *see also* Electrochemical capacitors (EC)
  electrostatic, 349
  pseudo-capacitor, 355
  vs. supercapacitors, 349–350
Capping, EL stabilization by, 26
Carbon, as electrode materials, 356
Carbon-based explosives, 388; *see also* Explosive devices
Carbon-coated materials, 338–339, 338f
Carbon materials deposition, on/in PSi, 200–202
  carbon nanotubes, 201
  fullerenes C60, 200–201
  graphite/graphene, 201–202
Carbon nanostructures; *see also specific types*
  hydrogen interaction, 274, 275
Carbon nanotubes (CNTs), 168, 178, 201, 364; *see also* Multiwalled carbon nanotubes (MWCNTs); Single wall carbon nanotubes (SWNT)
  optoelectronic properties, 46
  vs. high surface area graphite, 274
Carbon paste electrode (CPE), 304
Carboxymethylcellulose (CMC), 325
Catalysis, heterogeneous
  advantages, 300–301
  catalysts for microreactors, 299–300
  PSi matrix for, 300–307
    metal catalysts, 301–304, 302f
    photocatalysis, 304–307, 305f–306f
Catalyst deposition method, 312–313
Catalysts
  fuel cell, 250
  microreactors, 299–300
Cathode, 250, 251f; *see also* Cold cathodes
  Li-battery, 320
Cathode ray tubes (CRTs), 166, 166f
Cathodic polarization, of PSi with persulfate ions, 8–10, 9f
Cavities; *see also* Microcavities
  SMM for, 136–137, 136f–137f
CBE (chemical beam epitaxy) method, 153
CCD (charge-coupled device) arrays, 86
CdSe/CdS/ZnS QDs, 43
Cell phones, 250
Cetyl trimethyl ammonium chloride (CTAC), 329, 329f
Charge carrier transport
  in electrolytic systems, 10–11, 10f, 11f
  in solid-state devices, 12, 12f

Charge-discharge profile, of silicon-carbon based composite, 321–322, 321f
Chemical batteries, 241
Chemical beam epitaxy (CBE) method, 153
Chemical etching, 167, 299
  metal-assisted, 305–306
  for Si wafers texturization, 215, 216f
Chemically stimulated desorption, of hydrogen, 281–284, 282f, 284f
Chemical-mechanical polishing (CMP), 153
Chemical oxidation, 16, 16f
Chemical sensing applications, DOEs and, 93
Chemical storage, of hydrogen, 275–276; *see also* Hydrogen storage
Chemical vapor deposition (CVD), 112, 184
Chemical vapor deposition in ultra-high vacuum (UHV-CVD), 194
Chipfilm™ layer transfer process, 138, 138f
Chip-scale hybrid photodetectors, 46
Chronopotentiometry (CP), 354
CMOS circuitry, 36
CMOS-compatible processing in microelectronics, PSi for, 130–131
CMOS-compatible RF circuits, 139, 139f
CMOS (complementary metal oxide semiconductor) devices, 110, 376
CMOS technology, 52, 78, 119, 130, 173, 184
CNTs, *see* Carbon nanotubes (CNTs)
CNTs/PSi-based photodetectors, 46
Cobalt (Co), 185–189
Coherent porous silicon (CPS), 260–261
Cold cathodes, 165–179
  applications, 166–168
  approaches for, 168
  development of, 168
  disadvantages of, 168
  electric propulsion (EP) environments, 168
  electrode configurations, 168
  fabrication methods and, 166–168
  Fowler–Nordheim law, 166, 170
  gated FEA, cell in, 167f
  low voltage applications, 173
  overview, 166–168
  postemission acceleration schemes, 168
  principle of, 166
  PSi as substrate for new emitter materials, 178–179
  PSi-based, 169–173
    advantages, implementation of, 172–173
    background, 169
    band diagram, 169f
    BSD, 172–173
    electron emission process, 169–170, 169f
    energy distribution curves, 170f
    features, 172
    microscopic representation, 170, 171f
    multilayered and graded-multilayered PSi diodes, 171–172, 172f
    RTO process, 171
    structure and experimental configuration, 169f
    thickness of PSi layer and, 170–171, 171f
  Si-based
    examples, 167–168, 167f
    with Si-tips modified by porous layer, 173–177, 174f–177f
  Spindt-type, 167, 167f, 168
  work function, 166
Colloidal QDs, 42–43

Color-sensitive photodetectors, 45; *see also* Photodetectors
Compact PSi, between active PSi and top electrode, 23, 23f
Complementary metal oxide semiconductor (CMOS) devices, 110, 376
Complex hydrides, 276
Composite-based photodetectors, 42–43, 43f
Composite materials, 184; *see also* PSi-based composites
  vs. hybrid materials, 184
Concave mirrors
  applications, 88
  fabrication, 88–89, 89f
Conductive oxides, as top electrodes, 22, 22f–23f
Co_3O_4-ZnO catalyst, 310–311
Coplanar waveguides (CPW), 113, 114, 114f
  normalized power loss, 114, 115f
  RF electrical characteristics of, 114, 115t
  strips, 114, 114f
CO-PrOx microreactor, 312–313
CPS, *see* Coherent porous silicon (CPS)
CPW, *see* Coplanar waveguides (CPW)
CRTs, *see* Cathode ray tubes (CRTs)
Cryogenic liqui, 274
Cu/PSi composite powder, 303–304
Current gain, 36
Current–voltage curves, PSi-based photodetectors, 39–40, 39f
CVD, *see* Chemical vapor deposition (CVD)
Cyclic-voltammetry (CV) techniques, 354
Cyclo-1,3,5-trimethylene-2,4,6-trinitramine (RDX), 392

D

Dark current, 36
DBR, *see* Distributed Bragg reflector (DBR)
DC conductivity, of PSi, 122, 123f
Decay rate, radioisotopes, 246
Deep reactive ion etching (DRIE), 253, 256, 313, 314, 357
  silicon bipolar plate, 266–267, 266f–267f
Defects, in 2D photonic crystals, 63–65, 65f, 66f
Density functional theory, 380
Department of Energy (DOE), 274
Depth of discharge (DOD), 325
Desorption, of hydrogen from PSi
  chemically stimulated, 281–284, 282f, 284f
  photocatalytic evolution, 285
  thermally stimulated, 280–281, 281f
DHFC (direct hydrogen fuel cells), 251, 252t, 258, 258f
Diamond, 168
Diamond films
  application of, 158
  function and high electric strength, 158
  heteroepitaxial growth of, 157–158
Diamond-like carbon (DLC) layer, 178, 218–219
Dielectric permittivity, of PSi, 120, 121, 122t
Differential scanning calorimetry (DSC) techniques, 396
Diffraction-based biosensors, 93–94, 94f
Diffraction optical elements (DOE), 93–95, 93f–94f
  applications, 93
  chemical sensing applications, 93
  photolithography technique for, 93
  uses, 93

Dimethyl formamide (DMF), 329
Direct ethanol fuel cells (DEFC), 251, 260
Direct fuel cells, 251; *see also* Fuel cells (FC)
Direct hydrogen fuel cells (DHFC), 251, 252t, 258, 258f
Direct methanol fuel cells (DMFC), 251–252, 252t
Distributed Bragg reflector (DBR), 79–83, 79f
  electrochemical polishing process, 81
  FESEM image, 81, 81f
  IR mesoporous silicon filters, 81–82, 82f
  mesoporous, 81–82, 82f
  porous structure (categories), 81
  step-by-step preparation, 80, 80f
  vs. rugate filters, 82–83, 83f
DLC layer, *see* Diamond-like carbon (DLC) layer
DMFC (direct methanol fuel cell), 251–252, 252t
DOD (depth of discharge), 325
DOE, *see* Diffraction optical elements (DOE)
Double-layer capacitors, 348
Double-layer range/window, 351
Double-tank electrochemical etching, 87
Dry etching method, 89, 299
DSC (differential scanning calorimetry) techniques, 396

E

E-beam lithography, 94
ECO, *see* Electrochemical oxidation (ECO)
Eddy currents, 113, 116
EDLC, *see* Electrochemical double-layer capacitor (EDLC)
Effective optical thickness (EOT) distribution, 87, 87f
EIS, *see* Electrochemical impedance spectroscopy (EIS)
Electrical conductivity, 380
  isolating materials, 120
Electrical contacts, to PSi, organic substances for, 199–200, 200f
Electrical energy storage devices, Ragone plot for, 348f
Electrical isolation applications, of PSi, 109–123
  to active device isolation, 110–112, 111f, 112f
  electrical properties of PSi and, 120–123
    AC conductivity, 122–123
    DC conductivity, 122, 123f
    dielectric permittivity, 121, 122t
  overview, 109
  to RF passive devices, 113–120
    general considerations on device processing, 113
    hybrid substrates for, 119–120, 120f, 121f
    inductors, 116–119, 116f, 117t, 118f
    interconnect structures, 113–114, 114f, 115f, 115t
    RF functions, 119
  with sacrificial PSi layer, 139–140, 139f
Electrical properties, of PSi
  AC conductivity, 122–123
  DC conductivity, 122, 123f
  dielectric permittivity, 121, 122t
Electric propulsion (EP) environments, 168

"Electric-wire-in-cylinder" model, of capacitor, 356
Electroactive polymers, 356
Electrochemical anodization process, 88
Electrochemical capacitors (EC), 348; *see also* Electrochemical double-layer capacitor (EDLC); Supercapacitors
  background, 348–349
  electrode materials, 355–356
  vs. supercapacitors, 348
Electrochemical deposition method, 185
Electrochemical double-layer capacitor (EDLC), 348
  aqueous electrolytes in, 351
  charge density, 351
  cyclic-voltammetry curves, 354, 354f
  differential capacitance, 351
  electrode materials, 355–356
  electrolytes in, 351
  equivalent-circuit model of, 352–353, 352f–353f
  Gouy–Chapman diffuse layer, 350, 351f
  Helmholtz layer, 350–351, 351f
  inner Helmholtz plane (IHP), 350, 351f
  outer Helmholtz plane (OHP), 350, 351f
  with PSi electrodes, electrochemical performance of, 366t
  self-discharge, 353
  Stern–Grahame model, 350
  vs. supercapacitors, 350, 350f
Electrochemical etching methods, 54, 60, 80, 95, 178, 307, 310, 328, 357, 388; *see also specific types*
  double-tank, 87
  microlens and, 87
  parameters of explosive devices, 389, 389t
  single-tank, 87
  thin porous layer formation, 218, 219f
Electrochemical hydrogenation, 223
Electrochemical impedance spectroscopy (EIS), 339, 352
Electrochemical oxidation (ECO), 6–7, 16–17, 17f, 223
  effect on EL efficiency, 17, 17f
  EL stability and, 17
Electrochemical performance, of PSi-based supercapacitors, 364–371, 365t, 367t
  integrated on-perforated-silicon-chip micro-supercapacitors, 364–368, 366f, 367f
  with PSi electrodes, 368–371, 369f, 370f
Electrochemical polishing process, DBR, 81
Electrochemical texturization, of Si wafers, 214–215, 215f
Electrodes; *see also* Anode; Cathode; Top electrode
  cold cathodes, configurations, 168
  fuel cells, 250, 251f
  materials, 349
    for supercapacitors, 351, 355–356
  PSi-based, Li-ion batteries with, *see* Li-ion batteries, with PSi-based electrodes
  PSi integration in, 255–263
    macroporous silicon, 260–263
    mesoporous silicon, 256–259
  supercapacitors, fabrication, 360–364
    by anodic dissolution, 360
    by metal-assisted etching, 361–362, 361f–363f
    PSi nanowires as, 362–364, 363f

Electroluminescence (EL), of PSi, 4, 52
  background of, 5, 5t–6t
  ECO and, 17, 17f
  with persulfate ions, 8–10, 9f
  voltage-induced spectral shift, 9, 9f
Electroluminescence (EL) devices, 3–30
  background, 5, 5t–6t
  characteristics, 5
  charge carrier transport
    in electrolytic systems, 10–11, 10f, 11f
    in solid-state devices, 12, 12f
  electrolytic systems
    anodic polarization of PSi, 6–8
    cathodic polarization of PSi with persulfate ions, 8–10, 9f
  EL modulation speed, 28–29, 28f
  EL stabilization by
    PSi capping, 26
    PSi surface modification, 26–27, 26f
  EPE of, 5
  EQE of, 5
    requirements and status, 5t
  including as-formed PSi, 13, 13t
  integration of PSi EL, 29–30, 29f
  overview, 4–5
  partially oxidized PSi, 15–18, 15t
    chemical and thermal oxidation, 16, 16f
    electrochemical oxidation, 16–17, 17f
    high-pressure water vapor annealing, 18, 18f, 19f
    porosified p-n junctions, 13–15, 14f, 14t, 15f
  problems of, 5
  PSi impregnation, 18–21, 19t
    inert materials incorporation, 21
    metals incorporation, 20
    polymers incorporation, 20–21, 20f–21f
  top electrode, influence of, 21–25, 22t
    compact PSi between active PSi and, 23, 23f
    metals and conductive oxides as, 22, 22f–23f
    other materials between active PSi and, 24–25, 24f–25f
  tuning and narrowing EL spectrum with microcavities, 27–28, 27f
Electrolytes
  aqueous, 351
  in EDLC, 351
  fuel cell, 250
  in supercapacitors, 351
Electrolytic systems
  anodic polarization of PSi, 6–8
  cathodic polarization of PSi with persulfate ions, 8–10, 9f
  characteristics, 6t
  charge carrier transport in, 10–11, 10f, 11f
Electromagnetic interference (EMI) filter, 119–120, 121f
Electromagnetic waves, 53
Electromagnetism, 53
Electron emission process, PSi-based cold cathodes, 169–170, 169f
Electron flux density, of betavoltaic microbatteries, 241
Electron-hole pairs (EHPs), 241
  generation mechanism in solid-state PSi-based EL devices, 12, 12f
Electronic waves, 53

Electrostatic capacitors *vs.* supercapacitors, 349
Electrostatic discharge (ESD) protection diodes, 119–120
Electrostatic sensitivity, of explosion, 398
EL modulation speed, 28–29, 28f
EL spectrum with microcavities, tuning and narrowing, 27–28, 27f
EL stabilization
    by PSi capping, 26
    by PSi surface modification, 26–27, 26f
ELTRAN® (epitaxial layer transfer) technology, 112, 137, 144, 229
    development of, 146
    homoepitaxial growth on PSi layer and, 146
Emission spectrum, EL
    tuning and narrowing, 27–28, 27f
Emitters
    beta, 245, 245t
    homogeneous, 221
    nanoscale, 174, 175
    on-chip Si-based light, 4
    PSi as substrate for new materials, 178–179
    PSi-based, advantage of, 172
    selective, 221–222, 222f
    Si-based, 4
        comparison of turn-on fields for, 177, 178t
        visible light, 4
    thermal, 135
Energetic analysis, of PSi nanostructures, 291–292, 291t, 292f
Energetic materials; *see also* Explosive devices
    classification, 388
Energy yield, of explosion, 395–397, 396f
Environmental remediation, 304
Epitaxial films growth, PSi and, 143–159
    heteroepitaxial, 148–159
        diamond films, 157–158, 158t
        germanium and silicon-germanium alloys, 157
        of III-V compound semiconductors, 152–156
        of II-VI and IV-VI compound semiconductors, 148–152, 149f–150f, 151t, 152f
        other materials, 159
    homoepitaxial, 144–148, 144f
        APCVD and, 147
        buried Bragg reflector layer, 144, 147
        buried oxide (BOX) layer, 144, 145
        ELTRAN technology for, 146
        epitaxial layer detachment, 146
        IAD for, 147
        layer transfer technology for, 146
        LEPECVD and, 147
        low defect density in epitaxial films and, 145
        LPCVD and, 147
        LPE for, 147
        MBE, 145
        meso-PSi layers, 145
        numerical simulation methods, 146
        optimal conditions for, 145
        planar growth substrates for, 147
        PSi porosity and, 145, 147
        sacrificial layer, 144
        single-crystal silicon, 144–148, 144f
        sintering process, 145, 146
        substrate conductivity and, 145–146
        techniques used for, 145

thickness of epitaxial film and, 145, 147
thin-film silicon solar cells formation, 145–148
"zipper" layer, 146
overview, 144
Equivalent-circuit model, EDLC, 352–353, 352f–353f
Etching, *see* specific methods
Ethanol, 251, 392
Ethanol steam reformer, 312–313
Explosive devices, PSi, 387–399
    applications, 398–399, 399t
    device manufacture, 389–394
        internal surface stabilization, 391
        oxidants for, 391–392, 392t, 393f
        pore impregnation, 393–394
        properties, 389–391, 389t–390t, 390f
    history of technology, 388–389
    ignition of oxidants, 394–395
    overview, 388
    properties
        electrostatic sensitivity, 398
        energy yield, 395–397, 396f
        mechanical sensitivity, 398
        propagation velocity, 397–398, 397f–398f
Explosives, 388
External power efficiency (EPE), 5
External quantum efficiency (EQE), 4, 14–15
    defined, 5
    of photodiodes, 45
    of Si solar cells, porous Si ARC influencing, 220

F

Fabrication, 42; *see also specific methods*
    cantilevers, 131–133, 132f, 133f
    of capacitive microphone, 133–134, 134f
    concave mirrors, 88–89, 89f
    2D-photonic crystals, 60–63
        electrochemical etching method, 60
        proton beam writing method, 60–61, 61f
        using macroporous silicon, 62
        using macroporous silicon with standard lithography, 62–63, 63f
    3D photonic crystals, 68–71, 69f–70f
        lithographic prestructuring process, 68
        macroporous silicon method, 68–69, 69f–70f
    of light-emitting devices, 1D photonic crystals in, 58
    micro-probe card, 131–132, 132f
    microreactors, 299
    PSi-based supercapacitors, 357–364
        integrated on-perforated-silicon-chip micro-supercapacitors, 357–360, 358f–359f
        with PSi electrodes, 360–364
    PSi waveguide, 96–97, 97f
    silicon photodetectors, 43–44
    UV macroporous filter, 83–85, 84f–85f
Fabry–Perot interference filter, 83–85, 84f–85f, 103
Fabry–Perot resonator, 29
FC, *see* Fuel cells (FC)
FDTD (finite-difference time domain) algorithms, 53
FEAs, *see* Field emitter arrays (FEAs)

FEDs, *see* Field emission flat panel displays (FEDs)
Ferromagnetic metals, 185–189, 186f–188f
Fiber optical sensor, using PSi diffraction grating, 93, 93f
Field emission
    Fowler–Nordheim law of, 166
    phenomenon of, 166
Field emission devices, Si-based, 176t
Field emission flat panel displays (FEDs), 166, 166f
Field emitter arrays (FEAs), 166
Figures-of-merit (FoM), 114, 376, 396
Fill factor (FF), 217
Filtered photodetectors, 1D photonic crystals as, 59, 59f
Filters, 119
    optical, 79–86
        distributed Bragg reflector filters, 79–83, 79f–83f
        Fabry–Perot interference filter, 83–85, 84f–85f
        long wave pass filters, 85–86, 86f
        overview, 79
    PSi-based interference, 43–45, 44f–45f, 54
    rugate, 55, 82–83, 83f
Fine-resolution electron microscopy, 166
Finite-difference time domain (FDTD) algorithms, 53
Finite element method, inductors geometries and, 119
Finite 2D photonic crystals (slabs), 65–68, 66f–67f
    "air-bridge," 67
    light of cone, 66
FIPOS, *see* Full insulation porous oxidized silicon (FIPOS)
Flash ignition, 395
Flat panel displays (FPDs), 166, 172
Flow-through membrane PSi microreactors, 308–313, 308f–311f
Formic acid, 8
Formic acid fuel cells (FAFC), 251
Fourier transform infrared (FTIR) spectroscopy, 302, 306, 323
    hydrogen contents estimation in PSi by, 276–279, 277f–280f
Fowler–Nordheim law, 166, 170
FPDs, *see* Flat panel displays (FPDs)
Fraunhofer, 89–90
Free elements, SMM and, 135, 135f
Freestanding mechanical elements, SMM and, 131–135
    cantilevers, 131–133, 132f–133f
    free elements, 135, 135f
    microphones, 133–134, 134f
    thermal emitters, 135
Freestanding monocrystalline thin film silicon (FMS) process, 229
Fresnel equation, 89
FTIR spectroscopy, *see* Fourier transform infrared (FTIR) spectroscopy
Fuel cells (FC), 249–268; *see also* Micro-fuel cells
    advantages, 250
    anode, 250, 251f
    bipolar plates, 255
    catalysts, 250
    cathode, 250, 251f
    classification, 251–252, 252t
    described, 250–253

designing materials, 253–255
  metals, 253–254
  polymers, 254–255
direct, 251
electrodes, 250, 251f
electrolyte, 250
energy densities, 251t
gas diffusion layer, 255
gases used in, 251
H₂-O₂, 250
hydrogen in, 250
hydrogen oxidation, 250–251
indirect, 251
miniaturization of, 253
need for, 250
overview, 250
oxygen reduction, 251
power density *vs.* energy sources, 250t
PSi-based, 255–268
  integration in electrodes, 255–263
  macroporous silicon, 260–263
  mesoporous silicon, 256–259
  micro-machined Si-based supports for
    small fuel cells, 264–268, 266f–268f
  PEM, 264, 265f, 265t
  PSi integration in, 253–255, 254f
  supporting material, 255
  working principle, 250–251, 251f
Fullerenes C60, 200–201
Full insulation porous oxidized silicon (FIPOS),
  78, 110, 145

G

GaAs layer, heteroepitaxial growth, 152–153
Ga/Au contact, EL emission, 22, 23f
Gamma ray, 245
GaN layer, heteroepitaxial growth, 153–155,
  154f–156f
Gas diffusion layer (GDL), 255
Gases, in fuel cells, 251
Gasoline, 251
Gas-phase carbonization technique, 339
Gas sensor applications, of cold cathodes, 166
Gas sensors, 253
General Electric Company, 348
Germanium
  heteroepitaxial films of, 157
  and Si deposition into PSi, 194–195
Gettering process, of Si wafers, 213–214
Glucose monitoring microreactors, 298
Gold (Au), 13
  semitransparent, as top electrode, 22
Gold nanoparticles, 312
Gouy–Chapman diffuse layer, 350, 351f
Gradient refractive index (GRIN) microlens, 87
  effective optical thickness (EOT)
    distribution, 87, 87f
  interferogram showing Newton's rings, 88, 88f
  planar, 88
Graphene/PSi-based photodetectors, 46, 47f
Graphite/graphene, 46, 201–202
  CNTs *vs.* high surface area, 274
Ground signal ground (GSG) probes, 114
Guest substances, 184

H

Half-life, radioisotopes, 240
Heavily doped PSi *vs.* lightly doped PSi, 7

Helmholtz layer, EDLC, 350–351, 351f
Helmoltz capacitance, 9
Heteroepitaxial films, growth on PSi layer,
  148–159
  diamond films, 157–158, 158t
  germanium and silicon-germanium alloys,
    157
  of III-V compound semiconductors
    GaAs, 152–153
    GaN, 153–155, 154f–156f
    InSb, 153
  of II-VI and IV-VI compound
    semiconductors, 148–152, 149f–150f,
    151t, 152f
  other materials, 159
Heterogeneous catalysis/catalysts
  advantages, 300–301
  microreactors, 299–300
  PSi matrix for, 300–307
    metal catalysts, 301–304, 302f
    photocatalysis, 304–307, 305f–306f
HFCVD method, 157–158
High performance liquid chromatography
  (HPLC), 306
High-pressure gaseous hydrogen, 274
High-pressure water vapor annealing (HWA)
  advantages, 27
  oxidation by, 18, 18f, 19f
High resolution transmission electron
  microscopy (HRTEM), 323
High-sensitivity photodetectors, 36
High-speed photodetectors, 36
High-temperature applications, of cold
  cathodes, 166
H₂-O₂ fuel cells, 250
Hole injection, 6–8
Holey silicon, 380–381
Holography, 94–95
Homoepitaxial growth, on PSi layer, 144–148, 144f
  APCVD and, 147
  buried Bragg reflector layer, 144, 147
  buried oxide (BOX) layer, 144, 145
  ELTRAN technology for, 146
  epitaxial layer detachment, 146
  IAD for, 147
  layer transfer technology for, 146
  LEPECVD and, 147
  low defect density in epitaxial films and, 145
  LPCVD and, 147
  LPE for, 147
  MBE, 145
  meso-PSi layers, 145
  numerical simulation methods, 146
  optimal conditions for, 145
  planar growth substrates for, 147
  PSi porosity and, 145, 147
  sacrificial layer, 144
  single-crystal silicon, 144–148, 144f
  sintering process, 145, 146
  substrate conductivity and, 145–146
  techniques used for, 145
  thickness of epitaxial film and, 145, 147
  thin-film silicon solar cells formation,
    145–148
  "zipper" layer, 146
Homogeneous emitter, 221
Host material, PSi as, 184; *see also* Hybrid
  materials; PSi-based composites
HWA, *see* High-pressure water vapor annealing
  (HWA)

Hybrid heterojunction photodetectors, 45–46,
  46f–47f; *see also* Photodetectors
Hybrid materials, 183–202; *see also* PSi-based
  composites
  carbon materials deposition on/in PSi,
    200–202
    carbon nanotubes, 201
    fullerenes C60, 200–201
    graphite/graphene, 201–202
  for Li-ion battery anodes, 338–339
    carbon-coated materials, 338–339, 338f
    metal particle decoration, 339
  overview, 184
  by PSi infiltration with organic substances,
    195–200, 198f
    for electrical contacts to PSi, 199–200, 200f
    for optoelectronic devices, 197–198, 198f
    for passivation and functionalization of
      PSi surface, 198–199
  substrates, for active and passive devices,
    119–120, 120f
  vs. composite materials, 184
Hybrid PSi-glass, 103–104, 103f
Hydrogen, 274
  contents in PSi, FTIR method for
    estimating, 276–279, 277f–280f
  Van der Waals interactions, 274, 289
Hydrogenated Si nitride (SiNx:H), 220–221
Hydrogen-CNT interaction, 274
Hydrogen generation, 273–292
  energetic analysis PSi nanostructures and,
    291–292, 291t, 292f
  overview, 274
  from PSi, 280–285
    chemically stimulated desorption,
      281–284, 282f, 284f
    photocatalytic hydrogen evolution, 285
    thermally stimulated desorption,
      280–281, 281f
  in PSi composites, 290–291, 291f
  rate of, 285–289, 286f–288f, 288t
  in water solution, 285–286
Hydrogen ion implantation-porous silicon
  (HI-PSi) technique, 176–177, 177f
Hydrogenization, supplementary
  methods of, 289–290, 289f
Hydrogen oxidation, fuel cells, 250–251
Hydrogen storage, 273–292
  complex hydrides for, 276
  energetic analysis PSi nanostructures and,
    291–292, 291t, 292f
  interstitial hydrides for, 276
  metallic hydrides for, 276
  options, 274
  overview, 274
  in PSi composites, 290–291, 291f
  solid matrix method, 274–276
    chemical storage, 275–276
    physisorption, in porous matrixes,
      274–275, 275t
Hydrogen-terminated PSi, 302–303

I

IAD (ion assisted deposition), 147
Ignition, of oxidants, 394–395; *see also*
  Explosive devices
III-V compound semiconductors,
  heteroepitaxial films of
  GaAs, 152–153

GaN, 153–155, 154f–156f
InSb, 153
II-VI and IV-VI compound semiconductors, 191–194, 193f
heteroepitaxial films of, 148–152, 149f–150f, 151t, 152f
In/Au contact, EL emission, 22, 23f
Indigo carmine (IC), 305
Indirect fuel cells, 251; *see also* Fuel cells (FC)
Indium tin oxide (ITO), 13
as top electrode, 22
Inductors, 113, 116–119, 116f, 117t, 118f
electrical performances, 116, 117t
geometry, 116, 116f
finite element method, 119
quality factor (Q), 116–117
evolution of, 117, 118f
self-resonance frequency, 116
SGPS technique, 117, 118f
Inert materials, PSi impregnation and, 21
Inner Helmholtz plane (IHP), 350, 351f
InSb layer, heteroepitaxial growth, 153
Insertion losses (IL), 114
Insulation by porous oxidized silicon (IPOS), 110, 111f
Integrated circuits (ICs), 376
Integrated on-perforated-Si-chip micro-supercapacitors
3D trench-array, 357–358, 358f
electrochemical performance, 364–368, 366f, 367f
enhancement factor (AEF), 358
fabrication, 357–360
NiO/Si-MCP electrode, 360
preparation process stages, 358–359, 359f
Si-MCPs, 359–360, 359f
Integrated optoelectronics, 99–104, 100f–104f; *see also* Optoelectronics
Integration, of PSi EL, 29–30, 29f
Interconnects, role of, 78, 119
Interconnect structures, RF system, 113–114, 114f, 115f, 115t
Interdigitated back contacts (IBC) cell, 224, 224f
Internal quantum efficiency (IQE), 217, 219, 219f
of Si solar cells, porous Si ARC influencing, 219–221, 219f–220f
surface recombination velocity and, 222–223, 223f
Internal surface stabilization, PSi explosive devices and, 391
Interstitial hydrides, 276
Ion assisted deposition (IAD), 147
Ion beam, 60
Ion-irradiation process, 88
IPOS (insulation by porous oxidized silicon), 110, 111f
IR mesoporous silicon filters, 81–82, 82f
Iron (Fe), 185–189
Isolators, 78
I–V characteristics, 45
of MSM-based photodetectors, 40–41, 41f
of PSi LED devices, 101, 101f
surface recombination velocity and, 223

K

KClO$_4$, 391, 392
Kistler force sensor
KNO$_3$, 392

L

Laser ablation, 299
Laser beam induced current (LBIC) measurement, 224, 224f
LAST process, 229
Layer-transfer process, 146
SMM and, 137–139, 138f
solar cell applications, 228–229, 228f
ELTRAN process, 229
FMS process, 229
LAST process, 229
QMS process, 229
ψ-process, 229
for thin-film silicon solar cells, 146
Lenz's law, 113
LEPECVD (low energy plasma enhanced chemical vapor deposition), 147
Light beams, DOEs and, 93
Light-emitting devices; *see also* Electroluminescence (EL) devices
advantages, 58
1D photonic crystals in fabrication of, 58
Light emitting diodes (LEDs), 58
Light-induced passivation, on PSi/Si interface, 224, 224f
Lightly doped PSi *vs.* heavily doped PSi, 7
Light of cone, 66, 68
Li-ion batteries, with PSi-based electrodes, 319–339, 320f
hybrid structures performances, 338–339
carbon coating, 338–339, 338f
metal particle decoration, 339
overview, 320–321
synthesis and performances of anodes, 327–338
micro-particles, 337–338
SiNWs, 333–337, 334t, 335f–336f, 336t
thin films, 328–333
textured silicon material in anode
advantages, 326–327, 326t, 327f
binders in composite, 325–326
charge-discharge profile, 321–322, 321f
depth of discharge, 325
designs and reported performance, 326t
lithiation/de-lithiation mechanisms, 321–322, 322f
performance improvement, 324–326
principles and characterization, 321–324
SEI formation, 323–324, 323f
volume expansion of Si-based composites, 324, 325f
Line attenuation (α), 114
Liquid phase epitaxy (LPE), 147, 153
Liquid-phase infiltration, 338
Lithiation/de-lithiation mechanisms, 321–322, 322f
Lithography, 299
antireflection surfaces and, 90
macroporous silicon with, 2D-photonic crystals fabrication using, 62–63, 63f
Long wave pass filters, 85–86, 86f
applications, 85
variants, 85–86
Low energy plasma enhanced chemical vapor deposition (LEPECVD), 147
Low pressure chemical vapor deposition (LPCVD), 147
Low-pressure vapor-phase epitaxy (LPVPE), 112

LPCVD (low pressure chemical vapor deposition), 147
LPE (liquid phase epitaxy), 147, 153
Lucent technologies, 256

M

MaCE process, *see* Metal assisted chemical etching (MaCE) process
Macroporous silicon
as catalyst support structure, 310
in PSi-based fuel cells, 260–263
CPS, 260–261
3D substrates, 261, 261f
improved 3D structure, 261–262, 263f
micro-fuel cell prototypes, 263, 264t
multifunctional μ-FC, 262–263, 263f
RIPS, 260–261
for silicon anode batteries, 329, 329f
performances, 331–332, 331f, 332t
Macroporous silicon methods
3D photonic crystals fabrication, 68–69, 69f–70f
2D-photonic crystals fabrication using, 62
with standard lithography, 62–63, 63f
Macroporous silicon microreformer, 310–311, 310f
Macroporous silicon UV filters, 83–85, 84f–85f
Macroporous silicon with nanoporous surface (NP-MPSi), 306–307, 306f
Macroporous texturization, of Si wafers
alcaline *vs.*, 214–216, 214f–217f
influence on solar cell performance, 217
Magnetron sputtering, 184
Maxwell's equations, 53–54, 79–80
MBE, *see* Molecular beam epitaxy (MBE)
Mechanical sensitivity, of explosion, 398
Medical devices, 250
Membrane electrode assembly (MEA), 255
Membrane microreactors, 298–299, 299f
palladium reactors, 308–309, 308f, 309f
PSi thermal isolation ring, 309, 309f
MEMS (microelectromechanical systems) devices, 99–101, 130, 133, 139, 184, 253, 264, 388, 399
3-mercaptopropyltrimethoxysilane, 264
Mesoporous silicon
PEM performances, 265t
photocatalytic activity, 306–307, 306f
as proton exchange membrane, 253, 254f
in PSi-based fuel cells, 256–259, 256f–259f
channels, 256–257, 256f–257f
disadvantages, 259
micro-fuel cell prototypes, 258, 259t
performance factors, 258–259
process flow, 257–258, 257f, 258f
for silicon anode batteries, 329–330, 329f
performances, 332–333, 333f, 333t
Mesoporous silicon optical filters, 81–82, 82f
Meso-PSi layers, homoepitaxial growth and, 145
Metal assisted chemical etching (MaCE) process, 305–306, 328, 334–337
advantages, 328–329
Metal-assisted etching
PSi electrodes by, 361–362, 361f–363f
for Si wafers texturization, 215–216, 216f, 217f
Metal catalysts, PSi-supported
for heterogeneous catalysis, 301–304, 302f

Metallic hydrides, 276
Metallurgical grade Si, 337
 purification of, 213–214
Metal NPs, 339
Metal/PSi composites, 192t, 302–303
 applications, 184–191
 deposition method, 184–185
 electrochemical deposition method, 185
 ferromagnetic metals, 185–189, 186f–188f
 formation of, 184–191
 noble metals, 189–190, 190f
 properties, 184–191
Metals; *see also specific entries*
 antireflection coatings and, 91
 ferromagnetic, 185–189, 186f–188f
 in fuel cells, 253–254
 PSi impregnation and, 20
 as top electrodes, 22, 22f–23f
Metal–semiconductor–metal (MSM)-based
        photodetectors, 37–42, 37f, 43
 I–V characteristics of, 40–41, 41f
Methanol, 251, 392
 combustion of, 303
Methylviologen, 8
μ-FC, 262–263, 263f, 264
Microbatteries, betavoltaic devices, 240–241, 240f
 electron flux density, 241
 PSi-based, 242–244, 243f–244f
  radioisotopes for, 245–246, 245t
Micro-cathodes, 172, 175; *see also* Cold cathodes
Microcavities
 EL spectrum with, tuning and narrowing,
        27–28, 27f
 1D photonic crystals, 55–56, 57f
  linear reflection spectra, 58–59, 58f
Microdevices, 250
Micro-DMFC, 266
Microelectromechanical systems (MEMS)
        devices, 99–101, 133, 139, 184, 253,
        388, 399
Microelectronics, 4, 52
 CMOS-compatible processing in, PSi for,
        130–131
 electrical interconnect bottleneck in, 4
 electrical isolation applications of PSi in,
        109–123
  to active device isolation, 110–112, 111f,
        112f
  electrical properties of PSi and, 120–123
  overview, 109
  to RF passive devices, 113–120
 materials used in, 113
Micro-fuel cells, 246, 249–268, 276; *see also*
        Fuel cells (FC)
 described, 250–253
 overview, 250
 PSi-based, 254f
  mesoporous silicon in, 258, 259t
 self-regulating hydrogen generator for, 253
Microlens, 86–89, 87f–89f
 applications, 86
 concave mirrors, fabrication of, 88–89, 89f
 dry etching, 89
 electrochemical etching methods and, 87
 focal length, 86
 GRIN
  EOT distribution, 87, 87f
  interferogram showing Newton's rings,
        88, 88f
  planar, 88

optical oxidized PSi (OPS), 86
 planar, 87
Micro lithium batteries, 241
Microlithography, 299
Micro-machined Si-based supports, for small
        fuel cells, 264–268, 266f–268f
 micro-DMFC, 266
 silicon bipolar plate performed with DRIE,
        266–267, 266f–267f
Micromachining technique, 89
Microphones, 253
 fabrication of, 133–134, 134f
 SMM and, 133–134, 134f
Microphotonics, silicon, 78
Microporous silicon microreformer, 311–312,
        311f
Micro-probe card, fabrication process, 131–132,
        132f
Microreactors
 advantages, 298
 biosensing, 298
 conventional, 298, 298f
 fabrication methods, 299
 glucose monitoring, 298
 heterogeneous catalysis/catalysts
  metal catalysts, PSi supported, 301–304,
        302f
  photocatalysis, PSi supported, 304–307,
        305f–306f
  PSi matrix for, 300–307
 membrane, 298–299, 299f
 multichannel, 298, 298f
 overview, 298–300
 parameters, 299
 PSi-based, 297–316
  advantages of, 307
  fabrication process, 307, 307f
  flow-through membrane, 308–313,
        308f–311f
  PSi layer enhancement of, 313–316,
        314f–315f
 silicon for, 299
 silicon micromachining, 299, 301f
 single microchannel, 298, 298f
 studies on, 298
 testing and characterization, 299, 300f
Micro solar cell arrays, 241
Microstrips (MS), 113
Micro-supercapacitors
 basic principles, 357
 integrated on-perforated-Si-chip
  3D trench-array, 357–358, 358f
  electrochemical performance, 364–368,
        366f, 367f
  enhancement factor (AEF), 358
  fabrication, 357–360, 358f
  NiO/Si-MCP electrode, 360
  preparation process stages, 358–359,
        359f
  Si-MCPs, 359–360, 359f
Microsystems, 250
Microthrusters, on-chip PSi, 101
Microtips, 167
Microwave power amplifiers, 166
Military applications, 250
MOCVD method, 153, 154, 155, 157–158
Modulation speed, EL, 28–29, 28f
Modulators/switches, 78
Molecular beam epitaxy (MBE), 112, 145, 148,
        152–153

Molten carbonate fuel cells (MCFC), 252t
Molybdenum-100, 246
Monolithic microwave integrated circuits
        (MMIC), RF, 119
Moore's law, 388
"More than Moore" approach, 388
Multichannel microreactors, 298, 298f
Multiplexers/demultiplexers, 78
Multiwalled carbon nanotubes (MWCNTs), 178;
        see also Carbon nanotubes (CNTs)

N

NaBH$_4$, 303
NaClO$_4$, 398
Nanodots, 378
Nano-emitter, 174, 175
Nano-energetic composites, applications,
        398–399, 399t
Nano-explosions, 388, 393f; *see also* Explosive
        devices
 oxidants used for, 391–392, 392t
Nano-explosive devices, 391, 399; *see also*
        Explosive devices
Nanoparticles (NPs), 184, 325
 metal NPs, 339
Nanoporous carbons, 356
Nano PSi, as ARC
 advantages, 218
 background, 218
 disadvantages, 220
 PSi emission and quantum efficiency of
        solar cells, 221
 requirements for, on Si solar cells, 217–218
 SiNx:H layer, 220–221
 solar cell performance and, 219–221,
        219f–220f
 technology of, 218–219, 219f
Nanostructures, 4; *see also specific entries*
 applications, to hydrogen generation, 292,
        292f
 0D, 184
 1D, 184
 formation with PSi template, 184, 184f;
        see also Hybrid materials; PSi-based
        composites
 in photovoltaics, 211–233; *see also*
        Photovoltaics (PV)
 PSi, energetic analysis, 291–292, 291t, 292f
 TiO$_2$ nanoparticles, as photocatalysts, 304
Nanotubes (NTs), 184; *see also* Carbon
        nanotubes (CNTs)
Nanowires (NWs), 184; *see also* Silicon
        nanowires (SiNWs)
Natural gas, 251
NEAH power, 253
NEAH Power Company, 260
Near-infrared (NIR) optical applications, 36
Negative electron affinity (NEA) films, 168
NEMS devices, 184
New materials
 emitter, PSi as substrate for, 178–179
 search for, 178
Newton's rings, interferogram of GRIN lens
        showing, 88, 88f
Nexeon®, 337
NH$_4$ClO$_4$, 391, 392
Ni-63, radioisotope, 246
Nickel (Ni), 185–189

n-i-n-p-n device, 24
NiO/Si-MCP electrode, 360
n-i-p-n device, 24
Nippon Electric Company (NEC), 348
4-nitrophenol (4NP), 305
Noble metals, 189–190, 190f
Nonhygroscopic perchlorates, 391–392
Nonlinear optics, 1D photonic crystals
    applications in, 58–59, 58f
Nonthermal devices, 240; see also Betavoltaics/
    betavoltaic devices
Normalized photosensitivity spectra, of
    PSi-based photodetectors, 38f
Notebooks, 250
Novel carbon materials, 356
n-p-i-n device, 24
NP-MPSi (macroporous silicon with
    nanoporous surface),
    306–307, 306f
n-type process, active device isolation,
    110, 111f
n-type PSi, 9
(n-type) Si QD/(p-type) c-Si photovoltaic
    device, 233, 233f
n-type Si wafer, 91–92
Nuclear microbatteries, 240–241, 240f
    electron flux density, 241
    PSi-based, 242–244, 243f–244f
        radioisotopes for, 245–246, 245t
NXP, 357

O

On-chip PSi micro-thruster, 101
On-chip Si-based light emitters, 4
1D photonic crystals, 52f, 54–60, 54f–56f
    applications, 57–60, 59f
        as filtered photodetectors, 59, 59f
        light-emitting devices fabrication, 58
        in nonlinear optics, 58–59, 58f
        as optical sensors, 59–60
        PSi-based waveguides, 57
    distributed Bragg reflector (DBR), 79–83,
        79f–83f
    microcavities, 55–56, 57f
    stop-band, 54–55, 55f
Optical absorption, SiNWs, 23f, 231
Optical devices; see also specific entries
    disadvantages, 78
    in telecommunications, 78
Optical filters, 79–86
    applications, 83
    distributed Bragg reflector filters, 79–83,
        79f–83f
    Fabry–Perot interference filter, 83–85,
        84f–85f
    long wave pass filters, 85–86, 86f
    overview, 79
Optical ignition, 395
Optical lithography, 94
Optical nonlinearity, PSi, 29
Optical oxidized PSi (OPS) microlenses, 86
Optical reflector, PSi as
    Bragg reflector application, 225–226,
        225f–226f
    for thin film solar cell, 226–227, 226f, 227f
Optical sensors
    1D photonic crystals as, 59–60
    integrated PSi based, 102–103, 102f
    with PBG structure, 103

Optical waveguides, 95–99, 96f–97f
    applications, 97–98
    ARROW, 98
    background, 95
    Bragg grating fabrication, 96–97, 97f
    fabrication, 96–97, 97f
    multilayer structures, 98
    oxidized, 98
    properties, 95
Opto-coupler, silicon-integrated, 78
Optoelectronic devices, PSi-based
    organic substances for, 197–198, 198f
Optoelectronic integrated circuits (OEICs), 99
Optoelectronics
    antireflective coating, 89–92
    diffraction optical elements (DOE), 93–95
    integrated, 99–104
    microlens, 86–89
    optical filters, 79–86
    optical waveguides, 95–99
    overview, 78
Organic substances, hybrid materials formation
    and, 195–200, 198f
    for electrical contacts to PSi, 199–200, 200f
    for optoelectronic devices, 197–198, 198f
    for passivation and functionalization of PSi
        surface, 198–199
Outer Helmholtz plane (OHP), 350, 351f
Oxidants, for explosive devices
    choices for, 391–392, 392t, 393f
    ignition of, 394–395
Oxidation
    anodic, 6–8, 7f, 8f
    chemical reaction, 8
    ECO, 6–7
    hole injection, 6–8
    partially oxidized PSi, 15–18, 15t
        advantages, 15
        chemical and thermal oxidation,
          16, 16f
        electrochemical oxidation, 16–17, 17f
        high-pressure water vapor annealing,
          18, 18f, 19f
Ox-PSi-Sharpening, 176, 176f
Oxygen reduction, fuel cells, 251

P

Palladium membrane reactors, 308–309, 308f,
    309f
    PSi thermal isolation ring, 309, 309f
Palladium/silicon mesoporous composite,
    290–291, 291f
Partially oxidized PSi, 15–18, 15t
    advantages, 15
    chemical and thermal oxidation, 16, 16f
    electrochemical oxidation, 16–17, 17f
    high-pressure water vapor annealing, 18,
        18f, 19f
Passivation, surface, by porous layers
    light-induced, 224, 224f
    surface recombination velocity in PSi/Si
        interface, 222–224, 223f
Passive devices, RF, 113–120
    general considerations on processing, 113
    hybrid substrates for, 119–120, 120f, 121f
    inductors, 116–119, 116f, 117t, 118f
    interconnect structures, 113–114, 114f, 115f,
        115t
    RF functions, 119

PBS, see Photonic band structure (PBS)
PbS/PSi/Si heterostructures, heteroepitaxial, 149
PEM, see Proton exchange membrane (PEM)
PEMFC, see Proton exchange membrane fuel
    cells (PEMFC)
Perchlorates, as oxidants for explosive devices,
    391–392
Perforated PSi, 337
Persulfate ions, cathodic polarization of PSi
    with, 8–10, 9f
Philips, 357
Phosphoric acid fuel cells (PAFC), 252t
Phosphorous-doped silicon wafers, 81
Photocatalysis
    PSi-supported (heterogeneous), 304–307,
        305f–306f
    TiO_2 nanoparticles, 304–305
Photocatalytic/photoelectrochemical hydrogen
    evolution, 285
Photodetectors, PSi-based, 35–47
    advantages, 37
    antireflecting covering and filters, 43–45,
        44f–45f, 44t
    background, 36–37
    composite-based, 42–43, 43f
    current–voltage curves, 39–40, 39f
    dark current, 36
    development, 37
    disadvantages, 38–40
    fabricated, testing of, 43
    fabrication technology and, 42
    high-sensitivity, 36
    high-speed, 36
    hybrid heterojunction, 45–46, 46f–47f
    I–V characteristics of, 40–41, 41f
    MSM-based, 37–42, 37f
    normalized photosensitivity spectra, 38f
    overview, 36–37
    p-n junction-based, 37–42, 37f
    properties, 36
    quantum efficiency, 36
    response speed, 37
    responsivity of, 36
    Schootky barrier type, 37–42, 37f
    spectral sensitivity and, 38–39, 38f
    stabilization, 40–42, 41f, 42f
    "stop" ring, role of, 42
    temporal instability, 40
    thermal carbonization (TC), 40–41, 41f
    thermal oxidation, 41–42, 42f
Photoelectron spectroscopy (PES), 323
Photolithography technique, 96, 167
    for DOEs, 93
Photoluminescence (PL), 4, 52
Photonic-band, 53, 79
Photonic bandgaps, 53
Photonic band structure (PBS), 53
    calculation of, 53–54
    3D photonic crystals, 68, 69f
    2D photonic crystals, 61, 62f, 63, 64f
Photonic crystals, 51–72
    background, 52
    1D, 52f, 54–60, 54f–56f
        applications, 57–60, 58f, 59f
        microcavities, 55–56, 57f
    2D, 52f, 60–68
        applications, 71
        bulk, 60–63, 61f–64f
        defects in, 63–65, 65f, 66f
        finite (slabs), 65–68, 66f–67f

3D, 52f, 68–71, 69f–70f
    applications, 71
    examples, 52, 52f
    light propagation through (calculations), 53
    outlook, 71–72
    overview, 52–53
    principles, 53–54
    scalability, 53
Photonics, 253
Photons, 37
Photovoltaic effect, 240
Photovoltaics (PV)
    costs, 212
    gettering of Si wafers, 213–214
    macroporous texturization of Si wafers
        alcaline vs., 214–216, 214f–217f
        influence on solar cell performance, 217
    metallurgical grade Si, purification of,
        213–214
    nano PSi as antireflection coating
        influence on solar cell performance,
            219–221, 219f–220f
        PSi emission and quantum efficiency
            of solar cells, 221
        requirements for, on Si solar cells,
            217–218
        technology of, 218–219, 219f
    overview, 212–213
    PSi and Si nanostructures in, 211–233
    PSi as optical reflector
        rear Bragg reflector application,
            225–226, 225f–226f
        rear reflectors for thin film solar cell,
            226–227, 226f, 227f
    PSi for layer transfer processes, 228–229, 228f
    PSi selective emitter, 221–222, 222f
    solar cells with Si nanowires, 229–231,
        230f–231f
    surface passivation by porous layers
        light-induced, 224, 224f
        surface recombination velocity in PSi/Si
            interface, 222–224, 223f
    tandem Si solar cells with Si QDs, 231–233,
        232f–233f
Physical storage systems, of hydrogen, 274–275,
    275t
Physisorption, of hydrogen in porous matrixes,
    274–275, 275t
p-i-n device, 24
Pinnacle Research Institute (PRI), 348
PL, see Photoluminescence (PL)
Planar GRIN microlens, 88
Planar growth substrates, for homoepitaxial
    growth on PSi layer, 147
Planar inductors, 116; see also Inductors
Plane wave method, 53
Plasma enhanced chemical vapor deposition
    (PECVD), 113
Plasma-nitride treatment, 223
PL quenching, 10
Pm-147, radioisotope, 246
p-nitroaniline (p-NA), 303
p-n junction-based photodetectors, 37–42, 37f
p-n junctions, porosified, 13–15
    devices based on, 13–14, 14t
        efficiency and output of, 14–15, 15f
        PSi diode structure, 13–14, 14f
Polarization, of PSi
    anodic, 6–8
    cathodic, with persulfate ions, 8–10, 9f

Polyaniline (PANI), PSi impregnation and,
    20–21, 21f
Polycrystalline PSi, 170
Polydimethylsiloxane (PDMS), 254
Poly(3,4-ethylenedioxythiophene) (PEDOT)
    films, 364
Polymers; see also specific entries
    conducting, as electrode materials, 356
    electroactive, 356
    in fuel cells, 254–255
    PSi impregnation and, 20–21, 20f–21f
Polymethyl methacrylate (PMMA), 254
Polypyrrole (PPy), 356, 364
    PSi impregnation and, 20
Poly-Si-based switching TFT, 29
Pore-filling technique, 43
Pore impregnation, PSi explosive devices and,
    393–394
Porosified p-n junctions, 13–15
    devices based on, 13–14, 14t
        efficiency and output of, 14–15, 15f
        PSi diode structure, 13–14, 14f
Porosity, PSi, 145, 147
Porous silicon (PSi), 4; see also specific entries
    advantages, 52
    anodic polarization of, 6–8, 7f, 8f
    as antireflection coating material, 90
    capping, 26
    cathodic polarization with persulfate ions,
        8–10, 9f
    electrical properties
        AC conductivity, 122–123
        DC conductivity, 122, 123f
        dielectric permittivity, 121, 122t
    EL of, see Electroluminescence (EL), of PSi
    integration in fuel cells, 253–255, 254f;
        see also Fuel cells
    partially oxidized, 15–18, 15t
        advantages, 15
        chemical and thermal oxidation, 16, 16f
        electrochemical oxidation, 16–17, 17f
        high-pressure water vapor annealing,
            18, 18f, 19f
    properties, 52
    surface modification, 26–27, 26f
    thermal conductivity of, 379–381,
        379f–380f
    thermoelectric properties of, 379–381,
        381f–382f; see also Thermoelectric
        (TE) devices
Post-CMOS selectively grown porous silicon
    (SGPS) technique, 117, 118f
Potassium-40, 246
Power losses (PL), 114
Power splitters/combiners, 78
Pressure sensors, 253
Printed circuit boards (PCB), 254
Propagation velocity, of explosion, 397–398,
    397f–398f
Propellants, 388
Proton beam writing method, 60–61, 61f
Proton exchange membrane (PEM), 258
    mesoporous silicon-based, 265t
    PSi as, 264, 265f, 265t
Proton exchange membrane fuel cells
    (PEMFC), 251, 252t, 255, 307, 312
    described, 252–253, 252f
    performance factors, 258–259
    PSi-based, 254f
    working principle, 250–251, 251f

Pseudo-capacitive materials, 356, 357
Pseudo-capacitors, 355
    with PSi electrodes, electrochemical
        performance of, 366t
PSi, see Porous silicon (PSi)
PSi-based cold cathodes, 169–173; see also
        Cold cathodes
    advantages, implementation of, 172–173
    background, 169
    band diagram, 169f
    BSD, 172–173
    electron emission process, 169–170, 169f
    energy distribution curves, 170f
    features, 172
    microscopic representation, 170, 171f
    multilayered and graded-multilayered PSi
        diodes, 171–172, 172f
    RTO process, 171
    Si tips modified by, 173–177, 174f–177f
    structure and experimental configuration,
        169f
    thickness of PSi layer and, 170–171, 171f
PSi-based composites, 183–202; see also
        Hybrid materials
    formation methods, 184
    hydrogen generation in, 290–291, 291f
    hydrogen storage in, 290–291, 291f
    metal/PSi composites, 192t
        applications, 184–191
        deposition method, 184–185
        electrochemical deposition method, 185
        ferromagnetic metals, 185–189, 186f–188f
        formation of, 184–191
        noble metals, 189–190, 190f
        properties, 184–191
    overview, 184
    semiconductor/PSi composites, 192f, 196t
        applications, 191–195
        formation of, 191–195
        Ge and Si, 194–195
        II-VI and IV-VI compound
            semiconductors, 191–194, 193f
        properties, 191–195
PSi-based electrodes, Li-ion batteries with,
        319–339, 320f; see also Electrodes
    hybrid structures performances, 338–339
        carbon coating, 338–339, 338f
        metal particle decoration, 339
    overview, 320–321
    synthesis and performances of anodes,
        327–338
        micro-particles, 337–338
        SiNWs, 333–337, 334t, 335f–336f, 336t
        thin films, 328–333
    textured silicon material in anode
        advantages, 326–327, 326t, 327f
        performance improvement, 324–326
        principles and characterization, 321–324
PSi-based emitters; see also Emitters
    advantage of, 172
PSi-based fuel cells (FC), 255–268; see also
        Fuel cells (FC)
    integration in electrodes, 255–263
    macroporous silicon in, 260–263
        CPS, 260–261
        3D substrates, 261, 261f
        improved 3D structure, 261–262, 263f
        micro-fuel cell prototypes, 263, 264t
        multifunctional μ-FC, 262–263, 263f
        RIPS, 260–261

mesoporous silicon in, 256–259
  channels, 256–257, 256f–257f
  disadvantages, 259
  micro-fuel cell prototypes, 258, 259t
  performance factors, 258–259
  process flow, 257–258, 257f, 258f
micro-machined Si-based supports for
    small fuel cells, 264–268, 266f–268f
PEM, 264, 265f, 265t
PSi-based interference filters, 43–45, 45f, 54
PSi-based microreactors, 297–316; see also
    Microreactors
  advantages of, 307
  fabrication process, 307, 307f
  flow-through membrane, 308–313,
      308f–311f
  PSi layer enhancement of, 313–316,
      314f–315f
PSi-based supercapacitors, 356–364; see also
    Supercapacitors
  electrochemical performance, 364–371,
      365t, 367t
    integrated on-perforated-silicon-chip
        micro-supercapacitors, 364–368,
        366f, 367f
    with PSi electrodes, 368–371, 369f, 370f
  fabrication of, 357–364
    integrated on-perforated-silicon-chip
        micro-supercapacitors, 357–360,
        358f–359f
    with PSi electrodes, 360–364
PSi-diffraction based biosensor (PSi-DBB),
    93–94, 94f
PSi diffraction grating, with fiber optical
    sensor, 93, 93f
PSi electrodes, supercapacitors with
  by anodic dissolution, 360
  electrochemical performance, 368–370
  fabrication, 360–364
    by metal-assisted etching, 361–362,
        361f–363f
  PSi nanowires as, 362–364, 363f
PSi explosive devices, see Explosive devices, PSi
PSi impregnation, 18–21, 19t
  electrolytic systems
    anodic polarization of PSi, 6–8
    cathodic polarization of PSi with
        persulfate ions, 8–10, 9f
  inert materials incorporation, 21
  metals incorporation, 20
  polymers incorporation, 20–21, 20f–21f
PSi layer enhancement, of microreactors,
    313–316, 314f–315f
PSi matrix, for heterogeneous catalysis,
    300–307
  metal catalysts, 301–304, 302f
  photocatalysis, 304–307, 305f–306f
PSi micro-particles, for Li-ion battery anodes,
    337–338
PSi nanostructures, energetic analysis,
    291–292, 291t, 292f
PSi nanowires (PSiNWs)
  for Li-ion battery anodes, 335–337
  as supercapacitor electrodes, 362–364, 363f
PSi optical waveguides, 57, 95–96, 96f
  applications, 97–98
  fabrication, 96–97, 97f
  multilayer structures, 98
  oxidized, 98
  photoresist grating on, 97–98, 97f

PSi selective emitter, 221–222, 222f
  disadvantages, 222
  technologies, 221–222
PSi-Sharpening, 176, 176f
PSi/Si interface
  light-induced passivation on, 224, 224f
  surface recombination velocity in, 222–224,
      223f
PSi slab waveguides, 95, 96
PSi/sodium perchlorate system, 396
PSi-SWNT, 290–291
PSi template, nanostructures formation with,
    184, 184f; see also Hybrid materials;
    PSi-based composites
Pt/PSi composite, 302–303
p-type process, active device isolation, 110
p-type PSi, 9, 175
p-type Si wafer, 92
PV, see Photovoltaics (PV)
PVA/PDA (polyvinyl alcohol/phenol-2,4-
    disulfonic acid) electrolyte, 264, 265f
Pyrotechnics, 388

Q

QDs/PSi hybrid-MSM photodetector, 43, 43f
Quality factor (Q), planer inductors, 116–117
  evolution of, 117, 118f
Quantum confinement theory, 38
Quantum dots (QDs), 42–43
  tandem Si solar cells with, 231–233,
      232f–233f
Quantum efficiency; see also External quantum
    efficiency (EQE); Internal quantum
    efficiency (IQE)
  of photodetectors, 36
  of solar cells, PSi emission and, 221
Quasi-monocrystalline silicon (QMS) process,
    229

R

Radioactive batteries, 240, 240f
Radio frequency (RF) devices, 110
  passive devices, 113–120
    general considerations on processing,
        113
    hybrid substrates for, 119–120, 120f,
        121f
    inductors, 116–119, 116f, 117t, 118f
    interconnect structures, 113–114, 114f,
        115f, 115t
    RF functions, 119
Radioisotope generator, 240; see also
    Betavoltaics/betavoltaic devices
  advantages, 240
Radioisotopes; see also Betavoltaics/betavoltaic
    devices
  decay rate, 246
  half-life of, 240
  molybdenum-100, 246
  Ni-63, 246
  Pm-147, 246
  potassium-40, 246
  for PSi-based microbatteries, 245–246,
      245t
  safety concerns, 245–246
  stronium-90, 246
  tritium (T), 246
  zinc-70, 246

Random initiated porous silicon (RIPS),
    260–261
Rapid thermal oxidation (RTO), 171, 223
Rare earth metals, antireflection coatings and,
    91
Rayleigh, Lord, 89
RDX (cyclo-1,3,5-trimethylene-2,4,6-
    trinitramine), 392
Reactive ion etching, 264
Reflectance spectra
  of Bragg reflector, 55, 56f
  of SiO_2 coating, 90, 91f
  of ZnO/TiO_2 and PSi coating, 90, 91f
Reflection coefficient, Bragg mirror, 225–226,
    226f
Reflectors
  Bragg reflector, 54, 60, 100f
    application, 59
    concave PSi-based, 89
    reflectance spectra of, 55, 56f
    for Si solar cells, 225–226, 225f–226f
    in solar cells, PSi as, 147
  distributed Bragg reflector (DBR), 79–83,
      79f
    electrochemical polishing process, 81
    FESEM image, 81, 81f
    IR mesoporous silicon filters, 81–82, 82f
    mesoporous, 81–82, 82f
    porous structure (categories), 81
    step-by-step preparation, 80, 80f
    vs. rugate filters, 82–83, 83f
  optical, PSi as
    Bragg reflector application, 225–226,
        225f–226f
    for thin film solar cell, 226–227, 226f,
        227f
Reformed hydrogen fuel cells (RHFC), 251
Refractive index
  of dielectric layer, 217–2118
  of PSi, 43, 52, 54, 55f, 59–60
    volume fraction and, 90
Resonance frequency, inductors, 116
Response speed, photodetectors, 37
Responsivity, of of photodetectors, 36
RF functions, 119
RHFC (reformed hydrogen fuel cells), 251
RIPS, see Random initiated porous silicon
    (RIPS)
Rochow reaction, 339
RTO, see Rapid thermal oxidation (RTO)
Rugate filters, 55
  vs. DBR filters, 82–83, 83f
Ruthenium oxide (RuO_2), 356

S

Sacrificial layers, 129–140; see also Surface
    micromachining (SMM)
  homoepitaxial growth, 144
Sb/Au contact, EL emission, 22, 23f
Scalability, photonic crystals, 53
Scanning electron microscopy (SEM), 322
Scheme-aligning PSi technique, 221
Schootky barrier type photodetectors, 37–42,
    37f
Screen-printed multicrystalline silicon solar
    cell, 92
Screen-printing process, 222
Secondary ion mass spectrometry (TOF-SIMS),
    323

Seebeck coefficient, 376–377, 380
Seebeck effect, 376–377, 377f
Selective emitter, 221–222, 222f
    disadvantages, 222
    technologies, 221–222
Self-aligning PSi technique, 221
Self-regulating hydrogen generator, 253
Self-resonance frequency, inductors, 116
Semiconductor/PSi composites, 192f, 196t
    applications, 191–195
    formation of, 191–195
    Ge and Si, 194–195
    II-VI and IV-VI compound semiconductors,
        191–194, 193f
    properties, 191–195
Semiconductors; see also specific entries
    III-V compound, heteroepitaxial films of
        GaAs, 152–153
        GaN, 153–155, 154f–156f
        InSb, 153
    II-VI and IV-VI compound, heteroepitaxial
        films of, 148–152, 149f–150f, 151t,
        152f
Semitransparent gold, as top electrode, 22
Sensors, 166
Si-based cold cathodes; see also Cold cathodes
    comparison between devices, 176
    examples, 167–168, 167f
    with Si-tips modified by porous layer,
        173–177
        HII–PSi process, 176–177, 177f
        Ox-PSi-Sharpening, 176, 176f
        porosification technology, 173–174, 175
        PSi-Sharpening, 176, 176f
        spaghetti-like structures, 175, 175f
        threshold field for, 174–175, 174t
Si-based emitters; see also Emitters
    comparison of turn-on fields for, 177, 178t
    visible light, 4
Si-based field emission devices, comparison
        between, 176t
Si-based photodetectors; see also
        Photodetectors
    average reflectance values, 44t
    disadvantages, 43
    with PSi-based antireflecting covering,
        43–45, 44f–45f
    with PSi-based interference filters, 43–45,
        44f–45f
Si-based visible light emitter, 4
Silicon (Si)
    advantages, 52
    bandgap of, 4
    as dominant semiconductor material, 148
    metallurgical grade, purification of, 213–214
    for microreactors, 299
    optical properties, 52
    properties, 299
Silicon-carbon based composite, charge-
        discharge profile of, 321–322, 321f
Silicon-germanium alloys, heteroepitaxial films
        of, 157
Silicon-integrated opto-coupler, 78
Silicon microchannel plates (Si-MCPs),
        359–360, 359f
Silicon microphotonics, 78
Silicon nanowires (SiNWs)
    bottom up approach, 229
    cycling stability, enhancing, 335
    fabrication, 305, 305f

    for Li-ion battery anodes, 333–337, 334t,
        335f–336f, 336t
    NP-MPSi, 306
    optical absorption, 23f, 231
    photocatalytic activity, 305–306
    preparation method, 230, 230f
    SEI formation in, 323
    solar cells with, 229–231
    thermal conductivity of, 378–379, 378f
    top up approach, 229–230, 230f
Silicon nanowires arrays, vertically aligned,
        90, 91f
Silicon on an insulator (SOI) waveguide, 95–96,
        96f
Silicon on insulator (SOI) substrate, 78, 87
    uses of, 110
Silicon-on-insulator (SOI) technology, 29, 144
Silicon-on-nothing technology (SON), 139
Silicon on sapphire technology (SOS), 78
Si-material (textured), in Li-battery anode,
        321–327
    advantages, 326–327, 326t, 327f
    binders in composite, 325–326
    charge-discharge profile, 321–322, 321f
    depth of discharge, 325
    designs and reported performance, 326t
    lithiation/de-lithiation mechanisms,
        321–322, 322f
    performance improvement, 324–326
    principles and characterization, 321–324
    SEI formation, 323–324, 323f
    volume expansion of Si-based composites,
        324, 325f
Si nanocrystals, 18
Si nanoparticles, applications to hydrogen
        generation, 292, 292f
Single-chip capacitive microphone, 133, 134f
Single-crystal Si films, homoepitaxial growth
        of, 144–148, 144f
Single-crystal Si homoepitaxial growth, on PSi
        layer, 144–148, 144f
Single microchannel microreactors, 298, 298f
Single-tank electrochemical etching, 87
Single wall carbon nanotubes (SWNT),
        290–291
Sintered PSi (SPS) process, 146
    homoepitaxial growth on PSi layer and, 145
SiNWs, see Silicon nanowires (SiNWs)
SiO_2 coating, 90–91
    reflectance spectra of, 90, 91f
SiO_2/Si interface, 222
Si photovoltaics (PV); see also Photovoltaics
        (PV)
    modern tendencies in
        solar cells with Si nanowires, 229–231,
            230f–231f
        tandem Si solar cells with Si QDs,
            231–233, 232f–233f
Si porosification technology, 173–174, 327
Si/PSi-Ge/Ni MSM photodetectors, 43
Si QDs
    formation, 232–233, 233f
    properties, 232–233
    tandem Si solar cells with, 231–233,
        232f–233f
Si solar cells; see also Photovoltaics (PV);
        Solar cells
    advantages, 212
    basic design, 212, 213f
    Bragg reflector for, 225–226, 225f–226f

    disadvantages, 213
    requirements for ARC on, 217–218
    tandem, with Si QDs, 231–233, 232f–233f
Si TEG device, 382–383
Si wafers
    boron-doped, 81
    electrochemical texturization of, 214–215,
        215f
    gettering of, 213–214
    macroporous texturization of
        alcaline vs., 214–216, 214f–217f
        influence on solar cell performance, 217
    n-type, 91–92
    phosphorous-doped, 81
    p-type, 92
Slabs (finite 2D photonic crystals), 65–68,
        66f–67f
    "air-bridge," 67
    light of cone, 66, 68
SMM, see Surface micromachining (SMM)
Sn/Au contact, EL emission, 22, 23f
SOI wafers, formation of, 144–145
Solar cells; see also Photovoltaics (PV); Si solar
        cells
    basic design, 212, 213f
    layer-transfer process applications,
        228–229, 228f
        ELTRAN process, 229
        FMS process, 229
        LAST process, 229
        QMS process, 229
        ψ-process, 229
    macroporous texturization of Si wafers
        and, 217
    nano PSi as ARC and, 219–221, 219f–220f
    PSi as Bragg reflector in, 147
    PSi emission and quantum efficiency of, 221
    Si, requirements for ARC on, 217–218
    with Si nanowires, 229–231, 230f–231f
    tandem Si, with Si QDs, 231–233, 232f–233f
    thin-film silicon, see Thin-film silicon solar
        cells
Solar cells by liquid phase epitaxy over PSi
        (SCLIPS) process, 146
Solid electrolyte interface (SEI) formation
    Li-batteries silicon anodes, 323–324, 323f
Solid matrix method, hydrogen storage, 274–276
    chemical storage, 275–276
    physisorption, in porous matrixes, 274–275,
        275f
Solid oxide fuel cells (SOFC), 252, 252t
Solid-state devices
    charge carrier transport in, 12, 12f
    electron-hole pair generation mechanism, 12
    electron-hole pair generation mechanism
        in, 12, 12f
Spacer, 55
Spectral sensitivity
    PSi-based photodetectors, 38–39, 38f
    of sensor with PSi diffraction grating, 93
Spindt-type cathodes, 167, 167f, 168
Spiral inductors, 116; see also Inductors
Stabilization
    EL
        by PSi capping, 26
        by PSi surface modification, 26–27, 26f
    of PSi-based photodetectors, 40–42,
        41f, 42f
Stain etching process, 147, 302
    thin porous layer formation, 218

Stern–Grahame model, 350
"Stop" ring, role of, 42
Stronium-90, 246
SU8, 254
Substrate conductivity, homoepitaxial growth
     on PSi layer and, 145–146
Sulfuric acid, 307
Supercapacitors, 347–371
    AC impedance of, 352
    background, 348–349
    charge density, 351
    charge/discharge options, 354–355
    cyclic-voltammetry curves, 354, 354f
    definitions, 349–355
    differential capacitance, 351
    electrode materials for, 351, 355–356
    electrolytes in, 351
    equivalent-circuit model of, 352–353,
        352f–353f
    Gouy–Chapman diffuse layer, 350, 351f
    Helmholtz layer, 350–351, 351f
    inner Helmholtz plane (IHP), 350, 351f
    outer Helmholtz plane (OHP), 350, 351f
    overview, 348–349
    PSi-based, 356–364
        electrochemical performance, 364–371,
            365t, 367t
        fabrication of, 357–364
        integrated on-perforated-silicon-chip
            micro-supercapacitors, 357–360,
            358f–359f, 364–368
        with PSi electrodes, 360–364, 368–371
    specific energy density, 355
    specific power density, 355
    Stern–Grahame model, 350
    vs. batteries, 348
    vs. electrostatic capacitors, 349–350
    vs. electrochemical double-layer capacitor,
        350, 350f
Supplementary hydrogenization methods,
    289–290, 289f
Supporting material, 255
Surface micromachining (SMM), 129–140;
        see also Bulk micromachining
        technique; Sacrificial layers
    cavities and wafer-through holes, 136–137,
        136f–137f
    CMOS-compatible processing in
        microelectronics production line,
        PSi for, 130–131
    electrical isolation, 139–140, 139f
    freestanding mechanical elements, 131–135
        cantilevers, 131–133, 132f–133f
        free elements, 135, 135f
        microphones, 133–134, 134f
        thermal emitters, 135
    layer-transfer process, 137–139, 138f
    overview, 130
Surface passivation
    EL stabilization by, 26–27, 26f
    by porous layers
        light-induced, 224, 224f
        surface recombination velocity in PSi/Si
            interface, 222–224, 223f
Surface recombination velocity
    IQE and, 222–223, 223f
    I–V characteristics and, 223
    in PSi/Si interface, 222–224, 223f
Surface recombination velocity, in PSi/Si
    interface, 222–224, 223f

T

Tandem Si solar cells; see also Si solar cells;
    Solar cells
    "all-Si" tandem solar cells, 232, 232f
    with Si QDs, 231–233, 232f–233f
TE devices, see Thermoelectric (TE) devices
TEG, see Thermoelectric generator (TEG)
Telecommunications, optical devices in, 78
Temperature-programmed desorption (TPD)
    method, 280–281
Temporal instability, PSi-based photodetectors,
    40
Tetraethylammonium tetrafluoroborate
    (TEATFB) salt, 351
Tetramethylammonium hydroxide (TMAH)
    etching, 310
Textured Si-material, in Li-battery anode,
    321–327
    advantages, 326–327, 326t, 327f
    binders in composite, 325–326
    charge-discharge profile, 321–322, 321f
    depth of discharge, 325
    designs and reported performance, 326t
    lithiation/de-lithiation mechanisms,
        321–322, 322f
    performance improvement, 324–326
    principles and characterization, 321–324
    SEI formation, 323–324, 323f
    volume expansion of Si-based composites,
        324, 325f
Texturization, of Si wafers
    alcaline vs. macroporous, 214–216, 214f–217f
    chemical etching, 215, 216f
    electrochemical, 214–215, 215f
    metal assisted etching for, 215–216, 216f, 217f
    solar cell performance and, 217
Thermal carbonization (TC) technique
    PSi-based photodetectors and, 40–41, 41f
Thermal conductivity, 376
    of PSi, 379–381, 379f–380f
    of Si nanowires, 378–379, 378f
Thermal emitters
    SMM and, 135
    thermal isolation for, 135
Thermal evaporation, 43, 184
Thermally stimulated desorption, of hydrogen,
    280–281, 281f
Thermal management, 376
Thermal oxidation, 16, 16f, 90, 146; see also
    Rapid thermal oxidation (RTO)
    PSi-based photodetectors and, 41–42, 42f
Thermoelectric (TE) devices, 375–384
    advantages, 376
    applications, 376
    future perspectives, 383–384
    general considerations, 376–378
    overview, 376
    power factor, 377
    PSi-based, 378–381
        thermal conductivity and other
            parameters, 378–381, 378f–380f
        thermoelectric properties, 379–381,
            379f–382f
    Seebeck coefficient, 376–377
    Seebeck effect, 376–377, 377f
    thermoelectric generator (TEG), 376, 377,
        377f
        using PSi, 382–383, 382f–383f
    ZT factor, 376

Thermoelectric generator (TEG), 376, 377, 377f
    using PSi, 382–383, 382f–383f
Thin films, PSi, 328–333
    general considerations, 328–331, 329f–330f
    macroporous silicon anode performances,
        331–332, 331f, 332t
    mesoporous silicon anodes performances,
        332–333, 333f, 333t
Thin-film silicon solar cells, 144
    formation of, 144–145
    layer-transfer process for, 146
    low-grade starting silicon wafers for, 147–148
    processes for, 146–147
    PSi reflectors for, 226–227, 226f, 227f
    solar cells by liquid phase epitaxy over PSi
        (SCLIPS) process, 146
3D photonic crystals, 52f, 68–71, 69f–70f
    applications, 71
    fabrication, 68–71, 69f–70f
        lithographic prestructuring process, 68
        macroporous silicon method, 68–69,
            69f–70f
    light guiding and, 71
    PBS of, 68, 69f
3D PSi particles, 337
Threshold field, Si-based cold cathodes,
    174–175, 174t
Through silicon via (TSV) substrates, 260
TiO$_2$ nanoparticles, as photocatalysts, 304
Top electrode, device, 21–25, 22t
    compact PSi between active PSi and, 23, 23f
    metals and conductive oxides as, 22, 22f–23f
    other materials between active PSi and,
        24–25, 24f–25f
Top up approach, SiNWs, 229–230, 230f
TPD (temperature-programmed desorption)
    method, 280–281
Transfers matrix method, 53–54
Transition-metal oxides, as electrode materials,
    356
Transmission electron microscopy (TEM), 304,
    322, 326
Transversal electric (TE) polarization, 60, 98
Transversal magnetic (TM) polarization, 60, 98
Trinitrotoluene (TNT), 388
Tritium (T), 246
Turnover frequency (TOF), 303
2D photonic crystals, 52f, 60–68
    applications, 71
    bulk, 60–63, 61f–64f
    characteristics, 60
    defects in, 63–65, 65f, 66f
    fabrication
        electrochemical etching method, 60
        proton beam writing method, 60–61, 61f
        using macroporous silicon, 62
        using macroporous silicon with
            standard lithography, 62–63, 63f
    finite (slabs), 65–68, 66f–67f
    light guiding and, 71
    PBS for, 61, 62f, 63, 64f

U

Ultracapacitor, 348; see also Supercapacitors
Ultrafast processing of signals, 166
Ultraviolet-visible (UV-VIS) spectroscopy,
    303
UV–CVD, 224
UV photodetectors, 46

V

Vacuum-based methods, 184; *see also* specific methods
Vacuum fluorescent display (VFD), 166
Vacuum-less cathode-ray tube, 172
Valence electron energy loss spectrometry (VEELS), 322
Van der Waals interactions, of hydrogen molecules, 274, 289
Vapor compression techniques, 376
Vapor liquid solid (VLS) method, 229, 231, 326
Vegard's law, 121
Velocity, propagation, of explosion, 397–398, 397f–398f
VFD, *see* Vacuum fluorescent display (VFD)
VLSI technologies, 78
Voltage-induced spectral shift, of EL, 9, 9f

Volume expansion, of Si-based composites, 324, 325f

W

Wafer-through holes, SMM for, 136–137, 136f–137f
Waveguides, 78
    optical, 95–99, 96f–97f
    PSi-based, 57
Wet etching method, 89, 299
Wireless communications, 250
Work function, 166
Working principles, photonic crystals, 53–54

X

Xenon (Xe) flash lamp, 395
X-ray diffraction (XRD), 304, 322, 326

X-ray photoelectron spectroscopy (XPS), 306, 323

Y

ψ-process, 229

Z

Zeolites, 275
Zinc-70, 246
ZnO nanofilms, 104
ZnO nanofilms/PSi-based UV detectors, 104, 104f
ZnO/PSi/c-Si photodiode, 46f
ZnO/TiO$_2$ and PSi coating, 90–91
    reflectance spectra of, 91f
ZnSe/PSi/Si heteroepitaxial structure, 150, 152, 152f